Zoology: The Study of Animals

Zoology: The Study of Animals

Editor: Toby Simpson

STATES
ACADEMIC PRESS
www.statesacademicpress.com

States Academic Press,
109 South 5th Street,
Brooklyn, NY 11249, USA

Visit us on the World Wide Web at:
www.statesacademicpress.com

ISBN: 978-1-63989-576-2 (Hardback)

Cataloging-in-Publication Data

Zoology : the study of animals / edited by Toby Simpson.
p. cm.
Includes bibliographical references and index.
ISBN 978-1-63989-576-2
1. Zoology. 2. Animals. I. Simpson, Toby.
QL45.2 .Z663 2022
590--dc23

Table of Contents

Preface

In my initial years as a student, I used to run to the library at every possible instance to grab a book and learn something new. Books were my primary source of knowledge and I would not have come such a long way without all that I learnt from them. Thus, when I was approached to edit this book; I became understandably nostalgic. It was an absolute honor to be considered worthy of guiding the current generation as well as those to come. I put all my knowledge and hard work into making this book most beneficial for its readers.

Zoology is the branch of biology that studies the members of animal kingdom as well as their related aspects. It focuses on the constituent parts of animals, animal populations, and the interaction of animals with each other as well as with other biotic and abiotic components of the environment. The various facets studied under zoology are anatomy and morphology, taxonomy, physiology, embryology and evolutionism. Anatomy and morphology are the branches of biology which describe the internal organizations and external forms of living organisms respectively. The scientific study of naming and grouping organisms on the basis of shared characteristics is known as taxonomy. Embryology deals with the analysis of gametes, fertilization, and development stages of embryos and fetuses. Some of the diverse topics covered in this book address the varied branches that fall under this category. The topics covered in it deal with the core subjects of zoology. This book will provide comprehensive knowledge to the readers.

I wish to thank my publisher for supporting me at every step. I would also like to thank all the authors who have contributed their researches in this book. I hope this book will be a valuable contribution to the progress of the field.

Editor

1

Six new species of *Poeciloderrhis* from Brazil (Blattaria: Blaberidae) with a new record and a new combination

Leonardo de Oliveira Cardoso da Silva[1], Sonia Maria Lopes[1]

[1]*Department of Entomology, Museu Nacional, Universidade Federal do Rio de Janeiro, Quinta da Boa Vista, São Cristóvão, 20940-040 Rio de Janeiro, RJ, Brazil.*
Corresponding author: Sonia Maria Lopes (sonialfraga@gmail.com)

ABSTRACT. In this contribution six new species of *Poeciloderrhis* Stål, 1874 are described: *P. aureolatus* **sp. nov.** (Holotype male in MNRJ: Brazil, Rio de Janeiro State), *P. mediansclerostylatus* **sp. nov.** (Holotype male in MNRJ: Brazil, Rio de Janeiro State), *P. diamantinensis* **sp. nov.** (Holotype male in MNRJ: Brazil, Mato Grosso State), *P. itatiaiensis* **sp. nov.** (Holotype male in MNRJ: Brazil, Rio de Janeiro State), *P. vanzolinii* **sp. nov.** (Holotype male in MNRJ: Brazil, Rio de Janeiro State), and *P. penduloides* **sp. nov.** (Holotype male in MNRJ: Brazil, Rio de Janeiro State). The genus is recorded for the first time from the state of Rondônia, Brazil based on *P. diamantinensis* **sp. nov.** A new combination is proposed for *Poeciloderrhis bicolorata* Rocha e Silva & Lopes, 1977 that is transferred to *Epilampra* Burmeister, 1838. For specific delimation, the tergal modification in the abdomen, the genital plates and their respective structures were analyzed. A key for males including all species of *Poeciloderrhis* is also given.

KEY WORDS. Atlantic Forest, Epilamprinae, nomenclature, Serra da Mantiquera.

INTRODUCTION

The species of *Poeciloderrhis* Stål, 1874 differ from *Epilampra* Burmeister, 1838 by the tergal modifications in the first and second abdominal segments, L2d pointed and fused solidly to L2vm; prepuce membranous, not clearly defined by dense setae; R2 short and stout, without apical incision; L1 without a setal brush and the sclerotized cleft fused together (Roth 1970).

Until now 15 species of *Poeciloderrhis* are known, all endemic to Brazil, except *P. verticalis* (Burmeister, 1838), which also occurs in northeastern Argentina (Roth 1970, Rocha e Silva-Albuquerque and Lopes 1976, 1977, Rocha e Silva-Albuquerque and Jurberg 1978, Lopes and Oliveira 2006, Silva and Lopes 2015).

Recently, Silva and Lopes (2015) presented a synopsis of the genus and Anisyutkin (2016) redescribed the type of the species, *P. verticalis* (Burmeister, 1838), in detail, increasing our knowledge on this group of roaches.

The goals of this paper are to describe six new species of *Poeciloderrhis* and provide a new record of the genus from state of Rondônia, northern Brazil.

MATERIAL AND METHODS

The genital plates were removed after dissection of the posterior part of the abdomen, using standard techniques (e.g., Lopes and Oliveira 2000). After analysis, the genital plates and genital pieces were stored in microvials containing glycerin and were attached to the respective exemplar (Gurney et al. 1964). The male tergal modifications of the abdomen were observed from the base of the wings after preparation with diluted detergent over the wing insertions, using a brush. The terminology for the genitalia and the taxonomic classification follow Roth (2003). Digital images of the habitus, pronotum, head, tergal modification and genitalia were taken with a camera mounted on a Leica M205C stereoscopic microscope, with software LAS version 4.8. The holotypes are deposited in the collection of the Department of Entomology, Museu Nacional, Universidade Federal do Rio de Janeiro, Rio de Janeiro, Brazil (MNRJ). The images of *P. verticalis* (Figs 1–3), sent from the Zoologisches Institut der Martin-Luther Universität and the Naturhistorisches Museum Wien, have no scales.

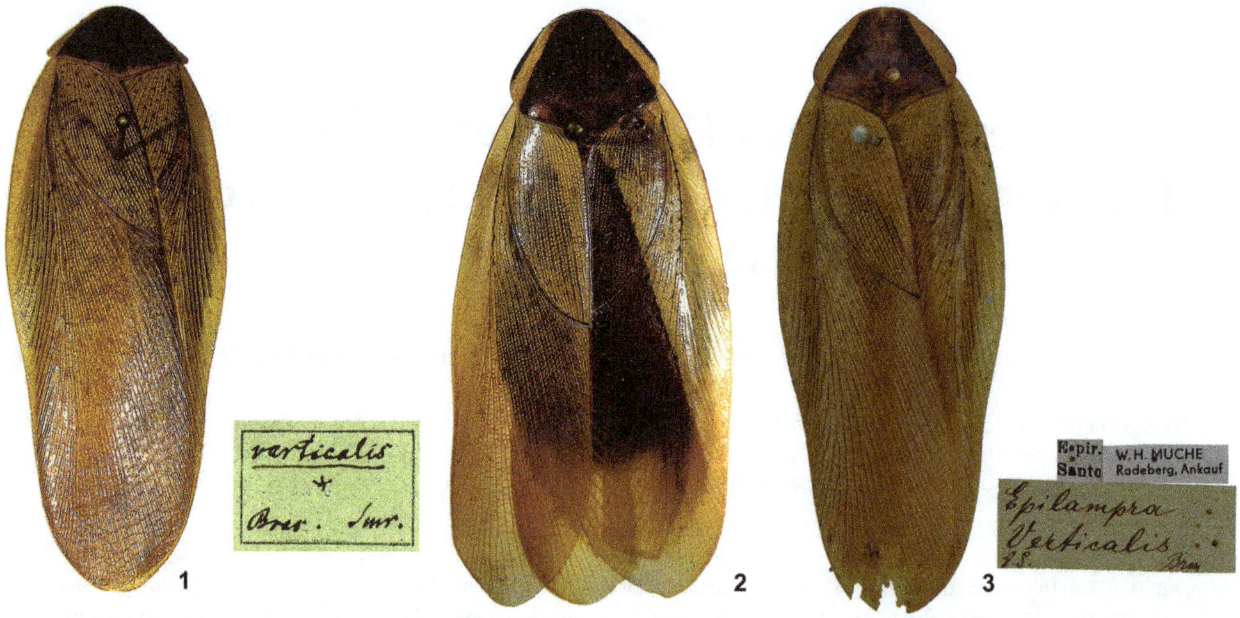

Figures 1–3. *Poeciloderrhis verticalis*: (1) habitus, dorsal view, female deposited in Halle Museum, length 40.5 mm; (2) habitus, dorsal view, female from deposited in Halle Museum, lenght 40.5 mm; (3) habitus, dorsal view, male deposited in Senkenbergisches Institut, lenght 40.5 mm.

TAXONOMY

Poeciloderrhis aureolatus sp. nov.

Figs 4–19

Diagnosis. General coloration. Shiny yellowish-brown (Fig. 4). Head with brown vertex (Fig. 5), with four light brown circular spots; interocular space, interantennal and central portion of clypeus brown. Maxillary palps with brown apical segment, with golden cilia. Eyes black. Pronotum semi-transparent, with central brown mark extending from posterior end nearly to the anterior end (Fig. 6). Tegmen semi-transparent, light brown, with brown spots. Legs brown with dark brown spines, arolia and claws. Abdomen light brown with dark brown punctuations.

Dimensions (mm). Holotype male. Total length: 42.8; total length of pronotum: 8.1; width of pronotum: 9.1; length of tegmen: 37.3; width of tegmen: 9.7. Paratype males. Total length: 41.6–45.7; total length of pronotum: 7.3–9.6; width of pronotum: 9.3–11.1; length of tegmen: 36.5–41.3; width of tegmen: 10.0–11.8. Paratype females. Total length: 45,4–45,7; total length of pronotum: 9.4–9.6; width of pronotum: 10.5–11.1; length of tegmen: 40.2–41.3; width of tegmen: 10.7–11.8.

Description of the holotype male. Head triangular with rounded borders, vertex barely visible in dorsal view; interocular space about 1.20 mm wide; antennae long, threadlike and tomentose, surpassing apex of cerci. Maxillary palps with first and second segments reduced, third segment 25% larger than fourth and 25% smaller than fifth segment, fifth segment slightly more swollen and densely tomentose.

Thorax. Pronotum ample, convex, with curved edges, base with small median projection. Legs with half of femur I with anteroventral surface bearing five robust spines followed by series of small spines up to apex, apex with two strong spines; posteroventral surface with three robust spines, one apical; femora II and III bearing few strong spines on ventral surface. Pulvilli on four tarsal segments, claws symmetrical, with two rows of small spines on ventral surface, similar to two spines on legs. Tegmen surpassing apex of abdomen. Marginal field elongated, slightly concave, scapular field elongated with oblique venules, discoidal field curved and convex apically, anal field convex and well defined.

Abdomen. Tergal modifications differentiated, with longitudinal humps on first segment; two lateral humps followed by two at middle of plate (Fig. 7). Supranal plate with small median reentrance; cerci surpassing length of plate (Fig. 8). Subgenital plate asymmetric (Fig. 9) with two thread-like styles, one long and one short. Left phallomere sclerotized (Fig. 10). Median sclerite spiniform apically, strongly sclerotized, with thin membranous structure subapically (Figs 11 and 12). Right phallomere with small short apex, bearing small spines near apex, membrane of phallomere with nine spines (Fig. 13).

Female larger than male (Fig. 14); front light brown (Fig. 15); pronotum brown with light brown lateral flaps (Fig. 16); supranal plate with median reentrance between cerci (Fig. 17);

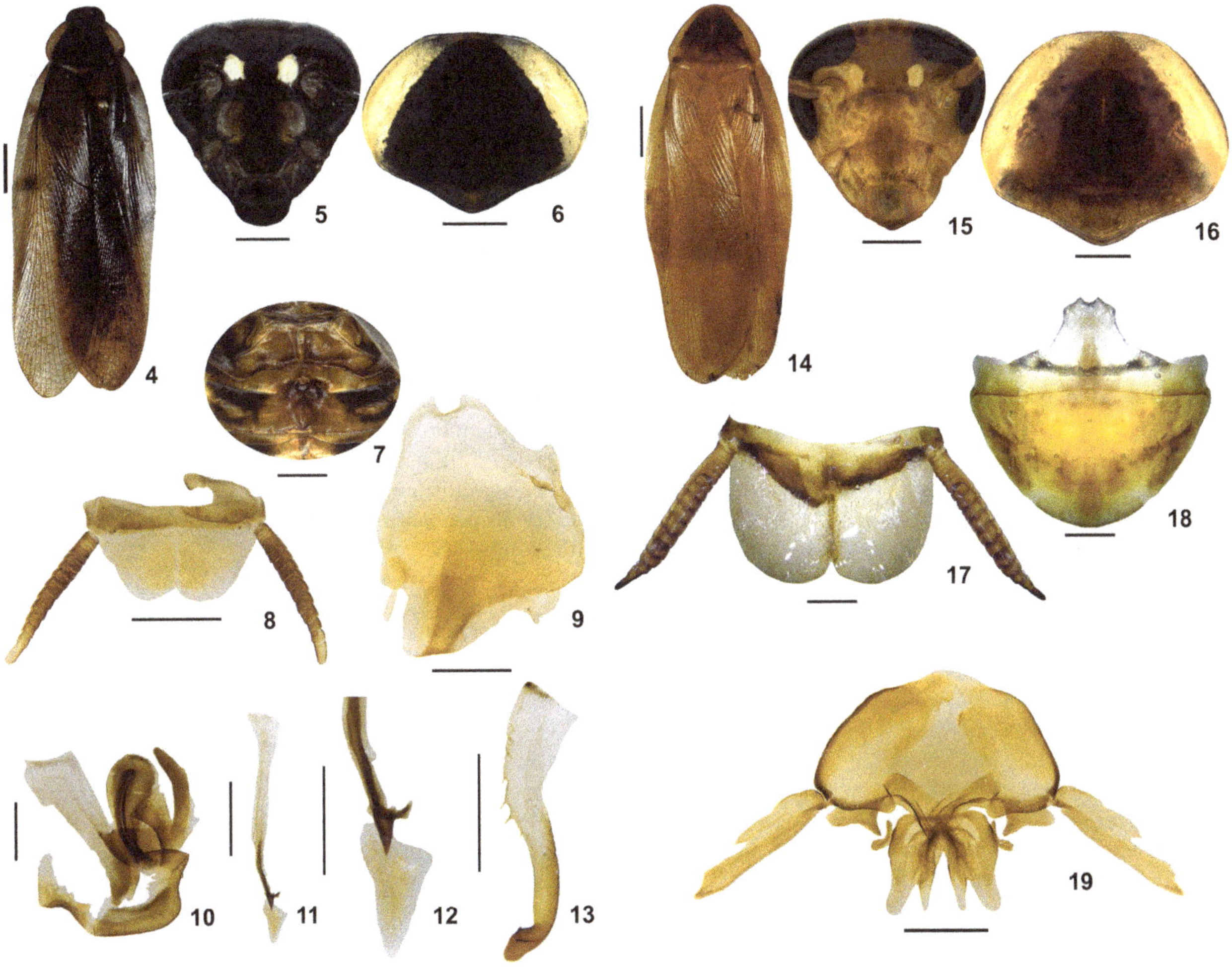

Figures 4–19. (4–13) *Poeciloderrhis aureolatus* sp. nov. holotype male: (4) habitus, dorsal view, (5) head, ventral view, (6) pronotum, dorsal view, (7) tergal modification of first and second abdominal segments, dorsal view, (8) supranal plate, dorsal view, (9) subgenital plate, ventral view, (10) left phallomere, dorsal view, (11 and 12) median sclerite, dorsal view, (13) right phallomere, dorsal view. (14–19) *Poeciloderrhis aureolatus* sp. nov. paratype female: (14) habitus, dorsal view, (15) head, ventral view, (16) pronotum, dorsal view, (17) supranal plate, dorsal view, (18) subgenital plate, ventral view, (19) valves, ventral view. Scale bars: 4 = 5.6 mm; 5, 9, 13, 15 = 2.0 mm; 6–8, 16 = 3.0 mm; 10 = 0.7 mm, 11, 19 = 1.0 mm, 12 = 0.7 mm; 14 = 6.0 mm, 17 = 1.2 mm, 18 = 1.9 mm.

subgenital plate triangular (Fig. 18); first pair the most developed, second smaller and tapered, third wider than second (Fig. 19).

Material examined. Holotype male, BRAZIL, Rio de Janeiro State, Itatiaia, PNI, Setor Lago Azul, Rio Campo Belo, 22°26′49″S, 44°36′45″W, 854 m a.s.l., 03/VIII–1/IX/2015, Biota Faperj, coleta 02. Paratypes: 5 males and 1 female, same data as holotype; 9 males and 1 female, same data as holotype but Complexo do Maromba, Cachoeira Véu da Noiva, PNI – M2A, 22°25′36″S, 44°37′05″W, 1153 m a.s.l., 2/X–2/XI/2015, Biota/Faperj, coleta 04; 1 male, Parada da Ribeira, R. Mangi, 9-II to 2-X-1959, Coelho & Castro cols; 1 male, Grajaú, collected in a water tank, 27-XII-1952, Santos & Machado cols; 8 females, Cachoeiras de Macacu (Rio São João), 5-I-1981, Santos & Neto cols; 1 male, 2/X/1959; 1 male, Minas Gerais, Oiticica col. All in MNRJ.

Etymology. The species name *aureolatus* is derived from the yellowish-brown coloration of the specimens.

Remarks. *Poeciloderrhis aureolatus* sp. nov. is similar to *Poeciloderrhis proxima* (Brunner von Wattenwyl, 1865) in coloration (both shiny brown). The new species might be distinguished from *P. proxima* by its tergal modification (the second abdominal segment bears two lateral humps followed by two at middle of plate in *P. aureolatus* sp. nov. while *P. proxima* have two lateral humps followed by two at middle of plate, being that second elevation of the middle of the plate ends in a hook-like structure) and shape of the genital parts (right phallomere with small short apex in *P. aureolatus* sp. nov., while *P. proxima* have right phallomere with rounded apex). Besides that, pre-apical region in *P. aureolatus* sp. nov. have an expansion, which is absent in *P. proxima*.

Poeciloderrhis mediansclerostylatus sp. nov.

Figs 20–29

Diagnosis. General coloration. Shiny brown (Fig. 20). Head with dark brown vertex; interocular and interantennal spaces dark brown, other head structures brown; maxillary palps with brown apical segment, with golden cilia. Eyes dark brown (Fig. 21). Pronotum light brown laterally, semi-transparent with dark brown trapezoidal spot (Fig. 22). Tegmen ranging from light brown to brown, semi-transparent with dark brown spots. Legs light brown, spines, pulvilli, arolia and claws brown. Abdomen light brown.

Dimensions (mm). Holotype male. Total length: 30.0; length of pronotum: 6.0; width of pronotum: 7.7; length of tegmen: 23.5; width of tegmen: 7.3.

Description of the holotype male. Head triangular with rounded corners, vertex barely exposed in dorsal view; interocular space about 1.24 mm. Antennae long, threadlike and tomentose, surpassing apex of cerci. Maxillary palps with first and second segments reduced, third segment longer than fourth, fifth segment slightly more dilated and densely tomentose.

Thorax. Pronotum wide, convex, angles curved, base with small median projection. Legs with femur I with anteroventral surface with six strong spines up to middle, followed by series of small spines toward apex, apex bearing two strong spines; posteroventral surface with three to four strong spines, one apical; femora II and III bearing few strong spines on their ventral surfaces. Pulvilli present on four tarsal segments, claws symmetrical and specialized, with two rows of small teeth, similar to spines on legs. Tegmen surpassing apex of abdomen. Marginal field elongated, slightly concave, scapular field elongated with oblique venules, discoidal field convex, curved apically, anal field convex and conspicuous.

Abdomen. Tergal modification consisting of longitudinal hump on first segment and complex of four humps on second segment (Fig. 23). Supranal plate with cerci surpassing size of plate, with conspicuous cilia in ventral view (Fig. 24). Subgenital plate asymmetric, left style, in ventral view, larger than right style (Fig. 25). Left phallomere sclerotized (Fig. 26). Median sclerite with apex spiniform (Figs 27 and 28). Right phallomere apically resembling hockey stick, with small spines near apex. Membrane of right phallomere with eight spines (Fig. 29).

Material examined. Holotype male, BRAZIL, Rio de Janeiro State, Conceição do Jacareí, 18-XI-1975, Prazeres col. (MNRJ).

Etymology. The species name, *mediansclerostylatus* is derived the styliform median sclerite, which is similar to *P. imperialis*.

Remarks. *Poeciloderrhis mediansclerostylatus* sp. nov. is similar to *Poeciloderrhis imperialis* Rocha e Silva & Jurberg, 1978 in coloration of the pronotum (semi-transparent with dark brown trapezoidal spot) and habitus (shiny brown). The new species might be distinguished from *P. imperialis* by its smaller size 30.0–31.1 mm (31.6–33.0 mm in *P. imperialis*), shape of the tergal modification consisting of longitudinal hump on first segment and a complex of four humps on second segment, while in *P. imperialis* is composed by a slight elevation, trianguliform near the apex and two elevations next from the base which follows from the center of the structure towards the sides in the first segment and another slight elevation in the base of the second segment. In addition, *P. mediansclerostylatus* sp. nov. can be distinguished too by its shiny-brown general coloration, while it is shiny light brown in *P. imperialis*, and genital morphology that presents the median sclerite with an apical spine and right phallomere with an irregular margin in the apex while the median sclerite presents a rounded structure and right phallomere with rounded apex in *P. imperialis*.

Poeciloderrhis diamantinensis sp. nov.

Figs 30–39

Diagnosis. General coloration. Shiny, light brown (Fig. 30). Head with front ranging from brown to light brown; interocular and interantennal spaces brown, other head structures light brown; ocelli brown; eyes dark brown (Fig. 31). Pronotum light brown, semi-transparent with brown punctuations (Fig. 32); tegmen light brown, semi-transparent, with dark brown spots. Legs light brown, spines, pulvilli, arolia and claws light brown. Abdomen brown.

Dimensions (mm). Holotype male Total length: 22.3; length of pronotum: 5.1; width of pronotum: 6.3; length of tegmen: 20.2; width of tegmen: 5.8. Paratype males. Total length: 22.1–24.5; length of pronotum: 4.9–6.0; width of pronotum: 6.3–7.0; length of tegmen: 20.2–22.0; width of tegmen: 5.5–6.5.

Description of the holotype male. Head triangular with rounded corners, vertex barely exposed in dorsal view; interocular space about 0.62 mm. Antennae long, threadlike and tomentose, surpassing apex of cerci. Eyes anterolateral; maxillary palps with first and second segments reduced, third segment longer than fourth, fifth segment slightly more dilated and densely tomentose.

Thorax. Pronotum ample, convex, curved laterally, punctuations lying in small depressions, base with small median projection. Legs with femur I with anteroventral surface bearing 4–6 strong spines followed by series of small spines toward apex, apex with two strong apical spines; posteroventral surface with four strong spines, one apical; femora II and III bearing few strong spines on ventral surface. Pulvilli present on four tarsal segments, claws symmetrical and specialized, with two rows of

Figures 20–39. (20–29) *Poeciloderrhis mediansclerostylatus* sp. nov. holotype male, dorsal view: (20) habitus, dorsal view, (21) head, ventral view, (22) pronotum, dorsal view, (23) tergal modification on first and second abdominal segments, dorsal view, (24) supranal plate, dorsal view, (25) subgenital plate, ventral view, (26) left phallomere, dorsal view, (27 and 28) median sclerite, dorsal view, (29) right phallomere, dorsal view. (30–39) *Poeciloderrhis diamantinensis* sp. nov. holotype male, dorsal view: (30) habitus; (31) Head, ventral view, (32) pronotum, dorsal view, (33) tergal modification of first and second abdominal segments, dorsal view, (34) supranal plate, dorsal view, (35) subgenital plate, ventral view, (36) left phallomere, dorsal view, (37-38) median sclerite, dorsal view, (39) right phallomere, dorsal view. Scale bars: 20 = 5.6 mm; 21 = 1.3 mm; 22 = 2.7 mm; 23-25, 32, 33 = 2.0 mm; 26 = 0.7 mm; 27, 29, 31, 34, 35 = 1.0 mm; 28, 38 = 0.3 mm; 30 = 3.4 mm; 36, 37, 39 = 0.7 mm.

small teeth, similar to spines on legs. Tegmen surpassing apex of abdômen. Marginal field elongate, slightly concave, scapular field elongated with oblique venules, discoidal field convex, curved apically, anal field convex and conspicuous.

Abdomen. Tergal modification consisting of two large transverse humped structures (Fig. 33) with apical cilia on first segment, and another transverse, arched and ciliated structure apically, on second segment. Supranal plate with cerci surpassing size of plate in length (Fig. 34); cerci with ventral cilia. Subgenital plate slightly asymmetric, left style, in ventral view, slightly larger than right (Fig. 35). Left phallomere sclerotized (Fig. 36). Median sclerite with apex aculeate, with small, strongly sclerotized curved (Figs 37 and 38). Right phallomere with curved, club-shaped apex, and small spines near apex (Fig. 39).

Material examined. Holotype male, BRAZIL, Mato Grosso State: Diamantino, Fazenda São João, Km 200, BR 163, without collector. Paratypes: Mato Grosso State: 1 male, Barra do Bugres, 19–21/XI/1983, J. Becker, O. Roppa, Belmiro, S. cols.; 7 males, Sinop, 12°31'S, 55° 37'W, BR 163, Km 500–600, 350 m, X-1974, Alvarenga & Roppa cols; Rondônia State: 1 male, Nova União, Ouro Preto, XI-1983, without collector; 1 male, Vilhena, XI-1973, Alvarenga col.; Goiás State: 1 male, Jataí, XII-1972, Oliveira col. All in MNRJ.

Etymology. The species name, *diamantinensis*, is derived from the name of the locality where this species was collected.

Remarks. *Poeciloderrhis diamantinensis* sp. nov. is similar to *Poeciloderrhis ferruginea* (Brunner von Wattenwyl, 1865) in coloration (shiny light brown in both). The new species might be distinguished from its smaller size 22.1–24.5 mm (25.0–32.0 mm in *P. ferruginea*), tergal modification (two large transverse humped structures with apical cilia on first segment, and another transverse, arched and ciliated structure apically, on second segment, while *P. ferruginea* presents a piramidal, long, with cilia, located in the first tergite and a cavity followed by a curved rod in the second tergite) and shape of genital parts with apex of right phallomere golf club-like and an pre-apical region of median sclerite without expansion in *P. diamantinensis* sp. nov., while in *P. ferruginea* the right phallomere bears an ax-shaped apex and median sclerite with a pre-apical expansion. *Poeciloderrhis diamantinensis* sp. nov. was determined by Roth (1970) as *Poeciloderrhis* A (pg. 108, figs 28–33).

Poeciloderrhis itatiaiensis sp. nov.

Figs 40–49

Diagnosis. General coloration. Light brown, shiny (Fig. 40). Head with brown front; interocular, interocellar and interantennal spaces brown. Eyes black (Fig. 41). Pronotum light brown, semi-transparent, with dark brown punctuations (Fig. 42). Tegmen light brown, semi-transparent with dark brown spots. Legs light brown, with spines, pulvilli, arolia and claws brown. Abdomen brown.

Dimensions of holotype male (mm). Total length: 48.9; length of pronotum: 7.6; width of pronotum: 9.7; length of tegmen: 42.6; width of tegmen: 10.0.

Description of the male holotype. Head triangular with rounded corners, vertex slightly exposed in dorsal view; interocular space about 1.22 mm. Maxillary palps with first and second segments reduced and fifth segment slightly densely tomentose.

Thorax. Pronotum ample, convex, curved angles, base with small median projection. Legs with femur I with anteroventral surface bearing 4–6 strong spines followed by series of small spines toward apex, apex with two strong apical spines; posteroventral surface with two or three strong spines, one apical; femora II and III bearing few strong spines on their ventral surfaces. Pulvilli present on all four tarsal segments, claws symmetrical and specialized, with two rows of small teeth, similar to spines on legs. Tegmen surpassing apex of abdomen. Marginal field wide, slightly concave, scapular field elongated with oblique venules, discoidal field convex, curved apically, anal field convex and conspicuous.

Abdomen. Tergal modification consisting of three transverse humps with cilia on hump closest to base, base of first segment with circular recess, base of second segment with arcuate hump (Fig. 43). Supranal plate with cerci surpassing size of plate; cerci with ventral cilia (Fig. 44). Subgenital plate asymmetric, left style, in dorsal view, larger than right (Fig. 45). Left phallomere sclerotized (Fig. 46). Median sclerite triangular, apex markedly sclerotized (Figs 47 and 48). Right phallomere with curved, club-shaped apex, bearing spines near apex (Fig. 49).

Material examined. Holotype male, BRAZIL, Rio de Janeiro State, Itatiaia, Parque Nacional de Itatiaia, 750 m, 04-IV-1960, Gouvêa col. (MNRJ).

Etymology. The species name *itatiaiensis* is derived from the locality where the species was collected Itatiaia.

Remarks. *Poeciloderrhis itatiaiensis* sp. nov. might be distinguished from *Poeciloderrhis verticalis* (Burmeister, 1838) by its larger size 48.9 mm (36.8–38.8 mm in *P. verticalis*), tergal modification consisting of three transverse humps with cilia on hump closest to base, base of first segment with circular recess, base of second segment with arcuate hump in *P. itatiaiensis* sp. nov., while in *P. verticalis* the tergal modification have an elevation at the middle of first segment followed by two lateral elevations at the lateral of the second segment and two elevation at the middle of the segment; the first of those two elevations is an hook-like shaped. The new species might be distinguished too by the shape of the genital pieces that have the median sclerite with triangular sclerotized apex and right phallomere with a fishhook-like at apex, while in *P. verticalis* bears a median sclerite with a spiked apex and right phallomere with irregular margin at the apex.

Poeciloderrhis vanzolinii sp. nov.

Figs 50–59

Diagnosis. General coloration. Shiny, light brown (Fig. 50). Head with light brown vertex (Fig. 51); interocular and interantennal spaces brown; ocelli brown; maxillary palps with light brown apical segment, with golden cilia. Eyes dark brown. Pronotum light brown (Fig. 52), semi-transparent, with brown punctuations. Tegmen light brown, semi-transparent, with brown spots. Legs light brown, with spines, pulvilli, arolia and claws brown. Abdomen brown, with middle portion of sclerite region dark brown.

Dimensions of holotype male (mm). Total length: 27.5; length of pronotum: 6.2; width of pronotum: 8.5; length of tegmen: 22.7; width of tegmen: 8.1

Description of the male holotype. Head triangular with rounded corners, vertex slightly exposed in dorsal view; interocular space about 1.34 mm long. Antennae threadlike and tomentose, surpassing apex of cerci. Maxillary palps with first and second segments reduced, third segment larger than fourth and fifth segments, slightly more dilated and densely tomentose.

Thorax. Pronotum ample, convex, with curved angles, base bearing small median projection. Legs with femur I with anteroventral surface with four strong spines up to median region, followed by series of small spines up to apex, apex with

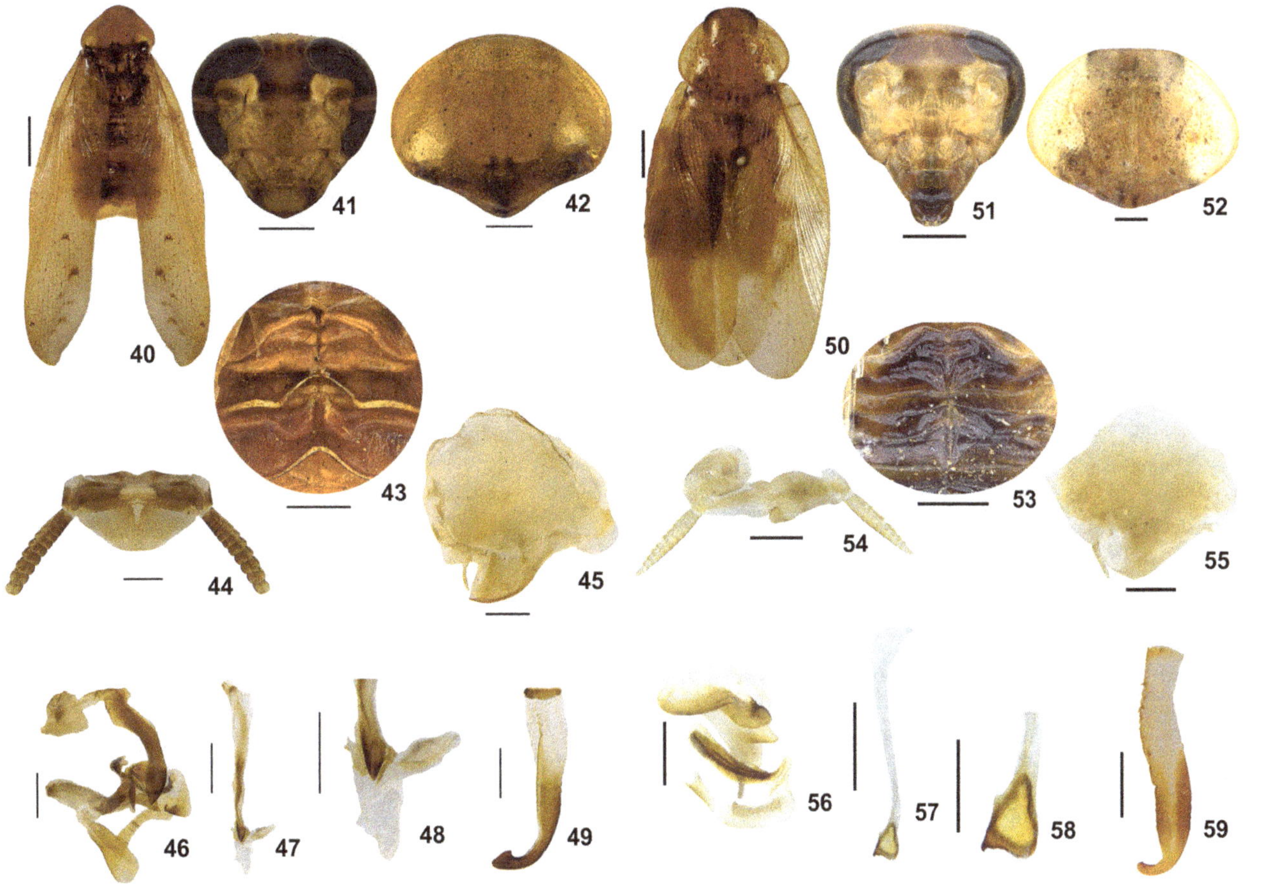

Figures 40–59. (40–49) *Poeciloderrhis itatiaiensis* sp. nov. holotype male, dorsal view: (40) habitus, dorsal view, (41) head, ventral view, (42) pronotum, dorsal view, (43) tergal modification on first and second abdominal segments, dorsal view, (44) supranal plate, dorsal view, (45) subgenital plate, ventral view, (46) left phallomere, dorsal view, (47-48) median sclerite, dorsal view, (49) right phallomere, dorsal view. (50–59) *Poeciloderrhis vanzolinii* sp. nov. holotype male: (50) habitus, dorsal view: (51) head, ventral view, (52) pronotum, dorsal view, (53) tergal modification of first and second abdominal segments, dorsal view, (54) supranal plate, dorsal view, (55) subgenital plate, ventral view, (56) left phallomere, dorsal view, (57-58) median sclerite, dorsal view; (59) right phallomere, dorsal view. Scale bars: 40 = 6.7 mm; 41, 43-45, 51-54, 57 = 2.0 mm; 42 = 3.0 mm; 46 = 0.8 mm; 47, 55 = 1.0 mm; 48, 49, 56, 58, 59 = 0.7 mm; 50 = 3.5 mm.

two strong apical spines; posteroventral surface with three strong spines, one apical; femora II and III with few strong ventral spines. Pulvilli present on all four tarsal segments, claws symmetrical and specialized, ventrally with two rows of small teeth, similar to those on legs. Tegmen surpassing apex of the abdomen. Marginal field wide, slightly concave, scapular field elongated with oblique venules, discoid field convex, curved apically, anal field convex and well-marked.

Abdomen. Tergal modification comprising three humped structures on first segment and one on second (Fig. 53). Supranal plate with cerci surpassing plate in length, with ventral cilia (Fig. 54). Subgenital plate asymmetric, with left style, in ventral view, larger than right (Fig. 55). Left phallomere with strongly sclerotized middle region (Fig. 56). Median sclerite with club-shaped apex, bearing small spine (Figs 57 and 58). Right phallomere with apex curved and small spine near apex (Fig. 59).

Material examined. Holotype male, BRAZIL, São Paulo State, Salesópolis, 22-V1-1947, Travassos & Vanzolini cols, MNRJ.

Etymology. The species name, *vanzolinii*, is named is of eminent researcher Dr. Paulo Emilio Vanzolini, who collected of the specimen.

Remarks. *Poeciloderrhis vanzolinii* sp. nov. is similar to *Poeciloderrhis ferruginea* (Brunner von Wattenwyl, 1865) in the coloration of the pronotum (light brown, semi-transparent with brown punctuations in both), and size 27.5 mm (25.0–32.0 mm in *P. ferruginea*). The new species might be distinguished from *P. ferruginea* by its tergal modification (three humped structures on first segment and one on second in *P. vanzolinii* sp. nov., while in *P. ferruginea* bears presents a piramidal, long, with cilia, located in the first tergite and a cavity followed by a curved rod in the second tergite, general coloration shiny light brown in *P. vanzolinii* sp. nov., while in *P. ferruginea* its presents shiny brown. The new species also might be distinguished by its shape of the genitalia (median sclerite with club-shaped apex and right phallomere with a small curved apex in *P. vanzolinii* sp. nov., while *P. ferruginea* bears a mediansclerite with spiked apex and a right phallomere with an ax-shaped apex.

Poeciloderrhis penduloides sp. nov.

Figs 60–69

Diagnosis. General coloration. Shiny, light brown (Fig. 60). Head with light brown vertex (Fig. 61); interocular and interocellar spaces, and center of clypeus brown; ocelli brown; maxillary palps with light brown apical segment, with golden cilia. Eyes brown. Pronotum semi-transparent, with dark brown punctuations (Fig. 62). Tegmen semi-transparent, with dark brown marks. Legs light brown, with spines, pulvilli, arolia and claws light brown. Abdomen brown with brown punctuations.

Dimensions of holotype male (mm). Total length: 26.9; length of pronotum: 6.9; width of pronotum: 8.7; length of tegmen: 22.1; width of tegmen: 9.0.

Description of the holotype male. Head triangular with rounded angles, vertex slightly exposed in dorsal view; interocular space about 1.45 mm long. Antennae threadlike and tomentose, surpassing apex of cerci. Maxillary palps with first and second segments reduced, third segment larger than fourth, and fifth segment slightly more dilated and densely tomentose.

Thorax. Pronotum ample, convex, with angles curved, base bearing small median projection. Legs with femur I with anteroventral surface bearing 3–4 strong spines up to median region, followed by series of small spines toward apex; apex with two strong apical spines; posteroventral surface with three strong spines, one apical; femora II and III bearing few strong spines on ventral surface. Claws symmetrical and specialized, ventrally with two rows of small teeth, similar to those on legs. Tegmen surpassing apex of abdomen. Marginal field ample, slightly concave; scapular field elongated with oblique venules; discoidal field convex, curved apically; anal field convex and well marked.

Abdomen. Tergal modification comprising three humped structures on first segment and one on second (Fig. 63). Supranal plate round (Fig. 64) with cerci surpassing length of plate; cerci with ventral surface bearing many cilia. Subgenital plate asymmetric (Fig. 65); left style, in ventral view, long, threadlike and well sclerotized; right style, in ventral view, tiny, slightly sclerotized. Left phallomere with sclerotized median structure (Fig. 66). Median sclerite with club-shaped apex, strongly sclerotized, with tiny spine (Figs 67 and 68). Right phallomere with curved apex and small spine near apex (Fig. 69).

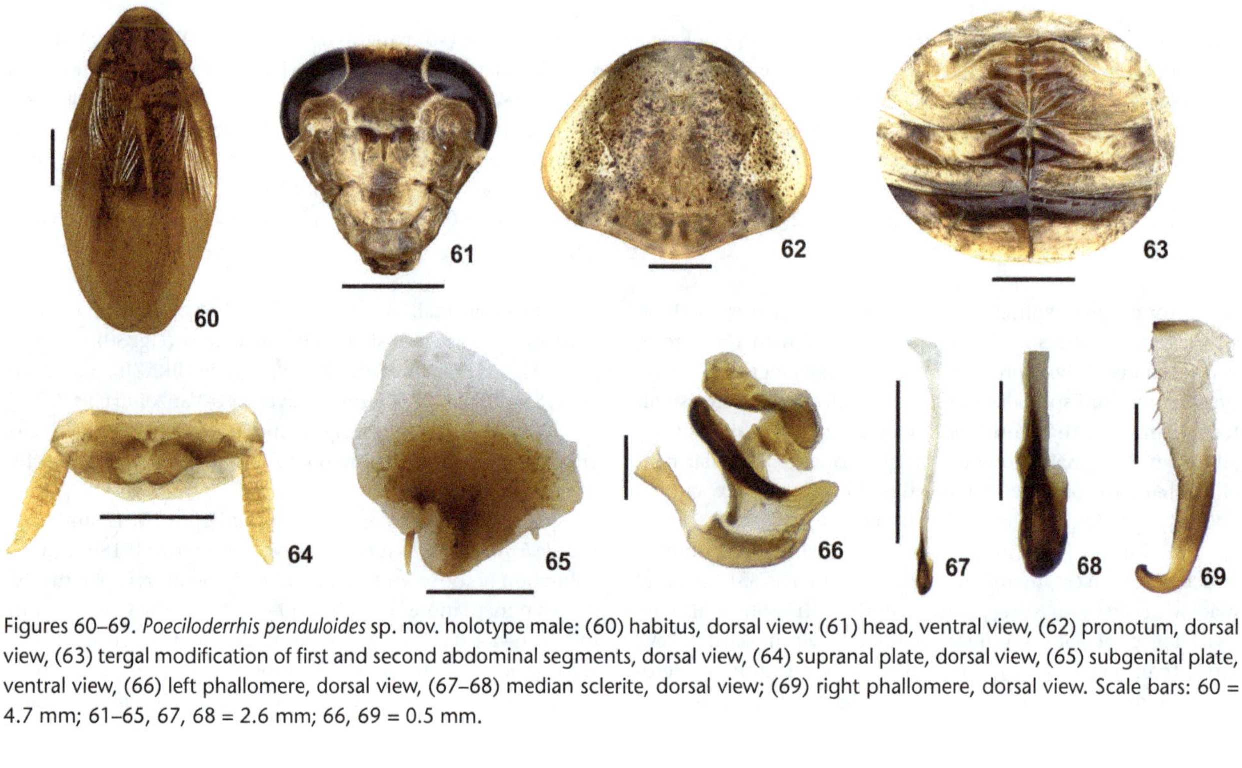

Figures 60–69. *Poeciloderrhis penduloides* sp. nov. holotype male: (60) habitus, dorsal view: (61) head, ventral view, (62) pronotum, dorsal view, (63) tergal modification of first and second abdominal segments, dorsal view, (64) supranal plate, dorsal view, (65) subgenital plate, ventral view, (66) left phallomere, dorsal view, (67–68) median sclerite, dorsal view; (69) right phallomere, dorsal view. Scale bars: 60 = 4.7 mm; 61–65, 67, 68 = 2.6 mm; 66, 69 = 0.5 mm.

Material examined. Holotype male, BRAZIL, Rio de Janeiro State, Itamanjuba, 09-IX-1981, Reys & Santos cols, MNRJ.

Etymology. The species *penduloides*, derived from to the pendulum-shaped apex of the median sclerite of the male.

Remarks. *Poeciloderrhis penduloides* sp. nov. is similar to *Poeciloderrhis tijucana* Silva & Lopes, 2015 in the coloration (light brown and semi-transparent in both). The new species might be distinguished from *P. tijucana* by its smaller size 26.9 mm (33.0 mm in *P. tijucana*) and pigmentation of the tegmen (*P. penduloides* sp. nov. bears small light brown spots, while *P. tijucana* have bigger spots), tergal modification (*P. tijucana* bears two latero-apical elevations converging toward thorax on first segment, while *P. penduloides* sp. nov. bears tree humped structures). Besides that, the new species might be also distinguished by the shape of genital parts that presents a clubbed-like median sclerite (in *P. tijucana* the structure have a rounded finger-shaped apex).

New combination

We analyzed the type material of *Poeciloderrhis bicolorata* Rocha e Silva Albuquerque & Lopes, 1977, and determined that it must be transferred to *Epilampra*. The specimens do not have the tergal modification and configuration of the genital pieces typical of *Poeciloderrhis*, with the right phallomere having a complete hook and the median sclerite with an accessory sclerite, termed the dorsal sclerite (Rocha e Silva Albuquerque and Lopes 1977: 509–510 figs 31–37). Thus the valid name is *Epilampra bicolorata* (Rocha e Silva Albuquerque & Lopes, 1977).

Key to males of *Poeciloderrhis*

1 Total length less than or equal to 19 mm .... *P. minoris* Silva & Lopes, 2015
1′ Total length more than 19 mm 2
2 Pronotum with a central dark brown to black mark 3
2′ Pronotum with pigmentation scattered on surface 8
3 Basal portion of discoidal and anal fields of tegmen covered more than 40% with black pigmentation .... *P. imperialis* Rocha e Silva & Jurberg, 1978
3′ Basal portion of discoidal and anal fields of tegmen covered less than 30% with dark brown pigmentation 4
4 The base of abdominal tergite II with dark brown hook-shaped projection .... *P. verticalis* (Burmeister, 1838)
4′ The base of abdominal tergite II bearing humps, cavity or pendulum 5
5 Head with region underneath ocelli black; second abdominal tergite with middle part presenting two humps 6
5′ Head with region underneath ocelli light brown to brown; second abdominal tergite with middle part presenting only one hump, or a cavity, a pendulum or a transversal elevation or a longitudinal elevation 7
6 Head with median region of clypeus brown; length of tegmen bigger than 35 mm .... *P. aureolatus* sp. nov.
6′ Head with median region of clypeus light brown; length of tegmen smaller than 28 mm .... *P. mediansclerostylatus* sp. nov.
7 Male genitalia with rounded pre-apical extension of median sclerite .... *P. atriventris* (Saussure, 1895)
7′ Male genitalia without pre-apical extension of median sclerite .... *P. paulistensis* Lopes & Oliveira, 2006
8 Head with vertex with four brown marks 9
8′ Head with vertex with fewer than four brown marks ... 12
9 Head with sub-ocular region without dark brown to black pigmentation; tegmen not reaching the apical third of abdomen .... *P. catharina* (Shelford, 1910)
9′ Head with sub-ocular region with dark brown to black pigmentation, tegmen surpassing the apical third of abdomen 10
10 Abdomen with tergite III bearing basal hump .... *P. tijucana* Silva & Lopes, 2015
10′ Abdomen with tergite III without basal hump 11
11 Male genitalia with shaft of median sclerite straight .... *P. vanzolinii* sp. nov.
11′ Male genitalia with shaft of median sclerite sinous .... *P. caracensis* Silva & Lopes, 2015
12 Male genitalia with right style, in ventral view, subequal to half of left style .... *P. diamantinensis* sp. nov.
12′ Male genitalia with right style, in ventral view, mesuring less than half of the left style 13
13 Tegmen not reaching the apex of abdomen .... *P. boraceiana* Lopes & Oliveira, 2006
13′ Tegmen reaching the apex of abdomen 14
14 Cerci not reaching the apex of supra-anal plate .... *P. santosi* (Rocha e Silva & Lopes, 1976
14′ Cerci reaching the apex of supra-anal plate 15
15 The base of abdominal of tergite II with a cavity followed by pendulum shape .... *P. ferruginea* (Brunner von Wattenwyl, 1865)
15′ The base of abdominal of tergite II without cavities or pendulum 16
16 Genitalia with apex of median sclerite clavate .... *P. penduloides* sp. nov.
16′ Genitalia with apex of median sclerite tapering .... *P. itatiaiensis* sp. nov.

ACKNOWLEDGEMENTS

We thank Karla Schneider of the Zoologisches Institut der Martin-Luther Universität, Halle, Germany; Klaus Klass and Christian Schmidt of the Senckenbergisches Institut, Frankfurt, Germany; Harald Bruckner of the Naturhistorisches Museum Wien, Vienna, Austria; Ellen Sandström of the Biological Museum, Lund, Sweden; George Beccaloni of the British Museum (Natural History), London, for the information and photographs sent. Also to Janet W. Reid by reviewing the language.

LITERATURE CITED

Anisyutkin LN (2016) New data on the subfamily epilamprinae (Dictyoptera, Blaberidae) of the new world, with description of a new genus and a new species from Ecuador. Entomological review 96(2): 199–217. https://doi.org/10.1134/S001387381602001X

Gurney AB, Kramer JP, Steyskal GC (1964). Some techniques for the preparation, study and storage in microvials of insect genitalia. Annals of the Entomological Society of America 57: 240–242. https://doi.org/10.1093/aesa/57.2.240

Lopes SM, Oliveira EH (2000) Espécie nova de *Eublaberus* Hebard, 1919 do Estado de Goiás, Brasil e notas sobre *E. marajoara* Rocha e Silva-Albuquerque, 1972 (Blaberidae, Blaberinae). Boletim do Museu Nacional de Zoologia 433: 1–5.

Lopes SM, Oliveira EH (2006). Duas espécies novas de *Poeciloderrhis* (Stål, 1874) do Estado de São Paulo, Brasil (Blattaria, Blaberidae, Poeciloderrhini). Arquivos do Museu Nacional, Rio de Janeiro 64(3): 223–226.

Rocha e Silva-Albuquerque I, Lopes SM (1976) Blattaria de Bromélia (Dictyoptera). Revista Brasileira de Biologia 36:873–901.

Rocha e Silva-Albuquerque I, Lopes SM (1977) Blattaria (Dictyoptera) do Alto da Mosela, Petrópolis, RJ. Revista Brasileira de Biologia 37:499–520.

Rocha e Silva-Albuquerque I, Jurberg J (1978). Uma nova espécie do Dictyoptera). Revista Brasileira de Biologia 38: 91–97.

Roth LM (1970) The male genitalia of Blattaria. II. *Poeciloderrhis* spp. (Blaberidae: Epilamprinae). Psyche 77: 104–119. https://doi.org/10.1155/1970/37214

Roth LM (2003) Systematics and Phylogeny of Cockroaches (Dictyoptera: Blattaria). Oriental Insects 37: 1–186. https://doi.org/10.1080/00305316.2003.10417344

Silva LCO, Lopes SM (2015) Synopsis of *Poeciloderrhis* Stål, 1874, with the description of three new species, and a redescription of the male and female of *Poeciloderrhis ferruginea* (Brunner von Wattenwyl, 1865) from southeast Brazil (Blattodea, Blaberidae, Epilamprinae) 545: 53–65. https://doi.org/10.3897/zookeys.545.6172

Author Contributions: LOCS and SML contributed equally to this article.

Competing Interests: The authors have declared that no competing interests exist.

2

Redescription of *Epiperipatus edwardsii*, and descriptions of five new species of *Epiperipatus* from Brazil (Onychophora: Peripatidae)

Cristiano Sampaio Costa[1], **Amazonas Chagas-Junior**[2], **Ricardo Pinto-da-Rocha**[1]

[1]*Departamento de Zoologia, Instituto de Biociências, Universidade de São Paulo. Rua do Matão-Travessa 14, 321, Cidade Universitária, 05508-090 São Paulo, SP, Brazil.*
[2]*Departamento de Biologia e Zoologia, Instituto de Biociências, Universidade Federal de Mato Grosso. Avenida Fernando Correa da Costa 2367, Boa Esperança, 78060-900 Cuiabá, MT, Brazil.*
Corresponding author: Cristiano Sampaio Costa (csampaioc@gmail.com)

ABSTRACT. *Epiperipatus edwardsii* (Blanchard, 1847) is redescribed based on fresh material collected near the type locality. Additionally, five new species of velvet worms of the genus *Epiperipatus* Clark, 1913 from Brazil are described: *Epiperipatus hyperbolicus* **sp. nov.**, *Epiperipatus lucerna* **sp. nov.**, *Epiperipatus titanicus* **sp. nov.** (Holotypes deposited in MNRJ: Murici, Alagoas State), *Epiperipatus beckeri* **sp. nov.** (Holotype female deposited in MNRJ: Camacan, Bahia State) and *Epiperipatus marajoara* **sp. nov.** (Holotype male deposited in MZUSP: Breves, Marajó island, Pará State). The peculiar shape of the primary papillae (artichoke-like) of *E. titanicus* **sp. nov.** and *E. beckeri* **sp. nov.** is documented for the first time. *Epiperipatus titanicus* **sp. nov.** has the largest number of pairs of oncopods (38 for males and 39 for females) among the Brazilian Onychophora. We also provide an identification key for Brazilian species of *Epiperipatus*.

KEY WORDS. French Guyana, neotropics, onychophorans, *Peripatus*, taxonomy.

INTRODUCTION

Epiperipatus Clark, 1913 is the most diverse genus of Neotropical Peripatidae. Currently, 26 species are known in Central and South America (Oliveira et al. 2012a). Clark (1913a) proposed *Epiperipatus* as a subgenus of *Peripatus* Guilding, 1826, and included the following species: *P.* (*E.*) *edwardsii* (Blanchard, 1847), *P.* (*E.*) *imlhurni* (Sclater, 1888), *P.* (*E.*) *trinidadensis* (Sedgwick, 1888), *P.* (*E.*) *brasiliensis* (Bouvier, 1899), *P.* (*E.*) *simoni* (Bouvier, 1899), *P.* (*E.*) *nicaraguensis* (Bouvier, 1900), *P.* (*E.*) *biolleyi* (Bouvier, 1902), *P.* (*E.*) *isthmicola* (Bouvier, 1902), *P.* (*E.*) *evansi* (Bouvier, 1904), and *P.* (*E.*) *barbouri* (Brues, 1911). The subgenus was delimited based on subjective characters, for instance shape, size and pattern of distribution of the dorsal papillae (Clark 1913a: 17). Many years later, Peck (1975) elevated *Epiperipatus* to genus, without performing a detailed morphological study. More recently, after evaluating scanning electron microscopy images of the integument (especially the primary papillae) of the Neotropical Peripatidae, Read (1988a) suggested that the patterns of scales ranks of the primary papillae can be used to delimit genera within peripatids.

Neopatida is the largest group of Peripatidae from the Neotropics. This taxonomically obscure group comprises the widespread genus *Macroperipatus* Clark, 1913, type species *Peripatus torquatus* von Kennel, 1883, from Trinidad, and *Epiperipatus*, type species *Peripatus edwardsii* Blanchard, 1847, from French Guiana. Those genera were included in a phylogenetic analyses of Costa (2016) and Giribet et al. (2018). Although phylogenetic studies on onychophorans have been seldom conducted including Brazilian species, the number of species described in *Epiperipatus* has increased considerably in the past few years. Brito et al. (2010) described *Epiperipatus cratensis* Brito et al., 2010 from Crato, Ceará State, Northeast Brazil. Oliveira et al. (2011) described three cryptic species of *Epiperipatus*, i.e., *E. diadenoproctus* Oliveira et al., 2011, *E. adenocryptus* Oliveira et al., 2011 and *E. paurognostus* Oliveira et al., 2011, from the Minas Gerais State, Southeast Brazil, using a molecular and morphological approach to delimit the new species. Oliveira et al. (2010) revised the taxonomic status of two species combined with *Macroperipatus*, *M. acacioi* (Marcus & Marcus, 1955) and *M. machadoi* Oliveira & Wieloch, 2005, and examined some characters of the integument and the crural tubercles, concluding that the two species are better combined with *Epiperipatus*. Recently, Chagas-Jr and Costa (2014) also revised a species of *Macroperipatus*, *M. ohausi* (Bouvier, 1900), and concluded that it equally belongs to *Epiperipatus*, not to *Macroperipatus*. Therefore, fourteen species of *Epiperipatus* are currently known to occur in Brazil.

Epiperipatus edwardsii is widespread in the Neotropics, distributed on a large area from Panama, Trininad & Tobago, Venezuela, French Guiana, Suriname, Colombia and Brazil (Read 1988b, Sampaio-Costa et al. 2009, Oliveira et al. 2012a). This contrasts with the narrow distribution of other species of onychophorans restricted to less than 100 km² (Marcus and Marcus 1955, Peck 1975). Blanchard (1847) described *E. edwardsii* based only on the number of oncopods (30 pairs). Bouvier (1905) proposed an emended description, adding, for instance, the diamond-shape pattern on the dorsal body, two incomplete folds, and base of primary papillae roundish, contiguous with dorsal papillae. These characters are shared by several species of *Epiperipatus*, which makes it impossible to identify any particular species using them.

Based on findings of Costa (2016), in this paper we analyzed a specimen of *E. edwardsii* from the Nouragues Field Station, French Guyana, a few kilometers from the species type locality, in Cayenne. We redescribe *E. edwardsii* and provide images documenting the dorsal color and the scale ranks of the primary papillae. Additionally, we describe five new species in *Epiperipatus*, all from Brazil. Two new species, *Epiperipatus titanicus* sp. nov. and *E. beckeri* sp. nov., have symmetrical apical piece with one scale rank, a feature that had not been observed before in Peripatidae.

MATERIAL AND METHODS

In total, we examined 65 specimens collected between 2008 and 2014. The material examined is deposited in the following institutions: Museu Nacional, Universidade Federal do Rio de Janeiro, Rio de Janeiro, Brazil (MNRJ); Museu de Zoologia da Universidade de São Paulo, São Paulo, Brazil (MZUSP); Universidade Federal de Minas Gerais, Belo Horizonte, Brazil (UFMG); Natural History Museum, London, United Kingdom (NHM); Collection of the Evolution et Diversité Biologique laboratory at Université Toulouse III Paul Sabatier, Toulouse, France (GF). Live specimens were photographed (mainly for color and body patterns) with SONY Cybershot DSC-HX1 with built-in flash, or Canon EOS Rebel XS with macrolens and flash circular cameras. The color of live specimens was described following the standard names of the 267 Color Centroids of NBS/ISCC Color System (Mundie 1995). Respective code numbers are given between parentheses following the color name.

We obtained the stacked images using a Leica MZ125 Stereomicroscope with built-in Leica DFC 290 camera, Leica Light source CLS100X series 00373 and assembled through of software Leica Application Suite version 3.3.0. We compiled descriptions, and dissection of dead specimens using a stereomicroscope. The morphological descriptive nomenclature follows the terminologies of Read (1988a), Morera-Brenes and Monge-Nájera (2010) and Oliveira et al. (2010).

Dorsal integument, jaws, legs and some antennae were photographed using Scanning Electron Microscopy (SEM). All tissues were cleaned, subjected to critical point and sputter coating. In the cleaning process, the soft parts were washed in a solution of hypochlorite 3% diluted in distilled water. The hard parts (for example jaws) were cleaned in distilled water and Ethylenediaminetetraacetic acid (or EDTA) heat solution in sonicator processor. For critical point, the tissues were dehydrated using an increasing ethanol series: 30%, 50%, 70%, 80%, 90% and 100%.

These tissues were critical point dried and mounted in SEM stubs with biadhesive carbon tape, in the critical point machines Baltec CPD 030. The sputter coating was completed using the machines Denton Desk IV gold sputter model and Balzer SCD 50 sputter coater. Generally, on average, 5 to 10 nm of gold/paladium or platinum/palladium were deposited over the tissues. We took SEM images using microscopes JEOL JSM-6390LV and DSM 940 Zeiss scanning electron. The Backscatter electron detector (BSD) function, when available on the microscope, was used to avoid the charging effects on the tissues.

The descriptions of new species are based on type series instead of just in the holotype (compound description). A full description is only provided for the first described species, *E. titanicus*, subsequent descriptions are comparative and complementary to that of *E. titanicus*.

TAXONOMY

Peripatidae Evans, 1901

Epiperipatus Clark, 1913

Peripatus (Epiperipatus) Clark, 1913: 17; Cockerell 1913: 87; Scorza 1953: 783; Froehlich 1968: 163, 165–168.

Epiperipatus Clark, 1937: 2; Arnett 1961: 218; Peck 1975: 334; Sampaio-Costa et al. 2009: 556; Morera-Brenes and Monge-Nájera 1990: 449, 2010: 1134; Morera-Brenes and León 1986: 278; Röhlig et al. 2010: 229; Chagas-Jr and Costa 2014: 979; Oliveira et al. 2010: 16, 2012a: 7, 2012b: 18, 2015: 227.

Type-species. *Peripatus edwardsii* Blanchard, 1847 by original designation.

Diagnosis. Dorsal papillae with roundish insertion. Bases of primary papillae with four to eighteen scale ranks. Apical pieces are conical, cylindrical, or rarely flat. Usually four complete and sometimes, rarely vestiges of a fifth spinous pad on the fourth and fifth legs. Nephridial tubercle on fourth and fifth pairs of legs, between third and fourth spinous pads. One to tree pregenital legs with crural papillae (male only).

Epiperipatus edwardsii (Blanchard, 1847)

Figs 1–6

Peripatus edwardsii Blanchard, 1847: 140.

Peripatus (Epiperipatus) edwardsii: Clark 1913: 18.

Epiperipatus edwardsii: Peck 1975: 345; Oliveira et al. 2012a: 10.

Figures 1–6. *Epiperipatus edwardsii*,GF180312HC003-06: (1) dorsal papillae arrangement; (2) dorsal view of the distal half of the body depicting diamond shaped pattern over the dorsomedian furrow; (3) foot in ventral view; (4) primary papillae in posterior view depicting differences in the number and shape of the scales (B and P); (5) outer blade in first plan of right jaw; (6) right fourth leg in ventral view depicting spinous pad undivided (I) and divided (IV) into two unqueal parts. (AT) Accessory tooth, (B) basal piece, (DR) series of denticles, (I) first spinous pads, (IJ) inner blade, (IV) last spinous pads, (nt) nephridial tubercle, (OJ) outer blade, (P) apical piece, (PP) prolateral foot papillae, (RP) retrolateral foot papillae. Scale bars: 1, 5 = 100 μm, 2 = 5 mm, 3 = 30 μm, 4 = 20 μm, 6 = 200 μm.

Diagnosis. The background color of the body of this species is in the purple spectrum; dorsal papillae round, primary papillae bearing asymmetrical apical piece, four complete spinous pads on the fourth and fifth legs (the last broken in some specimens).

Redescription. Measurements (mm): Type: length 35.0; width 4.0; height 3.0 (based on Bouvier 1905). Additional specimen (hereafter): length 30.0, width 3.0, height 3.0. Color (living specimen). Background color of body dark purplish gray (234), dorsal surface dark grayish reddish brown (47) with purplish gray (233) diamonds overlapping light brownish gray (63) dorsomedian furrow (Fig. 2). Background color of ventral body purplish gray (233) with light gray and yellowish-brown (79) median band. Legs white (263) and ventral organs pale pink (7). Description of body. Conspicuous dorsomedial furrow and hyaline organs on entire length of body. Twelve plicae per segment, two incomplete (not extending to insertion of legs), and seven crossing over to ventral side. Uniform width among to the dorsal plicae as mentioned by Bouvier (1905). Dorsal papillae on all plicae (Fig. 1). Primary papillae as the largest dorsal papillae, with roundish dome insertion and asymmetrical spherical apical piece (Figs 1, 4). Basal piece larger than apical piece, with a range of at least seven scale ranks. Apical piece with three posterior scale ranks (Fig. 4). Bristle directed posteriorly (Fig. 4). Accessory papillae usually the smallest, but some as large as the basal piece of a primary papilla (Fig. 1). Accessory papillae with roundish insertion. Accessory papillae more abundant than primary papillae per plica (Fig. 1). Frequently, primary papillae separated by one to four accessory papillae. Dorsal papillae always on the plicae, although only and rarely accessory papillae present on flanks.

Head. No evident structures or color patterns on head. Antennae with 44 rings: antennal tip composed of seven broad rings, excluding the terminal disc on top. Antennal body with alternating sequence of narrow and broad rings at least up to the eighteenth. Eyes and frontal organs present on ventrolateral region of antennal base. Frontal organs as long as four fused antennal papillae. Mouth opening surrounded by small, anterior, unique lobe, and seven flanked lobes decreasing in size from anterior to posterior ends of the mouth. Dental formula of inner and outer jaws, respectively: 1/1 and 1/1/10, with second accessory tooth of inner jaw vestigial (Fig. 5).

Legs. Male with 30 pairs of legs. Ventrally, fourth and fifth pairs of legs with four complete spinous pads (the fourth are sometimes broken) without vestige of fifth spinous pad (Fig. 6). Nephridial tubercle on fourth and fifth pairs of legs, between third and fourth spinous pads, and connected at tip with third spinous pad (Fig. 6). Two pro-lateral and one retro-lateral foot papillae the feet of fourth and fifth legs (Fig. 3). Conspicuous ventral and preventral organs, gonopore close to penultimate pairs of legs in both sexes.

Sexual dimorphism. Male pregenital leg without crural papillae (these organs had not been previously described for this species). Anal glands inconspicuous (male); represented by two pores on ventral portion of anal aperture.

Material examined. 1♂, French Guiana, Cayenne, Nouragues Field Station, no further data (GF180312HC003-06). The holotype, deposited at the Muséum National d'Histoire Naturelle, Paris, France (MNHN), was not examined because the curator did not reply to our loan requests.

Distribution. French Guyana: Cayenne and adjacency of Nouragues Field Station (Bélizon, Bita, Coco).

Remarks. The type of *Epiperipatus edwardsii*, deposited at the Muséum National d'Histoire Naturelle, Paris, France (MNHN) was not examined. We tried to contact the curator to arrange a loan of this material, but the curator did not reply. *Epiperipatus edwardsii* is the most frequently misidentified species of *Epiperipatus* and for this reason some records of its distribution (e.g., Brues 1914, Clark 1914, Peck 1975, Sampaio-Costa et al. 2009) are not accepted here. The shape and distribution of the dorsal papillae, dorsal diamond pattern and number of legs are frequently misinterpreted.

According to our observations, the diamond patterns are a general feature in most species of *Epiperipatus*. Morphologically, the diagnostic features to delimit *E. edwardsii* are (1) dorsal papillae over the plicae and never on the folds, and (2) primary papillae usually separated by one or two accessory papillae (however in some cases they are close to each other). Also, *E. edwardsii* is supported by molecular data as demonstrated in Maximum Parsimony and Maximum Likelihood analyses (e.g., Costa (2016) and Giribet et al. 2018).

The other records of *E. edwardsii* found in the literature, for instance from Panama, Venezuela, Colombia and Brazil (Brues 1914, Clark and Zetek 1946, Read 1988a, b, Sampaio-Costa et al. 2009, see discussion below) are probably based on undescribed species. Moreover, records of *E. edwardsii* from Aripo in Trinidad (Read 1988a) and Haut Sarare, Venezuela (Read 1988b) also correspond to an undescribed species. In conclusion, the distribution of *E. edwardsii* is likely limited to French Guyana.

Epiperipatus titanicus sp. nov.

Figs 7, 13–16, 37

Epiperipatus sp. 1: Sampaio-Costa et al.: 2009: 556, 557.

Diagnosis. *Epiperipatus* species with symmetrical, reduced or hidden apical piece. Body large, 29 up to 76 mm and 36 to 39 pairs of legs.

Compound description. Measurements. Males: length 29–41; width 1.0–3.0; height 1.0–2.5. Females: length 16–76; width 1.0–4.0; height 1.0–6.0. Background color of body in vivo is dark grayish red (20), with dorsal wavy band of light grayish red (18) and overlaid by diamond-shape marks, the same color as background color of body (Fig. 13). Dorsomedian furrow very dark purplish red (260) (Fig. 13). Anterior portion of head and antennae dark purple (224). Color of dorsal portion of legs the same as predominant dorsal body color. Largest primary papillae pale and distinguishable in vivo from light brown (57) color found on alternate plicae and forming longitudinal sequences on dorsal integument. The number of pale primary papillae increases near the legs. Legs and ventral surface displaying same color, pale pink (7). Description of body. Dorsal papillae aligned on top of folds, two primary papillae separated by one to three accessory papillae and rarely occurring close together. Both dorsal papillae have conical basal pieces, artichoke-like, composed of lanceolate scales that never overlap with one another at base of papillae (Figs 7, 14). The primary papillae are the largest and at their bases there are five to ten scale ranks. Apical piece can be reduced and there is just one scale rank (Fig. 7). Rank in some cases overlapped by scale rank of the base papilla, hiding the constriction between the two parts of the primary papillae (Fig. 7. See also Sampaio-Costa et al. 2009: 557, figs 6–8). The apical piece, when evident, is symmetrical and conical with narrow and elongated scales (Fig. 7). Needle-shaped sensory bristle curved or straight, placed at center of apical piece. Accessory papillae similar to primary papillae with respect to shape of scale and morphology of the base, but without an apical piece.

Head. Holotype and paratype with 44–45 antennal rings, respectively. In both specimens, antennal tip composed of seven broad rings (excluding the terminal disc on top) and followed by alternating narrow and broad rings until the eighteenth ring. Mouth opening surrounded by anterior oral lobe (unpaired lip) and seven pairs of oral lips. Jaws with two blades composed of one long curved main tooth followed by one accessory tooth (Figs 15, 16).

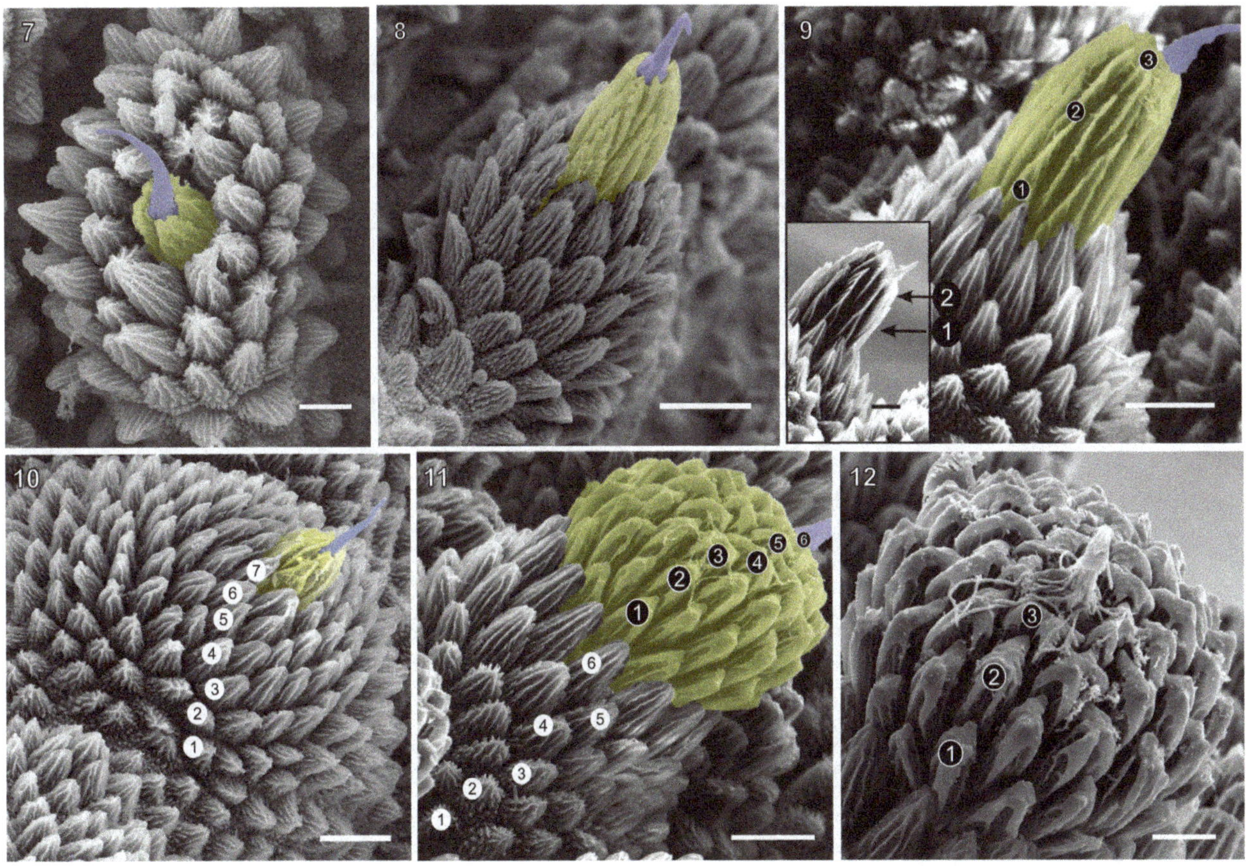

Figures 7–12. Shape of primary papillae (holotype, except when mentioned) of *Epiperipatus beckeri* sp. nov., *E.piperipatus lucerna* sp. nov., *Epiperipatus marajoara* sp. nov., *Epiperipatus hyperbolycus* sp. nov. and *Epiperipatus titanicus* sp. nov. (7) *Epiperipatus titanicus* sp. nov. (MNRJ 0057); (8) *Epiperipatus marajoara* sp. nov. (MZUSP 0022); (9) *Epiperipatus lucerna* sp. nov. MNRJ 0101; (10) *Epiperipatus beckeri* sp. nov. (MNRJ 0045); (11–12), *Epiperipatus hyperbolycus* sp. nov. (MNRJ 0104). Scales bars: 7, 12 (also in the small box of Fig. 9) = 10 μm, 8–11 = 20 μm.

Formulae of inner and outer jaw blades: 1/1 and 1/1/7–11 (Figs 15, 16), respectively. Legs. Males: 36–38 pairs of legs. Females: 36–39 pairs of legs. Nephridial tubercle on fourth and fifth pairs of legs, between third and fourth spinous pads (Fig. 17). On fourth and fifth pairs of legs, fifth spinous pad present and always reduced (Fig. 17).

Sexual dimorphism. Two or three pregenital legs with one crural papilla (male) each. Anal glands inconspicuous (male); represented by two pores on anterior margin of anal aperture.

Type material. Holotype. ♀, BRAZIL, Alagoas State: Murici, Estação Ecológica de Murici (Mata da Bananeira area), Lo-Man--Hung, N.F. leg. (MNRJ 0057). Paratypes. ♂, idem, 11–18.xiii.2009, Costa, C.S., Giupponi, A. and Chagas-Jr leg. (MNRJ 0035); 2 ♀, same date but 21.vii.1995, Freitas, L.M. and Natali, M.S. leg. (DZ-UFMG-ONY 0008 and 0009); 9♀, 4 ♂ and 1 unsexed, same data but, 11–18.xii.2009, Costa, C.S., Guipponi, A. and Chagas-Jr, A, leg (MNRJ 0053); 1♀, same data but, 26–29.vii.2012, Costa, C.S., Alcântara, D.M.C., Dias, P.S. and Nihei,S.S. leg (MZUSP 0010).

Distribution (Fig. 37). Only known from the type locality (Brazil, Alagoas state, Murici, Estação Ecológica de Murici).

Etymology. The specific name *titanicus*, *titanica*, *titanicum* (L., "titanic", adjective) refers to the large size of the females, and the great number of pairs of oncopods, more numerous than in any other Brazilian species of onychophorans.

Remarks. The number of leg pairs in this species varies from 36 to 39, and these numbers overlap in males and females, 36 to 38 in males and 36 to 39 in females.

Epiperipatus lucerna sp. nov.

Figs 9, 18–20, 37

Diagnosis. Apical piece cylindrical, asymmetric, have of two scale ranks on back side and three on front side; 29 to 30 pairs of legs.

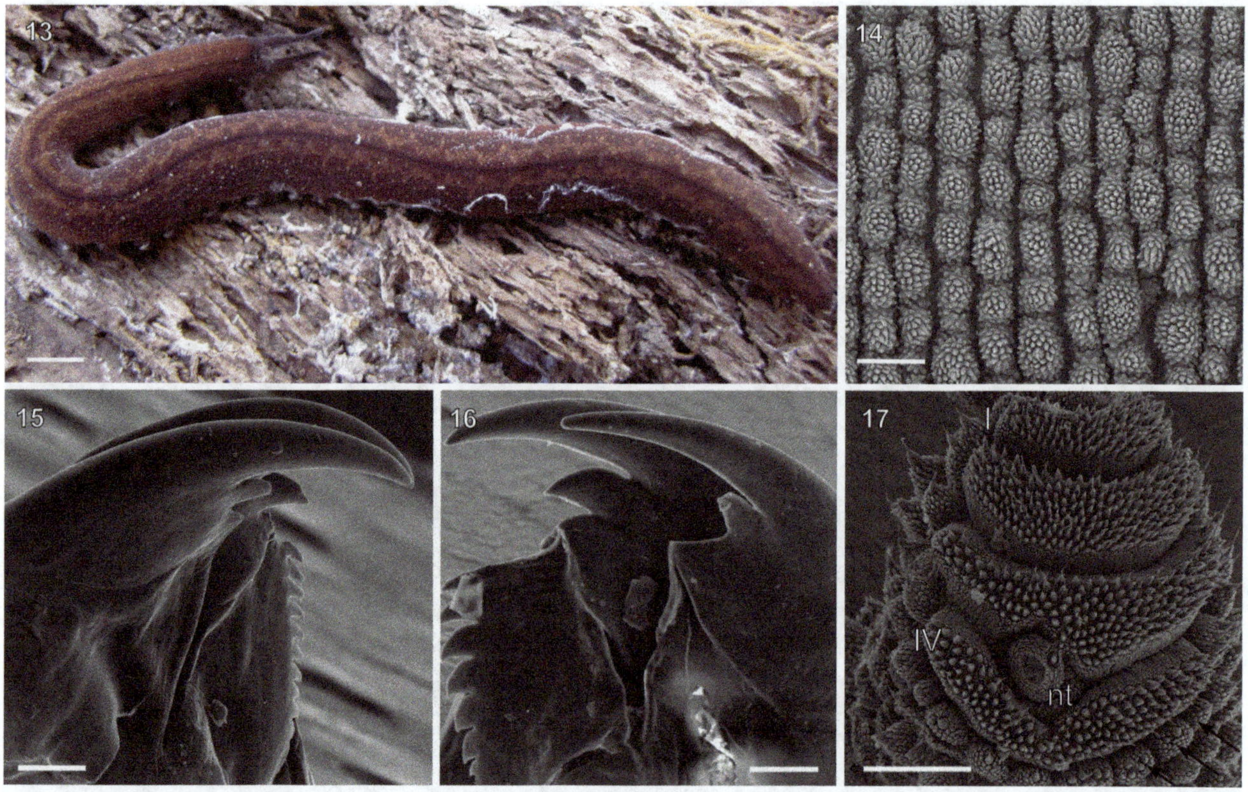

Figures 13–17. *Epiperipatus titanicus* sp. nov. habitus, dorsal plicae, spinous pad and jaws: (13) Living specimen in dorsal habitus, paratype; (14) dorsal papillae arrangement of the holotype; (15) paratype blades; (16) holotype blades with the accessory tooth broken in outer jaws; (17) spinous pad and nephridial tubercles of paratype. The Roman numbers indicate the first and last spinous pad of the leg. Black arrows point vestiges of a fifth spinous pad; (15–16) jaws with outer blade in first plane in the both figures. Scales bars: 13 = 2.7 mm, 17 = 200 µm, 14– 15 = 100 µm, 16 = 50 µm.

Compound description. Measurements. Male: length 13–29; width 1.0–3.0; height 1.0–1.5. Female: length 23–44; width 1.0–3.0; height 2.0–4.0. Color (living specimens). In vivo background color of body on dorsal portion, arrangement of pale primary papillae, and ventral region as described for *E. titanicus*. Description of body. Dorsal papillae aligned on top of folds, with two primary papillae close together or separated by one to five accessory papillae. Dorsal papillae with conical basal piece, composed of overlapping lanceolate scales. Primary papillae regular in size, at their bases four to eight scale ranks (Fig. 19). Apical piece cylindrical, asymmetrical, composed of two scale ranks on back side and three on front side (Fig. 9). This rank overlapped by scale rank of the base papilla and hiding constriction between the two parts of primary papillae (Figs 9, 19). Needle-shaped sensory bristle curved or straight and located on back side of apical piece (Fig. 9). Head. Number of complete antennal rings in holotype 29, alternating between narrow and large rings, from the eighth to the twentieth rings. Paratype with 46 rings and ring widths alternating from the eighth to the eighteenth ring. Antennal tips composed of seven large rings and terminal bud. Number of tooth in jaw blades as in *Epiperipatus titanicus* sp. nov. as in dental formula: (1/1; 1/1/8). Legs. Males: 27–29 pairs of legs. Females: 29–31 pairs of legs. Nephridial tubercle on fourth and fifth pairs of legs, between third and fourth spinous pads (Fig. 20). No evidence of fifth spinous pad on fourth and fifth pairs of legs.

Sexual dimorphism. One or two pregenital legs with crural papillae (male) with one crural papilla each (structure not identifiable in some adult males examined). Anal glands inconspicuous (male); represented by two pores on anterior board of anal aperture.

Type material. Holotype. MNRJ 0101, ♂, Brazil, Alagoas state, Murici, Estação Ecológica de Murici (Mata da Bananeira area), 11–18.xii.2009, Costa, C.S., Giupponi, A. and Chagas-Jr leg. Paratypes. MNRJ 0102, ♀, idem; MNRJ 0103, 13♀, 6 ♂ and 3 unsexed, Brazil, Alagoas state, Murici, Estação Ecológica de Murici (Mata da Bananeira area), 11–18.xii.2009, Costa, C.S., Giupponi, A. and Chagas-Jr leg; MZUSP 0011, 1♀, idem, 26–29.vii.2012, Costa, C.S., Alcântara, D.M.C., Dias, P.S. and Nihei,S.S. leg.

Figures 18–23. *Epiperipatus lucerna* sp. nov. and *Epiperipatus hyperbolicus* sp. nov.. *Epiperipatus lucerna* sp. nov., dorsal plicae, spinous pad and jaws: (18) dorsal papillae arrangement over the plicae (holotype); (19) primary papillae, holotype. Note the scales ranks on the basal piece; (20) spinous pad and nephridial tubercles of paratype. *Epiperipatus hyperbolicus* sp. nov., dorsal plicae, spinous pad and jaw, paratype images: (21) dorsal plicae on top view. Incomplete plicae position indicate by the arrow head; (22) spinous pad and nephridial tubercles. The Roman numbers indicate the first and last spinous pad of the leg; (23) jaw. The inner and outer blades are illustrated in the large and small boxes. Scales bars: 18, 21 = 500 µm, 19 = 20 µm, 20 = 200 µm, 22 = 100 µm, 23 (also in the small box) = 50 µm.

Distribution (Fig. 37). Only known from the type locality (Brazil, Alagoas State, Murici, Estação Ecológica de Murici).

Etymology. The specific name *lucerna* (L., "lamp", feminine noun in apposition) refers to the candle shape of the apical piece of the primary papillae of this species.

Epiperipatus hyperbolicus sp. nov.

Figs 11, 12, 21–23, 37

Diagnosis. Dorsal papillae with conical and small basal piece composed of truncated robust scales and primary papillae of regular size with six or seven scale ranks at base and robust and asymmetric spherical apical piece. Legs with 23 to 25 pairs.

Compound description. Measurements (based on a single male): length 26, width 2.0 and height 3.0. Females: length 22–33, width 17–2.0 and height 2.0–3.0. Color (living specimens). Background color of dorsal body, arrangement of pale primary papillae, and ventral region as in *Epiperipatus titanicus* sp. nov. Description of body. Dorsal papillae aligned on top of folds, two primary papillae occurring close together or separated by one to three accessory papillae. Dorsal papillae with conical basal piece, composed of truncated scales (Fig. 21). Primary papillae regular in size with six or seven scale ranks at base and robust and asymmetric spherical apical piece (Figs 11, 12). Apical piece composed of overlapped scale ranks, six or seven on front and three on back (Figs 11, 12). Constriction between the two portions of primary papillae hidden (Figs 11, 12). Needle-shaped sensory bristle curved or straight and dislocated to back side of apical piece.

Head. In holotype antennae with 27 rings (28 rings in paratype, some females with 36, 37 and 41 antennal rings). In both holotype and paratype the antennal tips composed of seven broad rings and terminal bud. Antennal tip followed by alternating narrow and broad rings until sixteenth ring. Jaw blades as in *Epiperipatus titanicus* sp. nov. with dental formula of holotype and paratype as follows: 1/1/7 and 1/1/11 (Fig. 23). Legs. Male holotype: 23 pairs of legs. Females: 24 and 25 pairs of legs. Nephridial tubercle on fourth and fifth pairs of legs, between third and the fourth spinous pads (Fig. 22). Fourth and fifth pairs of legs without evidence of fifth spinous pad (Fig. 22).

Sexual dimorphism. One or two pairs of pregenital legs present (male) with one crural papilla on each. Anal glands inconspicuous (male); represented only by two pores on anterior margin of anal aperture.

Type material. Holotype. MNRJ 0104, ♂, Brazil, Alagoas State, Murici, Estação Ecológica de Murici (Mata da Bananeira), 11–18.xii.2009, Costa, C.S., Giupponi, A. and Chagas-Jr leg. Paratypes. MNRJ 0105, ♀, idem; MNRJ 0106, 2♀, Brazil, Alagoas State, Murici, Estação Ecológica de Murici (Mata da Bananeira), 11–18.xii.2009, Costa, C.S., Giupponi, A. and Chagas-Jr leg.

Distribution (Fig. 37). Only known from the type locality (Brazil, Alagoas State, Murici, Estação Ecológica de Murici).

Etymology. The specific name *hyperbolicus*, *hyperbolica*, *hyperbolicum* (Latinized adjective "exaggerated", from Greek ὑπερβολή, plus suffix *-icus*) refers to the large spherical shape of the apical piece of the primary papillae of this species.

Epiperipatus beckeri sp. nov.

Figs 10, 24–27, 37

Diagnosis. Dorsal background color of body light orange, with dark orange dorsomedian furrow, and antennae purple brownish. Apical piece of primary papillae symmetrical, reduced, not hidden basally.

Compound description. Measurements: length 35–53, width 2.0–4.0, height 4.0. Holotype: length 53, width 4.0, height 2.0. Color (living specimens). Background color of body of the specimen is moderate yellowish pink (29) with a mid-longitudinal pale orange yellow (73) wavy stripe interrupted by well-defined background color cardioid patches (Fig. 24). Dorsomedian furrow thin, dark yellowish pink (30). Anterior portion of head strong yellowish pink (26) gradually turning to light violet (210) in antennae (Fig. 24). Arrangement of pale primary papillae as in *Epiperipatus titanicus*. Background color of body of juvenile light violet (210). Description of body. Dorsal papillae roundish and close together, rarely varying in the size. Larger, pale, primary papillae present, except in juveniles, marked only by faded aspect. Dorsal papillae aligned on top of folds. When separate there are 2 to 4 accessory papillae between two primary papillae (Fig. 25). Base of primary papillae conical and apical piece symmetrical and reduced, with some exceptions (see comments) (Fig. 10). Seven to twelve scale ranks at base and only one scale rank on apical piece (Fig. 10). Bristle on apical piece narrow and long, sometimes as long as apical piece (Fig. 10). Accessory papillae conical, generally the same size as primary papillae, but without apical piece (Fig. 26). Plicae regular, 12 per segment over legs, seven crossing to ventral side and incomplete plicae shorter than length of dorsal side (Fig. 25). Head. Holotype with 44 rings with seven broad rings on antennal tip and followed by sequence of alternating narrow and broad rings until fourteenth ring. In the other adult female there are 40 rings with the same aspect described for the holotype. Jaws (holotype) dental formula: 1/2 and 1/2/11 (Fig. 27).

Legs. Twenty-eight to thirty (juveniles included). Holotype with 30 pairs of legs. Nephridial tubercles between third and fourth spinous pads, connected by strait isthmus on top with third spinous pad. Two prolateral and one retrolateral foot papillae on feet of fourth and fifth legs.

Sexual dimorphism. Males unknown.

Type material. Holotype. ♀, Brazil, Bahia State, Camacan, Reserva Particular do Patrimônio Natural Serra Bonita, 11–15.vi.2009, Chagas-Jr, A., Dill, V., Giupponi, A., Pedroso, D. & Kury, A. Leg (MNRJ 0045),. Paratypes. 1 ♀ and 4 juveniles, Brazil, Bahia State, Camacan, Reserva Particular do Patrimônio Natural da Serra Bonita, 14.ii to 12.iii.2009, M. Teixeira Junior leg. (MZUSP 0017).

Distribution (Fig. 37). Only known from the type locality (Brazil, Bahia State, Camacan, Reserva Particular do Patrimônio Natural Serra Bonita.).

Etymology. The specific name *beckeri* (German surname, noun with Latin ending -i of the genitive case) honoring the naturalist and entomologist Vitor O. Becker, who created and is responsible for the administration of Reserva Particular da Serra Bonita, Camacan, Bahia State.

Remarks. The main differences between juveniles and adults are the color and the kind of dorsal papillae they display. A juvenile was photographed in vivo with background color of body very dark purplish red (260), without diamond areas. The other post-fixed juvenile darker than adults, with background color of body dark brown (55), diamond areas brownish orange (54) and legs on dorsal region pale yellow (89).

The regular primary papillae and accessory papillae have similar size in the adult. In the juveniles, the largest pale primary papillae are more diverse on the dorsal plicae. These primary papillae have a spherical everted and robust apical piece, their arrangement forming longitudinal lines similar to the condition found in the holotype.

Epiperipatus marajoara sp. nov.

Figs 8, 28–32, 37

Peripatus simoni: Bouvier, 1900: 753; 1905: 315; Froehlich 1968: 167 (misidentification).

Figures 24–30. *Epiperipatus beckeri* sp. nov. and *Epiperipatus marajoara* sp. nov. *Epiperipatus beckeri* sp. nov., dorsal body background and papillae and jaw, holotype, MNRJ 0045: (24) body background of dorsal side; (25) dorsal plicae and papillae arrangement. Note the incomplete plicae on the right side of the picture; (26) accessory papilla in posterior view. Note slight different shape between the accessory and primary papillae (see the Fig. 10); (27) teeth formula and shape. Note two accessory teeth for both blades. Note in the small box the denticles series of the outer blade, with eleven denticles. *Epiperipatus marajoara* sp. nov., body patterns: (28) body background of dorsal side; (29) body background of the ventral side. Note the ventral organs clearly visable. MZUSP 0022, holotype; (30) dorsal papillae arrangement around to the dorso-mediam furrow in anterior view. Note the elongate apical piece in the primary papillae and the small scale ranks (six) on the basal piece.Scale bars: 24, 28, 29 = 5.0 mm, 25 = 500 µm, 26 = 20 µm, 27 (in the small box = 50 µm), 30 = 100 µm.

Epiperipatus simoni: Clark and Zetek 1946: 211; Peck 1975: 346; Vasconcelos et al. 2004: 140; Sampaio-Costa et al. 2009: 556 (misidentification).

Diagnosis. This species has dorsal background color of body in the brown spectrum, dorsal diamond pattern inconspicuous or absent, roundish dorsal papillae, primary papillae with well-developed base, and a cylindrical apical piece.

Compound description. Measurements. Females. Length 40, 50 and 65 (specimens from NHM examined by Bouvier in his monograph). Males. Length 27–34, width 2.0–7.0, height 1.0–6.0. The single female is: length 33, width 2.8, height 1.2. Color (living specimens). Background color of body Light Grayish Brown (60) overlaid with brownish pink (33) diamond-shaped makers. Dorsalmedian furrow moderate brown (58), and antennae grayish reddish brown (46) (Figs 28, 29). Largest primary papillae pale and distinguishable in vivo by the light brown (57) color. Found on alternate plicae and forming longitudinal sequences on dorsal integument. Dorsal side of legs same color as dorsal background color of body. Color uniform on ventral surface, grayish yellowish pink (32), except for white spot splashed on integument over the ventral organs (Fig. 29). The number of pale primary papillae increases towards legs. Description of body. Dorsal papillae aligned on top of folds, with two primary papillae separated by one to three accessory papillae (Fig. 30). Both have conical and artichoke-like basal pieces, and are composed of overlapping

Figures 31–36. *Epiperipatus marajoara* sp. nov., *Epiperipatus cratensis*, *E. brasiliensis* and *E. paurognostus*. *Epiperipatus marajoara* sp. nov., MZUSP 0022, holotype: (31) spinous pad and the nephridial tubercle of the fifth leg; (32) inner and outer blade of right jaw. The accessory tooth is broken in the outer blade. (33, 35) *Epiperipatus cratensis*, MZUSP 0083, dorsal papillae arrangement with dorsomedian furrow on the top of image and primary papillae in lateral view, respectively. (34) *E. brasiliensis*, MZUSP 0121, primary papillae in lateral view. (36) *E. paurognostus*, UFMG 0184, dorsal papillae arrangement. Note all dorsal papillae are proeminent. Scale bars 31, 32 = 30 µm, 33 = 500 µm, 34 = 20 µm, 35 = 100 µm, 36 = 40 µm.

lanceolate scale ranks. Primary papillae the largest dorsal papillae, bearing five to six scale ranks basally. Apical piece conical and reduced (Fig. 30), composed of three scale ranks on front side and two scale ranks on back side. Constriction between base and apical piece hidden by basal scale ranks. Bristle slightly displaced to posterior region of apical piece.

Head. Antennae composed of 43 to 47 rings. Antennal tip composed of seven broad rings, except the terminal disc on top, followed by sequence of narrow and broad rings until at least the eighteenth antennal ring. Jaws with two blades and both composed of one long curved main tooth followed by one accessory tooth. Dental formula of outer and inner for holotype and paratype are: 1/1 and 1/1/8 (Fig. 32). Legs. Males: 27 pairs of legs. Single female with 31 pairs of legs. Nephridial tubercle on fourth and the fifth pairs of legs, between third and fourth spinous pads (Fig. 31), no evidence of fifth spinous pad.

Sexual dimorphism. One or two pregenital legs with one or two crural papillae (male), absent in females. Anal glands inconspicuous (male); represented only by two pores on anterior margin of anal aperture, absent in females.

Type material. Holotype. ♂, Brazil, Pará State, Island of Marajó, Breves, Reserva de Extrativista Mapuá, Breves; 4–8.xii.2012; Cristiano Sampaio Costa leg. (MZUSP 0022). Paratypes. Same data as holotype. ♀ (MZUSP 0027). 4 ♂ (MZUSP 0023-0026). 3♀, Breves, Marajó Island, Brazil, Amazon, 96.5.14.6-8 (NHM # 1004943, 1004944, 1004945).

Distribution (Fig. 37). Only known from the type locality (Brazil, Pará State, Breves, Reserva Extrativista Mapuá).

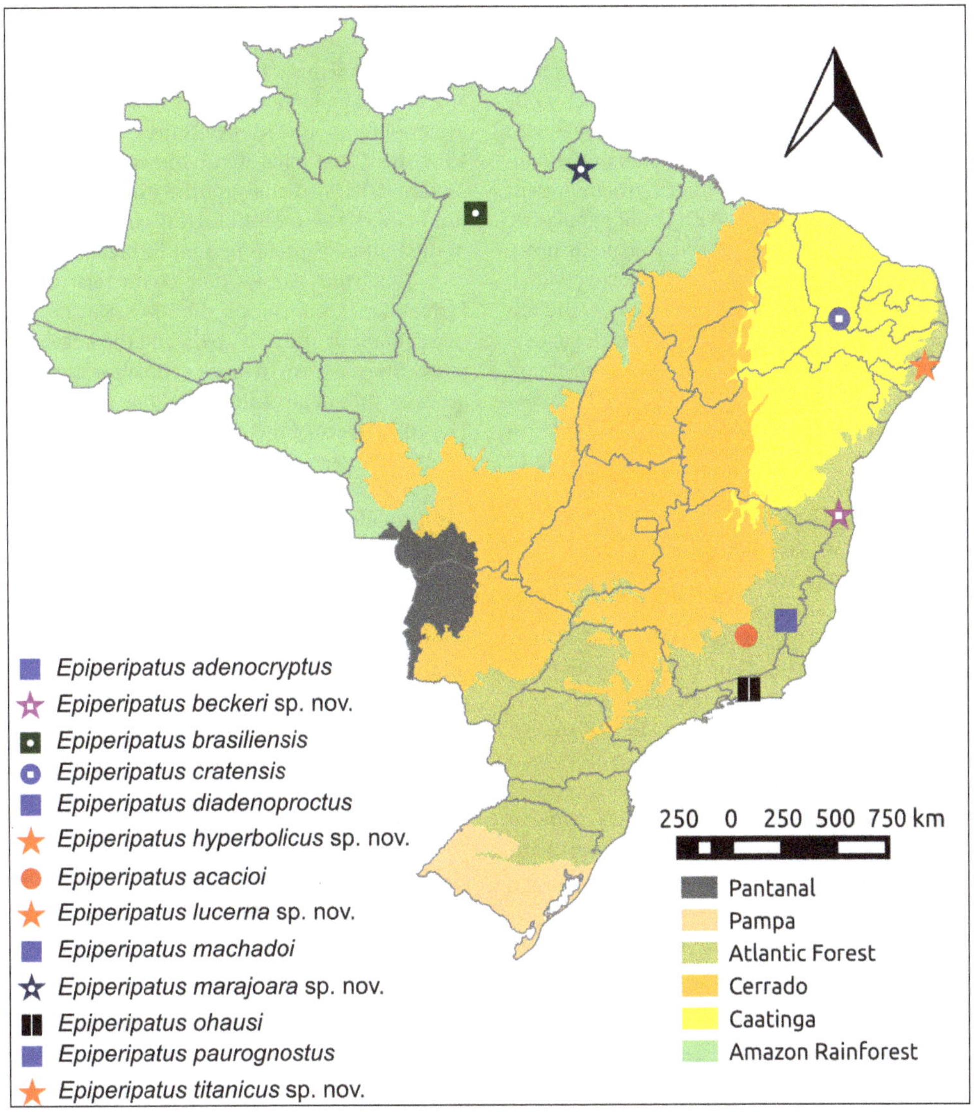

Figure 37. Distribution records of species of *Epiperipatus* in Brazil. Colors on the map correspond to biomes. The new species are represented by stars on the map. The absence of *E. tucupi*, is due to the inaccurate original label (the only data is "Pará state").

Etymology. The specific name *marajoara* (demonymic noun in apposition; invariable), honors the marajoara people, natives from the Island of Marajó.

Remarks. This species was discovered comparing the descriptions with *bona fide* specimens of *E. simoni* from NHM and specimens collected from the Island of Marajó, Pará State, Brazil. The type locality of *E. simoni* is Caracas in Venezuela, with records in Breves, Island of Marajó (Pará State) and Rio de Janeiro State (see Sampaio-Costa et al. 2009). We examined the specimens of *E. simoni* from Breves identified by Bouvier (1905) deposited in the NHM and we considered a misidentification of the new species here described. Also, we examined the integument of the specimens of *E. simoni* from Rio de Janeiro (Sampaio-Costa et al. 2009: 556) and we believe both records are misidentifications and belong to undescribed species.

DISCUSSION

Since 2010, the number of species of *Epiperipatus* has increased through descriptions of new species and new combinations. After the five new species erected in this work, the genus became the most diverse of Neotropical onychophorans, now including 31 species. In Brazil, *Epiperipatus* is also the most diverse genus of Peripatidae, with 14 species (excluding *E. edwardsii* and *E. simoni*, Fig. 37).

As stated by Bouvier (1905), the males of *Epiperipatus* have two pairs of pregenital oncopods with crural papillae, while males of *Peripatus* have at least two pregenital oncopods with crural papillae. Any new species described here fits in the Bouvier's definition for *Epiperipatus*, because three of them present one or two pairs of pregenital oncopods with crural papillae (*E. hyperbolicus* sp. nov., *E. lucerna* sp. nov. and *E. marajoara* sp. nov.), or two or three (*E. titanicus* sp. nov.). According to Read (1988a), *Epiperipatus* has three ranks of scales in the apical piece, and less than 12 ranks of scales in the basal piece. We also do not agreed with Read (1988a) because *E. beckeri* sp. nov. and *E. titanicus* sp. nov. have a short apical piece bearing one rank scale, *E. lucerna* sp. nov. and *E. marajaora* sp. nov. have two ranks of scales in the apical piece and *E. hyperbolicus* sp. nov. has three ranks of scales in the apical piece. The short apical piece bearing one rank scale is a feature never seen in other species of the genus and is described for the first time (see Figs 7, 10, Table 1).

However, the five new erected species are herein assigned to *Epiperipatus* because they share, with other species in the genus, the rounded insertion of the dorsal papillae, and distinct nephridial tubercles between the third and fourth spinous pads. They are also nested within of *Epiperipatus* based on molecular synapomorphyes of found in COI, 12S rRNA, 16S rRNA and 18S rRNA sequences (Costa 2016). Although it does not mean a phylogenetic relationship (see Costa 2016), the species of *Epiperipatus* could be divided into two groups according to the dorsal folds (complete vs. incomplete). The group with incomplete folds is formed by *E. acacioi*, *E. adenocryptus*, *E. barbadensis* (Froehlich, 1962), *E. barbouri*, *E. betheli* (Cockrell, 1913), *E. biolleyi*, *E. broadwayi* (Clark, 1913), *E. diadenoproctus*, *E. edwardsii*, *E. evansi*, *E. imthurni* (Sclater, 1888), *E. isthmicola*, *E. machadoi*, *E. nicaraguensis*, *E. ohausi*, *E. paurognostus*, *E. simoni*, *E. trinidadensis*, and *E. vespucci* Brues, 1914. While the group with complete folds is formed by *E. brasiliensis*, *E. cratensis*, *E. torrealbai* Scorza, 1953, *E. vagans* Brues, 1925 and *E. tucupi* Froehlich, 1968. Some species of *Epiperipatus* such as *E. hilkae* Morera-Brenes and Monge-Nájera, 1990 and *E. lewisi* Arnett, 1961 the dorsal folds are unknown. The five new species here present falls into the group with incomplete folds. Table 1 summarizes the morphological characteristics of *Epiperipatus* of the fourteen species present in Brazil.

Our study shows new characters to the taxonomy of Peripatidae. For instance, the few scales ranks on apical piece (see *E. titanicus* and *E. beckeri*), and the larger shape apical piece (see *E. hyperbolicus*). In resume, the new species differ from each other by the shape of the apical piece on the primary papillae. The apical piece of the primary papillae in *E. titanicus* is smaller than in *E. beckeri*. As for *E. lucerna* sp. nov. and *E. marajoara* sp. nov., they have conical apical piece, but *E. marajoara* sp. nov. has a smaller number of pairs of oncopods (27–31 pairs) and inconspicuous diamond pattern on the dorsal portion of the body. Among the five new species, *E. hyperbolicus* sp. nov. is distinguished by the particular shape of the apical piece of the primary papillae. Only in this species a robust and asymmetric spherical apical piece was observed.

Additionally, *E. lucerna* sp. nov., *E. hyperbolicus* sp. nov. and *E. titanicus* sp. nov. are the first species described in Brazil from the same locality, Estação Ecológica de Murici (Mata da Bananeira) in the Alagoas State, Brazil. This region is highland humid forest inserted in the realm of the Caatinga, and is really important to the understanding of the distribution and biogeographical history of onychophorans in the Neotropical region (Vasconcellos et al. 2004).

Due to the conservative morphology of Neotropical onychophorans, many specimens previously identified as *E. edwardsii* in the literature are not actually *E. edwardsii*. The

Table 1. The main characteristics present in the Brazilian *Epiperipatus* species.

Species	Morphology of the apical piece	Distribution of ranks of scales[1]	Range of scale ranks[2]	Incomplete plicae	Number of oncopods
E. acacioi	Spherical	Asymetrical	2 or 3	Present	24–26 (♂); 26–29 (♀)
E. adenocryptus	Conical	Asymetrical	2 or 3	Present	26 and 27 (♂); 28–30 (♀)
E. brasiliensis	Conical	Symetrical	2	Absent	29 (♂); 31 and 32 (♀)
E. cratensis	Spherical	Asymetrical	2 or 3	Absent	30–33 (♂); 33–34 (♀)
E. diadenoproctus	Conical	Asymetrical	2 or 3	Present	26–28 (♂); 29 and 30 (♀)
E. machadoi	Conical	Asymetrical	2 or 3	Present	27–29 (♂); 28–31 (♀)
E. ohausi	Spherical	Asymetrical	3	Present	26–28 (♂); 27–29 (♀)
E. paurognostus	Conical	Asymetrical	2 or 3	Present	26 and 27 (♂); 27–29 (♀)
E. tucupi	Conical	–	–	Absent	34 and 35 (♀)
E. beckeri sp. nov.	Conical reduced	Symetrical	1	Present	28–30 (♀)
E. hyperbolicus sp. nov	Spherical robust	Asymetrical	3	Present	23 (♂); 24 and 25 (♀)
E. lucerna sp. nov.	Conical	Asymetrical	2	Present	27–29 (♂); 29–31 (♀)
E. marajoara sp. nov.	Conical	Asymetrical	2	Present	27 (♂); 31 (♀)
E. titanicus sp. nov.	Conical reduced	Symetrical	1	Present	36 and 38 (♂); 36–39 (♀)

[1]Distribution of ranks of scales in the anterior and posterior portions of the distal end of papillae;
[2]Range of scale ranks on the posterior side of the apical piece.

characters that have been most often misinterpreted are the shape and distribution of the dorsal papillae, the dorsal diamond pattern, and the number of oncopods. A recent phylogenetic study revealed that *E. edwardsii* occurs only in Cayenne, or near this locality (Costa 2016). In addition, the specimens of *E. edwardsii* show diamond patterns on the dorsal side, rounded dorsal papillae, and basal and apical piece of the primary papillae covered by a small range of scale ranks. The broad distribution of *E. edwardsii* seems unlikely. Hence the records from Panama and Suriname (Bouvier 1905), Colombia (Brues 1914), Trinidad and Venezuela (Read 1988a, b) and Brazil (Clark and Zetek 1946, Froehlich 1968, Sampaio-Costa et al. 2009) belong to other species closely related to *E. edwardsii* and which share with it the similarities in the diamond shape on the dorsal pattern, and shape of the primary papillae. For the same morphological misinterpretation, we believe that all specimens from Brazil currently identified as *E. simoni* belong to undescribed species indeed.

Identification key for Brazilian species of *Epiperipatus*

1 Complete folds (Fig. 33) 2
1′ Incomplete folds (Fig. 21) 3
2 Symmetrical conical apical piece (Fig. 34), 29 pairs of legs (males) and 31 and 32 (females) .... *E. brasiliensis* Bouvier, 1899
2′ Asymmetrical conical apical piece (Fig. 35), 30 to 33 pairs of legs (males) and 33 to 34 (females) .... *E. cratensis* Brito et al., 2010
2″ Asymmetrical spherical well-developed apical piece (comparable with *E. cratensis*), 34 and 35 pairs of legs (females) .... *E. tucupi* Froehlich, 1968
3 Conical apical piece (Figs 7–10) 4
3′ Spherical apical piece (Fig. 11) 5
4 Conical normal-sized apical piece (Fig. 9) 6
4′ Conical/cylindrical reduced (comparable with *E. edwardsii*) apical piece (Figs 7, 10) 7
5 Spherical regular apical piece (Fig. 4) 8
5′ Spherical robust (comparable with *E. edwardsii*) apical piece (Fig. 11) .... *E. hyperbolicus* sp. nov.
6 Asymmetrical apical piece with cylindrical shape (Fig. 8), two primary papillae separated by one to three accessory (Fig. 30), accessory papillae occur on the flanks, 27 pairs of legs (males) and 31 (females) .... *E. marajoara* sp. nov.
6′ Asymmetrical apical piece with conical candle shape (Fig. 9), dorsal papillae aligned on top of folds (Fig. 18), with two primary papillae separated by one to five accessory, accessory papillae absent on flank of the folds (Fig. 18), 27 to 29 pairs of legs (males) and 29 to 31 (females) .... *E. lucerna* sp. nov.
7 Symmetrical reduced or hidden apical piece (Fig. 7), two primary papillae separated by one to three accessory (Fig. 14), reddish brown background color of body (*in vivo*, Fig. 13), dorsal papillae aligned on top of folds (Fig. 14), 36–38 pairs of legs (males) and 36–39 (females) .... *E. titanicus* sp. nov.
7′ Symmetrical reduced apical piece, dorsal papillae aligned on top of folds (Fig. 25), primary papillae when separated there are two to four accessory papillae between two of them (Fig. 25), yellowish pink background color of body (Fig. 24), 28–30 pairs of legs (females) .... *E. beckeri* sp. nov.
8 Peripheral accessory papillae frequently occur over the flank (Fig. 30) 9
8′ Peripheral accessory papillae rare over the flank (Oliveira et al. 2011: fig. 4A) 10
9 Background brownish, 27 to 29 pairs of legs and 28 to 31 (females) .... *E. machadoi* (Oliveira & Wieloch, 2005)
9′ Background dark red, 24 to 26 pairs of legs (males) and 26–29 (females) .... *E. acacioi* (Marcus & Marcus, 1955)
10 Conspicuous anal gland (only males, Oliveira et al. 2011: fig. 5C) 11
10′ Unconspicuous anal gland (only males, Oliveira et al. 2011: fig. 5E) 12
11 Male anal glands with two well-developed pores (like light spots) on the anterior board of anal aperture (Oliveira et al. 2011: fig. 5C), 26 to 28 pairs of legs (males) and 29 to 30 (females) .... *E. diadenoproctus* Oliveira et al., 2011
11′ Male unconspicuous anal gland, which was bean-shaped, hardly visible (Oliveira et al. 2011: fig. 5D), 26 and 27 pairs of legs (males) and 28 to 30 (females) without overlap between sexes .... *E. adenocryptus* Oliveira et al., 2011
12 Only primary papillae are prominent (Chagas-Jr & Sampaio 2014: fig. 6), primary papillae with robust apical piece (Chagas-Jr & Sampaio 2014: figs 8, 9), 26 to 28 pairs of legs (male) and 27 to 29 (female) .... *E. ohausi* Bouvier, 1900
12′ All dorsal papillae are prominent (Fig. 36), poorly developed anal gland (male feature) opened in two pores on the anterior board of anal aperture (Oliveira et al. 2011: fig. 5E), 26 and 27 pairs of legs (males) and 27 to 29 (females) .... *E. paurognostus* Oliveira et al., 2011

ACKNOWLEDGEMENTS

We are grateful to Jerome Murienne, who provided specimens of *E. edwardsii* for the present study. We also acknowledge partial support funding such as Labex CEBA (ANR-10-LABX-25-01) to collect *E. edwardsii*. This work was partially supported by Fundação de Amparo à Pesquisa do Estado de São Paulo (FAPESP 2011/20211-0, 2012/02969-6 and 2014/20557-2) to C.S.C. and R.P.R. We are grateful to Adriano B. Kury who provided support with the Latin grammatical forms. Beka Buckman, Brittany Damron and Denis Jacob provided help and comments to the manuscript. We acknowledge the support of the Center for Scanning Electron Microscopy of the Museu Nacional, Universidade Federal do Rio de Janeiro, the electron microscope laboratory at Instituto de Biociências of the Universidade São Paulo, the electron microscope laboratory at Museu de Zoologia da Universidade São Paulo, and the Center for Nanoscale Systems, Harvard University for the SEM images.

LITERATURE CITED

Arnett RH (1961) The Onychophora of Jamaica. Entomological News 72: 213–220.

Blanchard E (1847) Recherches sur l'organisation des Vers. Annales des Sciences Naturelles 8: 119–149.

Bouvier EL (1900) Nouvelles observations sur les Peripatus de la Collection du Musée Britannique. Quarterly Journal of Microscopical Science 43: 749–757. https://doi.org/10.5962/bhl.part.6551

Bouvier EL (1905) Monographie des Onychophores. Annales des Sciences Naturelles, Zoologie et Biologie Animale 2: 1–383.

Brito SV, Pereira JC, Ferreira FS, Vasconcellos A, Almeida WO (2010) *Epiperipatus cratensis* sp. nov. (Onychophora: Peripatidae) from Northeastern Brazil. Neotropical Biology and Conservation 5(1): 47–52. https://doi.org/10.4013/nbc.2010.51.07

Brues CT (1914) A new *Peripatus* from Colombia. Bulletin of the Museum of Comparative Zoology at Harvard College 58: 375–382. https://doi.org/10.5962/bhl.part.25124

Chagas-Jr A, Costa CS (2014) *Macroperipatus ohausi*: redescription and taxonomic notes on its status (Onychophora: Peripatidae). Revista de Biología Tropical 62: 977–985. https://doi.org/10.15517/rbt.v62i3.11643

Clark AH (1913) Piccole note su degli Onychophora. Zoologischer Anzeiger 42: 253–255.

Clark AH (1913a) Revision of the American species of *Peripatus*. Proceedings of the Biological Society Washington 26: 15–19.

Clark AH (1914) Notes on some specimens of a species of Onychophore (*Oroperipatus corradoi*) new to the fauna of Panama. Smithsonian Miscellaneous collections 63(2): 1–2.

Clark AH (1937) On some onychophores from the West Indies and Central America. Proceedings of the United States National Museum 85: 1–3. https://doi.org/10.5479/si.00963801.85-3027.1

Clark AH, Zetek J (1946) The Onychophores of Panama and the Canal Zone. Proceedings of the United States National Museum 96: 205–213. https://doi.org/10.5479/si.00963801.96-3197.205

Cockerell TDA (1913) A *Peripatus* from Guatemala. Proceedings of the Biological Society of Washington 26: 87–88.

Costa CS (2016) Systematic and phylogenetic analysis of *Epiperipatus* Clark, 1913 based on molecular and morphological data (Onychophora, Peripatidae). PhD Thesis, Universidade de São Paulo, São Paulo. https://doi.org/10.11606/T.41.2016.tde-25082016-100135

Froehlich CG (1968) On Some Brazilian Onychophores. Beitrage zur Neotropischen Fauna 5(3): 160–171. https://doi.org/10.1080/01650526809360404

Giribet G, Buckman-Young RS, Costa CS, Baker CM, Benavides LR, Branstetter MG, Daniels SR, Pinto-da-Rocha R. (2018) The "Peripatos" in Eurogondwana? Lack of evidence that southeast Asian onychophorans walked through Europe. Invertebrate Systematics 32: 842–865. https://doi.org/10.1071/IS18007

Marcus E, Marcus E (1955) A new *Peripatus* from Minas Gerais, Brazil. Anais da Academia Brasileira de Ciências 27(2): 189–193.

Morera-Brenes B, León MD (1986) Description of *Macroperipatus valerioi* sp. nov. from Costa Rica, and comments on the genus *Macroperipatus* (Onychophora: Peripatidae). Revista de Biología Tropical, 34(2): 277–282.

Morera-Brenes B, Monge-Nájera J (1990) *Epiperipatus hilkae*, sp. nov. from Costa Rica (Onychophora: Peripatidae). Revista de Biología Tropical 38: 449–455.

Morera-Brenes B, Monge-Nájera J (2010) A new giant species of placented worm and the mechanism by which onychophorans weave their nets (Onychophora: Peripatidae). Revista de Biología Tropical 58: 1127–1142. https://doi.org/10.15517/rbt.v58i4.5398

Mundie DA (1995) The NBS/ISCC Color System. Polymath Systems, Pittsburgh, 535.6 dc-20.

Oliveira IS, Wieloch AH, Mayer G (2010) Revised taxonomy and redescription of two species of the Peripatidae (Onychophora) from Brazil: a step towards consistent terminology of morphological characters. Zootaxa 2493: 16–34.

Oliveira IS, Lacorte GA, Fonseca CG, Wieloch AH, Mayer G (2011) Cryptic Speciation in Brazilian *Epiperipatus* (Onychophora: Peripatidae) Reveals an Underestimated Diversity among the Peripatid Velvet Worms. PloS One 6: e19973. https://doi.org/10.1371/journal.pone.0019973

Oliveira IS, Read VMSJ, Mayer G (2012a) A world checklist of Onychophora (velvet worms), with notes on nomenclature and status of names. ZooKeys 211: 1–70. https://doi.org/10.3897/zookeys.211.3463

Oliveira IS, Franke FA, Hering L, Schaffer S, Rowell DM (2012b) Unexplored Character Diversity in Onychophora (Velvet Worms): A Comparative Study of Three Peripatid Species. PLoS One 7(12): e51220. https://doi.org/10.1371/journal.pone.0051220

Oliveira IS, Lacorte GA, Weck-Heimann A, Cordeiro LM, Wieloch AH, Mayer G (2015) A new and critically endangered species and genus of Onychophora (Peripatidae) from the Brazilian savannah – a vulnerable biodiversity hotspot. Systematics and Biodiversity 13: 211–233. https://doi.org/10.1080/14772000.2014.985621

Peck SB (1975) A review of the New World Onychophora with the description of a new cavernicolous genus and species from Jamaica. Psyche 82: 341–358. https://doi.org/10.1155/1975/98614

Read VMSJ (1988a) The application of scanning electron microscopy to the systematics of the neotropical Peripatidae (Onychophora). Zoological Journal of the Linnean Society 93: 187–223. https://doi.org/10.1111/j.1096-3642.1988.tb01361.x

Read VMSJ (1988b) The Onychophora of Trinidad, Tobago and the Lesser Antilles. Zoological Journal of the Linnean Society 93: 225–257. https://doi.org/10.1111/j.1096-3642.1988.tb01362.x

Röhlig D, Dunlop JA, Ruhberg H, Friederichs A (2010) An annotated catalogue of the velvet worms (Onychophora) held in the Museum für Naturkunde Berlin. Zoosystematics and Evolution 86(2): 225–234. https://doi.org/10.1002/zoos.201000005

Sampaio-Costa C, Chagas-Jr A, Baptista RLC (2009) Brazilian species of Onychophora with notes on their taxonomy and distribution. Zoologia 26(3): 553–561. https://doi.org/10.1590/S1984-46702009005000004

Scorza JV (1953) Contribución al estudio de los Peripatus Caribes (Epiperipatus) de Venezuela, con adición de una nueva especie. Revista de Sanidad y Asistencia Social 18: 783–788.

Vasconcelos A, Almeida WO, Eloy ECC (2004) Onychophora de Florestas Úmidas do Complexo da Mata Atlântica do Nordeste Brasileiro e sua Importância para Conservação e Estudos Sistemáticos. In: Porto KC, Cabral JJP, Tabarelli M (Eds) Brejos de Altitude: história natural, ecologia e conservação. Ministério do Meio Ambiente, Brasília, 139–144.

Author Contributions: Contributions: CSC, ACJ, RPR wrote the paper; CSC did the descriptions and illustrations.
Competing Interests: The authors have declared no competing interest exist.

3

Areas of endemism of hummingbirds (Aves: Apodiformes: Trochilidae) in the Andean and Neotropical regions

Daiane Ouvernay[1], Ildemar Ferreira[1], Juan J. Morrone[2]

[1]*Departamento de Biologia Animal, Instituto de Biologia, Universidade Federal Rural do Rio de Janeiro. Rodovia BR-465 km 7, 23897-970 Seropédica, RJ, Brazil.*
[2]*Museo de Zoología "Alfonso L. Herrera", Departamento de Biología Evolutiva, Facultad de Ciencias, Universidad Nacional Autónoma de México. 04510 Mexico City, Mexico.*
Corresponding author: Daiane Ouvernay (ouvernay@gmail.com)

ABSTRACT. Using track analysis and cladistic biogeography, we identified areas of endemism of hummingbirds in the Andean and Neotropical regions. Our results point out that the current areas of endemism of hummingbirds occur in the Andes, Guiana Shield, the Lesser Antilles, western Central and North America and the Chiapas Highlands. The cladistic biogeographic analysis suggests a hummingbird distribution shaped mainly by dispersal events.

KEY WORDS. Historical biogeography, cladistic biogeography, dispersal events, PAE

INTRODUCTION

Biogeography is the science that aims at explaining the distribution of life on Earth and investigating the processes behind it. It also determines the historical relationship between the occurrence areas of a given taxon based on their distribution and phylogenetic analysis (Henderson 1991). According to Crother and Murray (2011), endemism is the concept most closely associated with distribution and results from both historical factors, as vicariance, and ecological factors, which are consistent with the current limits of the taxa (Morrone 2014b).

On the other hand, an area of endemism is a geographic region that has several taxa with their distribution restricted to it (Silva et al. 2004, Szumik and Goloboff 2004, Sirgrist and Carvalho 2009). According to the theory of vicariance, an area of endemism results from the fragmentation of the ancestral biota due to the emergence of a barrier that breaks the gene flow between populations and, consequently, results in allopatric speciation (Hausdorf and Hjenning 2003).

Based on the presence of endemic taxa, Morrone (2001a, b) delimited the Neotropical and Andean regions, which are natural biogeographic units. South America can be divided in two regions and, for some time it has been known that the Andes divides the continent into two very distinct areas. The tracks of several taxa that inhabit the western portion of South America connect this area with Australia and New Zealand; whereas the tracks of several taxa that inhabit the eastern portion of the continent connect this area to the Old World tropics. Hence, the western portion of the continent belongs to the Andean region and the rest belongs to the Neotropical region, which extends to Central America, the Antilles, and Mexico, where it limits with the Nearctic region.

Transition zones are also important elements, because they show a mixture of biotic elements of two regions (Sánchez-González et al. 2013). They are important from a biological perspective, as they represent areas of biotic interactions and can either be poor or show a considerable diversity (Morrone 2006). In the present study, we considered two transition zones: the Mexican transition zone, which comprises Neotropical and Nearctic elements, and the South American transition zone, which includes Neotropical and Andean elements (Morrone 2004, 2006, 2014a).

Both in the Neotropical and Andean regions, as well as in the transition zones, several areas of endemism have been identified for different taxa (Cracraft 1985, Vázquez-Miranda et al. 2007, Silva et al. 2004, Escalante et al. 2009, Echeverry and Morrone 2010, Prado et al. 2015).

With the objective of contributing to the knowledge of the biogeography of these regions, we undertook a track analysis and a cladistic biogeographic analysis, aimed at identifying areas of endemism of hummingbirds (Trochilidae) in the Andean and Neotropical regions. We also describe the distribution of hummingbirds in under the light of a current molecular phylogenetic hypothesis proposed for the taxon (McGuire et al. 2014).

MATERIAL AND METHODS

The Andean and Neotropical regions are natural biogeographic units delimited by the presence of endemic taxa (Morrone 2001a, 2014a).

We used as models 265 Trochilidae species (approximately 78% of hummingbirds' richness), which comprised the following clades proposed by Bleiweiss et al. (1997): Topazes, Hermits, Mountain Gems, Bees, Mangoes, Brilliants, Coquettes, Emeralds, and the species *Patagona gigas* (Vieillot, 1824). We obtained georeferenced records from the literature, consulted vouchers deposited in the ornithological collections of the Smithsonian Institution (http://www.si.edu/Collections) and the American Museum of Natural History (http://www.amnh.org/our-research/vertebrate-zoology/ornithology), and used the georeferenced point records available by BirdLife International (http://www.birdlife.org). Records without geographic coordinates or of doubtful origin were discarded from the study; therefore, it was not possible to sample 100% of the species of hummingbirds. Nomenclature follows Clements et al. (2012).

Track analysis is based on three basic concepts: individual tracks, generalized tracks, and nodes (Morrone and Crisci 1995). In the present study, we plotted hummingbird locality records on maps of the Andes and Neotropics using the shapefiles of the regions of Löwenberg-Neto (2014, 2015) in the software DIVA-GIS 7.5 (Hijmans et al. 2012).

We connected each locality to the nearest one, forming individual tracks. Superimposed individual tracks formed generalized tracks, which are areas where a former ancestral biota have fragmented due to vicariance events (Morrone and Escalante 2002, Carvalho 2011). Where two or more generalized tracks overlap, a node is identified, which is considered a high diversity zone (Morrone and Escalante 2002).

For this analysis, we used a parsimony analysis of endemicity (PAE) that classifies areas according to the species they share, which allows the identification of biotic relationships (Morrone and Crisci 1995, Urtubey et al. 2010). We built a presence/absence matrix (1/0) of each species based on individual tracks. We analyzed this matrix in the software WINCLADA (Nixon 1999) and built a cladogram. After the first analysis, we implemented a PAE-PCE or PAE with progressive character elimination, which consists in removing from the matrix synapomorphies (species) that define each clade, and successively carrying out parsimony analyses (Luna-Vega et al. 2000, Urtubey et al. 2010).

With this analysis, we sought to identify areas where ancestral populations of hummingbirds were fragmented by vicariance events. This analysis is possible through the generalized tracks (Morrone and Escalante 2002).

Cladistic biogeographic analysis

This analysis consists in replacing the name of the species with the area where it is distributed in the terminal branches of the taxon cladogram. The congruence between different area cladograms will allow obtaining a general area cladogram, whose sequence indicates its historical separation (Morrone and Escalante 2002). We used as a framework the molecular phylogeny of McGuire et al. (2014). The areas that replaced the species were the biogeographic provinces proposed by Morrone (2014a, 2015) for the Andean and Neotropical regions.

For each species of the cladogram we analyzed its individual track. This way, we obtained the provinces where each species was recorded. Based on this analysis, we built a presence/absence matrix (1/0), in which the rows were biogeographic provinces and the columns, species. Based on this matrix, we obtained the general cladograms of the areas using the software WinClada. Data from this study, the species used and the location of biogeographic provinces are presented in the supplementary material (Fig. S1 and Table S1).

RESULTS

Generalized tracks

Based on the superimposition of the individual tracks, we found 17 generalized tracks (Fig. 1), with the support of species of almost all clades. *Patagona gigas* and all the species of the clade that includes Topazes did not belong to any of the generalized tracks. Some generalized tracks coincide partially with tracks found for other plant and animal groups (Table 1).

Cladistic biogeography

The general area cladogram (Fig. 2) places the Ecuadorian province as the sister group of almost all the provinces of the Neotropical region, except Galapagos Islands, Atacaman, and Prepuna (Fig. 2 and Table 2).

Considering the area cladograms for each clade, the cladogram obtained in the biogeographic analysis for Topazes supports Araucaria Forest as the sister group of ((Atlantic, Parana Forest) Cerrado (Caatinga (Xingú-Tapajós (Pará (Roraima (Guiana Lowlands (Pantepui (Imeri, Madeira)))))))). The synapomorphic species of this clade, *Florisuga fusca*, was probably extinct in (Xingú-Tapajós (Pará (Roraima (Guiana Lowlands (Pantepui (Imeri, Madeira)))))) (see Fig. 2).

The Hermits clade shows a broad distribution, from the Mexican transition zone to the provinces of the Southern Brazil dominion, the South American transition zone and the Chacoan sub-region. The area cladogram implies several dispersal events and two synapomorphies, all in South America (Table 2). The first synapomorphies supports the dichotomy between the Atlantic and Parana Forest, and the second places Rondonia province as the sister group of (Ucayali (Napo (Páramo (Magdalena, Cauca)))).

Table 1. Support of the generalized track of Trochilidae and superimposition of other tracks.

Track	Species	Provinces (Morrone 2014a, 2015)	Overlapping tracks
A	*Eulampis holosericeus* and *Oreotrochilus estella*	Lesser Antilles	Echeverry and Morrone (2013) – (Animal and plants) and del Río et al. (2015) – Coleoptera
B	*Leucippus baeri* and *Myrmia micrura*	Ecuadorian + Cauca	Escalante et al. (2011) – (Nematoda)
D	*Atthis ellioti; Campylopterus rufus* and *Doricha enicura*	Chiapas Higlands + Veracruzan	Escalante et al. (2011) – (Nematoda)
E	*Chalcostigma ruficeps, Heliodoxa aurescens, H. branickii* and *Eriocnemis sapphiropygia*	Puna + Rondônia + Ucayali	Alzate et al. (2008) – (Alstroemeriaceae), Escalante et al. (2011) – (Nematoda) and Echeverry and Morrone (2010) – (Vascular plants)
F	*Adelomyia melanogenys, Amazilia viridicauda, Chalcostigma stanleyi* and *Metallura eupogon*	Puna + Rondônia + Ucayali	Alzate et al. (2008) – (Alstroemeriaceae) and Escalante et al. (2011) – (Nematoda)
G	*Aglaeactis aliciae* and *Metallura phoebe*	Puna	del Río et al. (2015) – Coleoptera
L	*Amazilia decora, Chlorostilbon assimilis, Elvira chionura, Eupherusa nigriventris, Lampornis calolaemus, L. hemileucus, Lophornis adorabilis, Selasphorus flammula* and *S. scintilla*	Guatuso-Talamanca + Puntarenas-Chiriquí	Escalante et al. (2011) – (Nematoda) and Escalante et al. (2011) – (Nematoda)
M	*Chaetocercus heliodor* and *Heliangelus mavors*	Guajira	Alzate et al. (2008) – (Alstroemeriaceae)
N	*Amazilia castaneiventris, A. saucerottei, Anthocephala floriceps,Boissonneaua flavescens* and *Coeligena bonapartei*	Magdalena + Cauca	Alzate et al. (2008) – (Alstroemeriaceae)
O	*Heliangelus exortis, Heliodoxa imperatrix* and *Metallura williami*	Cauca	Alzate et al. (2008) – (Alstroemeriaceae)
P	*Eriocnemis nigrivestris, Haplophaedia lugens, Heliangelus strophianus* and *Urochroa bougueri*	Cauca	del Río et al. (2015) – Coleoptera and Alzate et al. (2008) – (Alstroemeriaceae)
R	*Archilochus colubri* and *Hylocharis eliciae*	Guatuso-Talamanca + Pacific Lowlands + Chiapas Highlands	Escalante et al. (2011) – (Nematoda)

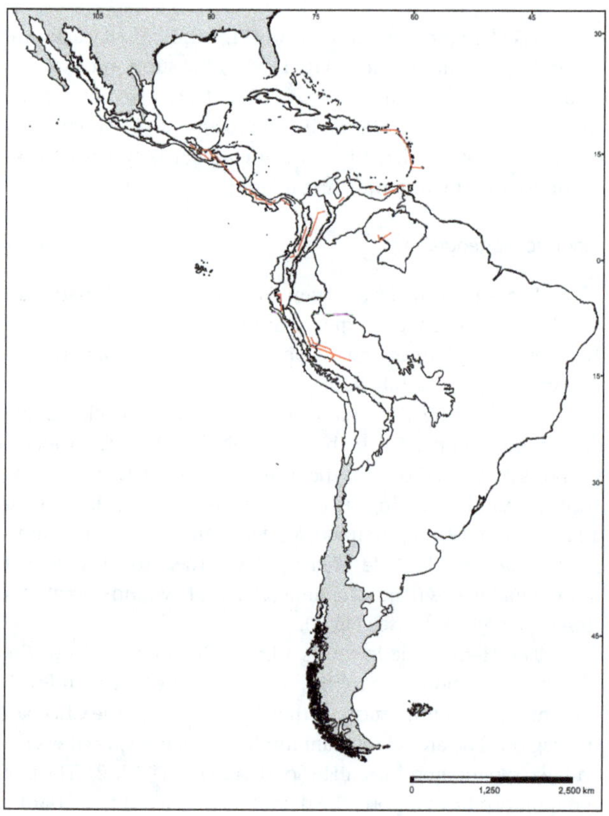

Figure 1. Generalized tracks for Trochilidae in the Neotropical and Andean regions.

The Mountain Gems clade shows a more restricted distribution than the previous clades. It occurs from the Mexican transition zone to the provinces in northern South America, in the Pacific dominion. The area cladogram supports with four synapomorphies the monophyly of the provinces Guatuso-Talamanca and Puntarenas-Chiquiri.

The distribution of the Bees clade is associated with the provinces close to the Pacific coast, in the western part of the American continent, from the Mexican transition zone to the South American transition zone. The area cladogram shows only one synapomorphy, which places the Transmexican Volcanic Belt as the sister group of (Sierra Madre del Sur, Balsas Basin). The other clade places the Desert province as the sister group of (Puna (Yungas (Rondônia (Ucayali (Napo (Paramo (Magdalena, Cauca))))))) and Rondônia as the sister group of (Ucayali (Napo (Paramo (Magdalena, Cauca)))).

The distribution of the Mangoes clade comprises the Mexican transition zone, the Antilles, all the Brazilian sub-region, part of the South American transition zone, and the Xingú-Tapajós province, in the South-eastern Amazonian dominion. The area cladogram shows groups supported by a single synapomorphy, in which the sister group of (Pará (Roraima (Guianan Lowlands (Pantepui (Imerí, Madeira))))) is followed by Pantepui of (Imerí, Madeira). In the Antilles sub-region, there is the clade Hispaniola and Puerto Rico, and in the Brazilian sub-region, the clade Guatuso-Talamanca and Puntarenas-Chiquiri.

The clade Brilliants is distributed from the Mesoamerican dominion to the South American transition zone, and some provinces of the Pacific, Boreal Brazilian and Chacoan domin-

Table 2. Events of dispersal, extinction and synapomorphies represented in the cladogram (Fig. 2). Each number represents the events described.

Number in the cladogram	Synapomorphy	Dispersal	Extinction
2	Ancestor of Brilliants, Coquettes, Montain Gems, Patagona gigas, Bees and Emeralds	Ancestor of Topazes and Hermits	–
3	Bees and Patagona gigas	–	–
4	Ancestor of Patagona gigas, Montain Gems, Bees and Emeralds	Ancestor of Brilliants, Coquettes, Montain Gems, Patagona gigas, Bees and Emeralds	–
5	–	Mangoes and ancestor of Topazes and Hermits	–
6	–	Emeralds and Mangoes	–
7	–	Coquettes, Emeralds, Hermits and ancestor of Hermits	–
8	–	Brilliants, Emeralds and Hermits	–
9	–	Brilliants, Mangoes and ancestor of Brilliants, Coquettes, Montain Gems, Patagona gigas, Bees and Emeralds	Mangoes
10	–	Brilliants, Coquettes, Mangoes, Hermits and ancestor of Coquettes	–
11	Brilliants and Coquettes	Brilliants, Coquettes, Emeralds, Montain Gems, Topazes, Patagona gigas and ancestor of Topazes and Bees	–
12	Coquettes	Brilliants and Emeralds	–
13	Brilliants, Coquettes and Emeralds	Brilliants, Coquettes, Hermits and Mangoes	–
14	Brilliants and Hermits	Bees, Brilliants, Emeralds, Hermits and Mangoes	–
15	Brilliants, Emeralds and Hermits	Emeralds and Hermits	Coquettes and Emeralds
16	Brilliants and Coquettes	Brilliants, Hermits, Mangoes and ancestor of Mangoes	Brilliants and Emeralds
17	Brilliants	Brilliants, Emeralds, Mangoes, Montain Gems and ancestor of Montain Gems and Bees	–
18	Brilliants, Coquettes and Emeralds	Coquettes, Emeralds, Hermits and Mangoes	Hermits
19	–	Ancestor of Montain Gems, Bees, Emeralds and Patagona gigas	–
20	–	Ancestor of Brilliants, Coquettes, Montain Gems, Bees, Emeralds and Patagona gigas	–
21	–	Brilliants and Mangoes	–
22	Coquettes	Brilliants	–
23	–	Ancestor of Montain Gems, Bees, Emeralds and Patagona gigas	–
24	–	Brilliants, Emeralds and Mangoes	–
25	–	–	Ancestor of Mangoes, Brilliants, Coquettes, Montain Gems, Emeralds and Patagona gigas
26	–	Emeralds	–
27	–	Emeralds	Emeralds
28	–	Emeralds	–
29	–	Emeralds	–
30	–	Emeralds	–
31	–	Coquettes, Montain Gems and ancestor of Montain Gems and Bees	Emeralds
32	–	Emeralds, Mangoes, Patagona gigas and ancestor of Bees, Brilliants, Coquettes, Emeralds, Hermits, Mangoes, Montain Gems and Topazes	–
33	–	Emeralds and Hermits	–
34	–	Bees, Coquettes, Emeralds, Mangoes, Montain Gems, Patagona gigas and ancestor of Topazes, Hermits, Coquettes, Brilliants, Mantoes, Montain Gems, Bees and Emeralds	–
35	–	Bees	–
36	–	Hermits and Montain Gems	–
37	–	Emeralds	–
38	Emeralds	Emeralds and Hermits	–
39	Ancestor of Topazes	Hermits, Topazes and ancestor of Hermits and Topazes	–
40	-	Hermits and Mangoes	–
41	Emeralds	Hermits and Emeralds	–
42	Brilliants, Emeralds and Hermits	Brilliants, Mangoes, Patagona gigas, and ancestor of Bees, Briliants, Coquettes, Emeralds, Mangoes and Montain Gems	–
43	–	Hermits and Montain Gems	–
44	Hermits	–	–
45	–	Emeralds and Hermits	Emeralds
46	–	Hermits	–

Number in the cladogram	Synapomorphy	Dispersal	Extinction
47	–	Hermits	Emeralds and Hermits
48	–	Emeralds	Hermits
49	Mangoes	Mangoes, Patagona gigas and ancestor of Bees, Brilliants, Coquettes, Emeralds, Mangoes and Montain Gems	Topazes
50	–	Emeralds and Hermits	Emeralds and Hermits
51	–	Mangoes	Ancestor of Hermits and Topazes
52	–	Emeralds, Hermits and Mangoes	–
53	Coquettes	Brilliants, Emeralds, Mangoes, Patagona gigas and ancestor of Bees, Brilliants, Coquettes, Emeralds and Montain Gems	–
54	–	–	Hermits and Mangoes
55	Mangoes	Mangoes	Emeralds
56	Emeralds	Emeralds	Hermits
57	Hermits	Brilliants, Emeralds, Hermits, Topazes and ancestor of Brilliants	Emeralds
58	–	Brilliants, Coquettes and Hermits	Mangoes and Topazes
59	–	Emeralds and Mangoes	Brilliants, Coquettes, Hermits, Patagona gigas and ancestor of Bees, Brilliants, Coquettes, Emeralds, Montain Gems
60	Emeralds and Mangoes	Emeralds	–
61	Mangoes	Emeralds	–
62	–	Emeralds	–
63	Emeralds and Mangoes	Mangoes	–
64	Emeralds and Mangoes	Bees, Emeralds, Mangoes and Montain Gems	–
65	–	Emeralds, Mangoes	–
66	–	Emeralds, Hermits and Montain Gems	Mangoes
67	Emeralds	Emeralds and Mangoes	–
68	–	Emeralds and Hermits	Coquettes and Hermits
69	–	Bees, Emeralds, Hermits, Mangoes and Montain Gems	–
70	–	–	Hermits
71	–	Bees, Emeralds and Mangoes	–
72	–	Emeralds	–
73	–	Emeralds, Hermits and Montain Gems	Emeralds
74	–	Brilliants, Emeralds, Hermits and Mangoes	Emeralds, Patagona gigas and ancestor of Bees, Brilliants, Coquettes, Emeralds, Hermits, Mangoes, Montain Gems and Topazes
75	–	Coquettes, Emeralds, Mangoes, Montain Gems and ancestor of Bees and Montain Gems	–
76	–	Brilliants, Coquettes, Emeralds and Hermits	Brilliants
77	–	Emeralds, Hermits and Mangoes	–
78	Emeralds	Brilliants, Coquettes, Emeralds, Hermits and Mangoes	–
79	Emeralds	Emeralds, Hermits and Mangoes	Brilliants and Coquettes
80	Brilliants	Brilliants, Coquettes, Emeralds and Montain Gems	Brilliants, Coquettes and Emeralds
81	–	Brilliants, Coquettes and Emeralds	Mangoes
82	Emeralds	Brilliants and Emeralds	Emeralds and Mangoes
83	Emeralds, Mangoes and Montain Gems	Bees, Emeralds, Mangoes and Montain Gems	Brilliants
84	Emeralds	Bees and Hermits	Montain Gems
85	Montain Gems	Emeralds and Montain Gems	Emeralds
86	Bees	Bees and Hermits	Emeralds
87	–	–	Patagona gigas, Emeralds and Montain Gems
88	Brilliants	Emeralds and Hermits	Brilliants and Coquettes
89	–	–	Brilliants and Patagona gigas
90	–	–	Coquettes and Brilliants
91	–	Bees	Emeralds and Hermits

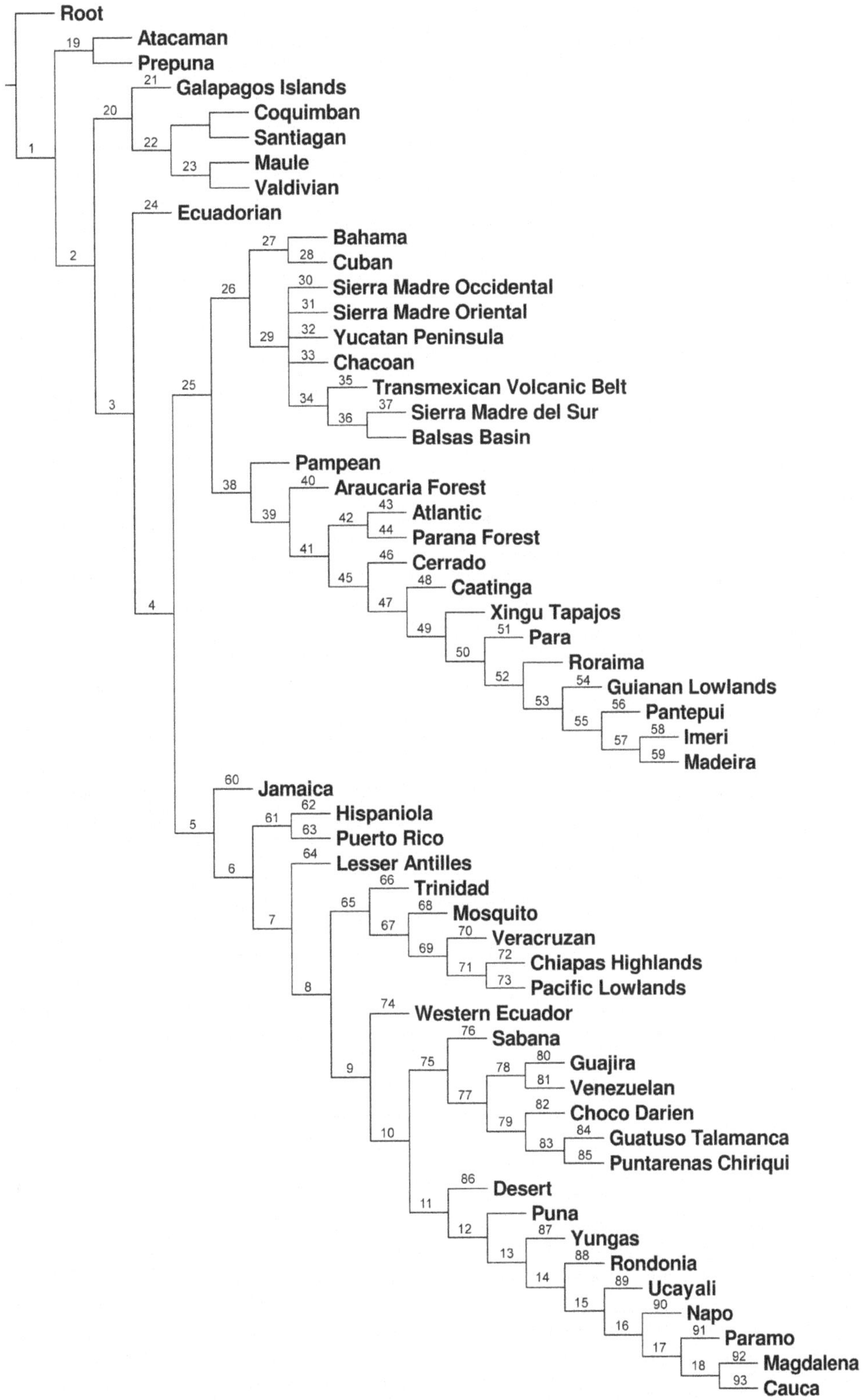

Figure 2. General area cladogram for Trochilidae. Events of dispersal, extinction, and synapomorphies detailed in the Table 2.

ions. The area cladogram shows nine synapomorphies; the best was supported by Yungas as the sister group of (Rondônia (Ucayali (Napo (Páramo (Magdalena, Cauca)))))), Ucayali as the sister group of (Napo (Páramo (Magdalena, Cauca))))), and Páramo as the sister group of (Magdalena, Cauca).

The clade Coquettes is distributed in western Neotropical provinces and some provinces of the Andes and South American transition zone. The area cladogram shows the extinction of the entire clade in the provinces Chocó-Darién, Guatuso-Talamanca, and Puntarenas-Chiquirí and strongly supports three groups. The first places Desert as the sister group of (Puna (Yungas (Rondônia (Ucayali (Napo (Páramo (Magdalena, Cauca))))))), the second places Puna as the sister group of (Yungas (Rondônia (Ucayali (Napo (Páramo (Magdalena, Cauca)))))), and the third supports the clade Magdalena + Cauca.

The clade Emeralds showed the largest number of records and also the largest distribution, which comprises the Mexican transition zone, the Antilles, and almost all the Neotropical provinces of South America, as well as part of the South American transition zone. The area cladogram showed 11 synapomorphies. The first places the Ecuadorian province as the sister group of the provinces in the Neotropical region and rhe Mexican transition zone except (Atacaman, Prepuna) and Galapagos Islands and Pampean as the sister group of (Araucaria forest (Atlantic, Parana forest) (Cerrado (Caatinga (Xingu-Tapajós (Pará (Roraima (Guianan Lowlands (Pantepui (Imeri, Madeira))))))))), also in the same clade between Atlantic and Parana forest, Pantepui and (Imeri, Madeira). In addition, Mosquito is the sister group of (Veracruzan (Chiapas Highlands, Pacific Lowlands)), Guajira and Venezuelan, Choco-Darién is the sister area of (Guatuso-Talamanca, Puntarenas-Chiquiri), Yungas is the sister group of (Rondônia (Ucayali (Napo (Paramo (Magdalena, Cauca))))), Ucayali is the sister group of (Napo (Paramo (Magdalena, Cauca))). The clade Magdalena + Cauca is supported, though weakly.

DISCUSSION

Generalized tracks

The track analysis suggests that current areas of endemism of hummingbirds occur in the Andes, Guiana Shield, Lesser Antilles, western Central and North America. In southern Mexico, more precisely in the Chiapas Highlands province, there is a biogeographic node: a zone of complex diversity for the family (Fig. 1). It is likely that vicariance events, especially of geological nature, occurred in these areas and involved all clades. In addition, the areas of endemism of Trochilidae in South America are within or partially superimpose the areas identified by Cracraft (1985) for the bird fauna. The generalized tracks corroborate some current hypotheses about the phylogeography of the family and reflect the geological, climatic, and tectonic events that occurred in the Neotropical and the Andean regions (McGuire et al. 2007, 2009, 2014, Bleiweiss et al. 1997, Bleiweiss 1998, 2008, Fogden et al. 2014).

Track "A" comprises the entire province of Lesser Antilles and part of Hispaniola, and is supported by Mangoes (*Eulampis holosericeus* (Linnaeus, 1758)) and Coquettes (*Oreotrochilus estella* (d'Orbigny & Lafresnaye, 1838)). This track suggests the fragmentation of the hummingbird population as had already occurred for other taxa (Ricklefs and Bermingham 2008), due to the effects of the geological history of Lesser Antilles. This track partially superimposes with the tracks obtained by Echeverry and Morrone (2013) and Del Río et al. (2015).

Tracks "B", "E", "F", "G", "M", "N", "O", and "P" are related to the uplift of the northern, southern, and central parts of the Andes and are supported by Bees (*Chaetocercus heliodor* (Bourcier, 1840) and *Myrmia micrura* (Gould, 1854)), Emeralds (*Amazilia castaneiventris* (Gould, 1856)), *A. saucerottei* (Delattre & Bourcier, 1846), *A. viridicauda* (von Berlepsch, 1883), *Anthocephala floriceps* (Gould, 1853) and *Leucippus baeri* Simon, 1901), Coquettes (*Adelomyia melanogenys* (Fraser, 1840), *Chalcostigma ruficeps* (Gould, 1846), *C. stanleyi* (Bourcier, 1851), *Heliangelus exortis* (Fraser, 1840), *H. mavors* Gould, 1848, *H. strophianus* (Gould, 1846), *Metallura eupogon* (Cabanis, 1874), *M. phoebe* (Lesson & Delattre, 1839) and *M. williami* (Delattre & Bourcier, 1846)) and Brilliants (*Aglaeactis aliciae* Salvin, 1896, *Boissonneaua flavescens* (Loddiges, 1832), *Coeligena bonapartei* (Boissonneau, 1840), *Eriocnemis mirabilis* Meyer de Schauensee, 1967, *E. nigrivestris* (Bourcier & Mulsant, 1852), *E. sapphiropygia* Taczanowski, 1874, *Haplophaedia lugens* (Gould, 1852), *Heliodoxa aurescens* (Gould, 1846), *H. branickii* (Taczanowski, 1874), *H. imperatrix* (Gould, 1856) and *Urochroa bougueri* (Bourcier, 1851) and comprise the Ecuadorian, Cauca, Desert, Ucayali, Rondônia, Yungas, Guajira, Magdalena, and Chocó-Darién provinces. The importance of the Andes for hummingbirds has been highlighted in several studies (Schuchmann et al. 2001, Weller and Schuchmann 2002, Schuchmann et al. 2003, Parra et al. 2009, Chaves et al. 2011) and corroborated by generalized tracks. According to Bleiweiss (1998), the main reason for the spectacular radiation of hummingbirds is their ecological capacity of colonizing high altitude habitats formed by the orogeny of the Andes, this fauna comprising several lineages with lowland origin. Moreover, the high potential for coevolution between hummingbirds and plants probably facilitated the success of the Andean taxa. On the other hand, McGuire et al. (2014) highlight the opening of new niches with the uplift of the Andes for well-succeeded colonization in that environment. Some species stand out in this habitat, especially the members of the clades Brilliants and Coquettes (McGuire et al. 2009). The tracks partially superimpose those obtained by Alzate et al. (2008), Echeverry and Morrone (2010), Escalante et al. (2011) and Del Río et al. (2015).

Tracks "H" and "J" comprise the Venezuelan and Trinidad provinces, in the Cordillera de La Costa and are supported by Emeralds (*Campylopterus ensipennis* (Swainson, 1822) and *Chalybura buffoni* (Lesson, 1832)), Bees (*Chaetocercus jourdanii* (Bourcier, 1839)) and Hermits (*Phaethornis longuemareus* (Lesson, 1832)).

Most of track "Q" is located in the Pantepui and part of Imerí, and is supported by Brilliants (*Heliodoxa xanthogonys* Salvin & Godman, 1882) and Emeralds (*Amazilia cupreicauda* Salvin & Godman, 1884). The area comprises a mountain complex whose importance and formation will be discussed later.

Track "I" is supported only by Emeralds (*Lepidopyga coeruleogularis* (Gould, 1851)) and *Juliamyia julie* (Bourcier, 1843)) and comprises the provinces of Chocó-Dárien and Guatuso-Talamanca. Therefore, we hypothesize that the closing of the Isthmus of Panama, at approximately 3-5 million years ago, affected the distribution of Trochilidae in the area.

Tracks "C", "D", "L" and "R" are supported by Mountain Gems (*Heliomaster constantii* (Delattre, 1843), *Lamprolaima rhami* (Lesson, 1839), *Lampornis calolaemus* (Salvin, 1865) and *L. hemileucus* (Salvin, 1865)), Bees (*Archilochus colubris* (Linnaeus, 1758), *Atthis ellioti* Ridgway, 1878, *Doricha enicura* (Vieillot, 1818), *Selasphorus flammula* Salvin, 1865, *S. scintilla* (Gould, 1851), *Tilmatura dupontii* (Lesson, 1832)), Emeralds (*Amazilia decora* (Salvin, 1891), *Campylopterus rufus* Lesson, 1840, *Chlorostilbon assimilis* Lawrence, 1861, *Elvira chionura* (Gould, 1851), *Eupherusa nigriventris* Lawrence, 1868 and *Hylocharis eliciae* (Bourcier & Mulsant, 1846) and Coquettes (*Lophornis adorabilis* Salvin, 1870). In the same area, there is also a biogeographic node. It comprises the provinces located in the Mexican transition zone and the Mesoamerican dominion. All tracks are located in provinces of the western coast, close to the Pacific Ocean. The tracks partially superimpose with those obtained by Escalante et al. (2011).

Paleogeographic events

Some important events that occurred in the Andean and Neotropical region during the Neogene equally influenced different clades of Trochilidae: uplift of the Andes (Gregory-Wodzicki 2000, Ramos 2009), the change of course of large rivers (Hoorn et al. 1995, Gamero 1996, Lovejoy et al. 1998, Lundberg et al. 1998, Nie et al. 2010, Sacek 2014), the closing of the Amazonian and Paranaense seas (Webb 1995), the closing of the Isthmus of Panama (Fortunato 2008, Farris et al. 2011), tectonic, volcanic, and climatic events in Central and North America (Pregill 1981, Briggs 1987, Pindell and Barret 1990, Hedges 1996, Ricklefs and Bermingham 2008), the Last Glacial Maximum (LGM) and, consequently, the rise in sea level and climate changes (Pregill and Olson 1981, Bourgouis et al. 1984, Codignotto et al 1992, Ferreira 2002, Filho et al. 2002), forest expansion and retraction (Filho et al 2002), volcanism (Fortunato 2008), and tectonism (Ramos 2009), and demonstrated in biogeographic analyses. Hence, we notice that the current distribution of Neotropical and Andean hummingbirds reflects the complex geological history of the region, which influenced the clades differently as we discuss as follows.

Hummingbirds and northern South America

Tectonic processes of the Andean orogeny reshaped South America nearly to the current form (Ramos 2009, Revollo 2015) and were crucial for the conformation of the current distribution of the family Trochilidae through vicariant events and the subsequent opening of new niches (McGuire et al. 2007, 2014). In addition, according to the generalized tracks, these tectonic processes divided the ancestral populations of hummingbirds. Uplift processes and all their resulting modifications in South America affected the populations of Emeralds, which can be corroborated by the presence of a generalized track (Fig. 1). Based on a molecular clock, the mininum age of hummingbirds has been postulated to be 65 Ma (Pacheco et al. 2011), at the K/T boundary.

The tectonism involved with the uplift of the central and northern parts of the Andes changed the drainage of large rivers, such as the Amazon, Orinoco, and Magdalena, and established the connection of the Amazon River with the Atlantic Ocean, completing the Amazon-Caribbean connection through the "Amazonian sea or Caribbean sea" (Hoorn et al. 1995, Webb 1995, Lundberg et al. 1998 Hernández et al. 2005, Nie et al. 2010, Sacek 2014, Revollo 2015). Trochilidae populations, in particular Hermits, may have been affected during this north-east movement of the Amazon River. The presence of a generalized track (Fig. 1) indicates that ancestral populations have been fragmented by vicariance events in this area. Our hypothesis is that the movement of the river has fragmented the population of hummingbirds.

Still in the context of the Andes, with the uplift of the Cordillera Oriental (East Andes) there was the formation of an arid valley (Magdalena Valley) and change in the drainage of the Magdalena River (Hoorn et al. 1995, Lundberg et al. 1998, Egbue and Kellogg 2012, Muñoz-Ortiz et al. 2015), which affected other taxa (Muñoz-Ortiz et al. 2015) and hummingbirds, in particular the clades Brilliants and Coquettes, as indicated by the presence of a generalized track (Fig. 1). This event may have acted in a similar way to the Amazon River, that is, this change in the course of the river may have fragmented ancestral hummingbird populations.

In addition to the Andes, the Pantepuis, located in northern South America also showed a significant relevance for Trochilidae, as a generalized track occurs in the region (Fig. 1). The Pantepuis are a mountain complex that belongs to the Precambrian Guiana Shield, which was separated from the African Shield by the opening of the South Atlantic. It contains the largest rivers in the world in terms of annual discharge (Désamoré et al. 2010, Costa et al. 2013). This area is particularly important for Mangoes, Brilliants, and Emeralds.

Generalized tracks indicate ancestral populations that have been fragmented by vicariance events (Morrone and Escalante 2002). Therefore, events involving Andean orogenic processes, changes in the Amazon and Magdalena Rivers, and the formation of the Pantepuis may have acted as barriers that fragmented the ancestral population of hummingbirds.

The relevance of the Andean orogeny for Trochilidae was pointed out by McGuire et al. (2014) and highlights the importance of the Neogene for the radiation of several taxa, including hummingbirds (Bleiweiss 1998). Our work corroborates the postulate by the authors, highlighting the importance of the upward "pulses" and all the geological changes that have taken place, the forms in South America, as the molding processes of the current distribution of the group in the continent.

Hummingbirds and the Last Glacial Maximum

Another important event in the Neotropical region was the Last Glacial Maximum, which produced profound changes in the physical geography of Earth (Hoorn et al. 1995, Ferreira 2002). Climatic fluctuations during the last million years led to long periods of global cooling, which changed the distribution of many species (Hewitt 2000). During the glacial period in the Pleistocene, the southern part of the Atlantic Forest became climatically unstable in comparison with its central part and served as refuge for several Neotropical species (Carnaval et al. 2009). The region also underwent with a drop in rainfall during the Mid-Holocene, when its climate was drier than today (Melo and Marengo 2008). The current Araucaria forest underwent a significant retraction during the Last Glacial Maximum, and it is thought to have been rare in high altitude areas (Behling 1998). The climatic gradient created by this event in the Neotropics, more precisely in the Atlantic Forest, was significant for Topazes, Brilliants, Emeralds, and Hermits.

Hummingbirds, North America and Mesoamerica

The closing of the Isthmus of Panama occurred at 3-5 million years ago and brought severe climatic and geological changes (Briggs 1987, Kellogg and Veja 1995, Coates et al. 2004, Fortunato 2008, Farris et al. 2011, Bacon et al. 2015). It comprised the tectonic collision between South America and Panama, which closed the connection between the Pacific Ocean and the Caribbean Sea and formed an important bridge for the exchange of fauna and flora (Farris et al. 2011, O'Dea et al. 2012). There was also a floristic diversification, which facilitated the dispersal of some birds, including hummingbirds (Ornelas et al. 2014). Concomitantly, there was an uplift of the Andes in northern Colombia (Kellogg and Veja 1995, Farris et al. 2011). The presence of generalized tracks (Fig. 1 and Suppl. material 1) demonstrates that this event was important for Mangoes, Mountain Gems, and Emeralds.

The origin of the Caribbean area is connected to the breakup of Pangea, when Laurasia began to separate from Gondwana (Hedges 1996). Current islands are classified as the Greater Antilles, which comprise old fragments of the continental crust, and the Lesser Antilles, which were formed by a volcanic arc and the current islands (Hedges 1996, Pindell and Barret 1990, Ricklefs and Bermingham 2008), where vicariant events could have generated endemism (Heads 2017). During the Last Glacial Maximum, the Bahamas were the most affected islands of the Antilles due to the sea level rise, which caused the loss of several habitats and a drier climate (Pregill and Olson 1981). In the period of a low sea level, the Lesser Antilles were probably connected, forming three large islands and their combination showed a different topography (Pregill and Olson 1981). This region stands out as a very important area for Trochilidae, in particular for Mangoes. The Lesser Antilles stand out due to the presence of a generalized track, which suggests the fragmentation of ancestral hummingbird populations (Fig. 1).

The Mexican transition zone and the provinces of the Mesoamerican dominion are equally important for hummingbirds, where, in addition to the generalized tracks that indicate endemism, there is a biogeographic node. Geological events that may have shaped the current Trochilidae distribution are volcanism (Mora et al. 2007) and tectonism (Bourgouis et al. 1984).

Hummingbirds and marine transgressions

During the Miocene, the rise in sea level and tectonic processes in South America produced a significant marine ingression into the continent (Lovejoy et al. 1998). These events contributed to the appearance of the Amazonian and Paranaense seas, which separated the north and south of the continent at 16 and 13 million years ago. Associated with Andean tectonic processes, these seas altered drastically the drainage of South American rivers (Lundberg et al. 1998, Hernández et al. 2005, Sacek 2014). Fossil evidence showed the existence of paleogeographic corridors formed by two large marine transgressions (Lundberg et al. 1998, Hernández et al. 2005). The area cladogram for hummingbirds shows signals of these events for the clades Emeralds and Coquettes.

Cladistic biogeography

Bleiweiss (1998) and McGuire et al. (2007, 2014) highlighted dispersal as a striking characteristic in hummingbird history, with some vicariance events. Considering the area cladogram of Trochilidae, dispersal is confirmed to be important in the family. However, this trend is not the same for all clades, i.e., some clades have a more restricted distribution, support several generalized tracks, and are synapomorphies of groups of some provinces. This divergence occurred mainly in the "Clade of the Andes".

The first synapomorphy found places the Ecuadorian province as the sister group of nearly all the other Neotropical provinces, except Galapagos Island, Prepuna, Atacaman (in the South America transition zone) and Coquimban, Santiagan, Maule and Valdivian. Prepuna and Atacaman would be more related to the provinces of the Andes than the Neotropical ones, as would be Galápagos (Fig. 2). Emeralds also support the group in which the Pampean province is the sister group of the provinces of the Chacoan, Parana, South-eastern, and Boreal Brazilian dominions, which occupy nearly the entire plains of South America, or the mid-northern area of the continent. The dichotomy between the Atlantic and Parana Forests is well supported by Brilliants, Emeralds, and Hermits. This result can be related to the floristic similarity between these two provinces, which are within the Atlantic Forest in southeastern and northeastern Brazil, respectively. These provinces have similar geological histories since the Last Glacial Maximum, which altered flora and fauna. The grouping of the provinces Guatuso-Talamanca and Puntarenas-Chiriquí is equally well supported by Mangoes, Mountain Gems, and Emeralds. Brilliants, Coquettes, and Mountain Gems appear as synapomorphies of the clade in which Desert is the sister group of the provinces of the South American transition zone and the South Brazilian and Boreal Brazilian dominions. In the cladogram, the same provinces are rearranged, with Puna as the sister group, followed by Yungas, Rondonia, Ucayali, Napo, and Paramo as sister groups of the provinces located to the west of South America on both sides of the Andes (Central and South). Provinces from different dominions were grouped, which reinforces the striking dispersal of the group. It is worth highlighting that, geographically, the cladogram reflects the grouping of provinces at the east and west of the continent (related to the Andes) and Central America.

Our results corroborate the hypothesis that the more basal group (Topazes clade) of Trochilidae was distributed in the plains of South America, and the dispersal to the Andes and North America were posterior events, before the Neogene phase of uplift of the Cordillera (McGuire et al. 2014). As shown in the area cladogram (Fig. 2), most grouped provinces belong to the Brazilian Shield and have the Araucaria forest region as sister group.

ACKNOWLEDGMENTS

The Coordenação de Aperfeiçoamento de Pessoal de Nível Superior (CAPES) funded the present study through its institutional program for sandwich Ph.D. (proc. BEX 10656/14-6), and the Universidad Nacional Autónoma de México (UNAM).

LITERATURE CITED

Alzate F, Quijano-Abril MA, Morrone JJ (2008) Panbiogeographical analysis of the genus *Bomarea* (Alstroemeriaceae). Journal of Biogeography 35(7): 1250–1257. https://doi.org/10.1111/j.1365-2699.2008.01896.x

Bacon CD, Silvestro D, Jaramilo C, Smith BT, Chakrabarty P, Antonelli A (2015) Biological evidence supports an early and complex emergence of the Isthmus of Panama. Proceedings of the National Academy of Sciences of the United States of America 112(19): 6110–6115. https://doi.org/10.1073/pnas.1423853112

Behling H (1998) Late Quaternary vegetational and climatic changes in Brazil. Review of Palaeobotany and Palynology 99(2): 143–156. https://doi.org/10.1016/S0034-6667(97)00044-4

Bleiweiss R (1998) Origin of hummingbird faunas. Biological Journal of the Linnean Society 65(1): 77–97. https://doi.org/10.1111/j.1095-8312.1998.tb00352.x

Bleiweiss R (2008) Tempo and mode of hummingbird evolution. Biological Journal of the Linnean Society 65(1): 63–76. https://doi.org/10.1111/j.1095-8312.1998.tb00351.x

Bleiweiss R, Kirsch JAW, Matheus JC (1997) DNA Hybridization Evidence for the Principal Lineages of Hummingbirds (Aves: Trochilidae). Molecular Biology and Evolution 14(3): 325–343. https://doi.org/10.1093/oxfordjournals.molbev.a025767

Bourgouis J, Azema J, Baumgartner PO, Tournon J, Desmet A, Aubouin J (1984) The geologic history of the Caribbean-Cocos plate boundary with special reference to the *Nicya ophiolite* Complex (Costa Rica) and D.S.D.P. results (legs 67 and 84 off Guatemala): A synthesis. Tecronophysics 108: 1–32. https://doi.org/10.1016/0040-1951(84)90152-5

Briggs JC (1987) Biogeography and Plate Tectonics. Elsevier, Amsterdam, 204 pp.

Carnaval AC, Hickerson MJ, Haddad CFB, Rodrigues MT, Moritz C (2009) Stability Predicts Genetic Diversity in the Brazilian Atlantic Forest Hotspot. Science 323(5915): 785–789. https://doi.org/10.1126/science.1166955

Carvalho CJB (2011) Áreas de Endemismo, p. 41–50. In: Carvalho CJB, Alemida EAB (Eds) Biogeografia da América do Sul: Padrões and Processos. Editora Roca Ltda, São Paulo, 308 pp.

Chaves JA, Weir JT, Smith TB (2011) Diversification in *Adelomyia* hummingbirds follows Andean uplift. Molecular Ecology 20(21): 4564–4576. https://doi.org/10.1111/j.1365-294X.2011.05304.x

Clements JF, Schlenberg TS, Iliff MJ, Roberson D, Fresweicks TA, Sullivan BL, Wood CL (2012) The eBird/Clements checklist of birds of the world. Available online at: http://www.birds.cornell.edu/clementschecklist/download [Accessed: 11/11/2013]

Coates AG, Collins LS, Aurbry MP, Berggren WA (2004) The Geology of the Darien, Panama, and the late Miocene-Pliocene collision of the Panama arc with northwestern South America. GSÄ Bulletin 116(11/12): 1327–1344. https://doi.org/10.1130/B25275.1

Codignotto JO, Kokot RR, Marcomini SC (1992) Neotectonism and sea-level changes in the Coastal Zone of Argentina. Journal of Coastal Research 8(1): 125–133.

Costa M, Viloria AL, Huber O, Attlal S, Orellana A (2013) Lepidoptera del Pantepui. Parte I: Endemismo y caracterización biogeográfica. Entomotropica 28(3): 193–217.

Cracraft J (1985) Historical Biogeography and Patterns of Differentiation within the South American Avifauna: Areas of Endemism. Ornithological Monographs 36: 49–84. https://doi.org/10.2307/40168278

Crother BI, Murray CM (2011) Ontology of areas of endemism. Journal of Biogeography 38(6): 1009–1015. https://doi.org/10.1111/j.1365-2699.2011.02483.x

DelRío MG, Morrone JJ, Lanteri AA (2015) Evolutionary biogeography of South American weevils of the tribe Naupactini (Coleoptera: Curculionidae). Journal of Biogeography 42(7): 1293–1304. https://doi.org/10.1111/jbi.12481

Désamoré A, Vanderpoorten A, Laenen B, Gradstein SR, Kok PJR (2010) Biogeography of the Lost World (Pantepui region, northeastern South America): Insights from bryophytes. Phytotaxa 9: 254–265. https://doi.org/10.11646/phytotaxa.9.1.14

Echeverry A, Morrone JJ (2010) Parsimony analysis of endemicity as a panbiogeographical tool: an analysis of Caribbean plant taxa. Biological Journal of the Linnean Society 101(4): 901–976. https://doi.org/10.1111/j.1095-8312.2010.01535.x

Echeverry A, Morrone JJ (2013) Generalized tracks, area cladograms and tectonics in the Caribbean. Journal of Biogeography 40(9): 1619–1637. https://doi.org/10.1111/jbi.12117

Egbue O, Kellogg J (2012) Three-dimensional structural evolution and kinematics of the Piedemonte Llanero, Central Llanos foothills, Eastern Cordillera, Colombia. Journal of South American Earth Sciences 39: 216–227. https://doi.org/10.1016/j.jsames.2012.04.012

Escalante T, Szumik C, Morrone JJ (2009) Areas of endemism of Mexican mammals: reanalysis applying the optimality criterion. Biological Journal of the Linnean Society 98(2): 468–478. https://doi.org/10.1111/j.1095-8312.2009.01293.x

Escalante T, Martínez-Salazar EA, Falcón-Ordaz J, Linaje M, Guerrero MR (2011) Análisis panbiogeográfico de Vexillata (Nematoda: Ornithostrongylidae) y sus huéspedes (Mammalia: Rodentia). Acta Zoológica Mexicana 27: 25–46. https://doi.org/10.21829/azm.2011.271732

Farris DW, Jaramillo C, Bayona G, Restrepo-Moreno SA, Montes C, Cardona A, Mora A, Speakman RJ, Glascock MD, Valencia V (2011) Fracturing of the Panamanian Isthmus during initial collision with South America. Geology 30(11): 1007–1010. https://doi.org/10.1130/G32237.1

Ferreira ADB (2002) Variabilidade climática e dinâmica geomorfológica. Publicações da Associação Portuguesa de Geomorfólogos 1: 7–15.

Filho AC, Schwartz D, Tatumi SH, Rosique T (2002) Amazonian Paleodunes Provide Evidence for Drier Climate Phases during the Late Pleistocene-Holocene. Quaternary Research 58(2): 205–209. https://doi.org/10.1006/qres.2002.2345

Fogden M, Taylor M, Williamson S (2014) Hummingbirds: A life-size guide to every species. Ivy Press, London, 400 pp.

Fortunato H (2008) The Central American land bridge: evolution at work. Naturwissenschaftliche Verein für Schleswig-Holstein 70: 56–72.

Gamero MLD (1996) The changing course of the Orinoco River during the Neogene: a review. Palaeogeography, Palaeoclimatology, Palaeoecology 123(1–4): 385–402. https://doi.org/10.1016/0031-0182(96)00115-0

Gregory-Wodzicki KM (2000) Uplift history of the Central and Northern Andes: A review. GSA Bulletin 112(7): 1091–1105. https://doi.org/10.1130/0016-7606(2000)112<1091:UHOTCA>2.0.CO;2

Hausdorf B, Hjenning C (2003) Biotic element analysis in biogeography. Systematic Biology 52(5): 717–723. https://doi.org/10.1080/10635150390235584

Heads M (2017) Metapopulation vicariance explains old endemics on young volcanic islands. Cladistcs 2017. https://doi.org/10.1111/cla.12204

Hedges SB (1996) Historical Biogeography of West Indian Vertebrates. Annual Review of Ecology and Systematics 27: 163–196. https://doi.org/10.1146/annurev.ecolsys.27.1.163

Hernández RM, Jordan TE, Farjat AD, Echavarría L, Idleman BD, Reynolds JH (2005) Age, distribution, tectonics, and eustatic controls of the Paranense and Caribbean marine transgressions in southern Bolivia and Argentina. Journal of South American Earth Sciences 19(4): 495–512. https://doi.org/10.1016/j.jsames.2005.06.007

Henderson IM (1991) Biogeography without Area? Australian Systematic Botany 4(1): 59–71. https://doi.org/10.1071/SB9910059

Hewitt G (2000) The genetic legacy of the Quaternary ice ages. Nature 405: 907. https://doi.org/10.1038/35016000

Hijmans RJ, Guarino L, Bussink C, Mathur P (2012) DIVA-GIS. v. 5.0. A geographic informationsystem for the analysis of species distribution data. http://www.diva-gis.org [Accessed: 17/01/2014]

Hoorn C, Guerrero J, Sarmiento GA, Lorente MA (1995) Andean tectonics as a cause for changing drainage patterns in Miocene northern South America. Geology 23(3): 237–240. https://doi.org/10.1130/0091-7613(1995)023<0237:ATAACF>2.3.CO;2

Kellogg JN, Veja V (1995) Tectonic development of Panama, Costa Rica, and the Colombian Andes: constraints from global positioning system geodetic studies and gravity. In: Mann P (Ed.) Geologic and Tectonic Development of the Caribbean plate boundary in Southern Central America. GSA Special Paper, Washington DC, 75–90. https://doi.org/10.1130/SPE295-p75

Lovejoy NR, Bermingham E, Martin AP (1998) Marine incursion into South America. Nature 396: 421–422. https://doi.org/10.1038/24757

Löwenberg-Neto P (2014) Neotropical region: a shapefile of Morrone's (2014) biogeographical regionalisation. Zootaxa 3802(2): 300. https://doi.org/10.11646/zootaxa.3802.2.12

Löwenberg-Neto P (2015) Andean region: a shapefile of Morrone's (2015) biogeographical regionalisation. Zootaxa 3936(4): 600. https://doi.org/10.11646/zootaxa.3985.4.9.

Luna-Vega I, Ayala OA, Morrone JJ, Organista DE (2000) Track analysis and conservation priorities in the cloud forests of Hidalgo, Mexico. Diversity and Distributions 6(3): 137–143. https://doi.org/10.1046/j.1472-4642.2000.00079.x

Lundberg JG, Marshal LG, Guerrero J, Horton B, Malabarba MCSL, Weddrlingh F (1998) The stage for neotropical fish diversification: A history of tropical South American rivers. In: Malabarba LR, Reis RE, Vari RP, Lucena ZM, Lucena CAS (Eds) Phylogeny and classification of Neotropical fishes. Edipucrs, Porto Alegre, 13–48.

McGuire JA, Witt CC, Altshuler DL, Rensen JR (2007) Phylogenetic Systematics and Biogeography of Hummingbirds: Bayesian and Maximum Likelihood Analyses of Partitioned Data and Selection of an Appropriate Partitioning Strategy. Systematic Biology 56(5): 837–856. https://doi.org/10.1080/10635150701656360

McGuire JA, Witt CC, Remsen JR, Dudley R, Altshuler DL (2009) A higher-level taxonomy for hummingbirds. Journal of Ornithology 150: 155–165. https://doi.org/10.1007/s10336-008-0330-x

McGuire JA, Witt CC, Remsen JV Jr, Corl VR, Rabosky A, Altshuler DL, Dudley R (2014) Molecular Phylogenetics and the Diversification of Hummingbirds. Current Biology 24(8): 1–7. https://doi.org/10.1016/j.cub.2014.03.016

Melo MLDD, Marengo JA (2008) Simulações do clima do Holoceno Médio na Amérca do Sul com o modelo de circulação geral da atmosfera do CPTEC. Revista Brasileira de Meteorologia 23(2): 191–205. https://doi.org/10.1590/S0102-77862008000200007

Mora JC, Jaimes-Viera MC, Guarduño-Monroy VH, Layer P, Pompa-Mera V, Godínez L (2007) Geology and geochemistry characteristics of the Chiapanecan Volcanic Arc (Central Area), Chiapas Mexico. Journal of Volcanology and Geothermal Research 162(1-2): 43–72. https://doi.org/10.1016/j.jvolgeores.2006.12.009

Morrone JJ (2001a) Biogeografía de América Latina y el Caribe. MandT – Manuales y Tesis SEA, Zaragoza, 55 pp.

Morrone JJ (2001b) A proposal concerning formal definitions of the Neotropical and Andean regions. Biogeographica 77: 65–82.

Morrone JJ (2006) Biogeographic areas and transition zones of Latin America and the Caribbean Islands based on panbiogeographic and cladstic analyses of the entomofauna. Annual Review of Entomology 51: 467–494. https://doi.org/10.1146/annurev.ento.50.071803.130447

Morrone JJ (2014a) Biogeographical regionalisation of the Neotropical region. Zootaxa 3782(1): 01–110. https://doi.org/10.11646/zootaxa.3782.1.1

Morrone JJ (2014b) Parsimony analysis of endemicity (PAE) revisited. Journal of Biogeography 41(5): 842–854. https://doi.org/10.1111/jbi.12251

Morrone JJ (2015) Biogeographical regionalisation of the Andean region. Zootaxa 3936(2): 207–36. https://doi.org/10.11646/zootaxa.3936.2.3

Morrone JJ, Crisci JV (1995) Historical Biogeography: Introduction to Methods. Annual Review of Ecology, Evolution and Systematics 26: 373–401. https://doi.org/10.1146/annurev.es.26.110195.002105

Morrone JJ, Escalante T (2002) Parsimony analysis of endemicity (PAE) of Mexican terrestrial mammals at different area units: when size matters. Journal of Biogeography 29(8): 1095–1104. https://doi.org/10.1046/j.1365-2699.2002.00753.x

Morrone JJ (2004) Panbiogeografía, componentes bióticos y zonas de transición. Revista Brasileira de Entomologia 48(2): 149–162. https://doi.org/10.1590/S0085-56262004000200001

Muñoz-Ortiz A, Velázquez-Álvarez AA, Guarnizo CE, Crawford AJ (2015) Of peaks and valleys: testing the roles of orogeny and habitat heterogeneity in driving allopatry in mid-elevation frogs (Aromobatidae: Rheobates) of the northern Andes. Journal of Biogeography 42(1): 193–205. https://doi.org/10.1111/jbi.12409

Nie J, Horton BK, Mora A, Saylor JE, Joush TB, Rubiano J, Naranjo J (2010) Tracking exhumation of Andean ranges bounding the Middle Magdalena Valley Basin, Colombia. Geology 38(5): 451–454. https://doi.org/10.1130/G30775.1

Nixon KC (1999–2002) WinClada. v. 1.0000, Published by the author, New York.

O'Dea A, Hoyos N, Rodríguez F, Gracia BD, Degracia C (2012) History of upwelling in the Tropical Eastern Pacific and the paleogeography of the Isthmus of Panama. Palaeogeography, Palaeoclimatology, Palaeoecology 348–349(1): 59–66. https://doi.org/10.1016/j.palaeo.2012.06.007

Ornelas JF, Gonzalez C, Espinosa de los Monteros A, Rodríguez-Gómez F, Garcia-Feria LM (2014) In and out of Mesoamerica: temporal divergence of *Amazilia* hummingbirds pre-dates the orthodox account of the completion of the Isthmus of Panama. Journal of Biogeography 41(1): 168–181. https://doi.org/10.1111/jbi.12184

Pacheco MA, Battistuzzi FU, Lentino M, Aguilar RF, Kumar S, Escalante AA (2011) Evolution of modern birds revealed by mitogenomics: timing the radiation and origino f major orders. Molecular Biology and Evolution 28(6): 1927–1942. https://doi.org/10.1093/molbev/msr014

Parra JL, Remsen JR, Alvarez-Rebolledo M, McGuire JA (2009) Molecular phylogenetics of the hummingbird genus *Coeligena*. Molecular Phylogenetics and Evolution 53(2): 425–434. https://doi.org/10.1016/j.ympev.2009.07.006

Pindell JL, Barret SF (1990) Geological evolution of the Caribbean region; a plate-tectonic perspective. In: Dengo G, Case JE (Eds) The Geology of North America. Geological Society of America, Washington, 583. https://doi.org/10.1130/DNAG-GNA-H.405

Prado JRD, Brennand PGG, Godoy LP, Libardi GS, Abreu-Junior EFD, Roth PRO, Chiquito A, Percequillo AR (2015) Species richness and areas of endemism of oryzomyine rodents (Cricetidae, Sigmodontinae) in South America: an NDM/VNDM approach. Journal of Biogeography 42(3): 540–551. https://doi.org/10.1111/jbi.12424

Pregill GK (1981) An Appraisal of the Vicariance Hypothesis of Caribbean Biogeography and Its Application to West Indian Terrestrial Vertebrates. Systematic Zoology 30: 147–155. https://doi.org/10.1093/sysbio/30.2.147

Pregill GK, Olson SL (1981) Zoogeography of West Indian Vertebrates in Relation to Pleistocene Climatic Cycles. Annual Review of Ecology and Systematics 12: 75–98. https://doi.org/10.1146/annurev.es.12.110181.000451

Ramos VA (2009) Anatomy and global context of the Andes: Main geologic features and the Andean orogenic cycle. The Geological Society of America Memoir 204: 31–65. https://doi.org/10.1130/2009.1204(02)

Revollo E (2015) Physical Geology and Geological History of South America. Editora Xlibris, Bloomington, 554 pp.

Ricklefs R, Bermingham E (2008) The West Indies as a laboratory of biogeography and evolution. Philosophical Transactions of the Royal Society B 363(1502): 2393–2413. https://doi.org/10.1098/rstb.2007.2068

Sacek V (2014) Drainage reversal of the Amazon River due to the coupling of surface and lithospheric processes. Earth and Planetary Science Letters 401: 301–312. https://doi.org/10.1016/j.epsl.2014.06.022

Sánchez-González LA, Navarro-Sigüenza AG, Ornelas, JF, Morrone JJ (2013) What's in a name?: Mesoamerica. Revista Mexicana de Biodiversidad 84(4): 1305–1308. https://doi.org/10.7550/rmb.34171

Schuchmann KL, Weller AA, Heynen I (2001) Systematics and biogeography of the Andean genus Eriocnemis (Aves: Trochilidae). Journal of Ornithology 142(4): 433–481. https://doi.org/10.1046/j.1439-0361.2001.01025.x

Schuchmann KL, Weller AA, Wulfmeyerm E (2003) Biogeography and taxonomy of Lafresnaya (Trochilidae) with a new subspecies from Colombia. Ornitologia Neotropical 14(2): 157–171.

Silva JMC, Sousa MC, Castelletti CHHM (2004) Areas of endemism for passerine birds in the Atlantic forest, South America. Global Ecology and Biogeography 13: 85–92. https://doi.org/10.1111/j.1466-882X.2004.00077.x

Sirgrist MS, Carvalho CJB (2009) Historical relationships among areas of endemism in the tropical South America using Brooks Parsimony Analysis (BPA). Biota Neotropica 9(4): 79–90. https://doi.org/10.1590/S1676-06032009000400009

Szumik CA, Goloboff PA (2004) Areas of endemism: An improved optimality criterion. Systematic Biology 53(6): 968–977. https://doi.org/10.1080/10635150490888859

Urtubey E, Stuessy TF, Tremetsberger K, Morrone JJ (2010) The South American Biogeographic Transition Zone: An analysis from Asteraceae. Taxon 59(2): 505–509. https://doi.org/10.2307/25677608

Vázquez-Miranda H, Navarro-Sigüenza AG, Morrone JJ (2007) Biogeographical patterns of the avifaunas of the Caribbean Basin Islands: a parsimony perspective. Cladistics 23(2): 180–200. https://doi.org/10.1111/j.1096-0031.2006.00133.x

Webb SD (1995) Biological implications of the middle Miocene Amazon seaway. Science 269(5222): 361–362. https://doi.org/10.1126/science.269.5222.361

Weller AA, Schuchmann KL (2002) Biogeography of *Ramphomicron* Bonaparte, 1850 (Aves: Trochilidae). Ecotropica 8: 47–58.

Author Contributions: DO and IF designed to the study; DO and JM analyzed the data; DO wrote the paper and IF and JM supervised the findings of this work. All authors discussed the results and contributed to the final manuscript.

Competing Interests: The authors have declared that no competing interests exist.

4

Revision of *Lineostethus* (Heteroptera: Pentatomidae: Discocephalini)

Thereza de Almeida Garbelotto[1], Ana Paula Leite Kochenborger[2], Luiz Alexandre Campos[2]

[1]*Departamento de Ciências Biológicas, Universidade do Sul Catarinense. Avenida José Acácio Moreira 787, 88704-900 Tubarão, SC, Brazil.*
[2]*Departamento de Zoologia, Universidade Federal do Rio Grande do Sul. Avenida Bento Gonçalves 9500, 91501-970 Porto Alegre, RS, Brazil.*
Corresponding author: Thereza de Almeida Garbelotto (therezagarbelotto@hotmail.com)

ABSTRACT. *Lineostethus* Ruckes, 1966 has four species, *L. clypeatus* (Stål, 1862), *L. graziae* Hildebrand & Becker, 1982, *L. marginellus* (Stål, 1872), and *L. tenebricornis* (Ruckes, 1957); mostly restricted to Central America, but with distribution records in the southern North America. *Lineostethus* is one of the discocephaline genera with discoid head. The last review of the genus, from the 1980's, focused on the genital morphology and failed to provide updated descriptions or a key to species. We had access to material from different collections, including several specimens representing new species of *Lineostethus*, and males of *L. marginellus* which shows important variations in genital morphology. Here we update the generic description of *Lineostethus* and its known species and describe the new species *Lineostethus acuminatus* **sp. nov.** (holotype male in DZUP: Costa Rica, Guanacaste), *Lineostethus auritus* **sp. nov.** (holotype male in AMNH: Mexico, Oaxaca), *Lineostethus otarus* **sp. nov.** (holotype male in AMNH: Mexico, Michoacán) and *Lineostethus sinuosus* **sp. nov.** (holotype male in DZUP: Mexico, Yucatán). Additionally, we provide identification keys and distribution maps.

KEY WORDS. *Discocephala*, Discocephalinae, Neotropical, *Platycarenus*, taxonomy.

INTRODUCTION

Members of Discocephalinae are recognized primarily by the labium inserted usually on or posterior to an imaginary transverse line crossing the head on the anterior limit of eyes, and by the abdominal trichobothria usually placed lateral to the spiracles (Grazia et al. 2015, Rider et al. 2017). Two tribes restricted to the Americas, Discocephalini and Ochlerini, are classified in Discocephalinae (Rolston and McDonald 1979, Rolston 1981). A generic key is available for the genera of Ochlerini (Rolston 1992) and another for the genera of broadheaded discocephalines (Rolston 1990), and though both are outdated, *Lineostethus* Ruckes, 1966 can be identified using the latter.

Discocephalini comprises 44 genera and 193 species (Grazia et al. 2015, Rosso and Campos 2017) of light-brown to brown hues, often mottled with black or shiny black, distributed mostly in the Neotropical region.

Ruckes (1966) revised *Platycarenus* Fieber, 1861, raising the subgenus *Discocephalessa* Kirkaldy, 1909 to genus, and describing five new genera to include species removed from both *Discocephalessa* and *Discocephala* Laporte, 1832, along with new species. *Lineostethus* Ruckes, 1966 was proposed to include three species, *Platycarenus* (*Discocephalessa*) *tenebricornis* Ruckes, 1957, *Platycarenus* (*Discocephalessa*) *clypeatus* (Stål, 1862), its type species, and *Platycarenus* (*Discocephalessa*) *marginellus* (Stål, 1872). Ruckes (1966) considered *P.* (*D.*) *marginellus* a subspecies of *P.* (*D.*) *clypeatus* and provided a key to the species of *Lineostethus*. Along with a morphological study of the genitalia for the genus, Hildebrand and Becker (1982) considered *P.* (*D.*) *marginellus* a valid species, and described *Lineostethus graziae* Hildebrand & Becker, 1982. However, although the authors characterized the genitalia of both sexes for all species, the general body morphology was not described nor an updated key for the species was provided.

Looking through specimens of Discocephalini available from different collections, we found specimens matching the characteristics of *Lineostethus* that did not fit in any of the described species, as well as several males of *L. marginellus* showing important variations on the genital morphology. Here *Lineostethus* is revised with the redescription of the four known species, and the description of four new species from Central America. Identification keys for males and females are presented, and a distribution map is given.

MATERIAL AND METHODS

Specimens were observed and measured under stereomicroscope. The following morphometric parameters were taken: length and width of the head, pronotum and scutellum; length of the antennal and labial segments; total length, from the apex of clypeus to the apex of segment VII of connexivum; and maximum abdominal width. Measurements (mean ± standard deviation, minimum, and maximum) are given in millimeters. Internal genitalic structures were studied after boiling in 10% KOH aqueous solution and staining in Congo red. Specimens and genital structures were photographed using a stereomicroscope Nikon AZ100M and pictures stacked with Nikon NIS-Elements AR Microscope Imaging Software. Line drawings were made with a vectorial image processor on the pictures taken and checked under stereomicroscope. For the phallus, the dorsal and ventral views have no significant variations in morphology between the species and so, these views are illustrated only for the type species of *Lineostethus*.

The terminologies of Baker (1931), Dupuis (1970), Campos and Grazia (2006) and Garbelotto et al. (2013, 2016) were adopted for genital structures, and of Kment and Vilímová (2010) for cuticular structures of the external scent efferent system.

Site collection data were georeferenced using the Global Gazetteer Version 2.2 (Falling Rain Genomics Inc 2010) for those specimens without geographic coordinates informed in the respective labels, and the distribution map for the studied specimens was made with the software QGIS 2.10.1.

The specimens examined are deposited in the following collections: AMNH – American Museum of Natural History, New York, USA; CAS – California Academy of Sciences, San Francisco, USA; DARC – David A. Rider Collection, Fargo, USA; DZUP – Museu de Entomologia Pe. Jesus Santiago Moure, Universidade Federal do Paraná, Curitiba, Brazil; EMEC – Essig Museum of Entomology, University of California Berkeley, Berkeley, USA; NHRS – Naturhistoriska riksmuseet, Stockholm, Sweden; UCRC – Entomology Research Museum, University of California Riverside, Riverside, USA; UFRG – Departamento de Zoologia da Universidade Federal do Rio Grande do Sul, Porto Alegre, Brazil; USNM – National Museum of Natural History, Washington D.C., USA.

TAXONOMY

Lineostethus Ruckes, 1966

Lineostethus Ruckes, 1966: 22–26, figs 2, 9, 12, 17–18; Hildebrand and Becker (1982: 773–784); Rolston (1990: 30).

Type species. *Discocephala clypeata* Stål, 1862 – by original designation.

Redescription. Body oval, general color ivory with brown punctures. Head discoid, flat, wider than long. Mandibular plates longer than clypeus, overlapping at apex; lateral margins of mandibular plates outlined in brown close to the eyes. Ocelli minute. Area between eyes and ocelli with 1+1 elliptical spots without punctures (Figs 1–8). Antennae 5-segmented. Head ventrally tumescent between eyes and along bucculae. Bucculae weakly developed. Labium reaching the abdominal segment III.

Pronotum trapezoidal, wider than long; cicatrices flat and outlined by punctures. Anterolateral margins straight. Scutellum longer than wide, reaching connexival segment VI; apex usually uniformly convex to angulate. Corium longer than scutellum, reaching connexival segment VII; unpunctured area lateral to the apex of each radial vein subequal to or smaller than the width of one eye. Marginal area of anterolateral margins of pronotum and coria delimited by a line of punctures. Membrane fumose, reaching or surpassing the apex of abdomen, with 6 to 9 veins. Prosternum furrowed; mesosternum with 1+1 tumescent lateral areas, xyphus carinated; metasternum lozenge, carinated medially. Mesopleural evaporatorium sinuous, extending laterad beyond half of mesopleura; lateral margin of mesopleura without evaporatorium, except its posterior angle. Metapleural evaporatorium extending laterad beyond half of metapleura and with lateral margin sinuous; both meso- and metapleural evaporatorium concolorous with the body. Peritreme spout-shaped, scalpel-like. Legs with brown spots; tibiae with dorsal furrow.

Connexivum narrowly to widely exposed; tergite VIII of females usually visible from above. Abdomen furrowed ventrally at least on base and with punctures concentrated on lateral thirds; spiracles brown.

Male genitalia. Pygophore globose. Dorsal rim with 1+1 subtriangular to subrectangular foliaceous expansions projected on each side of segment X; cutting of dorsal rim U-shaped or subquadrangular around the segment X (Figs 9, 19, 27, 37, 45, 54, 62, 70: fdr, dr). Basal impression of the foliaceous expansions subrectilinear to sinuous (Figs 9, 19, 27, 37, 45, 54, 62, 70: bi). Posterolateral angles of pygophore reduced, obtuse or acute, usually not covered by the foliaceous expansions (Figs 9, 19–20, 27, 37, 45, 54, 62–63, 70: pa). Segment X elliptical, twice longer than wide and surpassing ventral rim margin. Margin of superior layer of ventral rim sinuous to concave, median excavation reduced and U-shaped to drop-shaped (Figs 10, 20, 28, 38, 46, 55, 63, 71: vr). Margin of the inferior layer of ventral rim with one or two lines of short bristles below the projections of the superior layer of ventral rim, bristles associated or not with punctures (Figs 10–11, 20–21, 28–29, 38–39, 46–47, 55–56, 63–64, 71–72: b). Superior layer of ventral rim projected ventrally, forming 1+1 projections varying in contour, length, width, and shape of apex, these projections forming an acute to obtuse angle with the posterolateral angles (Figs 10–11, 20–21, 28–31, 38–39, 46–47, 55–56, 63–64, 71–72: psl, pa). Parameres placed laterally to segment X, at least reaching the expansions of the dorsal rim; base of parameres circular and head subtriangular to sublozenge (Figs 12–13, 22–23, 32–33, 40–41, 48–49, 57–58, 65–66, 73–74). Phallotheca longer than wide, narrowing toward vesica; ventral surface of phallotheca subrectilinear to concave,

Figures 1–8. *Lineostethus*, dorsal aspect: (1) *L. clypeatus* (female); (2) *L. tenebricornis* (male paratype); (3) *L. marginellus* (male); (4) *L. graziae* (female holotype, photo by Mike Narahara – CAS); (5) *L. sinuosus* sp. nov. (male holotype); (6) *L. acuminatus* sp. nov. (male holotype); (7) *L. otarus* sp. nov. (male holotype); (8) *L. auritus* sp. nov. (male holotype). Scale bars: 1mm.

with a ventro-apical projection (Figs 14–16, 24, 34, 42, 50, 59, 67, 75: ph, ap). Vesica cap-shaped, apex concave to acute, apical margin following or not the ductus seminis distalis (Figs 14–16, 24, 34, 42, 50, 59, 67, 75: v, av, ds).

Female genitalia. Posterior margin of urosternite VII uniformly concave, or sinuous with median third subrectilinear (Figs 17, 25, 35, 43, 51–52, 60, 68, 76: VII). Gonocoxites VIII subrectangular to subtriangular; posterior margin subrectilinear to convex; sutural angles not juxtaposed (Figs 17, 25, 35, 43, 51–52, 60, 68, 76: gcVIII). Laterotergites VIII subtriangular; posterior margin subrectilinear to concave; spiracles visible (Figs 17, 25, 35, 43, 51–52, 60, 68, 76: la8, s). Gonocoxites IX subtrapezoidal. Laterotergites IX digitiform to subtriangular. Segment X rectangular, surpassing half of laterotergites IX; posterior margin concave to subrectilinear. Thickening of gonapophyses IX paired, varying in shape (Figs 18, 26, 36, 44, 53, 61, 69, 77: tgIX). Ring sclerites present and circular. Thickening of vaginal intima circular to subquadrangular (Figs 18, 26, 36, 44, 53, 61, 69, 77: vi). Ductus receptaculi with median duct of vesicular area enlarged distally; anterior annular crest parallel or turned to distal ductus receptaculi; pars intermedialis completely sclerotized or just on basal and apical thirds; posterior annular crest perpendicular to or directed toward capsula seminalis; capsula seminalis globose, without processes.

Distribution (Fig. 78). USA: *Arizona*, *Texas*. Mexico: *Nuevo León*, *Tamaulipas*, *San Luis Potosi*, *Veracruz*, *Jalisco*, *Hidalgo*,

Michoacán, Morelos, Puebla, Oaxaca, Tabasco, Chiapas, Campeche, Yucatán, Quintana Roo. Belize (**new record**). Guatemala: *Alta Verapaz, Baja Verapaz, Sacatepequez*. Honduras: *Cortes, Santa Barbara, Comayagua, El Paraiso*. Costa Rica: *Guanacaste*. Panama: *Los Santos*.

Remarks. *Lineostethus* was named by Ruckes (1966) after the thin carina both in xyphus and metasternum. The genus is possibly related to the other Discocephalini with a discoid head (e.g., *Acclivilamna* Ruckes, 1966, *Alveostethus* Ruckes, 1966, *Discocephala*, *Ischnopelta* Stål, 1868, *Platycarenus* and *Phoeacia* Stål, 1862) grouped by Rolston (1990) for identification purposes as the 'broadheaded' discocephalines. *Lineostethus* can be differentiated from the remainder of broadheaded genera by the body shape, especially the head uniformly convex, with an inconspicuous anteocular process, the pronotum subtrapezoidal, with anterolateral margins slightly convex, and the scutellum reaching the abdominal segment VI with lateral margins narrowing evenly toward the concave apex.

The genitalic morphology in *Lineostethus* is also diagnostic, both the female genital plates and the male pygophore. The female well exposed gonocoxites IX and segment X (Figs 17, 25, 35, 43, 51–52, 60, 68, 76), and the male foliaceous expansions of the dorsal rim and the ventral projections of the superior layer of ventral rim of pygophore (Figs 9–11, 19–21, 27–31, 37–39, 45–47, 54–56, 62–64, 70–72: fdr, psl) are features seen only in *Lineostethus* within the Discocephalini. The males of *Acclivilamna* also have paired expansions of the dorsal rim, but extremely sclerotized, not foliaceous, and short in length if compared to *Lineostethus*. Undeveloped posterolateral angles of pygophore are also seen in *Acclivilamna* and in some undescribed species of *Ischnopelta*; though in *Ischnopelta* the dorsal and ventral rims have neither projections nor expansions.

Lineostethus have a homogeneous general morphology making it difficult to identify the species based on characteristics other than genitalic. In addition, we examined a large number of specimens from different localities and were unable to confirm the characters of general morphology treated by Ruckes (1966) as diagnostic for some species (e.g., antennae uniformly tenebrous in *L. tenebricornis*, and the distribution of punctures on anterolateral pronotal margins and on basal costal margins in *L. clypeatus* and *L. marginellus*). The male genitalia are clearly diagnostic for the genus, and species can be promptly separated by the morphology of pygophore, especially the foliaceous expansions of dorsal rim (Figs 9, 19, 27, 37, 45, 54, 62, 70) and the projections of the superior layer of ventral rim (Figs 10–11, 20–21, 28–31, 38–39, 46–47, 55–56, 63–64, 71–72). However, differences between female genitalia are subtle making the determination of female specimens difficult. Exceptions are *L. sinuosus* Kochenborger & Garbelotto sp. nov., *L. auritus* Garbelotto & Kochenborger sp. nov. and *L. otarus* Garbelotto & Kochenborger sp. nov., which have sinuous posterior margin in gonocoxites VIII. Since non-genital characters are insufficient for distinguishing species of *Lineostethus*, the keys presented below, one for males and another one for females, use only genital characters.

Lineostethus is the Discocephalinae genus with the northernmost distribution and the only so far recorded in the Nearctic Region. Though this Nearctic occurrence was previously known (Ruckes 1957, Hildebrand and Becker 1982), the discocephalines have been regarded as occurring exclusively in the Neotropical region [restricted to the Neotropics sensu Grazia et al. (2015), endemic to the Neotropics sensu Rider et al. (2017)]. Following classic as well recent biogeographic regionalization systems (Morrone 2014, 2015) the Mexican Transition Zone and the Nearctic region should also be considered as part of the distribution of the subfamily.

Key for males of *Lineostethus* Ruckes, 1966

1 Foliaceous expansions of dorsal rim of pygophore longer than wide, surpassing the posterior limit of ventral rim by at least one third of the expansion length (Figs 62, 70: fdr)........ 2

1′ Foliaceous expansions of dorsal rim of pygophore subequal in length and width or wider than long, sometimes surpassing the posterior limit of ventral rim but for much less than one third of the expansion length (Figs 9, 19, 27, 37, 45, 54, 62, 70: fdr)........ 3

2(1) Basal impression of the foliaceous expansions subrectilinear; proximal portion of the lateral margin of foliaceous expansions sinuous; apex of foliaceous expansions rounded; posterolateral angles usually hidden dorsally by the foliaceous expansions (Figs 62–63: bi, dr, fdr, pa)........ *L. otarus* Garbelotto & Kochenborger sp. nov. (Fig. 7)

2′ Basal impression of the foliaceous expansions sinuous; proximal portion of the lateral margin of foliaceous expansions straight; apex of foliaceous expansions truncate; posterolateral angles visible dorsally (Figs 70–71: bi, dr, fdr, pa)........ *L. auritus* Garbelotto & Kochenborger sp. nov. (Fig. 8)

3(1′) Foliaceous expansions of dorsal rim of pygophore subequal to posterolateral angles in dorsal view (Figs 37, 54: fdr, pa)........ 4

3′ Foliaceous expansions of dorsal rim of pygophore either longer or shorter than posterolateral angles in dorsal view (Figs 9, 19, 27, 45: fdr, pa)........ 5

4(3) Basal impression of the foliaceous expansions clearly defined; cutting of dorsal rim subrectangular around the segment X; posterolateral angles of pygophore with acute apex (Fig. 37: bi, fdr); projections of the superior layer of ventral rim with rounded apex (Fig. 39: psl)........ *L. graziae* Hildebrand & Becker, 1982 (Fig. 4)

4′ Basal impression of the foliaceous expansions less defined; cutting of dorsal rim U-shaped around the segment X; posterolateral angles of pygophore with rounded apex (Fig. 54: bi, fdr); projections of the superior layer of ventral rim with hook-shaped apex (Fig. 56: psl)........ *L. acuminatus* Kochenborger & Garbelotto sp. nov. (Fig. 6)

5(3′) Foliaceous expansions of dorsal rim shorter than both the posterior limit of ventral rim and the posterolateral angles (Fig. 9: fdr, pa); posterior limit of ventral rim subrectilinear (Fig. 10)...... *L. clypeatus* (Stål, 1862) (Fig. 1)

5′ Foliaceous expansions of dorsal rim at least as long as both the posterior limit of ventral rim and the posterolateral angles, hiding or not the posterolateral angles in dorsal view (Figs 19, 27, 45: fdr, pa); posterior limit of ventral rim sinuous (Figs 20, 28, 46) 6

6(5′) Foliaceous expansions of dorsal rim subquadrangular and expanded, hiding the posterolateral angles dorsally; cutting of dorsal rim subrectangular around the segment X (Fig. 19: fdr, pa, dr); margin of inferior layer of ventral rim with two lines of short bristles (Fig. 20: b)................. .... *L. tenebricornis* (Ruckes, 1957) (Fig. 2)

6′ Foliaceous expansions of dorsal rim subrectangular and not expanded, leaving the posterolateral angles visible dorsally; cutting of dorsal rim U-shaped around the segment X (Figs 27, 45: fdr, pa, dr); margin of inferior layer of ventral rim with one line of short bristles (Figs 28, 46: b).. 7

7(6′) Foliaceous expansions of dorsal rim longer than both the posterior limit of ventral rim and the posterolateral angles (Fig. 45: fdr); apex of the projections of the superior layer of ventral rim spatulate (Fig. 47: psl) .... *L. sinuosus* Kochenborger & Garbelotto sp. nov. (Fig. 5)

7′ Foliaceous expansions of dorsal rim at maximum as long as both the posterior limit of ventral rim and the posterolateral angles (Fig. 27: fdr); apex of the projections of the superior layer of ventral rim acute (Fig. 28: psl) .... *L. marginellus* (Stål, 1872) (Fig. 3)

Key for females of *Lineostethus* Ruckes, 1966

1 Posterior margin of gonocoxites VIII sinuous (Figs 51–52, 68, 76: gcVIII) 2

1′ Posterior margin of gonocoxites VIII convex or subrectilinear (Figs 17, 25, 35, 43, 60: gcVIII) 4

2(1) Gonocoxites VIII subtriangular; posterior margin of gonocoxites VIII strongly sinuous; sutural angles of gonocoxites VIII apart from each other and little projected over the laterotergites IX; lateral angles of gonocoxitesVIII little projected over the base of the laterotergites VIII (Figs 51–52: gcVIII, la8, la9, s) .... *L. sinuosus* Kochenborger & Garbelotto sp. nov. (Fig. 5)

2′ Gonocoxites VIII subrectangular; posterior margin of gonocoxites VIII weakly sinuous; sutural angles of gonocoxites VIII juxtaposed; lateral angles of gonocoxites VIII uniformly convex or truncated (Figs 68, 76: gcVIII, la8, la9, s)... 3

3(2′) Posterior margin of urosternite VII uniformly concave; lateral angles of gonocoxites VIII uniformly convex (Fig. 76: VII, gcVIII) ....*L. auritus* Garbelotto & Kochenborger sp. nov. (Fig. 8)

3′ Posterior margin of urosternite VII subrectilinear medially; lateral angles of gonocoxites VIII truncated (Fig. 68: VII, gcVIII) ....*L. otarus* Garbelotto & Kochenborger sp. nov. (Fig. 7)

4(1′) Spiracles of laterotergites VIII anterior to an imaginary transversal line through the posterior margin of gonocoxites VIII and placed close to the lateral angle of gonocoxites VIII (Figs 43, 60: s, la8, gcVIII) 5

4′ Spiracles of laterotergites VIII on or posterior to an imaginary transversal line through the posterior margin of gonocoxites VIII, and placed posterior to the lateral angle of gonocoxites VIII (Figs 17, 35: s, la8, gcVIII); if the spiracles are slightly anterior to the imaginary line, then each spiracle is placed away from the lateral angle of gonocoxites VIII (Fig. 25: s, la8, gcVIII).......................... 6

5(4) Posterior margin of urosternite VII uniformly concave; gonocoxites VIII subquadrangular (Fig. 43: VII, gcVIII).... ....*L. graziae* Hildebrand & Becker, 1982 (Fig. 4)

5′ Posterior margin of urosternite VII subrectilinear over gonocoxites VIII; gonocoxites VIII subtriangular (Fig. 60: VII).. *L. acuminatus* Kochenborger & Garbelotto sp. nov. (Fig. 6)

6(4′) Posterior margin of urosternite VII uniformly concave (Fig. 17: VII)..................... *L. clypeatus* (Stål, 1862) (Fig. 1)

6′ Posterior margin of urosternite VII subrectilinear medially (Figs 25, 35: VII) 7

7(6′) Posterior margin of urosternite VII sinuous; gonocoxites VIII subrectangular (Fig. 25: VII, gcVIII).......................... .... *L. tenebricornis* (Ruckes, 1957) (Fig. 2)

7′ Posterior margin of urosternite VII concave; gonocoxites VIII subtriangular (Fig. 35: VII, gcVIII) .... *L. marginellus* (Stål, 1872) (Fig. 3)

Lineostethus clypeatus (Stål, 1862)

Figs 1, 9–18, 78

Discocephala clypeata Stål, 1862: 96; Stål (1868: 18); Stål (1872: 6); Distant (1880: 45, plate 6, fig. 1); Uhler (1886: 5); Lethierry and Severin (1893: 83); Banks (1910: 92).

Discocephala inobtrusa: Walker, 1867: 183 (synonymy by Kirkaldy, 1909).

Platycarenus (*Discocephalessa*) *clypeatus*: Kirkaldy (1909: 215).

Lineostethus clypeatus clypeatus: Ruckes (1966: 24–26, figs 2, 5, 17, 18).

Lineostethus clypeatus: Hildebrand and Becker (1982: 778–780).

Diagnosis. Males of *L. clypeatus* differ from the other species by the shorter and more rounded foliaceous expansions of the dorsal rim of pygophore, and by the projections of the superior layer of ventral rim short and tapering to its apex (Figs 9–11: fdr, psl). The female genital plates are very similar to *L. marginellus* but can be differentiated by the slightly shorter gonocoxites VIII. The helicoidal distal portion of ductus receptaculi, the fusiform thickening of gonapophysis IX, and the elliptical thickening of vaginal intima also help to differentiate *L. clypeatus* from the other species.

Redescription. Antennal segments I–III with longitudinal brown lines. Anterolateral margins of pronotum and proximal portion of lateral margin of corium punctured in some specimens. Scutellum sometimes with a spot of brown punctures on disc and on post-frenal lobe. Unpunctured area laterad to the apex of radial vein of coria subequal to the width of one eye.

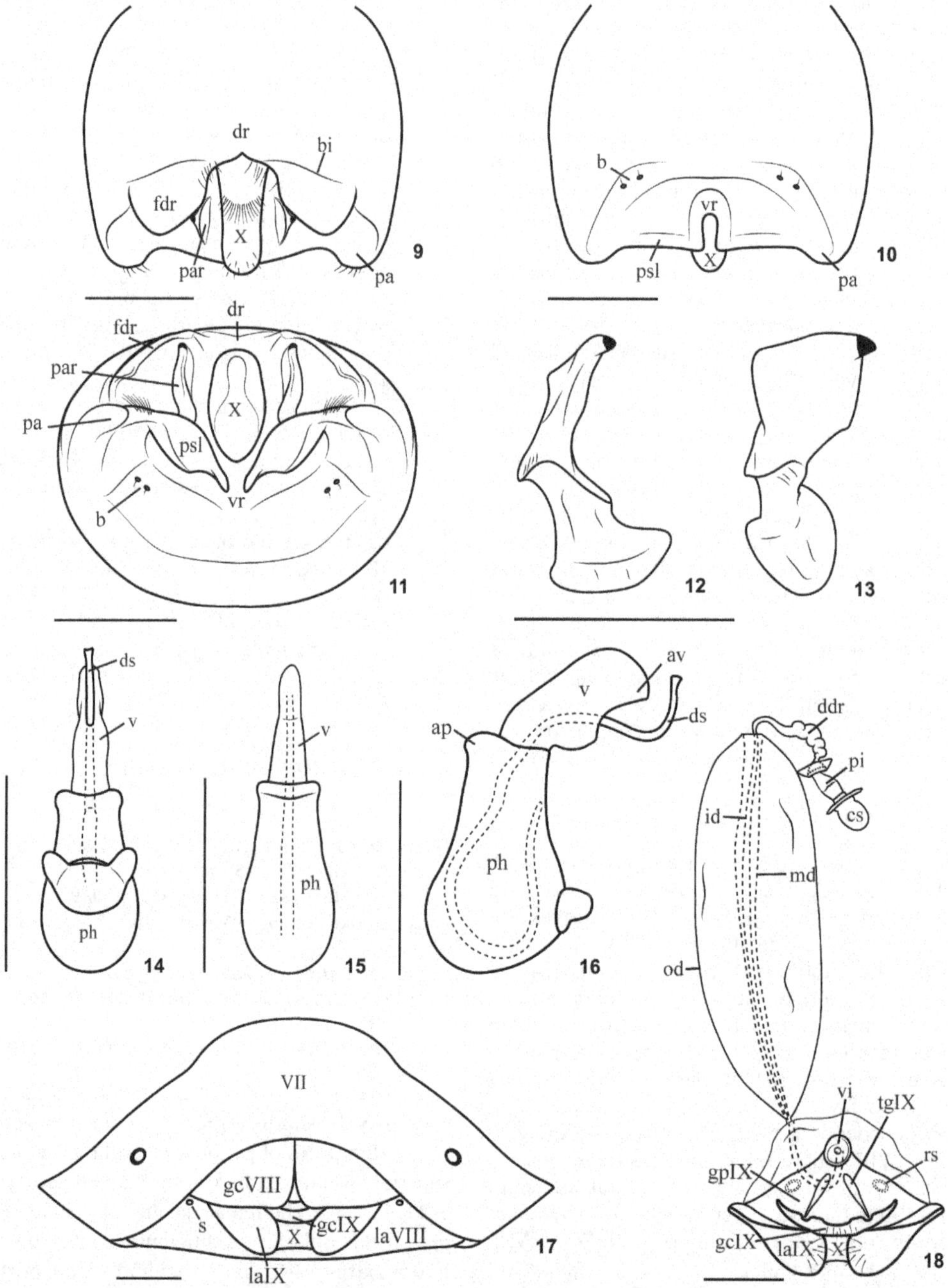

Figures 9–18. *Lineostethus clypeatus*: (9–11) male pygophore, dorsal, ventral and posterior views respectively; (12–13) left paramere, dorsal and lateral views; (14–16) phallus, dorsal, ventral and lateral view; (17) female genital plates; (18), female receptaculum seminis. Abbreviations: ap, ventral apical projection of phallotheca; av, apical margin of vesica; b, bristles; bi, basal impression of foliaceous expansions; cs, capsula seminalis; ddr, distal ductus receptaculi; dr, dorsal rim; ds, ductus seminis distalis; fdr, foliaceous expansions of dorsal rim; gcVIII, gonocoxites VIII; gcIX, gonocoxites IX; gpVIII, gonapophyses VIII; gpIX, gonapophyses IX; id, internal duct of vesicular area; la8, laterotergites VIII; la9, laterotergites IX; md, median duct of vesicular area; od, outer duct of vesicular area; pa, posterolateral angle; par, paramere; ph, phallotheca; pi, pars intermedialis; psl, projection of superior layer of ventral rim; rs, ring sclerites; s, spiracle; tgIX, thickening of gonapophyses IX; v, vesica; vi, thickening of vaginal intima; vr, ventral rim; VII, urosternite VII; X, segment X. Scale bars: 9–13, 17 = 1 mm; 14–16, 18 = 0.5 mm.

Hemelytral membrane slightly surpassing the apex of abdomen, with 6–7 veins. Ventrally, punctures sparse and concentrated laterally. Femora with sparse brown circular spots on apical half.

Male genitalia. Foliaceous expansions of dorsal rim subrectangular; apex of expansions not surpassing the posterior limit of ventral rim; posterolateral angles visible dorsally. Basal impression of the foliaceous expansions concave, cutting of dorsal rim U-shaped (Fig. 9: bi, fdr, dr). Posterolateral angles with rounded apex (Figs 9–11: pa). Margin of the superior layer of ventral rim flatly concave, median excavation drop-shaped (Fig. 10: vr). Margin of the inferior layer of ventral rim with one arc of short bristles (Figs 10, 11: b). Projections of the superior layer of ventral rim forming an obtuse angle with the posterolateral angles; projections tapering toward apex and covered by setae (Fig. 11: psl, pa). Apex of parameres diverging, not reaching the foliaceous expansions of dorsal rim; head lozenge (Figs 9, 11–13: par). Ventral surface of phallotheca subrectilinear (Figs 14–16: ph, ap). Vesica with rounded apex; apical margin concave extending to mid of ductus seminis distalis (Figs 14–16: v, av, ds). Distal portion of ductus seminis distalis subequal to vesica in length, tapering towards apex (Fig. 16: ds).

Female genitalia. Posterior margin of urosternite VII concave (Fig. 17: VII). Gonocoxites VIII subtriangular; posterior margin slightly convex, sutural margins juxtaposed at basal-two thirds; sutural angles in open "V" (Fig. 17: gcVIII). Posterior margin of laterotergites VIII slightly convex; limit between laterotergites VIII and tergite VIII sometimes not visible (Fig. 17: la8). Posterior margin of gonocoxites IX subrectilinear (Fig. 17: gcIX). Laterotergites IX digitiform, lateral margins convex and mesial margins rectilinear to slightly convex, subparallel or divergent (Fig. 17: la9). Posterior margin of segment X subrectilinear (Fig. 17: X). Thickening of gonapophyses IX fusiform, surpassing the lateral margin of the thickening of vaginal intima (Fig. 18: gpIX, tgIX). Thickening of vaginal intima elliptical and flat (Fig. 18: vi). Distal portion of ductus receptaculi helicoid apically (Fig. 18: ddr); anterior annular crest turned to distal ductus receptaculi; pars intermedialis subequal in length to capsula seminalis; middle third of pars intermedialis not sclerotized; posterior annular crest perpendicular to pars intermedialis (Fig. 18: pi). Capsula seminalis with a slight basal constriction (Fig. 18: cs).

Measurements. Male (n = 5): head length, 1.55 ± 0.08 (1.45–1.63); width, 3.20 ± 0.18 (2.98–3.36); length of antennal segments: I, 0.44 ± 0.04 (0.38–0.48); II, 0.83 ± 0.06 (0.80–0.90); III, 0.78 ± 0.05 (0.72–0.82); IV, 0.91 ± 0.01 (0.90–0.92); V, 1.06 ± 0.03 (1.04–1.08); length of labial segments: I, 0.72 ± 0.06 (0.64–0.80); II, 1.55 ± 0.21 (1.22–1.80); III, 0.65 ± 0.13 (0.62–0.82); IV, 0.47 ± 0.05 (0.40–0.50); pronotum length, 1.78 ± 0.09 (1.70–1.92); width, 4.32 ± 0.18 (4.10–4.50); scutellum length, 3.56 ± 0.19 (3.40–3.84); width, 2.84 ± 0.19 (2.64–3.00); total length, 7.25 ± 1.02 (6.75–8.50); abdominal width, 4.76 ± 0.22 (4.50–5.00).

Female (n = 5): head length, 1.55 ± 0.04 (1.5–1.63); width, 3.05 ± 0.11 (2.98–3.24); length of antennal segments: I, 0.41 ± 0.01 (0.40–0.42); II, 0.72 ± 0.06 (0.64–0.78); III, 0.71 ± 0.09 (0.60–0.80); IV, 0.87 ± 0.10 (0.78–1.00); V, 1.01 ± 0.01 (1.00–1.02); length of labial segments: I, 0.65 ± 0.05 (0.62–0.70); II, 1.38 ± 0.20 (1.10–1.58); III, 0.64 ± 0.06 (0.58–0.70); IV, 0.50 ± 0.03 (0.46–0.52); pronotum length, 1.86 ± 0.07 (1.76–1.92); width, 4.08 ± 0.23 (3.80–4.40); scutellum length, 3.45 ± 0.11 (3.32–3.56); width, 2.68 ± 0.11 (2.54–2.85); total length, 7.05 ± 0.54 (6.67–8.00); abdominal width, 4.49 ± 0.18 (4.35–4.80).

Material examined. Holotype (female): MEXICO, Sallé, Type, Typus, 80 81, NHRS-GULI 000011382 (NHRS – photographs examined). Other specimens: USA, *Texas*: Cameron (Co. Sabal Palm Grove Sanct.), 1 female, 3-VI-1984, Marlin Rice leg. (DARC); Progreso (L. Fife 2260, On cotton, Lot No 44-14405), 1 female, 7-VI-1944 (USNM); Brownsville, 1 female, 03-V — Brooklyn Museum Coll 1929 Catal No. 28 (USNM); Brownsville, 1 female, VI-1901, H.G. Barber Coll leg. (USNM); Browsville, 1 female, 5-VI-1932, J.O. Martin leg. (CAS). MEXICO, *Nuevo León*: (15 mi W. Linares), 1 female, 2-3-VII-1973, Mastro & Schaffner leg. (DARC); *Tamaulipas*: (Nac. del Rio Frio 7 km SSE Gomes Farías), 3 females, 15-16-VI-1983, T. Riley leg. (DARC); Bocatoma (7 km SSE Gomes Farías), 1 female, 5-7-I-1981, E.G. Riley leg. (DARC); (5 mi. SSE of Gomes Farías), 1 female, 19-20-VII-1970, Murray Phelps Hart Schaffener leg. (DARC); *San Luis Potosi*: Rio Verde (Media Luna, 1000), 1 female, 20-VI-1982, M.A. Ivie leg. (DARC); (Hwy 85, SW Tamazunchale, 600), 1 female, 24-VII-1982, C.W. & L.O'Brien & G.Wibmer leg. (DARC); (10 mi N Tamazunchale), 1 female, 24-XII-1940, G.E. Bohart Collector leg. (CAS); (10 mi N Tamazunchale), 1 female, 24-XII-1940, G.E. Bohart Collector leg. (AMNH); (7 mi N Tamazunchale), 1 female, 19-IX-1972, W.Hanson J.Poff leg. (DARC); *Veracruz*: Acayucan, 1 female, VII-1966, J. & W. Ivie leg. (AMNH); (Hwy. 140, 4mi. NW Rinconada, 800), 1 female, 21-VIII-1982, C.W. & L.O'Brien & G.Wibmer leg. (DARC); *Jalisco*: (Hwy. 80, 17 mi SW Autlan, 1800), 1 male, 11-VIII-1982, C.W. & L.O'Brien & G.Wibmer leg. (DARC); (6 mi. NE Barra de Navidad, 500), 1 female, 11-VIII-1982, C.W. & L.O'Brien & G.Wibmer leg. (DARC); *Guerrero*: Acapulco, 2 males, 6-7-X-1961, Pereirra S. Halffter leg. (UFRG); *Campeche*: (22 mi NE Campeche), 1 female, 5-VIII-1974, C.W. & L. O'Brien & Marshall leg. (AMNH); *Oaxaca*: (Hwy.147, 54 mi SW Tuxtepec, 800), 1 female, 27-V-1983, C.W. & L.O'Brien & G.B.Marshall leg. (DARC); (5 mi N Palomares) 1 female, 31-VIII-1964, J. & W. Ivie Collectors leg. (AMNH); La Ventosa (26 mi N) 2 males, 19-VII-1963, J. Doyen Collector leg. (UFRG); La Ventosa (26 mi N), 3 females, 19-VII-1963, J. Doyen Collector leg. (UFRG); La Ventosa (26 mi N), 1 female, 19-VII-1963, Coll.: W.A. Foster leg.; (23 mi E. Zanatepec, 1600), 1 male, 3-VI-1974, C.W. & L. O' Brien & Marshall leg. (AMNH). HONDURAS, *Comayagua*: (5 mi NW Comayagua, 1600), 2 females, 18-19-VII-1974, C.W. & L.O'Brien & Marshall leg. (DARC). COSTA RICA, *Guanacaste*: (13 mi SE Libería, 400), 1 female, 12-VII-1974, C.W. & L.O'Brien & Marshall leg. (DARC); Santa Rosa National Park, 1 female, 16-18-VII-1980, D.H.J. Janzen & W. Hallwachs leg. (DARC).

Distribution (Fig. 78). USA: *Texas* (**new record**). Mexico: *Nuevo León* (**new record**), *Tamaulipas* (**new record**), *San Luis*

Potosi (Ruckes 1966), *Veracruz* (**new record**), *Jalisco* (**new record**), *Oaxaca* (Ruckes 1966, Hildebrand and Becker 1982), *Campeche* (**new record**), *Yucatán* (Ruckes 1966). Guatemala: *Alta Verapaz*, *Baja Verapaz* (Ruckes 1966). Honduras - *Santa Barbara* (Ruckes 1966), *Comayagua* (**new record**). Costa Rica: *Guanacaste* (Ruckes 1966). Panama (Ruckes 1966).

Remarks. From the data retrieved of specimens' labels, cotton (*Gossypium hirsutum* L., Malvaceae) is a possible host plant.

Lineostethus tenebricornis (Ruckes, 1957)

Figs 2, 19–26, 78

Platycarenus (*Discocephalessa*) *tenebricornis* Ruckes, 1957: 16–17.
Lineostethus tenebricornis: Ruckes (1966: 24); Hildebrand and Becker (1982: 778–780).

Diagnosis. The specimens of *L. tenebricornis* may have brown antennae, also observed in *L. auritus* sp. nov. and *L. otarus* sp. nov. The males of *L. tenebricornis* are recognized by the subquadrangular foliaceous expansions of the dorsal rim of pygophore, covering dorsally the posterolateral angles (Figs 19–21: fdr, pa); the projections of the superior layer of ventral rim elongate and rounded at apex (the last feature similar to *L. graziae*) (Figs 21, 39: psl); the presence of two lines of short bristles below the projections of the superior layer of ventral rim (Figs 20–21: b); and the apical margin of vesica along with the ductus seminis distalis (Fig. 24: av, ds). In the females, diagnostic features are the posterior margin of urosternite VII subrectilinear on its median third along with subrectangular gonocoxites VIII (Fig. 25: VII, gcVIII); and the thickening of gonapophysis IX subrectangular with a median constriction, combined with a subtriangular thickening of vaginal intima (Fig. 25: tgIX, vi). The median third of the posterior margin of urosternite VII and the subtriangular thickening of vaginal intima are also found in *L. otarus* sp. nov. However, *L. tenebricornis* has shallower genital plates than *L. otarus* sp. nov. (Figs 25, 68), and the thickenings of gonapophysis IX have different shapes (Figs 26, 69: tgIX, vi).

Redescription. Antennal segments usually brown, sometimes ivory with longitudinal brown line on segments I–II; segment III usually longer than II. Anterolateral margins of pronotum and proximal portion of lateral margin of corium punctured. Unpunctured area laterad to the apex of radial vein of coria smaller than the width of one eye. Hemelytral membrane at least reaching the apex of abdomen, with 8 veins. Ventrally, punctures uniformly distributed. Femora with brown circular spots on the entire length or concentrated on distal half.

Male genitalia. Foliaceous expansions of dorsal rim subquadrangular; apex of expansions surpassing the posterior limit of ventral rim, hiding the posterolateral angles (Fig. 19: fdr). Basal impression of the foliaceous expansions subrectilinear, cutting of dorsal rim subrectangular, filled by the parameres and segment X (Fig. 19: bi, dr, pa, X). Posterolateral angles with sinuous apex (Figs 20–21: pa). Margin of the superior layer of ventral rim concave, median excavation U-shaped (Fig. 20: vr). Margin of the inferior layer of ventral rim with two subrectilinear lines of short bristles (Figs 20–21: b). Projections of the superior layer of ventral rim forming a right angle with the posterolateral angles; projections thin at base, median portion enlarged, and apex elongate; setae on the posterior margin of the projections (Fig. 21: psl, pa). Apex of parameres diverging, reaching or surpassing the foliaceous expansions of dorsal rim; head subtriangular with acute apex (Figs 21–23: par, X, fdr). Ventral surface of phallotheca concave (Fig. 24: ph, ap). Vesica more sclerotized on middle and over the ductus seminis distalis; apex of vesica concave; apical margin, concave, accompanying the ductus seminis distalis (Fig. 24: v, av, ds). Distal portion of ductus seminis distalis shorter than vesica (Fig. 24: ds).

Female genitalia. Posterior margin of urosternite VII slightly sinuous, middle third subrectilinear and lateral thirds convex (Fig. 25: VII). Gonocoxites VIII subrectangular; posterior margin slightly convex and sutural angles weakly divergent and basal two thirds of sutural margins juxtaposed (Fig. 25: gcVIII). Posterior margin of laterotergites VIII convex and limit between laterotergites VIII and tergite VIII indistinguishable (Fig. 25: la8). Posterior margin of gonocoxites IX slightly concave (Fig. 25: gcIX). Laterotergites IX digitiform (Fig. 25: la9). Posterior margin of segment X subrectilinear to concave (Fig. 25: X). Thickening of gonapophyses IX subrectangular narrowed medially, surpassing the lateral margin of the thickening of vaginal intima (Fig. 26: gpIX, tgIX). Thickening of vaginal intima subtriangular (Fig. 26: vi). Distal portion of ductus receptaculi medially sclerotized and enlarged (Fig. 26: ddr); anterior annular crest perpendicular to pars intermedialis; pars intermedialis sclerotized on basal two thirds and subequal to capsula seminalis in length; posterior annular crest turned to capsula seminalis (Fig. 26: pi, cs). Capsula seminalis with slight basal constriction (Fig. 26: cs).

Measurements. Male (n = 5): head length, 1.59 ± 0.05 (1.53–1.68); width, 2.96 ± 0.04 (2.88–3.01); length of antennal segments: I, 0.44 ± 0.04 (0.42–0.48); II, 0.76 ± 0.06 (0.72–0.82); III, 0.79 ± 0.06 (0.70–0.86); IV, 0.96 ± 0.05 (0.88–1.00); V, 1.17 ± 0.05 (1.08–1.22); length of labial segments: I, 0.76 ± 0.01 (0.76–0.78); II, 1.58 ± 0.05 (1.50–1.64); III, 0.68 ± 0.05 (0.62–0.74); IV, 0.51 ± 0.02 (0.50–0.54); pronotum length, 1.72 ± 0.06 (1.66–1.80); width, 4.24 ± 0.13 (4.10–4.45); scutellum length, 3.30 ± 0.20 (3.04–3.52); width, 2.71 ± 0.05 (2.67–2.79); total length, 6.82 ± 0.12 (6.67–7.03); abdominal width, 4.83 ± 0.12 (4.65–4.95).

Female (n = 5): head length, 1.56 ± 0.06 (1.5–1.63); width, 2.99 ± 0.08 (2.91–3.10); length of antennal segments: I, 0.43 ± 0.02 (0.40–0.46); II, 0.81 ± 0.04 (0.78–0.86); III, 0.82 ± 0.04 (0.80–0.86); IV, 1.02 ± 0.06 (1.00–1.12); V, 1.10 ± 0.20 (0.90–1.28); length of labial segments: I, 0.82 ± 0.04 (0.76–0.88); II, 1.60 ± 0.03 (0.02–1.64); III, 0.73 ± 0.03 (0.70–0.78); IV, 0.56 ± 0.03 (0.54–0.60); pronotum length, 1.89 ± 0.07 (1.82–1.98); width, 4.47 ± 0.08 (4.40–4.60); scutellum length, 3.66 ± 0.08 (3.60–3.80); width, 2.80 ± 0.16 (2.79–2.95); total length, 8.21 ± 0.74 (6.96–8.90); abdominal width, 5.10 ± 0.22 (5.00–5.48).

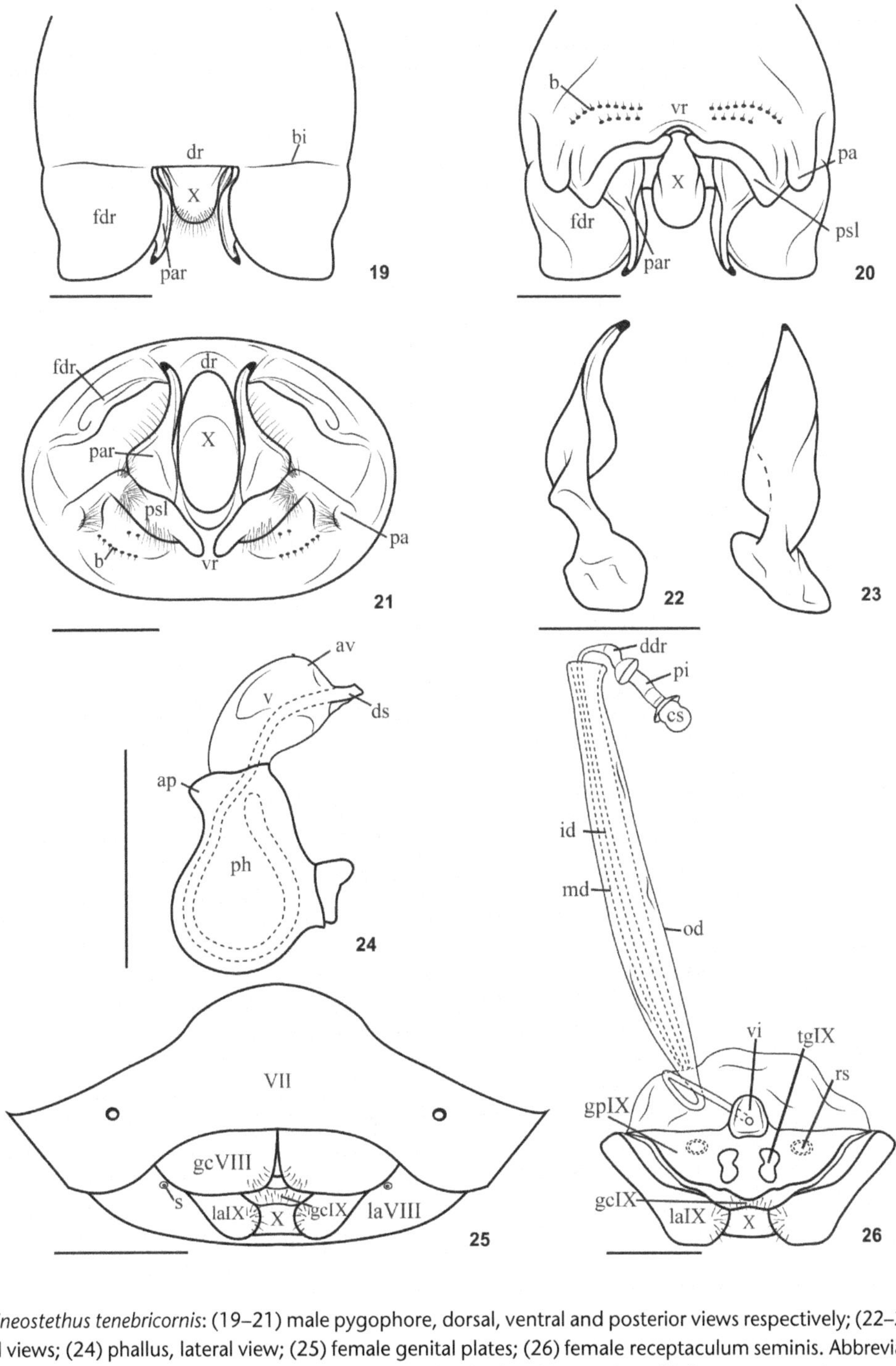

Figures 19–26. *Lineostethus tenebricornis*: (19–21) male pygophore, dorsal, ventral and posterior views respectively; (22–23) left paramere, dorsal and lateral views; (24) phallus, lateral view; (25) female genital plates; (26) female receptaculum seminis. Abbreviations: ap, ventral apical projection of phallotheca; av, apical margin of vesica; b, bristles; bi, basal impression of foliaceous expansions; cs, capsula seminalis; ddr, distal ductus receptaculi; dr, dorsal rim; ds, ductus seminis distalis; fdr, foliaceous expansions of dorsal rim; gcVIII, gonocoxites VIII; gcIX, gonocoxites IX; gpVIII, gonapophyses VIII; gpIX, gonapophyses IX; id, internal duct of vesicular area; la8, laterotergites VIII; la9, laterotergites IX; md, median duct of vesicular area; od, outer duct of vesicular area; pa, posterolateral angle; par, paramere; ph, phallotheca; pi, pars intermedialis; psl, projection of superior layer of ventral rim; rs, ring sclerites; s, spiracle; tgIX, thickening of gonapophyses IX; v, vesica; vi, thickening of vaginal intima; vr, ventral rim; VII, urosternite VII; X, segment X. Scale bars: 19–23, 25 = 1 mm; 24, 26 = 0.5 mm.

Material examamined. Paratypes: USA, *Arizona*: San Rita Mt, 1 female, 11-VII-1932, E.D. Ball leg. (AMNH); Madera Cyn. (Sta Rita Mts Pima Co.), 1 male, 13-VIII-1952, M. Cazier, R. Schrammel, C. & P. Vaurie leg. (CAS); Runaway Canyon Huachuca mts., 1 female, 15-I-1941 (AMNH); Huachuca Mts. (Rd to Petersens Rch), 1 female, 17-VIII-1940, E.C. Van Dyke leg. (CAS); Atascosa Mt., 1 female, 19-VII-1935, E.D. Ball leg. (AMNH). Other specimens: USA, *Arizona*: (Nr Oracle), 1 female, 14-VII-1940 (CAS); Sabino Cañon, Santa catalina Mts (4200 ft, on wild cotton), 1 male, 28-XI-1913, HSB leg.(USNM); (Cave Cr Chiricahua Mts.), 1 male and 2 females, 4-VII-1930, Pres. By. G. Linsley Collector leg. (CAS); (Cave Crk. Chiricahua Mts.), 1 male, 20-VI-1932, J.O. Martia leg. (CAS); Cochise Co., (Chiricahua Mts Horseshoe Cany), 6 males and 6 females, 4-VII-1977, Scott McCleve leg. (AMNH); Cochise Co., (Chiricahua Mts Horseshoe Cany), 1 male, 4-VII-1977, Scott McCleve, M. Cazier, R. Schrammel, C. & P. Vaurie Collectors leg. (AMNH); Cochise Co. Chiricahua Mts. (South Fork), 1 male and 1 female, 12-VIII-1978, Duane/Pam Flynn leg. (DARC); Cochise Co. Chiricahua Mts. (South Fork), 1 female, 12-VIII-1978, Duane/Pam Flynn leg. (DARC); Cochise Co. Huachuca Mts (Ash. Canyon, 5600-6000), 1 male, 29-VII-1979, W.F.Barr leg. (DARC); Cochise Co. Huachuca Mts (Ash. Canyon, 5600-6000), 1 female, 29-VII-1979, W.F.Barr leg. (DARC); Huachuca Mts, 1 female, 3-VII-2005 (CAS); Huachuca Mts, 1 male, 16-VII-2005 (CAS); Huachuca Mts. (Rd to Petersens Rch), 2 males and 1 female, 17-VIII-1940, E.C. Van Dyke leg. (CAS); Huachuca Mts. (Carr Cn), 2 males and 1 female, 7-VIII-1924, E.P. Van Duzee leg. (CAS); Huachuca Mts, 1 male, VII-1905, H.G. Barber Colln 1950 leg. (USNM); Cochise Co. (Cave Creek Cyn. 3 mi WE Portal), 1 male, 6-IX-1979, H.A.Hespenheide leg. (DARC); Cochise Co. (SW Res. Sta. 5mi W Portal), 5 females, 10-11-VIII-1983, E.G. & M. Riley leg. (DARC); Cochise Co. Chiracahua Mtns (Cave CK.), 1 female, 27-VIII-1983, D.B.Thomas leg. (DARC); Cochise Co. Chiracahua Mtns (Cave CK.), 1 male, 27-VIII-1983, D.B.Thomas leg. (DARC); Madera Cn., (Santa Rita Mts. Santa Cruz Co. 6200ft.), 2 males and 3 females, 1-2-VIII-1952, H.B. Leech & J.W. Green Collectors leg. (CAS); Florida Cn., 2 males, 10-VIII-1924, Santa Rita Mts, E.P. Van Duzee Collector leg. (CAS); Palmerlei, 1 female, July, H.G. Barber Collector leg. (USNM); (Camp Washington Patagonia Mts.), 1 male, 22-VI-1948, H.S.B. Barber leg. Request 1950 (USNM).

Distribution (Fig. 78). USA: Arizona (Ruckes 1957, 1966, Hildebrand and Becker 1982). Mexico: *Michoacán* (Hildebrand and Becker 1982).

Remarks. From the data retrieved of specimens' labels, cotton (*Gossypium hirsutum* L., Malvaceae) is a possible host plant.

Lineostethus marginellus (Stål, 1872)

Figs 3, 27–36, 78

Discocephala marginella Stål, 1872: 6; Distant (1880: 45, plate 5, fig.2); Uhler (1886: 5); Lethierry and Severin (1893: 84); Banks (1910: 92).
Platycarenus (*Discocephalessa*) *marginellus*: Kirkaldy (1909: 215).
Lineostethus clypeatus marginellus: Ruckes (1966: 26).
Lineostethus marginellus: Hildebrand and Becker (1982: 778–780).

Diagnosis. Males of *L. marginellus* differ from the other species by the subrectangular foliaceous expansions of the dorsal rim of pygophore (Fig. 27: fdr); and the projections of the superior layer of ventral rim acute at apex and forming an acute to right angle with the posterolateral angles (Figs 29–31: psl). In this species the enlargement of the median portion of the projections of the superior layer of the ventral rim is variable (Figs 29–31: psl), and consequently their apex length and shape also vary a little. The foliaceous expansions of the dorsal rim are subrectangular as in *L. acuminatus* sp. nov. but the species differ by the projections of ventral rim and by the phallus. The females are like *L. clypeatus* in the general shape of the genital plates, however in *L. marginellus* the gonocoxites VIII are longer (Figs 17, 35: gcVIII). Also, the sinuous thickening of gonapophysis IX, the subrectangular thickening of the vaginal intima, and the enlarged and straight ductus receptaculi distal are diagnostic and differ *L. marginellus* from *L. clypetus* (Figs 36: tgIX, vi, ddr).

Redescription. Antennal segments II–III with longitudinal brown line. Anterolateral pronotal margins and proximal portion of lateral margin of corium punctured in some specimens. Scutellum sometimes with a spot of brown punctures on disc and on post-frenal lobe. Unpunctured area laterad to the apex of radial vein of coria subequal to the width of one eye or inconspicuous. Hemelytral membrane slightly surpassing the apex of abdomen, with 6–7 veins. Ventrally, punctures sparse and concentrated laterally. Femora with sparse brown circular spots on apical half.

Male genitalia. Foliaceous expansions of dorsal rim subrectangular; apex of expansions not surpassing the posterior limit of ventral rim; posterolateral angles visible dorsally (Fig. 27: fdr, pa). Basal impression of the foliaceous expansions subrectilinear, cutting of dorsal rim U-shaped (Fig. 27: bi, dr). Posterolateral angles with rounded apex, sometimes with setae on the posterior margin (Fig. 27: pa). Margin of the superior layer of ventral rim sinuous, median excavation narrow and drop-shaped (Fig. 28: vr). Margin of the inferior layer of ventral rim with one arc of short bristles (Fig. 28: b). Projections of the superior layer of ventral rim forming an acute to right angle with the posterolateral angles; projections covered by setae on the posterior margin, base and median portions enlarged, apex acute (Figs 29–31: psl, pa); posterior surface of the projections usually depressed. Parameres parallel to segment X, usually projected over the cutting of the dorsal rim; head lozenge (Figs 29, 32–33: par, dr, X). Ventral surface of phallotheca subrectilinear (Fig. 34: ph, ap). Vesica apical margin sinuous, extending to mid of ductus seminis distalis (Fig. 34: v, av, ds). Distal portion of ductus seminis distalis half the length of vesica; uniformly wide (Fig. 34: ds).

Female genitalia. Posterior margin of urosternite VII concave (Fig. 35: VII). Gonocoxites VIII subtriangular; posterior margin

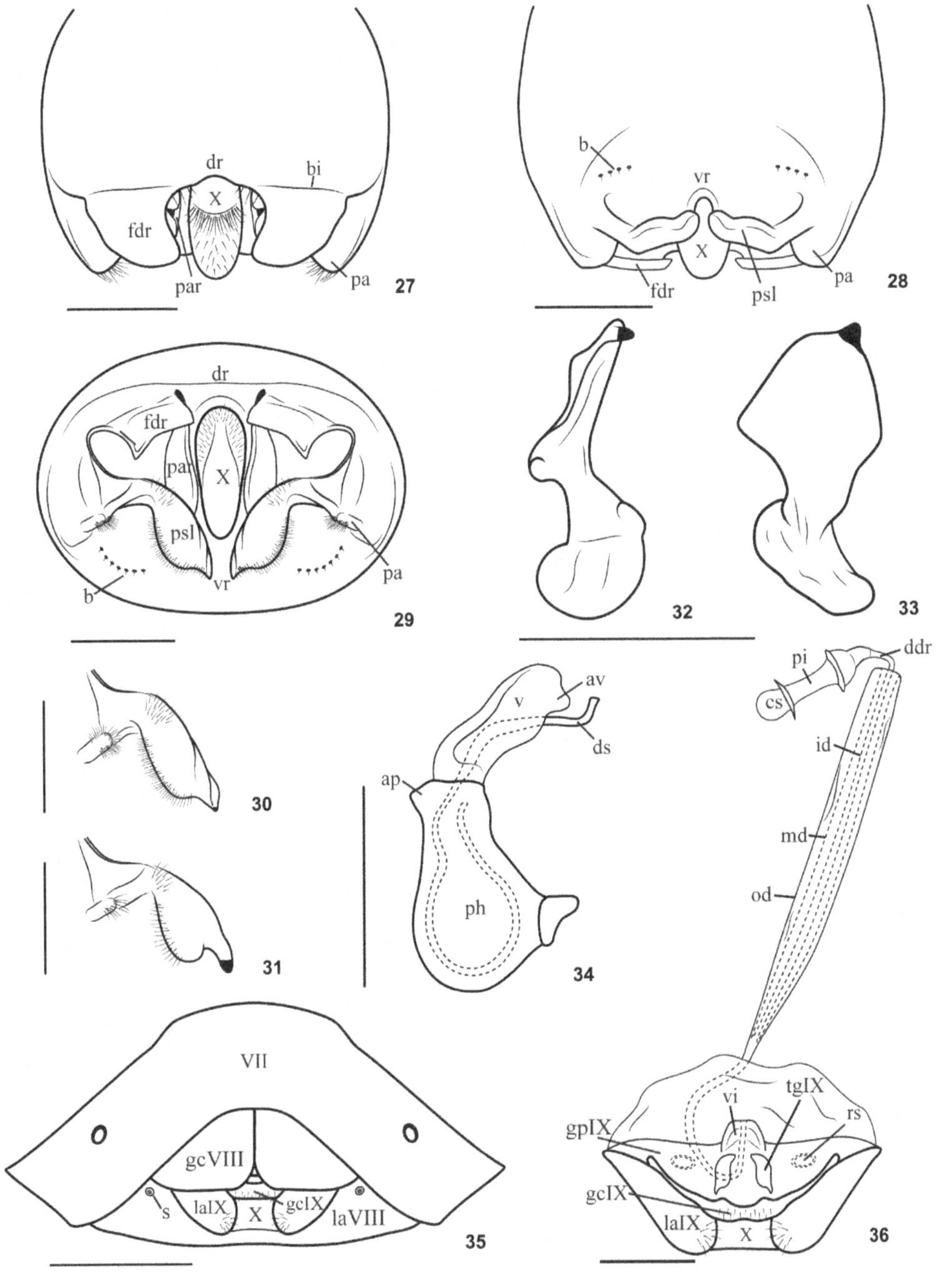

Figures 27–36. *Lineostethus marginellus*: (27–29) male pygophore, dorsal, ventral and posterior views respectively; (30–31) variations on the shape of the projections of superior layer of ventral rim; (32–33) left paramere, dorsal and lateral views; (34) phallus, lateral view; (35) female genital plates; (36) female receptaculum seminis. Abbreviations: ap, ventral apical projection of phallotheca; av, apical margin of vesica; b, bristles; bi, basal impression of foliaceous expansions; cs, capsula seminalis; ddr, distal ductus receptaculi; dr, dorsal rim; ds, ductus seminis distalis; fdr, foliaceous expansions of dorsal rim; gcVIII, gonocoxites VIII; gcIX, gonocoxites IX; gpVIII, gonapophyses VIII; gpIX, gonapophyses IX; id, internal duct of vesicular area; la8, laterotergites VIII; la9, laterotergites IX; md, median duct of vesicular area; od, outer duct of vesicular area; pa, posterolateral angle; par, paramere; ph, phallotheca; pi, pars intermedialis; psl, projection of superior layer of ventral rim; rs, ring sclerites; s, spiracle; tgIX, thickening of gonapophyses IX; v, vesica; vi, thickening of vaginal intima; vr, ventral rim; VII, urosternite VII; X, segment X. Scale bars: 27–33, 35 = 1 mm; 34, 36 = 0.5 mm.

subrectilinear (Fig. 35: gcVIII). Posterior margin of laterotergites VIII subrectilinear; limit between laterotergites VIII and tergite VIII indistinguishable (Fig. 35: la8). Posterior margin of gonocoxites IX subrectilinear (Fig. 35: gcIX). Laterotergites IX subtriangular, mesial margin slightly convex. Posterior margin of segment X slightly concave (Fig. 35: X). Thickening of gonapophyses IX sinuous, surpassing the lateral margin of the thickening of vaginal intima (Fig. 36: gpIX, tgIX). Thickening of vaginal intima subquadrangular (Fig. 36: vi). Distal portion of ductus receptaculi enlarged and more sclerotized; anterior and posterior annular crests perpendicular to pars intermedialis; pars intermedialis almost twice the length of capsula seminalis (Fig. 36: ddr, pi, cs).

Measurements. Male (n = 5): head length, 1.53 ± 0.04 (1.48–1.60); width, 3.03 ± 0.11 (2.88–3.16); length of antennal segments: I, 0.42 ± 0.02 (0.40–0.44); II, 0.70 ± 0.06 (0.64–0.78); III, 0.67 ± 0.09 (0.54–0.80); IV, 0.84 ± 0.03 (0.80–0.86); V, 1.01 ± 0.04 (0.98–1.06); length of labial segments: I, 0.66 ± 0.04 (0.60–0.70); II, 1.37 ± 0.08 (1.30–1.50); III, 0.64 ± 0.04 (0.60–0.68); IV, 0.43 ± 0.03 (0.40–0.46); pronotum length, 1.70 ± 0.08 (1.60–1.80); width, 4.07 ± 0.33 (3.80–4.65); scutellum length, 3.30 ± 0.08 (3.20–3.40); width, 2.61 ± 0.11 (2.48–2.76); total length, 6.48 ± 0.28 (6.11–6.82); abdominal width, 4.35 ± 0.17 (4.20–4.55).

Female (n = 5): head length, 1.58 ± 0.04 (1.52–1.64); width, 3.08 ± 0.12 (3.04–3.28); length of antennal segments: I, 0.40 ± 0.01 (0.38–0.42); II, 0.76 ± 0.07 (0.70–0.84); III, 0.75 ± 0.07 (0.64–0.82); IV, 0.90 ± 0.10 (0.80–1.00); V, 1.11 ± 0.09 (1.00–1.24); length of labial segments: I, 0.66 ± 0.08 (0.58–0.78); II, 1.35 ± 0.16 (1.30–1.52); III, 0.61 ± 0.08 (0.52–0.70); IV, 0.44 ± 0.04 (0.40–0.50); pronotum length, 1.80 ± 0.15 (1.60–1.94); width, 4.24 ± 0.38 (3.72–4.65); scutellum length, 3.50 ± 0.23 (3.20–3.76); width, 2.77 ± 0.23 (2.45–3.04); total length, 7.67 ± 0.95 (6.39–8.60); abdominal width, 4.80 ± 0.55 (4.30–5.54).

Material examined. Lectotype male: MEXICO, Sallé – type, *marginella* Stål – Paratypus, 82 81 – Lectotype *Discocephala marginella* Stål, 1872 – NHRS-GULI 000011383 (NHRS – photographs examined). Paralectotype female: MEXICO, Sallé – type – Typus, 81 81 – NHRS-GULI 000011382 (NHRS – photographs examined). Paralectotype female: MEXICO, Sallé – type – Typus, 80 81 – NHRS-GULI 000011384 (NHRS – photographs examined). Other specimens: USA, *Arizona*: Portal Cochise Co., 1 female, 14-19-VIII-1965, W.J. Gertsch Collector leg. (AMNH); Camp Washington Patagonia Mts, 4 females, 22-VI-1948, HSB Barber Request 1950 leg. (USNM); *Texas*: Star Co. (along Rio Grande at. Salineño), 1 female, 27-X-1991, E.G.Riley leg. (DARC); Cameron Co, 1 male, 8-III-1929, R.H. Beamer leg. (UFRG); Cameron Co. (Sabal Palm Grove Sanct.), 2 males, 28-29-III-1986, E.G. Riley leg. (DARC); Cameron Co. (Sabal Palm Groove Sanct.), 1 male, 4-X-1986, E.Riley & J.Negrón leg. (DARC); Cameron Co. (Sabal Palm Grove Audi. Sanct.), 2 males, 3-V-1987, E.Riley & Whitford leg. (DARC); Cameron Co. (Sabal Palm Grove Sanct.), 1 male, 5-IV-1987, E.Riley & D.Rider leg. (DARC); Cameron Co. (Sabal Palm Grove Sanct), 2 males and 2 females, 26-X-1991, E.G.Riley leg. (DARC); Cameron Co. (Paloma Blanca rd., nr. Sabal Palm Grove Sanct), 1 female, 26-X-1991, E.G.Riley leg. (DARC); Cameron Co., Brownsville (Sabal Palm Grove), 2 males, 14-IX-1982, C.W. & L.O'Brien & G. Wibmer leg. (DARC); Brownsville, 1 male, 3-VI-1932 (CAS); Brownsville, 1 female, 3-VI-1932, J.O. Martin leg. (CAS); Brownsville (Esper Rch), 1 male, no date (USNM); Brownsville, 1 male, June 1901, H.G. Barber Colln 1950 leg. (USNM). MEXICO, *Tamaulipas:* (Hwy. 80, 8mi W Jen. Hwys. 80 & 85, 1100), 1 male, 19-VI-1983, C.W. & L.O'Brien & G.B.Marshall leg. (DARC); (4 mi W Cd. Victoria, Canon del Navillo), 1 female, 14-XI-1985, P.Kovarik R.Jones & K.Haack leg. (DARC); *San Luis Potosi*: (10 Mi N. Tamazunchale), 1 female, 24-XII-1940, G.E. Bohart leg. (CAS); (Hwy. 85, 8mi SW Tamazunchale, 600), 1 female, 24-VII-1982, C.W. & L.O'Brien & G.Wibmer leg. (DARC); Vera Cruz Crawford, 1 male, no date (CAS); (7 mi N of San Juanito nr. Santos, Route 85, about 15mi S. of Ciudad Valles), 1 male, 23-IV-1965, H.V. Weems Jr. Cool leg. (DARC); *Veracruz*: Catemaco (Playa Azul), 1 male, 9-VIII-1966, J. & W. Ivie leg. (AMNH); (5mi E Catamaco), 1 female, 2-4-VII-1971, Clack Murray Hart Schaffner leg. (DARC); Acuyucan, 1 male, VIII-1966 J. & W. Ivie leg. (AMNH); 50' Montepio, 1 male and 2 females, 23-VIII-1982, C.W. & L.O'Brien & G. Wibmer leg. (DARC); (4 mi S Sontecomapan, 150), 2 females, 23-VIII-1982, C.W. & L.O'Brien & G.Wibmer leg. (DARC); *Hidalgo*: Otongo (3400), 1 female, 1-VIII-1982, O'Brien leg. (DARC); Otongo (3400, Minera Autlan), 1 male, 1-VIII-1982, C.W. & L.O'Brien & G.Wibmer leg. (DARC); *Colima*: (Hwy. 54, 24 mi W Manzanillo), 1 female, 9-VIII-1982, C.W. & L.O'Brien & G.Wibmer leg. (DARC); *Oaxaca*: (Hwy 147, 6 mi SE Jcn 175 & 147, 200), 1 male, C.W. & L.O'Brien & G.Wibmer leg. (DARC); *Tabasco*: (38 mi E Cardenas), 1 male, 4-X-1976, Cate & Clark leg. – beating flowers *Hampea nutricea* [sic] (male) (DARC); *Chiapas*: Hwy 307 (off Rd to Lacanja, 403m, 16.93°N 91.26°W), 3 males and 5 females, 07-VII-2009, D. Forero & G. Zhang leg. (UFRG); (in bananas, intercept N. Orleans US), 1 male, 19-II-1976 (USNM). HONDURAS, *Cortes*: (5 km N Cofradia), 1 male, 5-VIII-1977, C.W. & L.O'Brien & Marshall leg. PANAMA, *Los Santos*: Taboga, 1 female, XIII-1953, N.L.H. Kraus leg. (AMNH).

Distribution (Fig. 78). USA: *Arizona* (**new record**), *Texas* (Ruckes 1966; Hildebrand and Becker 1982). Mexico: *Nuevo León* (Ruckes 1966), *Tamaulipas* (**new record**), *San Luis Potosi* (**new record**), *Veracruz* (Ruckes 1966), *Hidalgo* (**new record**), *Tabasco* (**new record**), *Chiapas* (Hildebrand and Becker 1982). Honduras: *Cortes* (**new record**). Panama: *Los Santos* (**new record**).

Remarks. From the data retrieved of specimens' labels, banana (*Musa paradisiaca* L., Musaceae) and flowers of *Hampea nutricia* Fryxell (Malvaceae) are possible host plants.

Lineostethus graziae Hildebrand & Becker, 1982

Figs 4, 37–44, 78

Lineostethus graziae Hildebrand & Becker, 1982: 775–778.

Diagnosis. The males of *L. graziae* differ from other species by the subtriangular foliaceous expansions of the dorsal rim, the

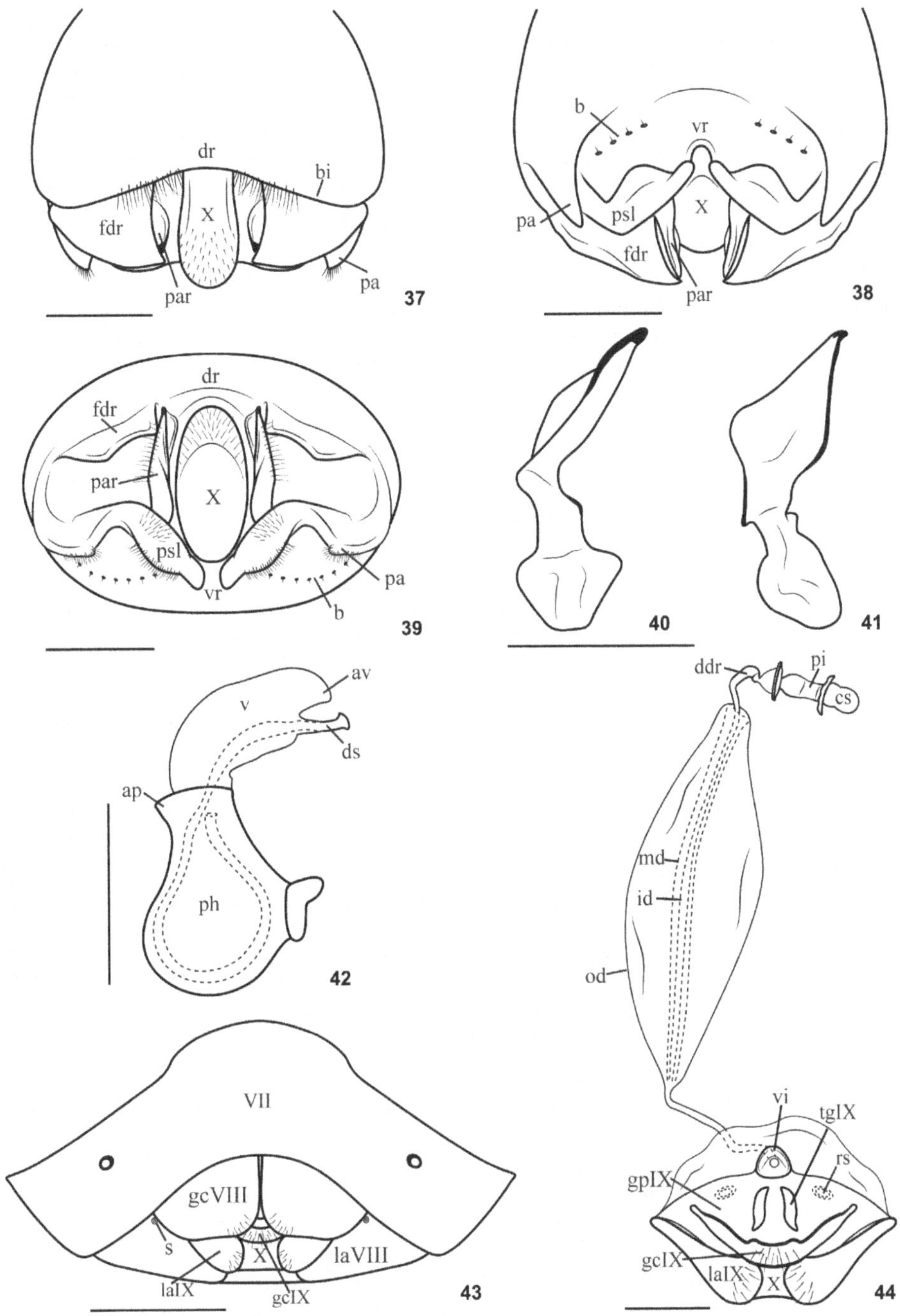

Figures 37–44. *Lineostethus graziae*: (37–39) male pygophore, dorsal, ventral and posterior views respectively; (40–41) left paramere, dorsal and lateral views; (42) phallus, lateral view; (43) female genital plates; (44) female receptaculum seminis. Abbreviations: ap, ventral apical projection of phallotheca; av, apical margin of vesica; b, bristles; bi, basal impression of foliaceous expansions; cs, capsula seminalis; ddr, distal ductus receptaculi; dr, dorsal rim; ds, ductus seminis distalis; fdr, foliaceous expansions of dorsal rim; gcVIII, gonocoxites VIII; gcIX, gonocoxites IX; gpVIII, gonapophyses VIII; gpIX, gonapophyses IX; id, internal duct of vesicular area; la8, laterotergites VIII; la9, laterotergites IX; md, median duct of vesicular area; od, outer duct of vesicular area; pa, posterolateral angle; par, paramere; ph, phallotheca; pi, pars intermedialis; psl, projection of superior layer of ventral rim; rs, ring sclerites; s, spiracle; tgIX, thickening of gonapophyses IX; v, vesica; vi, thickening of vaginal intima; vr, ventral rim; VII, urosternite VII; X, segment X. Scale bars: 37–41, 43 = 1 mm; 42, 44 = 0.5 mm.

subquadrangular cutting of the dorsal rim, the acute posterolateral angles (Fig. 37: fdr, dr, pa), the projections of the superior layer of ventral rim rounded at apex (Fig. 39: psl), and the ductus seminis distalis enlarged at apex (Fig. 42: ds). Subtriangular foliaceous expansions of the dorsal rim are also seen in *L. sinuosus* sp. nov., but the cutting of the dorsal rim, the posterolateral angles, and the projections of the ventral rim are different, as well as the phallus (Figs 37–39, 42, 45–47, 50: fdr, dr, pa). The females are distinguished by the subquadrangular gonocoxites VIII, the posterior margin of gonocoxites IX slightly concave (Fig. 43: gcVIII, gcIX), the enlarged distal portion of the ductus receptaculi, and the median constriction in the capsula seminalis (Fig. 44: ddr, cs). In the original description of *L. graziae*, Hildebrand and Becker (1982) described the connexival margins with large black strips, made of two lines of black punctures. However, the two specimens we found have thin strips, made of one line of black punctures.

Redescription. Antennal segments I–IV with longitudinal brown lines. Anterolateral margins of pronotum and proximal portion of lateral margin of corium without punctures. Unpunctured area laterad to the apex of radial vein of coria smaller than the width of one eye. Hemelytral membrane surpassing the apex do abdomen, with 7 veins. Anterior and posterior margins of connexival segments with one or two lines of black punctures (Fig. 4). Ventrally, punctures sparse and concentrated laterally or gradually dispersed toward the midline. Femora with sparse brown circular spots on distal half.

Male genitalia. Foliaceous expansions of dorsal rim subtriangular; apex of expansions not surpassing the posterior limit of ventral rim; posterolateral angles visible dorsally (Fig. 37: fdr, pa). Basal impression of the foliaceous expansions sinuous, cutting of dorsal rim subquadrangular, housing the apex of parameres (Fig. 37: bi, dr, par). Posterolateral angles with subtriangular apex (Figs 37–38: pa). Margin of the superior layer of ventral rim slightly concave, median excavation U-shaped (Fig. 38: vr). Margin of the inferior layer of ventral rim with one arc of short bristles (Fig. 38–39: b). Projections of the superior layer of ventral rim forming an acute angle with the posterolateral angles; projections with base and median portion enlarged and tapering to apex, covered by setae; apex rounded (Fig. 39: psl, pa). Parameres tangent to segment X, reaching the foliaceous expansions of dorsal rim; head lozenge (Figs 39–41: par, X; dr, fdr). Ventral surface of phallotheca concave (Fig. 44: ph, ap). Vesica with rounded apex, apical margin rounded and placed along ductus seminis distalis almost to its apex (Fig. 42: v, av, ds). Distal portion of ductus seminis distalis subequal to vesica in length, enlarged at apex (Fig. 42: ds).

Female genitalia. Posterior margin of urosternite VII concave (Fig. 43: VII). Gonocoxites VIII subquadrangular; apex slightly divergent; posterior margin convex (Fig. 43: gcVIII). Posterior margin of laterotergites VIII convex; limit between laterotergites VIII and tergite VIII visible (Fig. 43: la8). Posterior margin of gonocoxites IX convex (Fig. 43: gcIX). Laterotergites IX digitiform, mesial margins divergent (Fig. 43: la9). Posterior margin of segment X subrectilinear (Fig. 43: X). Thickening of gonapophyses IX sinuous, not surpassing the lateral margin of the thickening of vaginal intima (Fig. 44: gpIX, tgIX). Thickening of vaginal intima subtriangular, flat (Fig. 44: vi). Distal ductus recepitaculi sclerotized and enlarged; annular crests perpendicular to pars intermedialis; middle third of pars intermedialis less sclerotized (Fig. 44: ddr, pi, cs). Capsula seminalis subequal to pars intermedialis in length, bearing median constriction (Fig. 44: cs).

Measurements. Male (n = 1): head length, 1.60; width, 3.20; length of antennal segments: I, 0.44; II–V, missing; length of labial segments: I, 0.72; II, 1.56; III, 0.68; IV, 0.46; pronotum length, 1.72; width, 4.26; scutellum length, 3.56; width, 2.79; total length, 7.10; abdominal width, 4.50.

Female (n = 1): head length, 1.70; width, 3.40; length of antennal segments: I, 0.42; II, 0.78; III, 0.80; IV, 0.88; V, 0.98; length of labial segments: I, 0.78; II, 1.40; III, 0.70; IV, 0.58; pronotum length, 1.96; width, 3.56; scutellum length, 3.88; width, 3.80; total length, 8.70; abdominal width, 4.98.

Material examined. Holotype (female): MEXICO, *Oaxaca*: Totolapan (19 mi NW), 28-VII-1963, Collr: W. A. Foster leg. – California Academy of Sciences Type No. 15923 (CAS). Paratypes: MEXICO, *Morelos*: Cuernavaca (10mi E), 1 female, 31-VII-1963, W. A. Foster leg., CIS (EMEC – photographs examined); *Oaxaca*: Totolapan (19 mi NW), 2 males and 1 female, 28-VII-1963, W. A. Foster leg., CIS (EMEC – photographs examined). Other specimens: GUATEMALA, *Sacatepequez*: Capetillo (5000 ft), 1 male, 21-VIII-1947, F. Johnson Donor, Cols. C. & P. Vaurie leg. (AMNH); Antigua (1500-1600 mi), 1 female, June 1981, N.L.H. Krauss leg. (AMNH).

Distribution (Fig. 78). Mexico: *Morelos, Puebla, Oaxaca* (Hildebrand and Becker 1982). Guatemala: *Sacatepequez* (**new record**).

Remarks. Hildebrand and Becker (1982) indicated the Essig Museum of Entomology in the University of California Berkeley (EMEC) as deposit collection for the female holotype of *L. graziae*, but it is in the California Academy of Sciences under the type number 15923 and was examined by T.A.G.

Lineostethus sinuosus Kochenborger & Garbelotto, sp. nov.

Figs 5, 45–53, 78

Diagnosis. The males of *L. sinuosus* sp. nov. differ from other species by the subrectangular expansions of the dorsal rim of pygophore, though it is similar in shape with *L. marginellus* whose dorsal rim cutting is wider (Figs 27, 45: fdr). The median portion of the projections of the superior layer of ventral rim in *L. sinuosus* sp. nov. is also the widest among the species with projections spatulate at apex (Fig. 47: psl). The females of *L. sinuosus* sp. nov. can be recognized by the sinuous posterior margin of gonocoxites VIII (Figs 51–52: gcVIII), and although a sinuous outline is also observed in *L. otarus* sp. nov. and *L. auritus* sp. nov. (Figs 68, 76: gcVIII), the strongly divergent

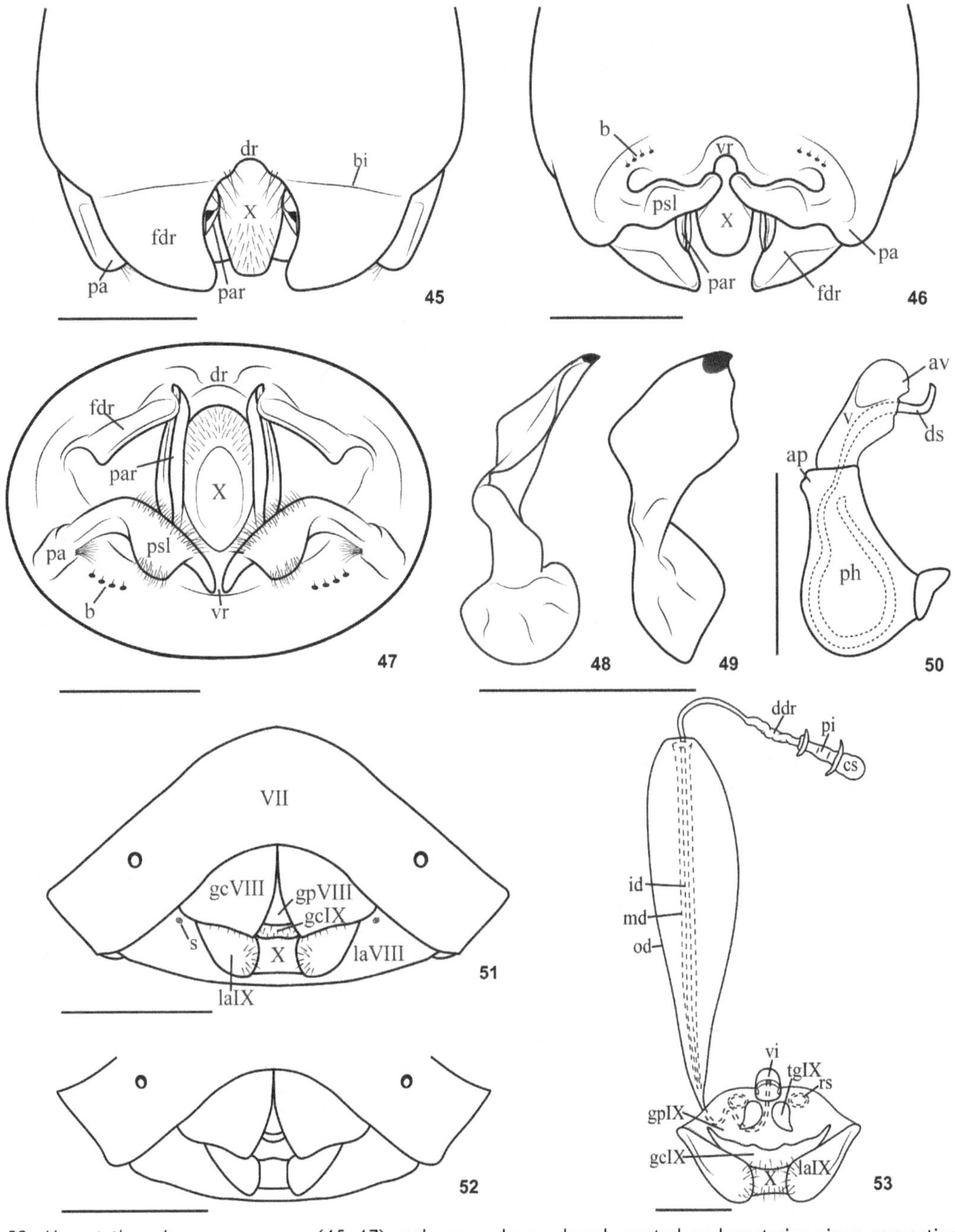

Figures 45–53. *Lineostethus sinuosus* sp. nov.: (45–47) male pygophore, dorsal, ventral and posterior views respectively; (48–49) left paramere, dorsal and lateral views; (50) phallus, lateral view; (51–52) female genital plates; (53) female receptaculum seminis. Abbreviations: ap, ventral apical projection of phallotheca; av, apical margin of vesica; b, bristles; bi, basal impression of foliaceous expansions; cs, capsula seminalis; ddr, distal ductus receptaculi; dr, dorsal rim; ds, ductus seminis distalis; fdr, foliaceous expansions of dorsal rim; gcVIII, gonocoxites VIII; gcIX, gonocoxites IX; gpVIII, gonapophyses VIII; gpIX, gonapophyses IX; id, internal duct of vesicular area; la8, laterotergites VIII; la9, laterotergites IX; md, median duct of vesicular area; od, outer duct of vesicular area; pa, posterolateral angle; par, paramere; ph, phallotheca; pi, pars intermedialis; psl, projection of superior layer of ventral rim; rs, ring sclerites; s, spiracle; tgIX, thickening of gonapophyses IX; v, vesica; vi, thickening of vaginal intima; vr, ventral rim; VII, urosternite VII; X, segment X. Scale bars: 45–49, 51–52 = 1 mm; 50, 53 = 0.5 mm.

sutural angles are exclusive to *L. sinuosus* sp. nov. The shape of the thickening of the gonapophyses VIII is also exclusive inside the genus, and along with the helicoid distal ductus receptaculi is diagnostic to the species.

Description. Antennae ivory; segments II–III with longitudinal brown lines. Anterolateral margins of pronotum and proximal portion of lateral margin of corium punctured in some specimens. Unpunctured area laterad to the apex of radial vein of coria smaller to subequal to the width of one eye. Hemelytral membrane at least reaching the apex of abdomen, with 6 veins. Ventrally, punctures sparse and laterally concentrated. Femora with brown circular spots on distal half.

Male genitalia. Foliaceous expansions of dorsal rim subrectangular; apex of expansions at least reaching the posterior limit of ventral rim; posterolateral angles visible dorsally (Fig. 45: fdr, pa). Basal impression of the foliaceous expansions subrectilinear; cutting of dorsal rim U-shaped (Fig. 45: bi, dr). Posterolateral angles rounded, apex obtuse, sometimes with setae on posterior margin (Figs 45–46: pa). Segment X visible on the cutting of dorsal rim (Fig. 45: dr, X). Margin of the superior layer of ventral rim sinuous, median excavation U-shaped (Fig. 46: vr). Margin of the inferior layer of ventral rim with one arc of short bristles (Figs 46–47: b). Projections of the superior layer of ventral rim spatulate at apex, forming a right angle with the posterolateral angles; dorsal margin convex and ventral margin sinuous, median third enlarged in relation to the lateral thirds; setae on the posterior margin of the projections (Fig. 47: psl). Parameres sinuous, parallel to segment X, apex reaching the mesial margin of the foliaceous expansions of the dorsal rim; head lozenge (Figs 47–49: par, X, dr). Ventral surface of phallotheca subrectilinear (Fig. 50: ph, ap). Vesica obtuse and more sclerotized at apex; apical margin concave, extending to basal 1/3 of ductus seminis distalis (Fig. 50: v, av, ds). Distal portion of the ductus seminis distalis about 2/3 of vesica length, narrowing posteriorly (Fig. 50: ds).

Female genitalia. Posterior margin of urosternite VII concave (Figs 51–52: VII). Gonocoxites VIII subtriangular; sutural margins diverging towards apex, sutural angles posteriorly projected; posterior margin sinuous, varying in intensity (Figs 51–52: gcVIII) with the sinuosity of the posterior margins, sometimes with lateral angles partially hiding the spiracles of the laterotergites VIII, and sutural angles projected over the base of the laterotergites IX, or not. Posterior margin of laterotergites VIII convex; limit between laterotergites VIII and tergite VIII indistinguishable (Figs 51–52: la8). Posterior margin of gonocoxites IX subrectilinear (Fig. 51: gcIX) to convex (Fig. 52: gcIX). Gonapophyses VIII visible and smooth (Figs 51–52: gpVIII). Laterotergites IX digitiform, with a median depression; varying in shape from lateral margins convex with apex rounded to lateral margins subrectilinear with apex acute (Figs 51–52: la9). Posterior margin of segment X subrectilinear (Figs 51–52: X). Thickening of gonapophyses IX sinuous, surpassing the lateral margin of the thickening of vaginal intima (Fig. 53: gpIX, tgIX). Thickening of vaginal intima subtriangular, flat (Fig. 53: vi). Distal portion of ductus receptaculi helicoid (Fig. 53: ddr); anterior annular crest turned to ductus receptaculi; pars intermedialis subequal in length to capsula seminalis; posterior annular crest turned to pars intermedialis (Fig. 53: pi, cs).

Measurements. Male (n = 5): head length, 1.54 ± 0,07 (1.48–1.63); width, 2.98 ± 0.13 (2.79–3.16); length of antennal segments: I, 0.42 ± 0.03 (0.38–0.44); II, 0.68 ± 0.02 (0.66–0.70); III, 0.69 ± 0.08 (0.62–0.78); IV, 0.80; V, 0.96; length of labial segments: I, 0.67 ± 0.06 (0.62–0.76); II, 1.50 ± 0.05 (1.44–1.56); III, 0.61 ± 0.04 (0.56–0.68); IV, 0.48 ± 0.07 (0.42–0.60); pronotum length, 1.75 ± 0.09 (1.64–1.84); width, 4.00 ± 0.16 (3.76–4.15); scutellum length, 3.38 ± 0.21 (3.07–3.64); width, 2.66 ± 0.12 (2.48–2.82); total length, 6.43 ± 0.31 (6.11–6.82); abdominal width, 4.45 ± 0.22 (4.20–4.70).

Female (n = 5): head length, 1.49 ± 0.10 (1.32–1.58); width, 2.99 ± 0.20 (2.67–3.20); length of antennal segments: I, 0.40 ± 0.02 (0.38–0.42); II, 0.73 ± 0.10 (0.62–0.82); III, 0.64 ± 0.06 (0.54–0.70); IV, 0.81 ± 0.09 (0.70–0.90); V, 0.98 ± 0.07 (0.88–1.04); length of labial segments: I, 0.68 ± 0.10 (0.58–0.78); II, 1.44 ± 0.10 (1.30–1.52); III, 0.64 ± 0.07 (0.56–0.70); IV, 0.47 ± 0.04 (0.44–0.52); pronotum length, 1.76 ± 0.09 (1.60–1.82); width, 4.05 ± 0.26 (3.64–4.25); scutellum length, 3.35 ± 0.13 (3.13–3.48); width, 2.63 ± 0.12 (2.45–2.79); total length, 6.68 ± 0.37 (6.04–7.03); abdominal width, 4.44 ± 0.31 (3.92–4.75).

Type material. Holotype (male): MEXICO, *Yucatán*: Puerto Chicxulub (Convolvulaceae), 25-XI-2011, Jean-Luc Dzido leg. (deposited in: DZUP). Paratypes: MEXICO, *Yucatán*: (12 km N Quintana Roo, Hwy 295), 1 male and 2 females, 20-V-1987, D.A. Rider E.G. & T.J. Riley leg. (DARC); Chichen-Itza, 1 female, 16-VII-1952, J. & D. Pallister leg. (AMNH); *Quintana Roo:* Chichén-Itzá, 2 males, 10-VIII-1990, D.A. Rider leg. (1 male UFRG and 1 male DARC); Coba, 1 male, 11-VIII-1990, D.A. Rider leg. (DARC); (45 mi W Chetumal), 3 females, 9-VIII-1974, C.W. & L.O'Brien & Marshall leg. (1 female UFRG and 2 females DARC). BELIZE: (H?25), 1 male, no date (CAS); (Belize Mile 36 Northern Road), 1 male and 1 female, 11-VIII-1977, C.W. & L. O'Brien & Marshall leg. (AMNH). HONDURAS, *Cortes*: (5 km N Cofradia), 1 male, 5-VIII-1977, C.W. & L.B. O'Brien & Marshall leg. (AMNH).

Distribution (Fig. 78). Mexico: *Yucatán*, *Quintana Roo*; Belize; Honduras: *Cortes*.

Etymology. The epithet refers to the shape of the females' gonocoxites VIII; Latin adjective masculine, *sinuosus (-a, -um)* = full of bendings, windings.

Remarks. From the data retrieved of specimens' labels, Convolvulaceae are possible host plants.

Lineostethus acuminatus Kochenborger & Garbelotto, sp. nov.

Figs 6, 54–61, 78

Diagnosis. Males of *L. acuminatus* sp. nov. differ from most species by the subtriangular expansions of the dorsal rim,

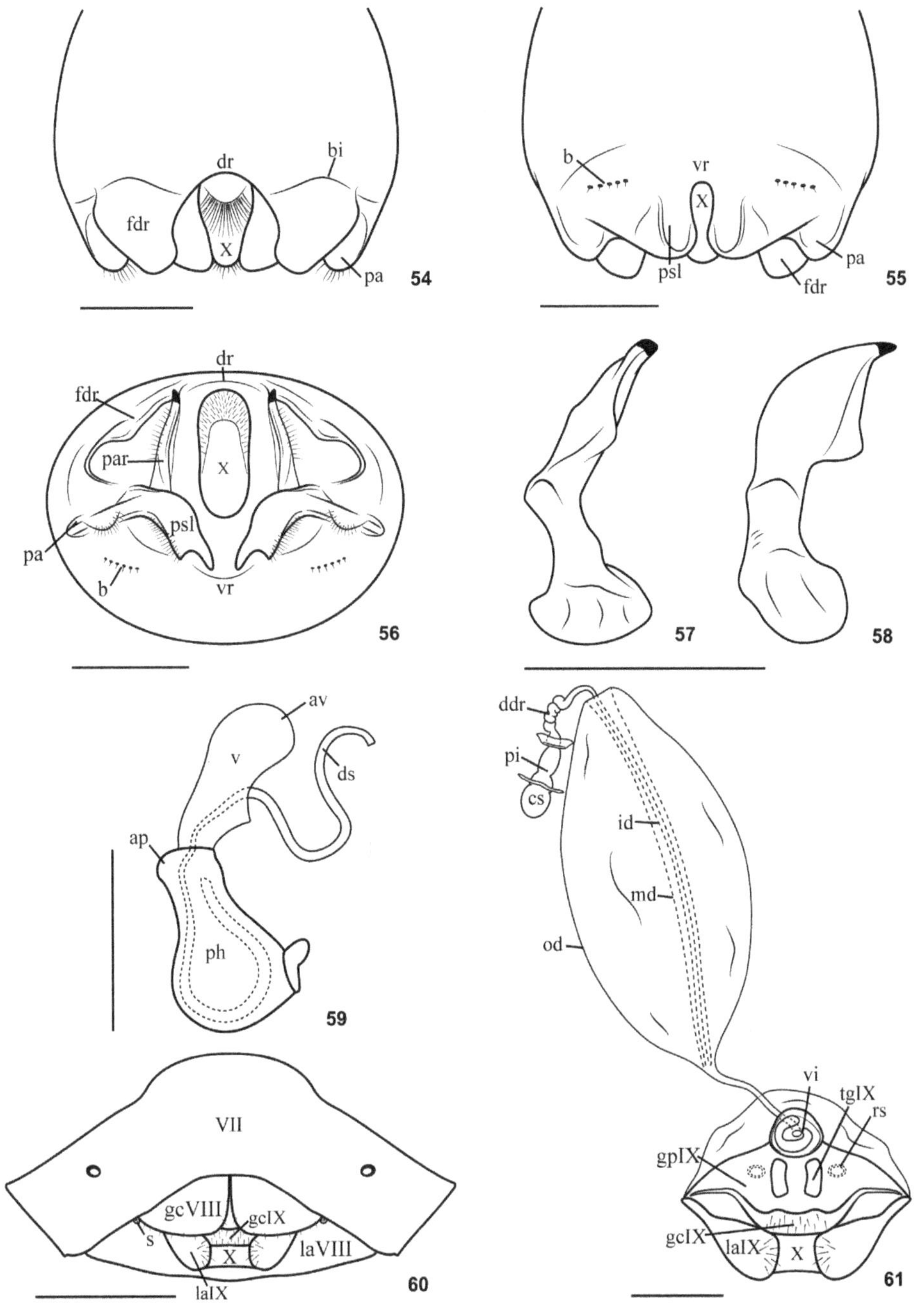

Figures 54–61. *Lineostethus acuminatus* sp. nov.: (54–56) male pygophore, dorsal, ventral and posterior views respectively; (57–58) left paramere, dorsal and lateral views; (59) phallus, lateral view; (60) female genital plates; (61) female receptaculum seminis. Abbreviations: ap, ventral apical projection of phallotheca; av, apical margin of vesica; b, bristles; bi, basal impression of foliaceous expansions; cs, capsula seminalis; ddr, distal ductus receptaculi; dr, dorsal rim; ds, ductus seminis distalis; fdr, foliaceous expansions of dorsal rim; gcVIII, gonocoxites VIII; gcIX, gonocoxites IX; gpVIII, gonapophyses VIII; gpIX, gonapophyses IX; id, internal duct of vesicular area; la8, laterotergites VIII; la9, laterotergites IX; md, median duct of vesicular area; od, outer duct of vesicular area; pa, posterolateral angle; par, paramere; ph, phallotheca; pi, pars intermedialis; psl, projection of superior layer of ventral rim; rs, ring sclerites; s, spiracle; tgIX, thickening of gonapophyses IX; v, vesica; vi, thickening of vaginal intima; vr, ventral rim; VII, urosternite VII; X, segment X. Scale bars: 54–58, 60 = 1 mm; 59, 61 = 0.5 mm.

although resembling *L. graziae*. *L. acuminatus* sp. nov. expansions are longer and narrower than *L. graziae* (Figs 37, 54: fdr), and the dorsal rim cutting is U-shaped in *L. acuminatus* sp. nov. while subquadrangular in *L. graziae* (Figs 37, 54: dr). Also, the projections of the superior layer of the ventral rim in *L. acuminatus* sp. nov. are uniformly wide and the apex is hook-shaped (Fig. 55: psl). The phallus of *L. acuminatus* sp. nov. bears the largest vesica and the longest ductus seminis distalis in the genus (Fig. 59: v, ds). The females of the new species are recognized by the truncated apex of laterotergites IX and the posterior margin of segment X concave to subrectilinear (Fig. 60: la9, X); the circular thickening of vaginal intima, the ductus receptaculi distal helicoid, and the pars intermedialis with an apical constriction help to identify this species (Fig. 61: vi, ddr, pi).

Description. Antennae ivory; segments II–III with longitudinal brown lines. Anterolateral margins of pronotum and proximal portion of lateral margin of corium punctured in some specimens. Scutellum may present a spot of brown punctures on disc and on post-frenal lobe. Unpunctured area laterad to the apex of radial vein of coria smaller than the width of one eye. Hemelytral membrane reaching the apex of abdomen, with 6–7 veins. Ventrally, punctures sparse and laterally concentrated. Femora with sparse brown circular spots on distal half.

Male genitalia. Foliaceous expansions of dorsal rim subtriangular; apex of expansions rounded and surpassing the posterior limit of ventral rim; posterolateral angles visible dorsally (Fig. 54: fdr, pa). Basal impression of the foliaceous expansions sinuous, cutting of dorsal rim U-shaped (Fig. 54: bi, dr). Posterolateral angles rounded (Figs 55–56: pa). Margin of the superior layer of ventral rim sinuous, median excavation drop-shaped (Fig. 55: vr). Margin of the inferior layer of ventral rim with one subrectilinear line of short bristles (Figs 55–56: b). Segment X with a line of setae on base (Figs 54–55: X). Projections of the superior layer of ventral rim forming a right angle with posterolateral angles, covered by setae; dorsal margin convex and ventral margin concave; narrow at base, apex acute and hook-shaped, median third uniformly enlarged towards the apex (Fig. 56: psl, pa). Parameres parallel or slightly directed to segment X, reaching the foliaceous expansions of dorsal rim; head lozenge (Figs 56–58: par, X, fdr). Ventral surface of phallotheca concave (Fig. 59: ph, ap). Vesica apex rounded; apical margin concave extending to mid of ductus seminis distalis (Fig. 59: v, av, ds). Distal portion of ductus seminis distalis sinuous, about 2/3 longer than vesica, narrowing posteriorly (Fig. 59: ds).

Female genitalia. Posterior margin of urosternite VII concave (Fig. 60: VII). Gonocoxites VIII subtriangular; sutural and posterior margins subrectilinear (Fig. 60: gcVIII). Posterior margin of laterotergites VIII slightly convex; limit between laterotergites VIII and tergite VIII indistinguishable (Fig. 60: la8). Posterior margin of gonocoxites IX subrectilinear (Fig. 60: gcIX). Laterotergites IX digitiform and truncate apex; mesial and lateral margins convex (Fig. 60: la9). Posterior margin of segment X concave to rectilinear (Fig. 60: X). Thickening of gonapophyses IX subrectangular, not surpassing the lateral margin of the thickening of vaginal intima (Fig. 61: gpIX, tgIX). Thickening of vaginal intima circular and flat (Fig. 61: vi). Distal portion of ductus receptaculi helicoid (Fig. 61: ddr); anterior annular crest turned to ductus receptaculi; pars intermedialis with a shallow constriction on its apex and subequal in length to capsula seminalis; posterior annular crest perpendicular to pars intermedialis (Fig. 61: pi, cs). Capsula seminalis shallowly constricted near base (Fig. 61: cs).

Measurements. Male (n = 5): head length, 1.54 ± 0.07 (1.43–1.63); width, 2.96 ± 0.16 (2.70–3.10); length of antennal segments: I, 0.40 ± 0.01 (0.40–0.42); II, 0.68 ± 0.04 (0.62–0.70); III, 0.67 ± 0.06 (0.60–0.74); IV, 0.89 ± 0.01 (0.88–0.90); V, 1.05 ± 0.07 (1.00–1.10); length of labial segments: I, 0.66 ± 0.05 (0.60–0.70); II, 1.34 ± 0.05 (1.30–1.40); III, 0.63 ± 0.05 (0.58–0.70); IV, 0.46 ± 0.05 (0.38–0.50); pronotum length, 1.66 ± 0.11 (1.52–1.80); width, 3.95 ± 0.18 (3.65–4.10); scutellum length, 3.30 ± 0.15 (3.08–3.48); width, 2.61 ± 0.12 (2.42–2.73); total length, 6.26 ± 0.30 (6.32–6.53); abdominal width, 4.28 ± 0.19 (4.00–4.45).

Female (n = 5): head length, 1.55 ± 0.06 (1.45–1.63); width, 3.09 ± 0.13 (2.95–3.24); length of antennal segments: I, 0.42 ± 0.02 (0.40–0.46); II, 0.71 ± 0.06 (0.64–0.78); III, 0.65 ± 0.04 (0.60–0.70); IV, 0.81 ± 0.07 (0.74–0.90); V, 0.96 ± 0.03 (0.94–1.00); length of labial segments: I, 0.67 ± 0.04 (0.60–0.70); II, 1.39 ± 0.08 (1.26–1.48); III, 0.65 ± 0.09 (0.60–0.80); IV, 0.43 ± 0.04 (0.40–0.48); pronotum length, 1.76 ± 0.08 (1.62–1.82); width, 4.07 ± 0.13 (3.88–4.20); scutellum length, 3.36 ± 0.08 (3.28–3.44); width, 2.68 ± 0.10 (2.64–2.79); total length, 6.67 ± 0.19 (6.60–6.96); abdominal width, 4.52 ± 0.14 (4.30–4.65).

Type material. Holotype (male): COSTA RICA, *Guanacaste*: Santa Rosa (NP, Research Station 308.1, 10.81N 85.62W), 7-XII-2008, Weirauch & Forero leg. (deposited in: DZUP). Paratypes: HONDURAS: *El Paraiso*: Zamorano, 1 male, 28-VII-1977, O'Brien leg. (DARC). COSTA RICA: (C.a. E.P.), 1 male, no date (CAS); *Guanacaste*: Las Cañas (Finca La Taboga), 1 female, 17-27-VI-1969, Toby Schuh, Janet Crane leg. (AMNH); Santa Rosa (NP, Research Station, 308m, 10.84°N 85.62°W), 2 males and 4 females, 06-XII-2008, Weirauch & Forero leg. (UCRC); Palo Verde National Park (Nr OTS research station 26m, 10.35°N 85.35°W), 3 males and 8 females, 4-XII-2008, Weirauch & Forero leg. (UFRG). PANAMA, *Los Santos*: Taboga, 1 male, XII-1953, Collector: N.L.H. Kraus leg. (AMNH).

Distribution (Fig. 78). Honduras: *El Paraiso*; Costa Rica: *Guanacaste*; Panama: *Los Santos*.

Etymology. The epithet refers to the shape of the apex of each projection of the superior layer of ventral rim of pygophore; Latin adjective masculine, *acuminatus* (*-a*, *-um*) = pointed, sharpened.

Lineostethus otarus Garbelotto & Kochenborger, sp. nov.

Figs 7, 62–69, 78

Diagnosis. *Lineostethus otarus* sp. nov. and *L. auritus* sp. nov. share similar features in both male and female genitalia: the long foliaceous expansions of the dorsal rim of pygophore; two lines of bristles on the ventral wall, one of them not associated with punctures; and broad gonocoxites VIII. *Lineostethus otarus* sp. nov. is distinguished by the wider foliaceous expansions of the dorsal rim, hiding the posterolateral angles of pygophore; the sinuous proximal portion of the lateral margin of the expansions; and the projections of the superior layer of ventral rim more elongated. Females of *L. otarus* sp. nov. are distinguished by the posterior margin of urosternite VII subrectilinear over laterotergites IX, and the sinuous posterior margin of gonocoxites VIII.

Description. General color of the specimens with some variation, in melanic specimens margins of connexival segments with lines of black punctures more evident. Antennae ivory; segments II–III with longitudinal brown lines dorsally and ventrally; segments IV–V usually darker than the previous. Anterolateral margins of pronotum and proximal portion of lateral margin of corium punctured. Anterior margin of scutellum with two to four brown spots, and one ivory spot medially; post-frenal lobe with 1+1 lateral brown maculae. Unpunctured area laterad to the apex of radial vein of coria inconspicuous to equivalent to the width of one eye. Hemelytral membrane at least reaching apex of abdomen, with 6 veins. Ventrally, punctures uniformly distributed, absent on sterna. Femora with sparse brown circular spots on distal half.

Male genitalia. Foliaceous expansions of dorsal rim subrectangular, elongated, longer than wide, narrowing towards the apex; proximal portion of the lateral margin of expansions sinuous; apex rounded, surpassing the posterior limit of ventral rim by half of the expansion length, and visible dorsally when fixed at the specimen (Fig. 62: fdr). Basal impression of the foliaceous expansions subrectilinear, cutting of dorsal rim subrectangular (Fig. 62: bi, dr). Posterolateral angles acute and dorsally hidden by the dorsal rim expansions, covered by setae (Figs 62–63: pa). Margin of the superior layer of ventral rim V-shaped, median excavation drop-shaped (Fig. 63: vr). Margin of the inferior layer of ventral rim with two lines of short bristles: one posterior, subrectilinear and not associated with punctures; and another anterior, concave and associated with punctures (Figs 63–64: b). Segment X with a line of setae on middle (Fig. 62: X). Projections of the superior layer of ventral rim forming a right angle with the posterolateral angles, covered by setae; dorsal margin convex and ventral margin subrectlinear, elongated and uniformly wide throughout its length, apex acute and hook-shaped (Fig. 64: psl, pa). Parameres parallel to segment X, at least reaching the foliaceous expansions of dorsal rim; head lozenge (Figs 64–66: par, X, fdr). Ventral surface of phallotheca concave (Fig. 67: ph, ap). Vesica apex subquadrangular; apical margin with a sclerotized contour, subrectilinear, and surpassing ductus seminis distalis (Fig. 67: v, av, ds). Distal portion of ductus seminis distalis placed in a concavity on the middle of vesical, short and bending down (Fig. 67: ds).

Female genitalia. Posterior margin of urosternite VII sinuous, middle third subrectilinear and lateral thirds convex (Fig. 68: VII). Gonocoxites VIII subquadrangular, broad; posterior margin slightly sinuous, subrectilinear over laterotergites IX, with external angles truncated and weakly projecting over laterotergites VIII (Fig. 68: gcVIII). Posterior margin of laterotergites VIII subrectilinear; limit between laterotergites VIII and tergite VIII indistinguishable (Fig. 68: la8). Posterior margin of gonocoxites IX subrectilinear to convex (Fig. 68: gcIX). Laterotergites IX digitiform, tumescent at apex; apex varying between rounded and slightly acute (Fig. 68: la9). Posterior margin of segment X subrectilinear (Fig. 68: X). Thickening of gonapophyses IX kidney-shaped, surpassing the lateral margin of the thickening of vaginal intima (Fig. 69: gpIX, tgIX). Thickening of vaginal intima subquadrangular (Fig. 69: vi). Distal portion of ductus receptaculi apically enlarged and more sclerotized (Fig. 69: ddr); anterior annular crest turned to ductus receptaculi; pars intermedialis membranous at apex, longer than capsula seminalis; posterior annular crest perpendicular to pars intermedialis (Fig. 69: pi, cs). Capsula seminalis shallowly constricted near base (Fig. 69: cs).

Measurements. Male (n = 5): head length, 1.75 ± 0.06 (1.68–1.82); width, 3.13 ± 0.06 (3.04–3.20); length of antennal segments: I, 0.48 ± 0.02 (0.46–0.50); II, 0.73 ± 0.03 (0.70–0.78); III, 0.80 ± 0.01 (0.80–0.82); IV, 0.97 ± 0.09 (0.90–1.10); V, 1.06 ± 0.07 (0.98–1.12); length of labial segments: I, 0.76 ± 0.04 (0.72–0.80); II, 1.63 ± 0.11 (1.52–1.82); III, 0.72 ± 0.07 (0.66–0.84); IV, 0.53 ± 0.02 (0.50–0.54); pronotum length, 1.79 ± 0.07 (1.78–1.86); width, 4.41 ± 0.33 (4.20–4.98); scutellum length, 3.46 ± 0.15 (3.28–3.68); width, 2.74 ± 0.07 (2.67–2.84); total length, 8.16 ± 0.25 (7.80–8.40); abdominal width, 4.82 ± 0.10 (4.70–4.95).

Female (n = 5): head length, 1.81 ± 0.08 (1.72–1.92); width, 3.22 ± 0.11 (3.10–3.36); length of antennal segments: I, 0.49 ± 0.02 (0.48–0.52); II, 0.77 ± 0.05 (0.70–0.80); III, 0.82 ± 0.09 (0.72–0.92); IV, 0.97 ± 0.05 (0.92–1.02); V, 1.16 ± 0.04 (1.10–1.12); length of labial segments: I, 0.77 ± 0.10 (0.60–0.86); II, 1.40 ± 0.45 (0.60–1.70); III, 0.71 ± 0.08 (0.62–0.84); IV, 0.52 ± 0.04 (0.46–0.56); pronotum length, 1.82 ± 0.09 (1.82–1.94); width, 4.35 ± 0.21 (4.10–4.68); scutellum length, 3.54 ± 0.18 (3.28–3.68); width, 2.86 ± 0.14 (2.70–3.07); total length, 8.58 ± 0.30 (8.20–9.00); abdominal width, 4.94 ± 0.30 (4.55–5.36).

Type material. Holotype (male): MEXICO, *Michoacán*: (Hwy. 15, 21 mi E Morelia, 7000'), 15-VIII-1982, C.W. & L.O'Brien & G.Wibmer leg. (AMNH). Paratypes: MEXICO,

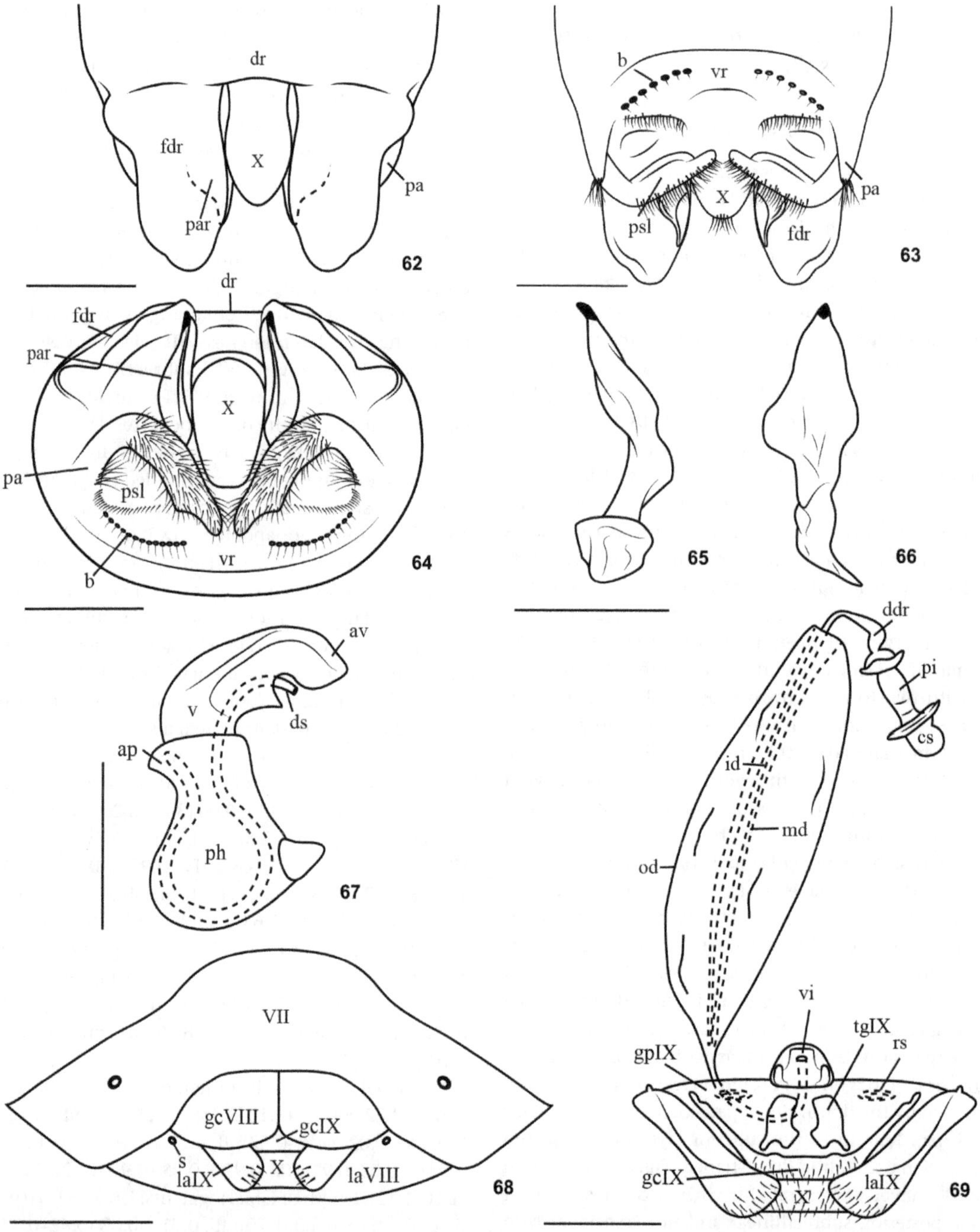

Figures 62–69. *Lineostethus otarus* sp. nov.: (62–64) male pygophore, dorsal, ventral and posterior views respectively; (65–66) left paramere, dorsal and lateral views; (67) phallus, lateral view; 68, female genital plates; (69) female receptaculum seminis. Abbreviations: ap, ventral apical projection of phallotheca; av, apical margin of vesica; b, bristles; bi, basal impression of foliaceous expansions; cs, capsula seminalis; ddr, distal ductus receptaculi; dr, dorsal rim; ds, ductus seminis distalis; fdr, foliaceous expansions of dorsal rim; gcVIII, gonocoxites VIII; gcIX, gonocoxites IX; gpVIII, gonapophyses VIII; gpIX, gonapophyses IX; id, internal duct of vesicular area; la8, laterotergites VIII; la9, laterotergites IX; md, median duct of vesicular area; od, outer duct of vesicular area; pa, posterolateral angle; par, paramere; ph, phallotheca; pi, pars intermedialis; psl, projection of superior layer of ventral rim; rs, ring sclerites; s, spiracle; tgIX, thickening of gonapophyses IX; v, vesica; vi, thickening of vaginal intima; vr, ventral rim; VII, urosternite VII; X, segment X. Scale bars: 62–66, 68 = 1 mm; 67, 69 = 0.5 mm.

Michoacán: (Hwy 15,21 mi E Morelia, 7000′), 8 males and 13 females, 15-VIII-1982, C.W. & L.O'Brien & G.B. Marshall leg. (1 male UFRG; 1 female AMNH; 7 males and 12 females DARC); *Puebla*: (Hwy 18 mi NW Azumbilla, 7700′), 1 female, 14-V-1983, 1100′, C.W. & L.O'Brien & G.B. Marshall leg. (UFRG).

Distribution (Fig. 78). México: *Michoacán, Puebla* (Fig. 78).

Etymology. The epithet refers to the shape of the foliaceous expansions of dorsal rim of pygophore; adjective *otarus* (*-a, -um*), latinized from Greek *otaros* = large-eared.

Lineostethus auritus Garbelotto & Kochenborger, sp. nov.

Figs 8, 70–78

Diagnosis. Males of *L. auritus* sp. nov. are distinguished by the long and narrow foliaceous expansions of dorsal rim truncate at apex, leaving the posterolateral angles of pygophore visible dorsally. The females have broad gonocoxites VIII with the posterior margin uniformly convex; the shape of the thickening of the gonapophyses IX and of the thickening of vaginal intima are also diagnostic.

Description. Antennae usually brown, with lateral lines ivory, sometimes antennae variegated. Anterolateral margins of pronotum and proximal portion of lateral margin of corium punctured. Anterior margin of scutellum with two to four brown spots, and at least one ivory spot medially; post-frenal lobe sometimes with lateral brown maculae. Connexival segments with lines of black punctures on anterior and posterior margin of each segment. Unpunctured area laterad to the apex of radial vein of coria at least equivalent to half of eye width. Hemelytral membrane slightly surpassing the apex of abdomen, with 9 veins. Ventrally, punctures concentrated on lateral margins of abdomen. Femora with sparse brown circular spots on distal half.

Male genitalia. Foliaceous expansions of dorsal rim subrectangular, elongated, longer than wide; lateral margins of expansions slightly concave; apex truncate, slightly wider than the median width, surpassing the posterior limit of ventral rim by half of the expansion length; not hiding the posterolateral angles (Fig. 70: fdr, pa). Basal impression of the foliaceous expansions sinuous, cutting of dorsal rim subrectangular (Fig. 70: bi, dr). Posterolateral angles acute, ventrally covered by setae (Figs 70–71: pa). Margin of the superior layer of ventral rim open V-shaped, median excavation drop-shaped (Fig. 71: vr). Margin of the inferior layer of ventral rim with two lines of short bristles: one posterior, subrectilinear and not associated with punctures; and another anterior, concave and associated with punctures (Figs 71–72: b). Segment X with dorsal setae on apex (Fig. 70: X). Projections of the superior layer of ventral rim forming a right angle with the posterolateral angles, covered by setae; dorsal and ventral margins subrectilinear, apex spatulate (Fig. 72: psl, pa). Parameres parallel to segment X, surpassing the foliaceous expansions of dorsal rim; head subtriangular (Figs 72–74: par, X, fdr). Ventral surface of phallotheca concave (Fig. 75: ph, ap). Vesica apex rounded; apical margin convex, extending to mid of ductus seminis distalis (Fig. 75: v, av, ds). Distal portion of ductus seminis distalis subequal to vesica in length (Fig. 75: ds).

Female genitalia. Posterior margin of urosternite VII uniformly concave (Fig. 76: VII). Gonocoxites VIII subquadrangular, broad; posterior margin uniformly convex (Fig. 76: gcVIII). Posterior margin of laterotergites VIII subrectilinear; limit between laterotergites VII and tergite VIII indistinguishable (Fig. 76: la8). Gonocoxites IX small, posterior margin subrectilinear (Fig. 76: gcIX). Laterotergites IX digitiform, tumescent and rounded at apex (Fig. 76: la9). Posterior margin of segment X concave (Fig. 76: X). Thickening of gonapophyses IX subcilindrical slightly expanded on base, surpassing the lateral margin of the thickening of vaginal intima (Fig. 77: gpIX, tgIX). Thickening of vaginal intima subtriangular (Fig. 77: vi). Distal portion of ductus receptaculi apically enlarged and more sclerotized (Fig. 77: ddr); anterior and posterior annular crests perpendicular to pars intermedialis; pars intermedialis membranous at apex, subequal to capsula seminalis (Fig. 77: pi, cs). Capsula seminalis shallowly constricted near base (Fig. 77: cs).

Measurements. Male (n = 2): head length, 1.78 ± 0.25 (1.60–1.96); width, 3.13 ± 0.21 (2.98–3.28); length of antennal segments: I, 0.46 ± 0.06 (0.42–0.50); II, 0.74; III, 0.52; IV, 0.78; V, 1.00; length of labial segments: I, 0.71 ± 0.10 (0.64–0.78); II, 1.50 ± 0.14 (1.40–1.60); III, 0.70 ± 0.03 (0.68–0.72); IV, 0.49 ± 0.07 (0.44–0.54); pronotum length, 1.8 ± 0.20 (1.66–1.94); width, 4.11 ± 0.43 (3.80–4.41); scutellum length, 3.38 ± 0.31 (3.16–3.60); width, 2.71 ± 0.29 (2.50–2.91); total length, 7.72 ± 1.68 (6.53–8.90); abdominal width, 4.53 ± 0.53 (4.15–4.90).

Female (n = 2): head length, 1.95 ± 0.04 (1.92–1.98); width, 3.49 ± 0.07 (3.44–3.54); length of antennal segments: I, 0.67 ± 0.18 (0.54–0.8); II, 0.89 ± 0.16 (0.78–1.00); III, 0.53 ± 0.04 (0.50–0.56); IV, 1.05 ± 0.04 (1.02–1.08); V, 1.25 ± 0.07 (1.20–1.30); length of labial segments: I, 0.74 ± 0.08 (0.68–0.80); II, 1.62 ± 0.03 (1.60–1.64); III, 0.77 ± 0.10 (0.70–0.84); IV, 0.54 ± 0.04 (0.51–0.56); pronotum length, 2.03 ± 0.07 (1.98–2.08); width, 4.78 ± 0.11 (4.70–4.85); scutellum length, 3.88 ± 0.11 (3.08–3.95); width, 2.92 ± 0.09 (2.85–2.98); total length, 9.40; abdominal width, 5.36.

Type material. Holotype (male): MEXICO, *Oaxaca*: (9 mi N Oaxaca), 31-V-1983, C.W. & L.O'Brien & Marshall leg. (deposited in: AMNH). Paratypes: MEXICO, *Oaxaca*: (5mi S Candelaria Loxicha), 1 male, 18-19-VII-1974, Clar Marray Ashe & Shaffner leg. (DARC); (9mi N Oaxaca), 2 females, 31-V-1983, C.W. & L.O'Brien & Marshall leg. (1 female AMNH; 1 female DARC).

Distribution (Fig. 78). Mexico: *Oaxaca*.

Etymology. The epithet refers to the shape of the foliaceous expansions of dorsal rim of pygophore; Latin adjective, *auritus* (*-a, -um*) = eared.

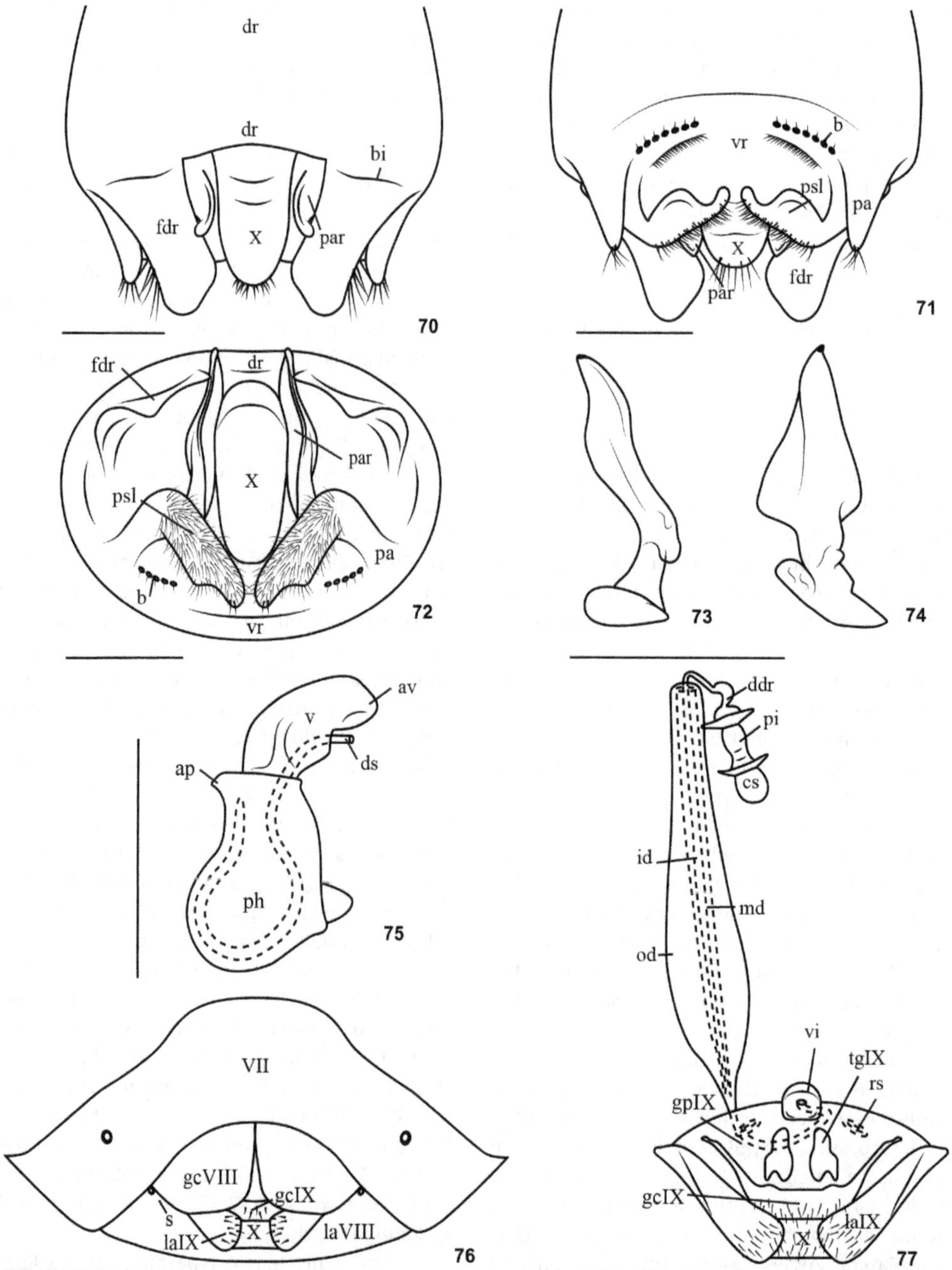

Figures 70–77. *Lineostethus auritus* sp. nov.: (70–72) male pygophore, dorsal, ventral and posterior views respectively; (73–74) left paramere, dorsal and lateral views; (75) phallus, lateral view; (76) female genital plates; (77) female receptaculum seminis. Abbreviations: ap, ventral apical projection of phallotheca; av, apical margin of vesica; b, bristles; bi, basal impression of foliaceous expansions; cs, capsula seminalis; ddr, distal ductus receptaculi; dr, dorsal rim; ds, ductus seminis distalis; fdr, foliaceous expansions of dorsal rim; gcVIII, gonocoxites VIII; gcIX, gonocoxites IX; gpVIII, gonapophyses VIII; gpIX, gonapophyses IX; id, internal duct of vesicular area; la8, laterotergites VIII; la9, laterotergites IX; md, median duct of vesicular area; od, outer duct of vesicular area; pa, posterolateral angle; par, paramere; ph, phallotheca; pi, pars intermedialis; psl, projection of superior layer of ventral rim; rs, ring sclerites; s, spiracle; tgIX, thickening of gonapophyses IX; v, vesica; vi, thickening of vaginal intima; vr, ventral rim; VII, urosternite VII; X, segment X. Scale bars: 70–74, 76 = 1 mm; 75, 77 = 0.5 mm.

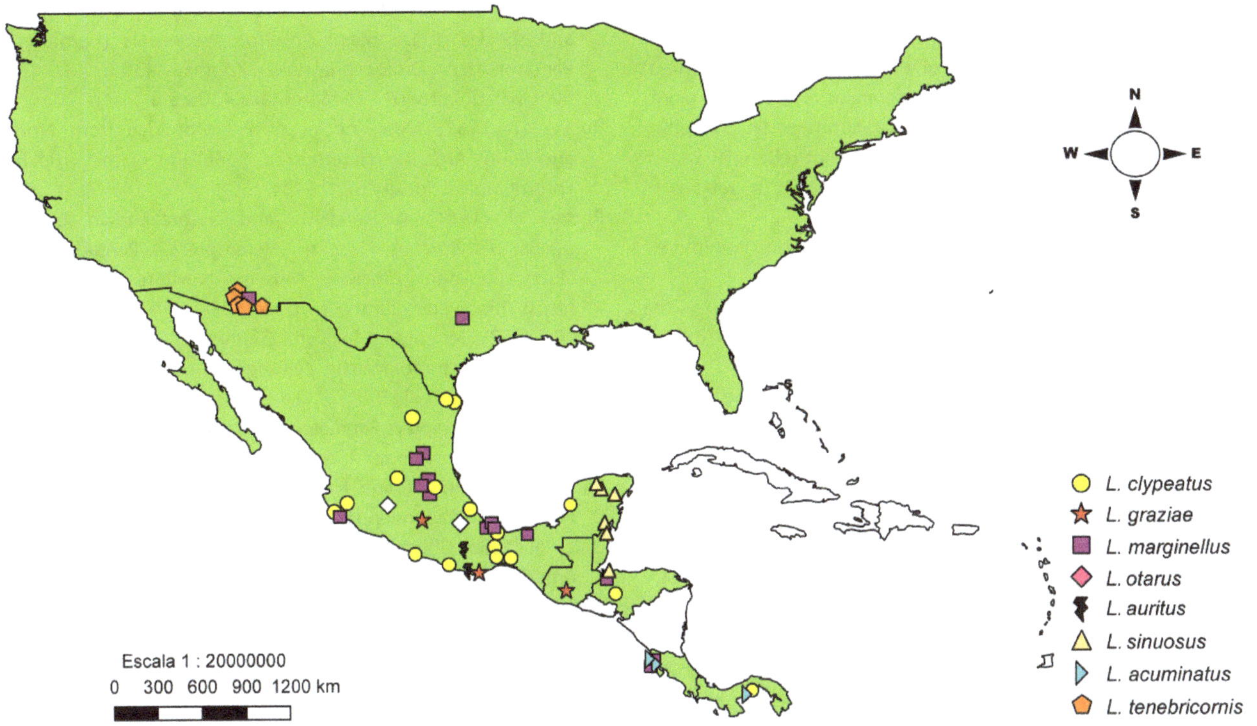

Figure 78. Distributional map for studied specimens of the species of *Lineostethus*. Countries with species records are in green.

ACKNOWLEDGMENTS

We are thankful to all the curators of the listed collections, who provided access to specimens, types or not, especially to: Mike Narahara and Vicent F. Lee (CAS) and Peter T. Oboyski (EMEC) who provided photos of L. graziae, and Gunvi Lindberg (NHRS) who provided photos of *L. clypeatus and L. marginellus*. To Christiane Weirauch for providing access to specimens of several species, including some new species. To D.A. Rider and P. Kment for the valuable suggestions on the original manuscript. Scholarships from Conselho Nacional de Desenvolvimento Científico e Tecnológico (CNPq) and Coordenação de Aperfeiçoamento de Pessoal de Nível Superior (Capes) to T.A.G. (CNPq 142448/2011-7; CAPES-PDSE 5788-13-7), and CNPq fellowship to L.A.C. (307204/2015-4) are also gratefully acknowledged. The manuscript did not undergo grammar revision and any improprieties of English proficiency are the authors' entire responsibility.

LITERATURE CITED

Baker AD (1931) A study of the male genitalia of Canadian species of Pentatomidae. Canadian Journal of Research 4: 148–220. https://doi.org/10.1139/cjr31-012

Banks N (1910) Catalogue of the Nearctic Hemiptera Heteroptera. Philadelphia, 103 pp. https://doi.org/10.5962/bhl.title.12353

Campos LA, Grazia J (2006) Análise cladística e biogeografia de Ochlerini (Heteroptera, Pentatomidae, Discocephalinae). Iheringia, Série Zoologia 96: 147–163. https://doi.org/10.1590/S0073–47212006000200004

Distant WL (1880–1893) Heteroptera. In: Godman F, Salvin O (Eds) Biologia Centrali-Americana. London, 462 pp.

Dupuis C (1970) Heteroptera. In: Tuxen SL (Ed.) Taxonomist's Glossary of Genitalia of Insects. Munksgaard, Copenhagen, 190–208.

Falling Rain Genomics Inc (2010) Global Gazetteer v. 2.2. http://www.fallingrain.com/world [Accessed: 20/03/2016]

Garbelotto TA, Campos LA, Grazia J (2013) Cladistics and revision of *Alitocoris* with considerations on the phylogeny of the *Herrichella* clade (Hemiptera, Pentatomidae, Discocephalinae, Ochlerini). Zoological Journal of the Linnean Society 168: 452-472. https://doi.org/10.1111/zoj.12032

Garbelotto TA, Campos LA, Grazia J (2016) *Stapecolis*, new genus of Ochlerini (Hemiptera: Heteroptera: Discocephalinae). Zootaxa 4137: 545-552. https://doi.org/10.11646/zootaxa.4137.4.6

Grazia J, Panizzi AR, Greve C, Schwertner CF, Campos LA, Garbelotto TA, Fernandes, JAM (2015) Stink bugs (Pentatomidae). In: Panizzi AR, Grazia J (Eds) True Bugs (Heteroptera) of the Neotropics. Springer, New York, (Entomology in focus 2), 681–756. https://doi.org/10.1007/978-94-017-9861-7_22

Hildebrand R, Becker M (1982) Uma nova espécie de *Lineostethus* Ruckes, 1966 com o estudo de genitália do gênero (Heteroptera,

Pentatomidae, Discocephalinae). Revista Brasileira de Biologia 42: 773–784.

Kirkaldy GW (1909) Catalogue of the Hemiptera (Heteroptera) with biological and anatomical references, lists of food plants and parasites, etc. Cimicidae. Felix L. Dames, Berlin, 392 pp.

Kment P, Vilímová J (2010) Thoracic scent efferent system of Pentatomoidea (Hemiptera: Heteroptera): a review of terminology. Zootaxa 2706: 1–77.

Lethierry L, Severin G (1893) Catalogue general des hemipteres. Brussels, vol. 1, 286 pp.

Morrone JJ (2014) Biogeographical regionalisation of the Neotropical region. Zootaxa 3782: 1–110. http://dx.doi.org/10.11646/zootaxa.3782.1.1

Morrone JJ (2015) Biogeographical regionalisation of the world: a reappraisal. Australian Systematic Botany 28: 81–90. https://doi.org/10.1071/SB14042

Rider DA, Schwertner CF, Vilimová J, Rédei D, Kment P, Thomas DB (2017) Higher systematics of the Pentatomoidea. In: McOherson JE (Ed.) Invasive stink bugs and related species (Pentatomoidea) – Biology, higher systematics, semiochemistry and management. CRC Press, Boca Raton, 25–176.

Rolston LH (1981) Ochlerini, a new tribe in Discocephalinae (Hemiptera: Pentatomidae). Journal of the New York Entomological Society 89: 40–42. http://www.jstor.org/stable/25009234

Rolston LH (1990) Key and diagnoses for the genera of 'Broadheaded' Discocephalines (Hemiptera: Pentatomidae). Journal of the New York Entomological Society 98: 14–31. http://www.jstor.org/stable/25009813

Rolston LH (1992) Key and diagnoses for the genera of Ochlerini (Hemiptera: Pentatomidae: Discocephalinae). Journal of the New York Entomological Society 100: 1–41. http://www.jstor.org/stable/25009944

Rolston LH, McDonald F (1979) Keys and diagnoses for the families of Western Hemisphere Pentatomoidea, subfamilies of Pentatomidae and tribes of Pentatominae (Hemiptera). Journal of the New York Entomological Society 87: 189–207. http://www.jstor.org/stable/25009160

Rosso P, Campos LA (2017) *Nigrisagitta distichus*, a new genus and species of Discocephalini from Venezuela (Hemiptera, Pentatomidae, Discocephalinae). Zootaxa 4303: 427–436. http://dx.doi.org/10.11646/zootaxa.4303.3.8

Ruckes H (1957) New species of Pentatomidae from North and South America (Heteroptera) I. Bulletin of the Brooklyn Entomological Society 52: 16–24.

Ruckes H (1966) An analysis and a breakdown of the genus *Patycarenus* Fieber (Heteroptera, Pentadomidae, Discocephalinae). American Museum Novitates 2255: 1–42. https://biodiversitylibrary.org/page/50589944

Stål C (1862) Hemiptera Mexicana Enumeravit Speciesque Novas Descripsit. Entomologische Zeitung entomologischen Vereine zu Stettin 23: 81–118.

Stål C (1868) Hemiptera Fabriciana. Kongelige Svenska Vetenskpas Akademien Handlingar 7(2): 1–48.

Stål C (1872) Enumeratio Hemipterorum II. Kongelige Svenska Vetenskpas Akademien Handlingar 10(4): 1–159.

Uhler PR (1886) Check-list of the Hemiptera-Heteroptera of North America. Brooklyn, 29 pp.

Walker F (1867) Catalogue of the specimens of heteropterous Hemiptera in the collection of the British Museum. London, pt. 1, 240 pp. https://doi.org/10.5962/bhl.title.118688

Author Contributions: TAG, APLK and LAC contributed equally to this article.

Competing Interests: The authors have declared no competing interest exist.

5

Proposal of *Tiddergasilus* gen. nov. (Ergasilidae: Cyclopoida) for *T. iheringi* comb. nov. from the gills of *Hoplias malabaricus* (Erythrinidae: Characiformes) from Brazil

Taísa Mendes Marques[1,2], Walter A. Boeger[1]

[1] *Laboratório de Ecologia Molecular e Parasitologia Evolutiva, Departamento de Zoologia, Universidade Federal do Paraná. Caixa Postal 19073, 81531-980 Curitiba, PR, Brazil.*
[2] *Departamento de Patologia, Universidade Federal do Paraná, Curitiba, PR, Brazil.*
Corresponding author: Taísa Mendes Marques (taisaorsi@gmail.com)

ABSTRACT. *Ergasilus* von Nordmann, 1832 is the type and the most speciose genus of Ergasilidae, with more than 150 valid species. The first species of this genus, *Ergasilus iheringi* Tidd, 1942, was described from the Neotropical Region, from the gills of *Hoplias malabaricus* (Bloch, 1794). Since then, 35 new species of Ergasilidae have been described from the Neotropics. However, a reanalysis of the available type specimens of *E. iheringi* revealed features that suggest that it represents a lineage independent from *Ergasilus*. Consequently, we propose *Tiddergasilus* **gen. nov.** (Ergasilidae: Cyclopoida) to accommodate *T. iheringi* (Tidd, 1942) **comb. nov.** The species was re-examined and redescribed based on available type specimens. *Tiddergasilus* **gen. nov.** is characterized by species with exopod of leg 4 2-segmented; antenna small and robust, with a short second segment (approximately three times shorter than the first segment), and claw with a unique morphology (short, recurved with a sub-proximal indentation in the inner margin). A phylogenetic analysis, based on a matrix of 15 taxa and 16 morphological characters in PAUP* software 4.0a152, resulted in 15 equally parsimonious trees. Based on this analysis, *Tiddergasilus* is closely related to a group composed of Neotropical species of *Ergasilus* – sharing with them antennule with 6-segments. When the highly homoplastic character "number of segments in the antennule" is removed from the analysis, *Tiddergasilus* forms a polytomy with the clade of genera of the former Vaigamidae, *Ergasilus*, and the clade *Rhinergasilus+Brasergasilus+Pindapixara*.

KEY WORDS. Copepoda, Ergasilidae, taxonomy.

INTRODUCTION

Ergasilus von Nordmann, 1832 is the oldest genus of Ergasilidae, being the type and the most species-rich genus of the family, including more than 150 valid species (Walter and Boxshall 2017) distributed on all continents except Antarctica (Lacerda et al. 2007; Boxshall and Defaye 2008). The Neotropical Region has the richest freshwater fauna of parasitic ergasilids (Boxshall and Defaye 2008). The first representative of Ergasilidae described from the Neotropics was *E. iheringi* Tidd, 1942, a gill parasite of *Hoplias malabaricus* (Bloch, 1794). Since then, 35 new species of Ergasilidae have been described from the Neotropical region.

Reanalysis of the available type specimens of *E. iheringi* revealed that the species belongs to an undescribed genus of Ergasilidae. In this study, we redescribe the species, re-evaluate its phylogenetic relationships and propose a new genus, *Tiddergasilus* gen. nov., to accommodate the species originally described by Tidd (1942).

MATERIAL AND METHODS

The specimens were obtained on loan from the United States National Museum, Washington, D.C. (USNM). Nine adult females preserved in ethanol were cleared with lactic acid, mounted on temporary slides, and examined under a BX51 DIC microscope. One specimen was dissected with glass microprobes and mounted in Hoyer's mounting medium (Humason 1979) on a semi-permanent slide. Illustrations were prepared with a drawing tube attached to the microscope. All measurements are expressed in micrometers. A parsimony analysis was performed with a matrix of 15 taxa and 16 morphological characters using PAUP 4.0a152 (Swofford 2002). The characters used were based on previous phylogenetic analyses (Amado et al. 1995, Tang and Kalman 2008); in addition, new characters were included in the analysis. The included taxa are those genera with species occurring in the freshwater habitats of the Neotropical Region. Species of these genera form a putative monophyletic group, as suggested by the following putative synapomorphy: endopod of leg 1 with two segments. Rooting was based on a functional outgroup determined on preliminary phylogenetic analysis of known genera of Ergasilidae. Three genera were chosen as functional outgroups: *Acusicola, Miracetyma,* and *Amplexibranchius.* Species of these genera are characterized by having a latching groove in the third segment for receiving the opposite claw (Thatcher 1984). Clade stability was measured using Bremer's support (Bremer 1994), calculated in PAUP according to Bremer (1994). All character-states were equally weighted, unordered, and non-applicable character states were coded as 'missing'. A heuristic search analysis consisted of 1,000 random stepwise-addition replicates. Character matrix and morphological character states are presented in Tables 1 and 2.

Table 1. Morphological characters used in the analysis of the genera of the Ergasilidae in the Neotropical Region; the number in parentheses indicates the code of each state.

#	Characters
1	Rostral spine absent (0), or present (1)
2	Retrostylets absent (0), or present (1)
3	Antennule 5-segmented (0), or 6-segmented (1)
4	Latching antenna absent (0), or present (1)
5	Second segment of antennae normal (0), reduced (1), or absent (2)
6	Antenna tipped with 2 claws (0), or 1 claw (1)
7	Claw short (0), or elongated (1)
8	Leg 1 endopod 3-segmented (0), or 2-segmented (1)
9	Leg 1 second endopodal segment normal (0), or modified (1)
10	Spines on Leg 1 endopod present (0), or absent (1)
11	Spines on Leg 2 exopod absent (0), or present (1)
12	Spines on Leg 2 endopod present (0), or absent (1)
13	Leg 4 biramous (0), represented by one seta (1), or absent (2)
14	Leg 4 exopod 2-segmented (0), or 1-segmented (1)
15	Leg 4 endopod 3-segmented (0), or 2-segmented (1)
16	Egg sac with multiple rows (0), or composed of single row (1)

Table 2. Character matrix for 16 morphological characters used in the analysis of the genera of the Ergasilidae in Neotropical Region.

Taxon	Character															
	1	2	3	4	5	6	7	8	9	10	11	12	13	14	15	16
Acusicola	0	0	1	1	0	1	0	1	0	0	1	0	0	0	0	0
Amplexibranchius	0	0	1	1	0	1	0	1	1	1	0	0	0	0	0	0
Brasergasilus	0	0	0	0	2	1	1	1	0	0	1	0	1	-	-	1
Ergasilus	0	0	0	0	0	1	0	1	0	0	1	0	0	0	1	0
Gamidactylus	0	1	1	0	0	0	0	1	0	0	1	1	0	1	1	1
Gamispatulus	1	1	1	0	0	0	0	1	0	0	1	1	0	1	1	1
Gamispinus	0	1	1	0	0	1	0	1	0	0	1	0	0	1	1	1
Gauchergasilus	0	0	1	0	0	1	0	1	0	0	1	0	0	0	0	0
Miracetyma	0	0	1	1	0	1	0	1	1	1	0	0	0	0	0	0
Pindapixara	0	0	0	0	1	1	1	1	0	0	0	0	0	0	1	1
Prehendorastrus	0	0	1	0	0	1	0	1	0	0	1	0	0	0	0	0
Pseudovaigamus	1	1	1	0	0	1	0	1	0	0	1	1	0	0	0	1
Rhinergasilus	0	0	1	0	1	1	1	1	0	1	1	0	1	-	-	1
Tiddergasilus gen. nov.	0	0	0	0	1	1	0	1	0	0	1	0	0	0	1	0
Vaigamus	1	1	0	0	0	1	0	1	0	0	1	0	0	1	1	1

TAXONOMY

Ergasilidae Burmeister, 1835

Tiddergasilus gen. nov.

Diagnosis. Adult female body comprising prosome and urosome; cephalosome well defined; antenna prehensile presenting coxobasis, two segments, hooked claw; antennule 6-segmented; mandible presenting two blades (median and posterior), basis of maxilla partially toothed on convex margin; four pairs of legs; leg 1 exopod 3-segmented, endopod 2-segmented; legs 2 and 3 differ in number of setae on endopod; legs 2 and 3 exopod and endopod 3-segmented; leg 4 exopod 2-segmented, endopod 2-segmented; abdominal somites with spinules ventrally; caudal rami armed with 4 setae, egg sac long with 2-3 rows of eggs.

Type species. *Tiddergasilus iheringi* (Tidd, 1942) comb. nov.

Etymology. The generic epithet is in honor of Wilbur M. Tidd, author of *Ergasilus iheringi.*

Tiddergasilus iheringi (Tidd, 1942), comb. nov.

Figs 1–10

Type host. *Hoplias malabaricus* (Bloch, 1974)

Site on host. Gill filaments.

Type locality. Campina Grande, Paraíba, Brazil.

Male. Unknown

Type specimens. Holotype and Paratype in USNM Catalog #79255.

Redescription (based on 9 females). Body comprising prosome and urosome; prosome consisting of non-inflated cephalosome, rostrum broad and well defined. Cephalosome well defined. Cephalosome and first pedigerous somite partially fused, followed by three free pedigerous somites. Cephalothorax length less than twice width, more than half

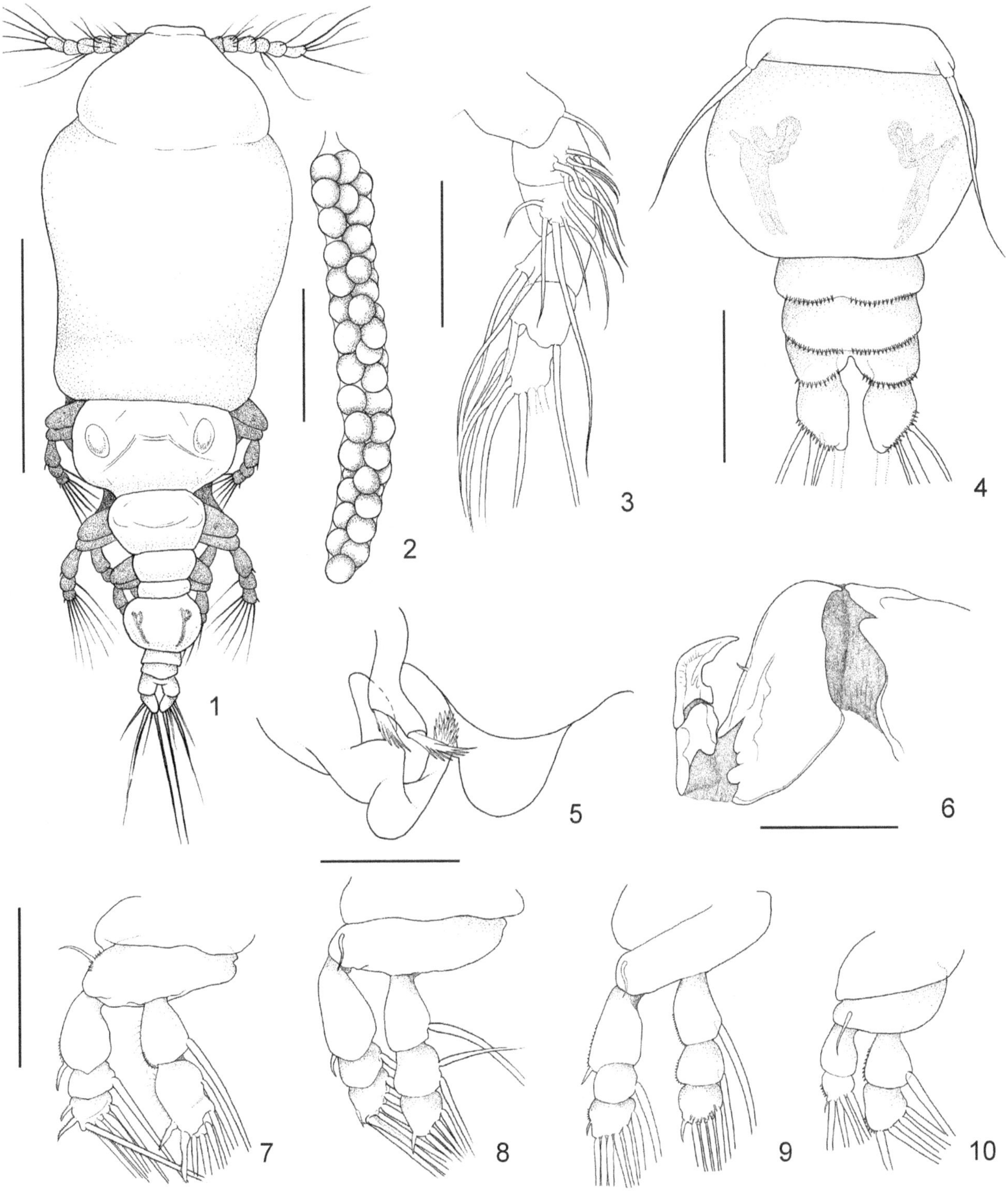

Figures 1–10. *Tiddergasilus iheringi* comb. nov. adult female. (1) dorsal view; (2) egg sac; (3) antennule; (4) urosome, caudal ramus; (5) mouth parts; (6) antenna; (7) leg 1; (8) leg 2; (9) leg 3; (10) leg 4. Scale bars: 1–2 = vertical 250 μm, 3–10 = 50 μm.

body length. Antennule 6-segmented, tapering distally, with 23 setae, aesthetascs not observed. Antenna small and robust, with coxobasis, 2-segmented endopod, and terminal claw. Coxobasis short, with small seta at inner distal corner, membrane between coxobasis and first endopodal segment not inflated; first endopodal segment lacking spinules on internal margin, submedian sensillum present, lacking cuticular elevation; second segment (approximately trhee times shorter than first), lacking sensillum; claw with unique morphology (short, recurved with sub-proximal indentation in inner margin). Maxillule not observed; mandible with middle and posterior blades, anterior blade absent, middle blade toothed along posterior edge. Maxilla lacking spinulate seta; basis partially toothed on convex margin. Swimming legs 1–4 biramous each with 2-segmented protopod comprising coxa and basis. Interpodal plates of all legs with two rows of spinules ventrally. The armature of legs (spines, Roman numerals; setae, Arabic numerals) as follows:

	Coxa	Basis	Exopod	Endopod
Leg 1	0-0	1-0	I-0; 0-1; II-4	0-1; II-5
Leg 2	0-0	1-0	I-0; 0-1; I-6	0-1; 0-2; I-4
Leg 3	0-0	1-0	I-0; 0-1; I-6	0-1; 0-1; I-4
Leg 4	0-0	1-0	0-0; I-4	0-1; 0-4

Leg 1. Coxa with smooth margins, lacking spinules; basis with spinules and outer seta. Exopod 2-segmented; first segment non-pilose, with spinules distally on outer margin, and a spinulate distal spine; second segment lacking spinules and distal spine, with single seta; third segment lacking spinules, lacking pectinate seta, with four pilose setae and one spinulate seta, plus two distal spinulate spines. Endopod 2-segmented; first segment with spinules distally on outer margin, pilose, with single seta; second segment with spinules on outer margin, with five setae, plus 2 spinulated spines, one long, one short. **Leg 2.** Coxa lacking spinules; basis lacking spinules, with outer seta. Exopod 3-segmented; first segment non-pilose, with spinules on outer margin and non-spinulate distal spine, lacking seta; second segment non-pilose, lacking spinules, with single seta; third segment with spinules distally on outer margin, with six setae, plus non-spinulated distal spine. Endopod 3-segmented; first segment non-pilose, with spinules on outer margin and with single seta; second segment non-pilose, with spinules on outer margin, with single seta, distal spine absent; third segment with 4 setae and spinules immediately anterior to setae, plus non-spinulated distal spine. **Leg 3.** Coxa lacking spinules; basis lacking spinules, with outer seta. Exopod 3-segmented; first segment non-pilose, with spinules on outer margin and non-spinulate distal spine, lacking seta; second segment non-pilose, lacking spinules, with single seta, distal spine absent; third segment with spinules distally on outer margin, with six setae, plus non-spinulate distal spine. Endopod 3-segmented; first segment non-pilose, with spinules on outer margin and with single seta; second segment non-pilose, with spinules on outer margin, with single seta, distal spine absent; third segment with 4 setae and spinules immediately anterior to setae, plus non-spinulate distal spine. **Leg 4.** Coxa lacking spinules; basis with outer seta, lacking spinules. Exopod 2-segmented; first segment non-pilose, lacking spinules and seta, distal spine absent; second segment with spinules on outer margin, with 4 setae, plus non-spinulate distal spine. Endopod 2-segmented; first segment non-pilose, with spinules on outer margin, with single seta; second segment with spinules distally on outer margin, with 4 setae, distal spine absent. **Leg V** reduced to single seta.

Urosome consisting of fifth somite, genital double-somite, and three free abdominal somites. Genital double-somite barrel-shaped narrowing posteriorly, ventral surface lacking spinules. Abdominal somites with a row of spinules on posteroventral margins; third abdominal somite bipartite. Caudal rami as long as wide, each armed with one long, one medium and two smaller (lateral and ventral) setae, rows of spinules present ventrally, immediately anterior to smaller setae. Egg sac many times longer than wide, composed of 2–3 rows of eggs.

Phylogeny

An initial hypothesis of the evolutionary relationships of Ergasilidae genera was constructed manually using Hennigian Argumentation (Hennig 1966, Wiley 1981). The topology of the resulting cladogram was tested with PAUP* to confirm that it was one of the most-parsimonious trees. This hypothesis is presented in Figure 11 (tree length 28; C.I. = 0.68 and R.I. = 0.76) – it is one of 15 equally parsimonious trees obtained with PAUP*. Bremer Support is presented in gray numbers on each respective node.

Three main clades were observed within the Neotropical Ergasilidae. One is comprised of the three outgroup genera: *Acusicola* Cressey, 1970, *Miracetyma* Malta, 1993 and *Amplexibranchius* Thatcher & Paredes, 1985. Another clade is composed of *Gamidactylus* Thatcher & Boeger, 1984, *Gamispatulus* Thatcher & Boeger, 1984, *Pseudovaigamus* Amado, Ho & Rocha, 1995, *Vaigamus* Thatcher & Robertson, 1984 and *Gamispinus* Thatcher & Boeger, 1984, which represents members of the clade formerly referred to as the Vaigamidae. Although this clade corroborates the results of past morphological analyses (Amado et al. 1995, Tang and Kalman 2008), the relationships between the included genera differ.

The same is true for *Acusicola*, *Amplexibranchius* and *Miracetyma*. As in the previous phylogenies (Amado et al. 1995, Tang and Kalman 2008), those genera are sister taxa, but their relative phylogenetic relationships differ in the present analysis, with the sister-group *Amplexibranchius* and *Miracetyma* supported the following synapomorphies: second endopod segment of leg 1 modified (Character 9), endopod of leg 1 lacking spines, and endopod of leg 2 lacking spines (Characters 10 and 12). Accordingly, this is the most robust clade in the phylogeny, presenting the highest Bremer support.

Tiddergasilus gen. nov. is a putative sister taxon of Neotropical species of *Ergasilus*. The two genera share antennule with 6-segments (Character 3). The hypothetical ancestor of this group of *Ergasilus* and *Tiddergasilus* gen. nov. forms a polytomy

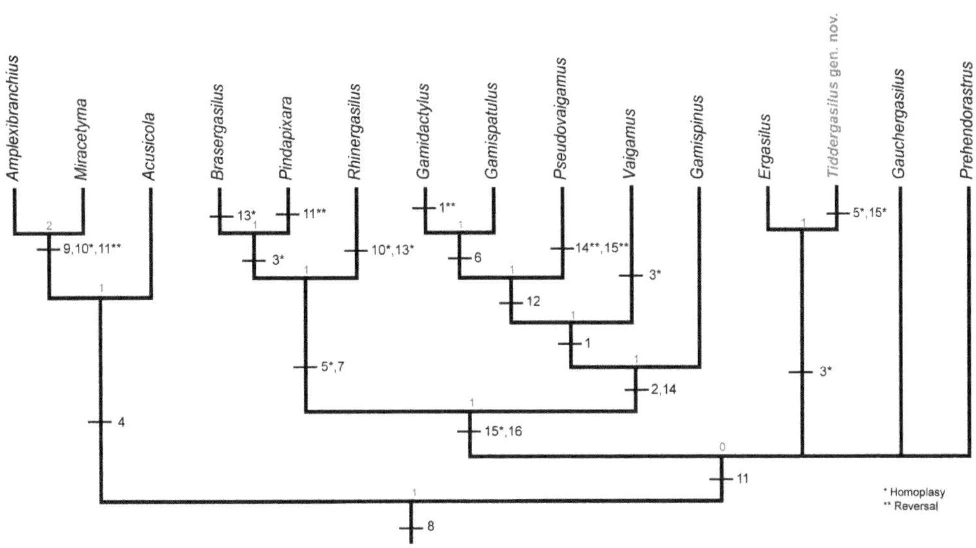

11

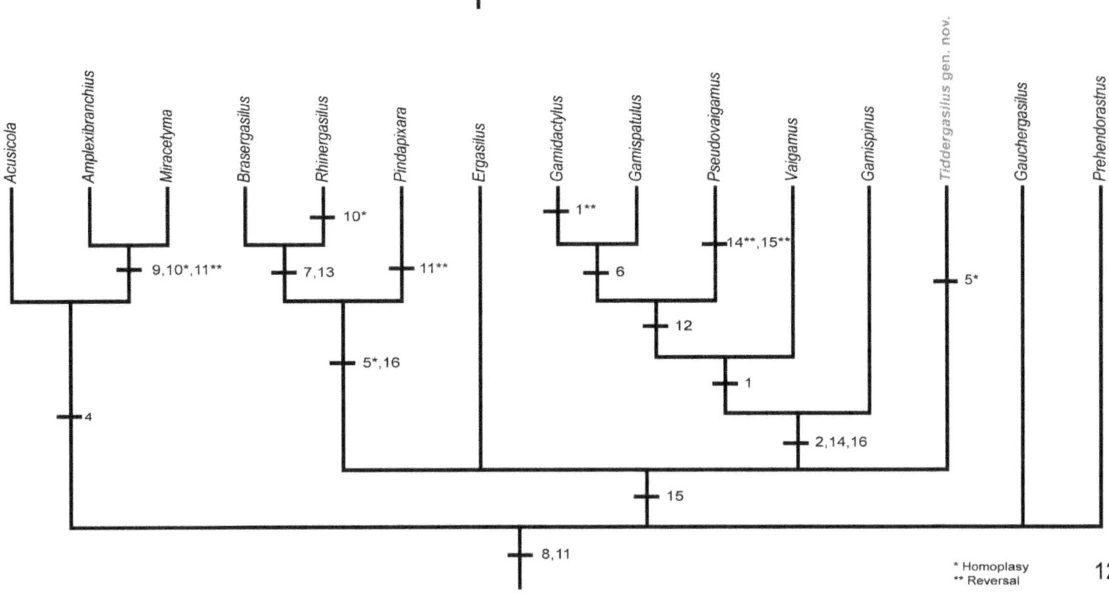

12

Figure 11–12. (11) Morphological phylogeny of 15 genera of Ergasilidae showing Bremer support values and character changes supporting each node. Bremer support values are indicated in gray numerals; characters that indicate a homoplasy or reversal events are indicated. (12) Morphological phylogeny of 15 genera of Ergasilidae without the Character 3, showing Bremer support values and character changes supporting each node. Bremer support values are indicated in gray numerals; characters that indicate a homoplasy or reversal events are indicated.

with *Gauchergasilus* Montu & Boxshall, 2002 and *Prehendorastrus* Boeger & Thatcher, 1990. Removal of the highly homoplastic character 3 from the analysis results in a polytomy with the clade of genera of the former Vaigamidae, *Ergasilus*, and the clade *Rhinergasilus+Brasergasilus+Pindapixara* (Fig. 12, treelength 24; C.I. = 0.70 and R.I. = 0.76).

Five homoplasious character states were observed: (1) antennule 6-segmented in *Ergasilus* + *Tiddergasilus*, *Vaigamus*, and *Brasergasilus* + *Pindapixara* (Character 3); (2) second segment of antenna reduced in *Brasergasilus* + *Pindapixara* + *Rhinergasilus* and *Tiddergasilus* (Character 5); (3) spines on second endopod absent in *Amplexibranchius* + *Miracetyma* and *Rhinergasilus* (Character 10); (4) leg 4 absent in *Rhinergasilus* and *Brasergasilus* (Character 13). (5) leg 4 endopod 2-segmented in *Brasergasilus* + *Pindapixara* + *Rhinergasilus* + *Gamidactylus* + *Gamispatulus* + *Vaigamus* + *Gamispinus* and *Tiddergasilus* (Character 15).

The analysis also shows four reversals: loss of rostral spine in *Gamidactylus* (Character 1); spines on exopod 2 absent in *Pindapixara* (Character 11); leg 4 with exopod 2-segmented and endopod 3-segmented in *Pseudovaigamus* (Characters 14 and 15).

DISCUSSION

The present redescription, based on the study of type specimens, revealed some features not reported in the original description made by Tidd (1942). These are: (1) number of setae on antennule (16 reported by Tidd, 23 reported here); (2) the presence of a submedian sensillum on the first endopodal segment of the antenna; (2) abdominal somites ornamented with spinules on their posteroventral margins; (4) caudal rami with four setae; (5) basis of all legs armed with outer seta; (6) leg 1 with pectinate seta on distal exopod segment; (7) leg 4 with exopod 2-segmented; distal spine on two exopod; and (8) interpodal plates between legs ornamented with two rows of spinules ventrally.

Based on the phylogenetic analysis, a combination of five features can be considered diagnostic of *Tiddergasilus*: (1) leg 1 endopod 2-segmented; (2) spines on exopod of leg 2 present; (3) antennule 6-segmented; (4) antenna small and robust, with a short second segment (approximately 3 times shorter than the first segment), claw with a unique morphology (short, recurved with a sub-proximal indentation in the inner margin); and (5) endopod of leg 4 2-segmented. Although a putative sister group of the Neotropical species of *Ergasilus*, the single species of *Tiddergasilus* gen. nov., *T. iheringi*, possesses morphological features that can easily distinguish it as a new genus. *Tiddergasilus iheringi* can be distinguished from *Ergasilus* spp. by the morphology of antennae and the well-defined rostrum.

Tiddergasilus gen. nov., like species of many other genera of Ergasilidae, presents exopod of leg 4 2-segmented; and antenna small and robust, with the short second segment, similar to species of *Majalincola* Tang & Kalman, 2008, *Paeonodes* Wilson, 1944, *Teredophilus* Rancurel, 1954, *Therodamas* Kroyer, 1862 and *Urogasilus* Rosim, Boxshall & Ceccarelli, 2012.

A more detailed redescription of certain features (e.g. number of setae on antennule) was hampered by the conditions of the preserved specimens. Some structures such as setae can be lost with time. For instance, the antennulemay presents more setae than those reported herein, and the maxillule could not be observed.

LITERATURE CITED

Amado MA, Ho J, Rocha CEF (1995) Phylogeny and biogeography of the Ergasilidae (Copepoda, Poecilostomatoida), with a reconsideration of the taxonomic status of the Vaigamidae. Contributions to Zoology 65: 222–242.

Boxshall GA, Defaye D (2008) Global diversity of copepods (Crustacea: Copepoda) in freshwater. Hydrobiologia 595(1): 195–207. https://doi.org/10.1007/s10750-007-9014-4

Bremer K (1994) Branch support and tree stability. Cladistics 10: 295–304. https://doi.org/10.1111/j.1096-0031.1994.tb00179.x

Hennig W (1966) Phylogenetic Systematics. University of Illinois Press, Urbana.

Humason GL (1979) Animal tissue techniques. W.H. Freeman and Company, San Francisco.

Lacerda ANF, Takemoto RM, Lizama MAP, Pavanelli GC (2007) Parasitic copepods in the nasal fossae of five fish species (Characiformes) from the upper Paraná river floodplain, Paraná, Brazil. Acta Scientiarum Biological Sciences 29(4): 429–425.

Swofford DL (2002) PAUP*. Phylogenetic Analysis Using Parsimony (*and other methods). Sunderland, Sinauer, version 4.

Tang D, Kalman JE (2008) A new genus and species of mesoparasitic ergasilid (Copepoda: Cyclopoida) from brackish water pufferfishes collected in northern Australian waters. Systematic Parasitology 69: 89–99. https://doi.org/10.1007/s11230-007-9109-3

Thatcher VE (1984) The parasitic crustaceans of fishes from the Brazilian Amazon, 7, *Acusicola tucunarense*, n. sp. (Copepoda: Cyclopidea) from *Cichla ocellaris*, with an emendation of the genus and the proposal of Acusicolinae subfam. nov. Revista Brasileira de Biologia 44(2): 181–189.

Tidd WM (1942) A new species of *Ergasilus* (parasitic Copepoda) from a characinid fish of Brazil. Transactions of the American Microscopical Society 61(1): 62–65. https://doi.org/10.2307/3222682

Walter TC, Boxshall G (2017) World of Copepods database. *Ergasilus* von Nordmann, 1832. http://www.marinespecies.org/aphia.php?p=taxdetails&id=128641 [Accessed: 25/05/2017]

Wiley EO (1981) Phylogenetics. The Theory and Practice of Phylogenetic Systematics. John Wiley, New York.

Author Contributions: TMM was responsible for writing of the manuscript, revision of the morphological characters and phylogenetic analysis. WAB was responsible for drawing and revision of morphology, analysis and revising of the manuscript.

Competing Interests: The authors have declared that no competing interests exist.

6

Metrics derived from fish assemblages as indicators of environmental degradation in Cerrado streams

Milton P. Ávila[1], Rafaela Neves Carvalho[1], Lilian Casatti[2], Juliana Simião-Ferreira[1], Lorrane Faria de Morais[1], Fabrício B. Teresa[1]

[1]*Campus de Ciências Exatas e Tecnológicas, Universidade Estadual de Goiás. Rodovia BR 153, 3105, Fazenda Barreiro do Meio, Caixa Postal 459, 75132-903 Anápolis, GO, Brazil.*
[2]*Instituto de Biociências, Letras e Ciências Exatas, Universidade Estadual Paulista. Rua Cristovão Colombo 2265, 15054-000 São José do Rio Preto, SP, Brazil.*
Corresponding author: Fabrício B. Teresa (fabricioteresa@yahoo.com.br)

ABSTRACT. The development of effective monitoring tools depends on finding sensitive metrics that are capable of detecting the most important environmental impacts at a given region. We assessed if metrics derived from stream fish assemblages reflect physical habitat degradation and changes in land cover. We sampled the ichthyofauna and environmental characteristics of 16 stream sites of first and second order in the Upper Tocantins River basin. The streams were classified according to their environmental characteristics into reference (n = 5), intermediate (n = 4), and impacted (n = 7). A total of 4,079 individuals in five orders, 12 families, and 30 species were collected. Of the 20 metrics tested, eight were non-collinear and were tested for their performance in discriminating among groups of streams. Three metrics were sensitive to the gradient of degradation: Berger-Parker dominance index, percentage of characiform fish, and percentage of rheophilic individuals. Some commonly used metrics did not reflect the disturbances and many others were redundant with those that did. These results indicate that the metrics derived from fish assemblages may be informative for identifying the conservation status of streams, with the potential to be used in biomonitoring.

KEY WORDS. Bioindicators, Brazilian savannah, fish fauna, multimetric index.

INTRODUCTION

Anthropogenic impacts, such as pollution, riparian zone logging, and landscape (including network) fragmentation (Barletta et al. 2010, Naiman and Dudgeon 2011) affect aquatic ecosystems almost everywhere. These impacts have negatively influenced the chemical, physical, and biological integrity of ecosystems (Barletta et al. 2010, Casatti et al. 2009). In this scenario, the development of environmental monitoring tools is fundamental to provide information for the management and protection of natural resources (Yoccoz et al. 2001).

Biological communities can provide valuable information on ecosystem functioning and environmental quality (Roset et al. 2007). Several descriptors (metrics) can be derived from biological communities, including taxonomic structure (e.g. diversity, evenness or dominance), taxonomic composition (e.g. percentage of individuals of a given taxonomic group), and functional composition (e.g. percentage of individuals of a given ecological guild). For example, it has been suggested that curves of relative abundances and biomass allow the assessment of the level of disturbance in an environment (Clarke and Warwick 2001). These curves are based on the assumption that small-sized species dominate in more disturbed environments, whereas large species dominate in minimally disturbed environments (Yemane et al. 2005). The number of species, the presence of rare species, and the distribution of the abundance of individuals among species (evenness) in an assemblage are also potentially informative metrics for the differentiation of assemblages that have been subjected to different levels of disturbance (Barman and Gupta 2015).

Environmental quality has often been measured and monitored using metrics that have not been thoroughtly evaluated, particularly with respect to how they respond to the disturbance gradient (Roset et al. 2007). And since metric responses can vary regionally and also accorging to the type of impact (Bonada et al. 2006, Teresa and Casatti 2017), it is necessary that they are evaluated before they can be used. Thus, studies that test the response of biological metrics to the gradients of anthropogenic disturbance provide an important step to find reliable regional indicators.

Fish are excellent indicators of the conditions of aquatic environments (Karr 1981, Ruaro and Gubiani 2013) and many different indicators based on fish have been developed worldwide (Schiemer 2000, Maire et al. 2013). In Brazil, most efforts to validate metrics extracted from fish assemblages have been concentrated in regions with high densities of humam populations, such as those in the south and southeast Brazil (e.g., Araújo et al. 2003, Casatti et al. 2009 and Costa and Schulz 2010). This asymmetry highlights the need to validate these metrics for other regions, such as the Cerrado (e.g. Carvalho et al. 2017), where the expansion of the agricultural frontier and degradation of aquatic environments have been of major concern for more than a decade (Wantzen et al. 2006, Hunke et al. 2015).

We studied the response of headwater fish assemblages according to contrasting conditions of physical habitat degradation associated with the deforestation of the riparian zone. We evaluated whether metrics based on fish assemblages can be used to differentiate streams according to their conservation status.

MATERIAL AND METHODS

Study area and general procedures

We described fish assemblage of headwater sites in relation to potential indicators (metrics) and tested whether these data would allow us to distinguish among streams with different levels of environmental degradation. Stream sites were classified according to their environmental characteristics, representing a gradient of land use and habitat degradation.

The studied streams are part of the Upper Tocantins basin, Santa Tereza sub-basin (northern Goiás state) (Fig. 1). The Upper Tocantins basin covers an area of 123,800 km^2, encompassing parts of the states of Tocantins, Goiás and Distrito Federal. It drains part of the Cerrado biome and the climate is classified as Cwa, according to the Köppen classification system. Annual rainfall is 1,100-1,700 mm and is concentrated between December and February (Ferreira and Tokarski 2007). In the studied region, the native vegetation comprises 85% of the land cover (Borges et al. 2016). Impacts resulting from removal of the riparian forest contribute to a loss of physical integrity in aquatic habitats, and are among the main antropogenic impacts in local streams (Borges et al. 2015). Minimally disturbed streams have streambeds with stable substrate (composed of rocks and pebbles) and forested riparian zones and catchments.

Sampling

We sampled fish license: ICMBio #34144-1 from 16 stream sites (sampling reach length of 80 meters) of 1st and 2nd order (sensu Sthraler 1957) (Fig. 1) during the dry period (September 2013), when sampling effectiveness is high due to low water volume. The stream sites were chosen to represent the gradient of environmental conditions in the region and the length of our sampling site was sufficient to estimate fish assemblage metrics based on abundance (Terra et al. 2013a). Sampling effort consisted of one passage of electrofishing (4A, 220V) performed by three people for 1 h in the upstream direction. Fishes were fixed in formaldehyde solution (10%) and transferred to 70 % ethanol after 72 hours. Voucher materials are stored in the Coleção Zoológica da Universidade Estadual de Goiás (UEG 291 to 300).

Streams were characterized based on their environmental variables: 1) depth, 2) wetted width, 3) water velocity, 4) percentage of stable (hard) substrate (rocks and pebbles), 5) riparian forest width, 6) proportion of trees and grass in the riparian zone (30 m from the stream channel in each margin), 7) percentage of native vegetation in a buffer of 100- and 500-m upstream of

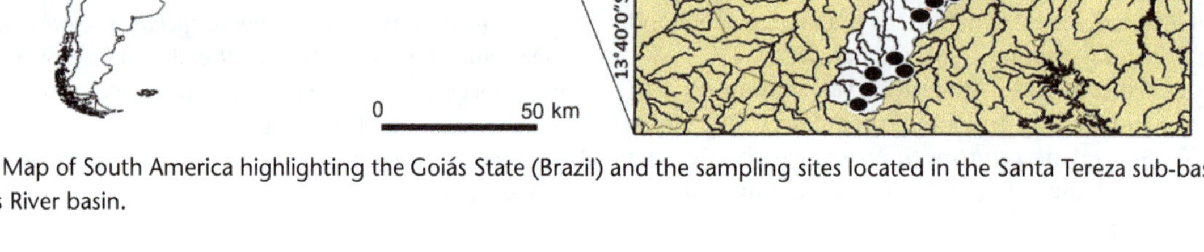

Figure 1. Map of South America highlighting the Goiás State (Brazil) and the sampling sites located in the Santa Tereza sub-basin, Upper Tocantins River basin.

the sampling sites, and 8) percentage of native vegetation in the entire catchment. These variables describe the natural environmental gradient (variables 1 to 3) and the level of stream degradation (variables 4 to 8) (Casatti et al. 2006, Leal et al. 2016, Teresa and Casatti 2017).

The physical habitat variables were assessed using a protocol adapted from Casatti et al. (2006). Each sampled site was divided into nine equidistant transversal transects. Depth and wetted width were measured using a meter ruler and measuring tape, respectively. Depth was measured at five equidistant points from one margin to the other in each transect. Water velocity was measured at ~ 0.6 times the total depth at the center of each transect with a mechanical flowmeter model 2030 (General Oceanics). The relative proportion of each type of substrate was visually estimated at each transect according to Cummins (1962). The proportion of pebbles, cobbles and boulders were grouped into a single category named stable substrate. The values of instream and riparian variables for the whole site were obtained by calculating mean values across transects. The percentage of native vegetation in the catchment and in a buffer of 100 and 500-meters upstream of the sampling site were obtained using high resolution images of the RapidEye satellite (spatial resolution of 5 m), made available by the Brazilian Ministry of the Environment, as well as SRTM (Shuttle Radar Topography Mission) images.

Assemblage-based metrics

We tested 20 metrics to describe fish assemblages. These metrics describe patterns of diversity and dominance (Shannon diversity index, Berger-Parker dominance index, and Pielou's evenness index), taxonomic composition (percentage of individuals in each family or order), size distribution (W statistic for ABC curve summarizing abundance and biomass accumulated among species), trophic ecology, and habitat use (percentage of individuals in each guild).

In order to identify trophic guilds, fish diets were evaluated. Ten individuals of each species had their stomachs removed and analyzed under a stereomicroscope. Food items were separated and identified in broad categories: detritus, terrestrial invertebrates, aquatic invertebrates, plant material (seeds and leaf fragments), and filamentous algae. The alimentary index (IAi) was calculated for each item following Kawakami and Vazzoler (1980). Species were grouped into trophic guilds according to the most frequent and dominant item (high values of IAi). The classification of species according to habitat use was based in the literature (e.g. Teresa et al. 2016, Teresa and Casatti 2017) and personal observations. Species were classified as being either water column dwellers, benthic or nektobenthic, and in relation to rheophily.

The relationship between species abundance and biomass within an assemblage was summarized by ABC curves (Warwick 1986). The abundance and biomass accumulated among the species of each assemblage were ranked in decreasing order of importance (Warwick 1986), using the software package 'PRIMER' (Clarke and Gorley 2001). From the ABC curve, it is possible to calculate the W-statistic, which indicates the relationship between the curves. When the biomass curve is positioned above the abundance curve (positive W-statistic), this indicates that large species are dominant in the assemblage, which is typical of better conserved environments (Clarke and Warwick 2001). When the abundance curve is positioned above the biomass curve (negative W-statistic), this indicates that the dominant species has a small contribution to the total biomass, which is typical of disturbed environments (Clarke and Warwick 2001).

Data analysis

Our data anaysis can be summarized in three steps. In step one, we removed colinear variables from environmental and fish-based metric data, performing Pearson pairwise correlations in each matrix and removing variables that were strongly correlated (R > |0.75|). When two or more metrics were redundant with one another, we retained the metric that could more be objectively identified or recorded by a technician. In step two, streams were classified as reference (minimally disturbed), intermediate, or impacted, according to their environmental variables. Only variables that describe the gradient of environmental degradation in agroecosystems (variables 4-8 in sampling section), and which are useful to explain the variation in the structure and composition of fish assemblages were considered for stream classification (Casatti et al. 2015, Leal et al. 2016, Teresa and Casatti 2017). We generated a matrix of environmental dissimilarity among stream sites (Euclidian distance) based on a standardized environmental matrix (z-score) and performed a cluster analysis using the UPGMA algorithm. We also compared stream groups in relation to their hydrological characteristics (depth, wet width, and water velocity) in order to ensure that stream groups were similar in relation to natural environmental gradients using a one-way ANOVA. Finally, in step three, the performance of the metrics in distinguishing stream groups (obtained in the previous step) was tested with a one-way ANOVA and, in case of significant result, we obtained the Least Significant Difference (LSD) using the software STATISTICA 7.0.

RESULTS

Two variables (width of riparian zone and the percentage of trees in the riparian zone) were removed prior to the clustering analysis because they were collinear with the percentage of grass in the riparian zone. Cluster analysis based on environmental variables associated with dirturbance resulted in four major groups (Fig. 2). One group, named reference hereafter, was composed of five stream sites that presented the highest proportion of consolidated substrate and with forested riparian zones with the highest amount of native vegetation in the landscape (Table 1). A second group, termed intermediate, was more similar to the reference group and was formed by four stream sites, with intermediate

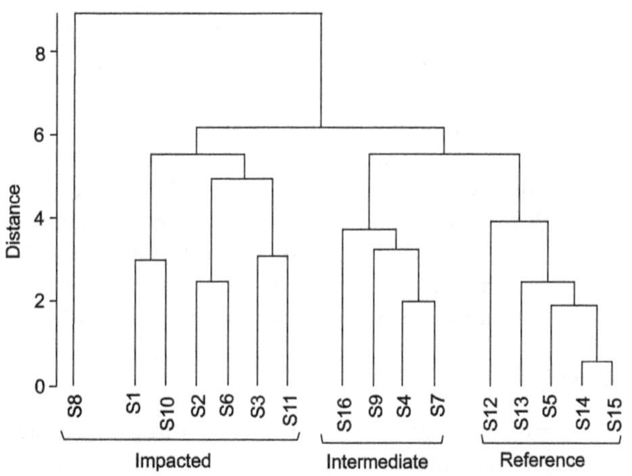

Figure 2. Stream sites grouped by their environmental characteristics by using UPGMA (Unweighted Pair Group Method with Arithmetic Mean). Height at y-axis refers to Euclidean distance.

Table 1. Environmental variables (mean ± standard deviation) from reference, intermediate and impacted stream groups.

Variables	Reference	Intermediate	Impacted
Depth (cm)	13.73 ± 1.98	13.82 ± 4.39	10.62 ± 2.41
Wet width (m)	3.16 ± 0.79	2.56 ± 0.78	2.18 ± 0.64
Water velocity (m/s)	0.22 ± 0.11	0.13 ± 0.08	0.22 ± 0.14
Hard substrate (%)	65.78 ± 10.10	41.11 ± 8.73	30.09 ± 20.03
Grass in the riparian zone (%)	18.00 ± 19.32	23.75 ± 20.16	72.92 ± 15.69
Native vegetation (% in 100 m buffer upstream)	94.23 ± 12.90	89.49 ± 11.20	45.38 ± 29.37
Native vegetation (% in 500 m buffer upstream)	95.45 ± 4.24	72.27 ± 14.82	81.85 ± 12.32
Native vegetation (% in the whole catchment)	88.25 ± 15.48	63.92 ± 8.88	57.20 ± 25.96

values for the majority of variables (Table 1). The third group was formed by six impacted stream sites that had unconsolidated substrate, grass covering most of the riparian zones, and less forested catchments (Table 1). The fourth group was formed by a single stream site (S8). Given the characteristics of this stream, which fit with the typicaly highly degraded streams of agroecosystems, we consider this stream as belonging to the third group (Fig. 2, Table 1). There were no differences among groups of streams with respect to the hydrological variables depth ($F(2,13) = 2.35$, $p = 0.13$), wet width ($F(2,13) = 2.73$, $p = 0.13$), and water velocity ($F(2,13) = 0.76$, $p = 0.48$).

A total of 4,079 individuals in five orders, 12 families, and 30 species were sampled. The species accumulation curves reached the asymptote, indicating that the species sampling was representative (see Suppl. material 1: Table S1 and Fig. S1). The order Characiformes was dominant, representing 81 % of all individuals, followed by Siluriformes, which represented 17 % of the total abundance. Richness varied from 7 to 21 and the most abundant species was *Knodus* cf. *chapadae* (46% of the total abundance).

Table 2. Performance of the metrics in distinguishing streams according to their conservation status, with the corresponding F statistic and p-value. For each sensitive metric, the redundant metrics (R > |0.75|) are shown. (+) positive correlation with the sensitive metric; (–) negative correlation with the sensitive metric.

Metrics	Redundant metrics	F(2,13)	p-value
Berger-Parker dominance index	W-statistic (–) Shannon diversity (–) Pielou evenness (–) % omnivorous individuals (+)	10.78	0.001
% individuals of Characiformes	% individuals of Siluriformes (–) % individuals of Characidae (+) % individuals of Loricariidae (–) % benthic individuals (–)	3.82	0.05
% rheophilic individuals	% individuals of Characidae (–) % individuals of Loricariidae (+) % water column individuals (–) % benthic individuals (+) % nektobenthic individuals (+) % aquatic invertivorous individuals (+) % omnivorous individuals (–) % detritivorous/periphytivorous individuals (+)	7.24	0.001
Abundance		1.15	NS
% individuals of Gymnotiformes		2.52	NS
% individuals of Heptapteridae		0.02	NS
% individuals of Characiformes + Siluriformes		1.17	NS
% terrestrial invertivorous		0.36	NS

Of the 20 metrics, only eight non-collinear metrics were retained for the assessment of monitoring potential (Table 2). Three metrics met the response criteria in distinguishing the stream groups: dominance of Berger-Parker ($F(2,13) = 10.8$, $p < 0.01$), percentage of individuals of Characiformes ($F(2,13) = 3.82$, $p = 0.05$), and percentage of rheophilic individuals ($F(2,13) = 7.24$, $p < 0.01$) (Table 2, Fig. 3). The pairwise comparison among the three groups of streams showed that the first two metrics distinguished impacted from intermediate and reference streams (Fisher's LSD, $p < 0.04$) (Fig. 3). However, these metrics were unable to distinguish reference and intermediate streams (Fisher's LSD, $p > 0.53$). The percentage of rheophilic individuals was higher in reference than in intermediate and impacted streams (Fisher's LSD, $p < 0.03$), and no differences occurred between the latter two groups (Fisher's LSD, $p = 0.34$) (Fig. 3). The three sensitive metrics (dominance of Berger-Parker, percentage of individuals of Characiformes, and percentage of rheophilic individuals) were redundant with 13 other metrics (Table 2).

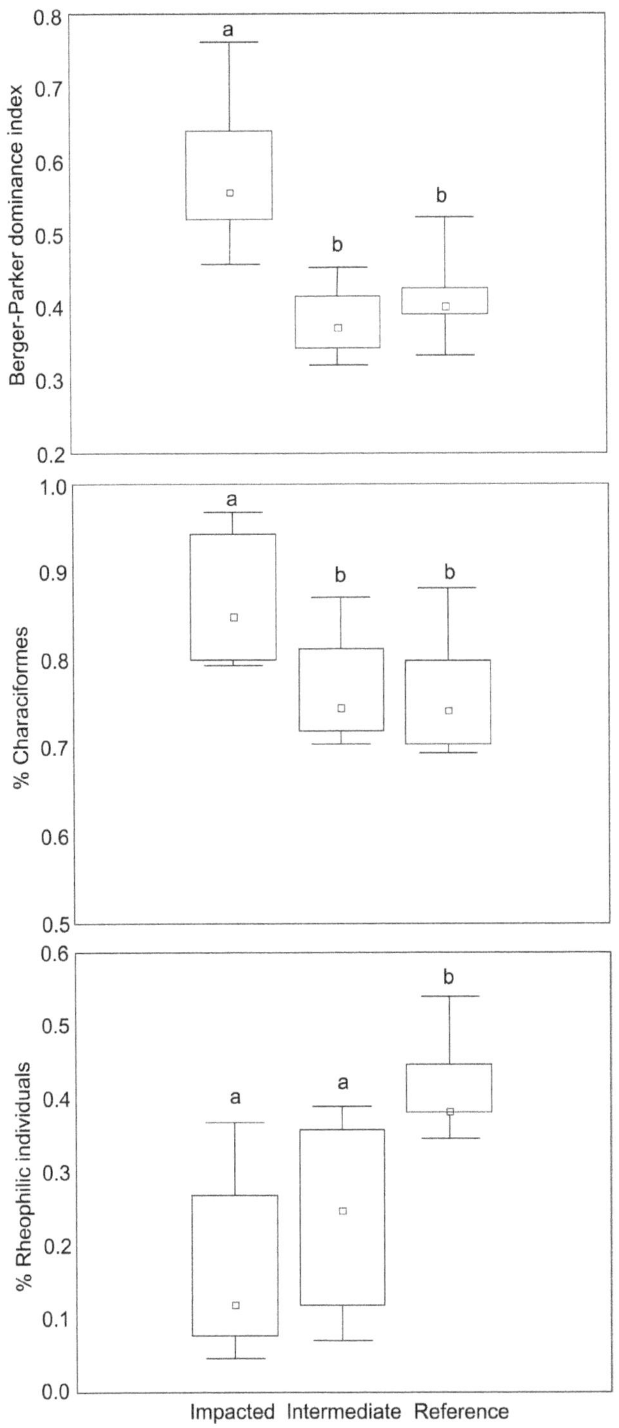

Figure 3. Box-and-Whisker plots of the three sensitive metrics. Rectangles represent the 1st and 3rd quartiles, small squares are medians, bars are maximum and minimum values. Different letters indicate statistically significant differences.

DISCUSSION

We found that at least three metrics were capable of distinguishing streams in contrasting conditions. Accordingly, they can be useful for identifying the conservation status of streams. Thus, our study assists with the fundamental step of incorporating bioindicators in environmental assessment frameworks, which is the identification of sensitive metrics (Bonada et al. 2006, Roset et al. 2007).

The stream groups identified by us represent the typical physionomies present in agroecosystems (Terra et al. 2013b, Teresa and Casatti 2017). The reference streams have well-preserved riparian forests, with a dominance of woody vegetation. These streams also have good physical integrity, with high environmental complexity, and hard substrate that maintains high species richness (Dala-Corte et al. 2016). On the other hand, impacted streams have more unconsolidated substrate, which is evidence of the siltation process that is currently underway in the region (Borges et al. 2015). These impacts, associated with riparian forest removal and loss of native vegetation surrounding streams, cause habitat homogeneity, favoring tolerant fish species that become abundant (Lorion and Kennedy 2009). The most dominant species was *Knodus* cf. *chapadae*, a species with high phenotypic plasticity (Teresa et al. 2016). Like its congener, *K. moenkahusii*, this species seems to be tolerant of environmental degradation, with a great ability to exploit niche opportunities in degraded streams (Ceneviva-Bastos et al. 2007).

Despite the high proportion of minimally disturbed landscapes (Borges et al. 2016), the region has been experiencing livestock and crop expansion, especially in lowlands, which threatens the conservation of rivers and streams. As exemplified by *Knodus* cf. *chapadae*, in impacted conditions, some opportunistic species may be favoured, while many specialized species tend to experience a reduction in their populations (Casatti et al. 2009). Our results suggest that other species of Characiformes may be dominant in impacted streams. In fact, many species (particulary Characidae) of this order are active water column swimmers (Souza et al. 2014). Their ability to exploit the entire water column makes them greatly flexible regarding feeding items (Moraes et al. 2013). Consequently, they are not dependent on physical structures that are lost in degraded streams, such as fallen leaves, logs or stable substrate (Bozzetti and Schulz 2004).

In streams with steep slopes, like those studied here, runs and riffles are prevalent along the longitudinal gradient. Thus, an abundance of rheophilic species is expected. Most of these species are dependent upon hard substrates, where they find suitable microhabitat and food resources (Teresa et al. 2016). However, with the increased runoff of terrestrial sediments, the stream bed becomes silted and less suitable for rheophilic species. In this context, the metric "percentage of rheophilic species" was informative enough to discriminate intermediate conditions from the reference. The importance of rheophilic species as bioindicators has been documented in other studies (Terra et al. 2013b, Mostafavi et al. 2015). Armored catfishes, which have benthic and periphytivorous habits (Teresa and Casatti 2012), compose this functional group. Indeed, in our study the percentage of rheophilic species was correlated with the percentage of Loricariidae individuals, of benthic individuals, and detritivorous/periphytivorous individuals.

Our sensitive metrics are complementary, as they allow to distinguish different stream conditions along the gradient of perturbation. Two metrics (dominance and percentage of Characiformes) are capable of distinguishing more impacted streams. However, at intermediate levels of disturbance these metrics should be uninformative. In these cases, the percentage of rheophilic individuals would allow the discrimination of reference from intermediate conditions. Therefore, these three metrics can be used jointly to improve the discriminatory power of assessment indices when the goal is to classify regional streams according to their environmental condition. In fact, the capacity to discriminate different conditions along a gradient of environmental perturbation is a desirable feature of bioindicators (Bonada et al. 2006).

Recently a new fish-based multimetric index has been developed for streams in the Brazilian savannah (Carvalho et al. 2017). However, we found that the response of at least one sensitive metric (i.e. percentage of characiformes fish) contradicted the findings of Carvalho et al. (2017). Other metrics (e.g. percentage of Trichomycteridae and percentage of *Poecilia reticulata* individuals) are not applicable in our study system, since these taxa were not recorded. Therefore, the validation of context-specific metrics are fundamental, given the fact that indicators cannot be reliably used across different regions (Teresa and Casatti 2017). In summary, our results suggest that physical habitat degradation and changes in land cover in the riparian zone and in the catchment can be detected by fish-based metrics. We expect that our results, coupled with recent iniciatives (Carvalho et al. 2017), represent part of the initial efforts to develop reliable bioindicators for the streams in the Brazilian savannah.

ACKNOWLEDGMENTS

We are grateful to the Laboratório de Ciências Biológicas (Universidade Estadual de Goiás/Porangatu) and the Laboratório de Biogeografia e Ecologia Aquática (Universidade Estadual de Goiás/Anápolis) for data collection; to Fernando R. Carvalho for fish identification; to Fundação de Apoio à Pesquisa do Estado de Goiás (201210267000703 and AUXPE 2036/2013) and Conselho Nacional de Desenvolvimento Científico e Tecnológico (482185/2012-0) for research funding. MPA and LFM received scholarships from FAPEG; FBT and LC receive research grants from CNPq (302158/2015-4, 301755/2013-2 respectively).

LITERATURE CITED

Araújo FG, Fichberg I, Pinto BCT, Peixoto MG (2003) A preliminary index of biotic integrity for monitoring the condition of the Rio Paraíba do Sul, southeast Brazil. Environmental Management 32: 516–526. https://doi.org/10.1007/s00267-003-3003-9

Barletta M, Jaureguizar AJ, Baigun C, Fontoura NF, Agostinho AA, Almeida-Val VMF, Val AL, Torres RA, Jimenes-Segura LF, Giarrizzo T, Fabré NN, Batista VS, Lasso C, Taphorn DC, Costa MF, Chaves PT, Vieira JP, Corrêa MFM (2010) Fish and aquatic habitat conservation in South America: a continental overview with emphasis on Neotropical systems. Journal of Fish Biology 76: 2118–2176. https://doi.org/10.1111/j.1095-8649.2010.02684.x

Barman B, Gupta S (2015) Aquatic insects as bio-indicator of water quality: a study on Bakuamari stream, Chakras hila Wildlife Sanctuary, Assam, North East India. Journal of Entomology and Zoology Studies 3: 178–186.

Bonada N, Prat N, Resh VH, Statzner B (2006). Developments in aquatic insect biomonitoring: a comparative analysis of recent approaches. Annual Review of Entomology 51: 495–523. https://doi.org/10.1146/annurev.ento.51.110104.151124

Borges PP, Teresa FB, Martins PTA, Nabout JC (2015) Relative influence of direct and indirect environmental effects on sestonic chlorophyll-a concentration in Cerrado streams. Acta Limnologica Brasiensia 27: 301–310. https://doi.org/10.1590/S2179-975X1815

Borges PP, Martins PTA, Ferreira AA (2016) Uso e ocupação do solo por meio de uma série histórica na bacia do rio Santa Teresa em Goiás. Revista Brasileira de Geografia Física 9: 296–304. https://doi.org/10.5935/1984-2295.20160020

Bozzetti M, Schulz UH (2004) An index of biotic integrity based on fish assemblages for subtropical streams in southern Brazil. Hydrobiologia 529: 133–144. https://doi.org/10.1007/s10750-004-5738-6

Carvalho DR, Leal CG, Junqueira NT, Castro MA, Fagundes DC, Alves CBM, Hughes RM, Pompeu PS (2017) A fish-based multimetric index for Brazilian savanna streams. Ecological Indicators 77: 386–396. https://doi.org/10.1016/j.ecolind.2017.02.032

Casatti L, Langeani F, Silva AM, Castro RMC (2006) Stream fish, water and habitat quality in a pasture dominated basin, southeastern Brazil. Brazilian Journal of Biology 66: 681–696. https://doi.org/10.1590/S1519-69842006000400012

Casatti L, Ferreira CP, Langeani F (2009) A fish-based biotic integrity index for assessment of lowland streams in southeastern Brazil. Hydrobiologia 623: 173–189. https://doi.org/10.1007/s10750-008-9656-x

Casatti L, Teresa FB, Zeni JO, Ribeiro MD, Brejão GL, Ceneviva-Bastos M (2015) More of the same: high functional redundancy in stream fish assemblages from tropical agroecosystems. Environmental Management 55: 1300–1314. https://doi.org/10.1007/s00267-015-0461-9

Ceneviva-Bastos M, Casatti L, Rossa-Feres DC (2007) Meso and microhabitat analysis and feeding habits of small nektonic characins (Teleostei: Characiformes) in Neotropical streams. Zoologia 27: 191-200. https://doi.org/10.1590/S1984-46702010000200006

Clarke KR, Gorley RN (2001) Software PRIMER. Plymouth, Primer-E Ltd, v. 5.

Clarke KR, Warwick RM (2001) Change in marine communities: an approach to statistical analysis and interpretation, 2nd ed., PRIMER-E, Plymouth, 172 p.

Costa PF, Schulz UH (2010) The fish community as an indicator of biotic integrity of the streams in the Sinos River basin, Brazil. Brazilian Journal of Biology 70: 1195–1205. https://doi.org/10.1590/S1519-69842010000600009

Cummins KH (1962) An evaluation of some techniques for the collection and analysis of benthic samples with special emphasis on lotic waters. The American Midland Naturalist 67: 477–504. https://doi.org/10.2307/2422722

Dala-Corte RB, Giam G, Olden JD, Becker FG, Guimarães TF, Melo AS (2016) Revealing the pathways by which agricultural land-use affects stream fish communities in South Brazilian grasslands. Freshwater Biology 61: 1921–1934. https://doi.org/10.1111/fwb.12825

Ferreira EAB, Tokarski DJ (2007) Bacia Hidrográfica do Alto Tocantins Retrato e Reflexões. ECODATA, WWF – Brasil, 102 pp.

Hunke P, Mueller EN, Schröder N, Zeilhofer P (2015) The Brazilian Cerrado: assessment of water and soil degradation in catchments under intensive agricultural use. Ecohydrology 8: 1154–1180. https://doi.org/10.1002/eco.1573

Karr JR (1981) Assessment of biotic integrity using fish communities. Fisheries 6: 21–27. https://doi.org/10.1577/1548-8446(1981)006<0021:AOBIUF>2.0.CO;2

Kawakami E, Vazzoler G (1980). Método gráfico e estimativa de índice alimentar aplicado no estudo de alimentação de peixes. Boletim do Instituto Oceanográfico 29: 205–207. https://doi.org/10.1590/S0373-55241980000200043

Leal CG, Pompeu PS, Gardner TA, Leitão RP, Hughes RM, Kaufmann PR, Zuanon J, Paula FR, Ferraz SFB, Thomson JR, Nally RM, Ferreira J, Barlow J (2016) Multi-scale assessment of human-induced changes to Amazonian instream habitats. Landscape Ecology 31: 1725–1745. https://doi.org/10.1007/s10980-016-0358-x

Lorion CM, Kennedy BP (2009) Riparian forest buffers mitigate the effects of deforestation on fish assemblages in tropical headwater streams. Ecological Applications 19: 468–479. https://doi.org/10.1890/08-0050.1

Maire A, Buisson L, Biau S, Canal J, Laffaille P (2013) A multi-faceted framework of diversity for prioritizing the conservation of fish assemblages. Ecological Indicators 34: 450– 459. https://doi.org/10.1016/j.ecolind.2013.06.009

Moraes M, Rezende CF, Mazzoni R (2013) The feeding ecology of stream-dwelling Characidae (Osteichthyes: Characiformes) from the upper Tocantins River, Brazil. Zoologia 30: 645–651. https://doi.org/10.1590/S1984

Mostafavi H, Schinegger R, Melcher A, Moder K, Mielach C, Schmutza S (2015) A new fish-based multi-metric assessment index for cyprinid streams in the Iranian Caspian Sea Basin. Limnologica 51: 37–52. https://doi.org/10.1016/j.limno.2014.10.006

Myers N, Mittermeier RA, Mittermeier CG, Fonseca GA, Kent J (2000) Biodiversity hotspots for conservation priorities. Nature 403: 853–858. https://doi.org/10.1038/35002501

Naiman RJ, Dudgeon D (2011) Global alteration of freshwaters: influences on human and environmental well-being. Ecological Research 26: 865–873. https://doi.org/10.1007/s11284-010-0693-3

Roset N, Grenouillet G, Goffaux D, Pont D, Kestemont PA (2007) Review of existing fish assemblage indicators and methodologies. Fisheries Management and Ecology 14: 393–405. https://doi.org/10.1111/j.1365-2400.2007.00589.x

Ruaro R, Gubiani EA (2013) A scientometric assessment of 30 years of the Index of Biotic Integrity in aquatic ecosystems: applications and main flaws. Ecological Indicators 29: 105–110. https://doi.org/10.1016/j.ecolind.2012.12.016

Schiemer F (2000) Fish as indicators for the assessment of the ecological integrity of large rivers. Hydrobiologia 422/423: 371–278. https://doi.org/10.1023/A:1017086703551

Souza MA, Fagundes DC, Leal CG, Pompeu PS (2014) Ecomorphology of Astyanax species in streams with different substrates. Zoologia 31: 42–50. https://doi.org/10.1590/S1984-46702014000100006

Sthraler AN (1957) Quantitative analysis of watershed geomorphology. Trans. American Geophysical Union 38: 913–920. https://doi.org/10.1029/TR038i006p00913

Teresa FB, Casatti L (2012) Influence of forest cover and mesohabitat types on functional and taxonomic diversity of fish communities in Neotropical lowland streams. Ecology of Freshwater Fish 21: 433–442. https://doi.org/10.1111/j.1600-0633.2012.00562.x

Teresa FB, Casatti L (2017) Trait-based metrics as bioindicators: responses of stream fish assemblages to a gradient of environmental degradation. Ecological Indicators 75: 249–258. https://doi.org/10.1016/j.ecolind.2016.12.041

Teresa FB, Souza LS, Silva DMA, Barbosa HO, Lima JD, Nabout JC (2016) Environmental constraints structuring fish assemblages in riffles: evidences from a tropical stream. Neotropical Ichthyology 14: e150185. https://doi.org/10.1590/1982-0224-20150185

Terra BF, Hughes, RM, Araújo, FG (2013a) Sampling sufficiency for fish assemblage surveys of tropical Atlantic forest streams, Southeastern Brazil. Fisheries 38: 150–158. https://doi.org/10.1080/03632415.2013.775572

Terra BF, Hughes RM, Francelino MR, Araújo FG (2013b) Assessment of biotic condition of Atlantic Rain Forest streams: a fish-based multimetric approach. Ecological Indicators 34: 136–148. https://doi.org/10.1016/j.ecolind.2013.05.001

Wantzen KM, Siqueira A, Cunha C, Sá MF (2006) Conservation scheme for forest-stream-ecosystems of the Brazilian Cerrado and similar biomes in the seasonal tropics. Aquatic Conservation 16: 713–732. https://doi.org/10.1002/aqc.807

Warwick RM (1986) A new method for detecting pollution effects on marine macrobenthic communities. Marine Biology 92: 557–562. https://doi.org/10.1007/BF00392515

Yemane D, Field JG, Leslie RW (2005) Exploring the effects of fishing on fish assemblages using Abundance Biomass Comparison (ABC) curves. ICES Journal of Marine Science 62: 374-379. https://doi.org/10.1016/j.icesjms.2005.01.009

Yoccoz NG, Nichols JD, Boulinier T (2001) Monitoring of biological diversity in space and time. Trends in Ecology & Evolution 16: 446–453. https://doi.org/10.1016/S0169-5347(01)02205-4

Author Contributions: MPA JSF and FBT designed the sampling design; MPA, RNC, LFM and FBT collected data; MPA RNC LFM and FBT analyzed the data; MPA RNC LC JSF LRM and FBT wrote the paper.

Competing Interests: The authors have declared that no competing interests exist.

7

Dactylogyrids (Platyhelminthes: Monogenoidea) from the gills of *Hassar gabiru* and *Hassar orestis* (Siluriformes: Doradidae) from the Xingu Basin, Brazil

Geusivam Barbosa Soares[1,3], **João Flor dos Santos Neto**[1,2], **Marcus Vinicius Domingues**[1,2,3]

[1]*Laboratório de Sistemática e Coevolução, Instituto de Estudos Costeiros, Universidade Federal do Pará. Alameda Leandro Ribeiro, 68600-000 Bragança, PA, Brazil.*
[2]*Programa de Pós-Graduação em Biologia Ambiental, Universidade Federal do Pará. Alameda Leandro Ribeiro, 68600-000 Bragança, PA, Brazil.*
[3]*Programa de Pós-Graduação em Biodiversidade e Conservação, Universidade Federal do Pará. Rua Coronel José Porfírio 2515, 68372-040 Altamira, PA, Brazil.*
Corresponding author: Marcus Vinicius Domingues (mvdomingues@ufpa.br)

ABSTRACT. Four species of *Cosmetocleithrum* (three new) and one new species of *Vancleaveus* are described or reported parasitizing the gills of doradid catfishes (Siluriformes) from Xingu River and related tributaries: *Cosmetocleithrum phryctophallus* **sp. nov.** and *Cosmetocleithrum bifurcum* Mendoza-Franco, Mendoza-Palmero & Scholz, 2016 from *Hassar orestis*; *Cosmetocleithrum leandroi* **sp. nov.** from *Hassar gabiru*; *Cosmetocleithrum akuanduba* **sp. nov.** and *Vancleaveus klasseni* **sp. nov.** from *Hassar orestis* and *H. gabiru*. *Cosmetocleithrum phryctophallus* **sp. nov.** differs from its congeners by possessing a male copulatory organ (MCO) with 2 ½ counterclockwise rings, and an accessory piece with an elongate torch-shaped blade. *Cosmetocleithrum leandroi* **sp. nov.** has a MCO comprising a coil of about 3 ½ rings, a sigmoid accessory piece with a cup-shaped distal portion, a single type of hooks, and anchors with poorly differentiated roots. *Cosmetocleithrum akuanduba* **sp. nov.** is characterized mainly by having a J-shaped MCO, an elongate accessory piece with sharp distal region, distal portion with a small gutter, and a heavily sclerotized vagina with short "S"-shaped vaginal canal. *Vancleaveus klasseni* **sp. nov.** differs from the other species of the genus occurring in doradids by possessing anchors with triangular to subtriangular superficial root, developed deep root, and a coiled MCO with 2 ½ counterclockwise rings. *Cosmetocleithrum bifurcum* was reported for the first time parasitizing doradids from Brazilian Amazon.

KEY WORDS. Amazon, *Cosmetocleithrum*, *Vancleaveus*.

INTRODUCTION

Freshwater catfish belonging to the Doradidae Bleeker, 1858 (Actinopterygii: Siluriformes) are endemic to the Neotropics, being reported in all basins of South America, mainly in Brazil, Peru, and Guianas (Birindelli 2014, Nelson et al. 2016). There are currently 65 species of Doradidae occurring in Brazilian inland waters, of which ~70% are reported from the Amazon Basin (Buckup et al. 2007, Birindelli 2014, Sabaj-Pérez and Hernández 2017). Within the Amazon Basin, 15 species of the following genera are reported to occur in the Xingu River: *Anadoras* Eigenmann, 1925 (1 species), *Doras* Lacepède, 1803 (1 species), *Hassar* Eigenmann & Eigenmann, 1888 (2 species), *Leptodoras* Boulenger, 1898 (2 species), *Megalodoras* Eigenmann, 1925 (1 species), *Nemadoras* Eigenmann, 1925 (1 species), *Ossancora* Sabaj-Pérez & Birindelli, 2011 (1 species), *Oxydoras* Kner, 1855 (1 species), *Platydoras* Bleeker, 1862 (2 species), *Rhinodoras* Bleeker, 1862 (1 species), *Rhynchodoras* Klausewitz & Rössel, 1961

Table 1. List of parasite species of *Vancleaveus* and *Cosmetocleithrum*, hosts species, host family, locality and references. AR = Argentina, BR = Brazil, PE = Peru.

Parasite	Host	Family	Locality	References
Vancleaveus janauacaensis Kritsky, Thatcher & Boeger, 1986	*Pterodoras granulosus*	Doradidae	BR	1
			AR	3
			PE	8, 10
	Hoplias malabaricus	Erythrynidae	BR	9
	Pseudoplatystoma reticulatum	Pimelodidae	BR	12
	Pseudoplatystoma. corrucans	Pimelodidae	BR	12
Vancleaveus cicinnus Kritsky, Thatcher & Boeger, 1986	*Phractocephalus hemiliopterus*	Pimelodidae	BR	1
	Fransciscodoras marmoratus	Doradidae	BR	4
	Pimelodus albicans	Pimelodidae	AR	3
	Pseudoplatystoma reticulatum	Pimelodidae	BR	12
	Pseudoplatystoma. corrucans	Pimelodidae	BR	12
Vancleaveus fungulus Kritsky, Thatcher & Boeger, 1986	*Pseudoplatystoma tigrinum*	Pimelodidae	BR	1
	Pseudoplatystoma fasciatus	Pimelodidae	BR	5
			PE	8
	Pseudoplatystoma reticulatum	Pimelodidae	BR	12
	Pseudoplatystoma. corrucans	Pimelodidae	BR	12
	Pseudoplatystoma. corrucans	Pimelodidae	PE	5
Vancleaveus platyrhynchi Kritsky, Thatcher & Boeger, 1986	*Hemisorubim platyrhynchos*	Pimelodidae	BR	1
			PE	8
Cosmetocleithrum bulbocirrus Kritsky, Thatcher & Boeger, 1986	*Pterodoras granulosus*	Doradidae	BR	1
	Ageneiosus ucayalensis	Auchenipteridae	BR	11
	Hoplias malabaricus	Erythrinidae	BR	9
			PE	8
Cosmetocleithrum confusus Kritsky, Thatcher & Boeger, 1986	*Oxydoras niger*	Doradidae	BR	1
			PE	2
Cosmetocleithrum gussevi Kritsky, Thatcher & Boeger, 1986	*Oxydoras niger*	Doradidae	BR	1, 6
			PE	2
Cosmetocleithrum parvum Kritsky, Thatcher & Boeger, 1986	*Oxydoras niger*	Doradidae	BR	1
Cosmetocleithrum rarum Kritsky, Thatcher & Boeger, 1986	*Oxydoras niger*	Doradidae	BR	1
Cosmetocleithrum sobrinus Kritsky, Thatcher & Boeger, 1986	*Oxydoras niger*	Doradidae	BR	1
			PE	2
Cosmetocleithrum longivaginatum Suriano & Incorvaia, 1995	*Pimelodus albicans*	Pimelodidae	AR	3
Cosmetocleithrum striatuli Abdallah, Azevedo & Luque, 2012	*Trachelyopterus striatulus*	Auchenipteridae	BR	7
	Auchenipterus nuchalis	Auchenipteridae	BR	15
	Trachelyopterus coriaceus	Auchenipteridae	BR	14
	Trachelyopterus galeatus	Auchenipteridae	BR	14
Cosmetocleithrum tortum Mendoza-Franco, Mendoza-Palmero & Scholz, 2016	*Nemadoras hemipeltis*	Doradidae	PE	13
Cosmetocleithrum bifurcum Mendoza-Franco, Mendoza-Palmero & Scholz, 2016	*Hassar orestis*	Doradidae	PE	13
Cosmetocleithrum laciniatum Yamada, Yamada, Silva & Anjos, 2017	*Trachelyopterus galeatus*	Auchenipteridae	BR	16
Paracosmetocleithrum trachydorasi Acosta, Scholz, Blasco-Costa, Alves & da Silva, 2018	*Trachydoras paraguayensis*	Doradidae	BR	17

1. Kritsky et al. (1986); 2. Iannacone and Luque (1991); 3. Suriano and Incorvaia (1995); 4. Santos and Brasil-Sato (2006); 5. Takemoto et al. (2009); 6. Silva et al. (2011); 7. Abdallah et al. (2012); 8. Mendoza-Palmero et al. (2012); 9. Graça et al. (2013); 10. Mendoza-Palmero et al. (2015); 11. Ferreira and Tavares-Dias (2016); 12. Jeronimo et al. (2016); 13. Mendoza-Franco et al. (2016); 14. Pantoja et al. (2016); 15. Tavares-Dias (2017); 16. Yamada et al. (2017); 17. Acosta et al. (2018).

(1 species) and *Trachydoras* Eigenmann, 1925 (1 species) (Eler et al. 2007, Buckup et al. 2007, Birindelli et al. 2011).

Thirty-five species of metazoan parasites have been reported to infect doradids (i.e., 2 Acanthocephala, 1 Crustacea, 12 Nematoda and 20 Platyhelminthes) (Thatcher 2006, Kohn et al. 2007, Luque et al. 2011, Cohen et al. 2013, Mendoza-Franco et al. 2016, Acosta et al. 2018). Among these, monogenoids correspond to 30% (11 spp.) of this diversity (Table 1) (Cohen et al. 2013, Mendoza-Franco et al. 2016, Acosta et al. 2018). However, in spite of this diversity, only ~9% of the doradid species have been investigated for monogenoidean parasites (i.e., *Franciscodoras marmoratus* [Lütken, 1874]; *Pterodoras granulosus* [Valenciennes, 1921]; *Oxydoras niger* [Valenciennes, 1821]; *Nemadoras hemipeltis* [Eigenmann, 1925], *Hassar orestis* [Steindachner, 1875]; *Trachydoras paraguayensis* [Eigenmann & Ward, 1907]) (Cohen et al. 2013, Mendoza-Franco et al. 2016, Acosta et al. 2018).

During a study of monogenoids infecting doradid fish from the Xingu River and related tributaries, three new species of *Cosmetocleithrum* and a new species of *Vancleaveus* (Dactylogyridae) were found parasitizing the gills of *Hassar orestis* and *H. gabiru* Birindelli, Fayal & Wosiacki,2011. Descriptions of the new species are presented herein. *Cosmetocleithrum bifurcum* Mendoza-Franco, Mendoza-Palmero & Scholz, 2016 is reported for the first time parasitizing *Hassar orestis* in Brazilian waters, and the first time parasitizing *H. gabiru.*

MATERIAL AND METHODS

Hosts were collected by trammel net from the Xingu River and related tributaries (Amazon basin: Amazonas, Xingu, Iriri, and Paru sub-basins) during 2015 (Table 2). Host scientific names follow Eschmeyer et al. (2017). Names of the basins and sub-basins nomenclature follow the Agência Nacional de Águas, Ministério do Meio Ambiente, Brazil (http://hidroweb.ana.gov.br).

Gill arches were removed and placed in vials containing heated water (~65 °C). Each vial was vigorously shaken and formalin was added to obtain a 5% solution. In the laboratory, the contents of each vial were examined under a Leica S6D dissecting microscope and helminths were removed from the gills or sediment using small probes. Some specimens were stained with Gomori's trichrome (Humason 1979, Boeger and Vianna 2006) and mounted in Damar Gum or Canada balsam to determine internal soft structures and others were mounted in Hoyer's medium or Gray & Wess medium (Humason 1979, Boeger and Vianna 2006) for study of sclerotized structures. The measurements, all in micrometers, were obtained according to the procedures of Mizelle and Klucka (1953). Dimensions of organs and other structures represent the greatest measurement in dorso-ventral view; lengths of bent structures (i.e., bars and accessory piece) represent the straight line distances between extreme ends; lengths of anchors follow those presented in Fig. 6; total lengths of the MCO were carried out using IMAGEJ (Rasband 1997–2016) on drawing tube images. The average measurement is followed by the ranges and the number (n) of specimens measured in parentheses. Illustrations were prepared with the aid of a drawing tube on a Leica DM 2500 microscope with differential interference contrast and phase contrast optics. Definitions of prevalence, mean intensity and mean abundance follow Bush et al. (1997). Type specimens and vouchers were deposited in the following collections: Helminthological Collection of the Instituto Oswaldo Cruz (CHIOC), Rio de Janeiro, RJ, Brazil; Invertebrate Collection of the Instituto de Pesquisas da Amazônia (INPA), Manaus, AM, Brazil; Invertebrate Collection of the Museu Paraense Emílio Goeldi (MPEG), Belém, PA, Brazil. Historical review of species containing relevant taxonomic contributions, such as description (descr), redescription (redes), synonymization (synon), new record (recor), citation (citat), figure (fig) are included after the valid species name.

Table 2. Host species, number of host specimens collected, localities and respective geographical coordinates.

Species	N	Locality	Coordinates
H. gabiru	2	Iriri River, Brazil	03°49'06,4"S, 52°41'25,8"W
	19	Ilha Grande – Xingu River, Brazil	03°35'50,2"S, 52°21'22,5"W
	6	Gorgulho da Rita Community – Xingu River, Brazil	03°21'15,7"S, 52°11'47,5"W
	15	Bacajá River, Brazil	03°33'47,1"S, 51°36'50,3"W
H. orestis	16	Belo Monte Community – Xingu River, Brazil	03°05'52,5"S, 51°43'18,0"W
	8	Vitória do Xingu – Xingu River, Brazil	02°47'27,1"S, 51°59'50,0"W
Total	66		

TAXONOMY

Class Monogenoidea Bychowsky, 1937
Subclass Polyonchoinea Bychowsky, 1937
Order Dactylogyridea Bychowsky, 1937
Dactylogyridae Bychowsky, 1933

Cosmetocleithrum phryctophallus sp. nov.

Figs 1–7

Type host. *Hassar orestis* (Steindachner, 1875), Doradidae

Site of infection. Gill filaments.

Type locality. Xingu River, Belo Monte Community, municipality of Vitória do Xingu, Pará (03°05'52.5"S, 51°43'18.0"W).

Prevalence. 100% of 16 hosts examined.

Mean intensity. 6.5 parasites per infected host.

Mean abundance. 6.5 parasites per host.

Other records. *Hassar orestis* (Prevalence: 100% of 8 hosts; Mean intensity: 6; Mean abundance: 6), Xingu River, municipality of Vitória do Xingu, Pará (02°47'27.1"S, 51°59'50.0"W).

Specimens deposited. Holotype 39055 a, and 9 paratypes, CHIOC 39055 b–g, INPA 770, MPEG 0135; 9 vouchers, CHIOC 39056 a–b, 39057 a–c, INPA 771–772, MPEG 0136.

Description (based on 10 type specimens, 5 mounted in Hoyer, 5 stained with Gomori's trichrome): Body fusiform, total length including haptor 356 (250–462; n = 10) long, 77 (62–95; n = 10) wide at level of germarium. Tegument smooth. (Fig. 1). Cephalic margin rounded, cephalic lobes inconspicuous; 3 or 4 bilateral pairs of head organs with rod-shaped secretion; cephalic glands unicellular, posterolateral to pharynx. Eyes, pigment

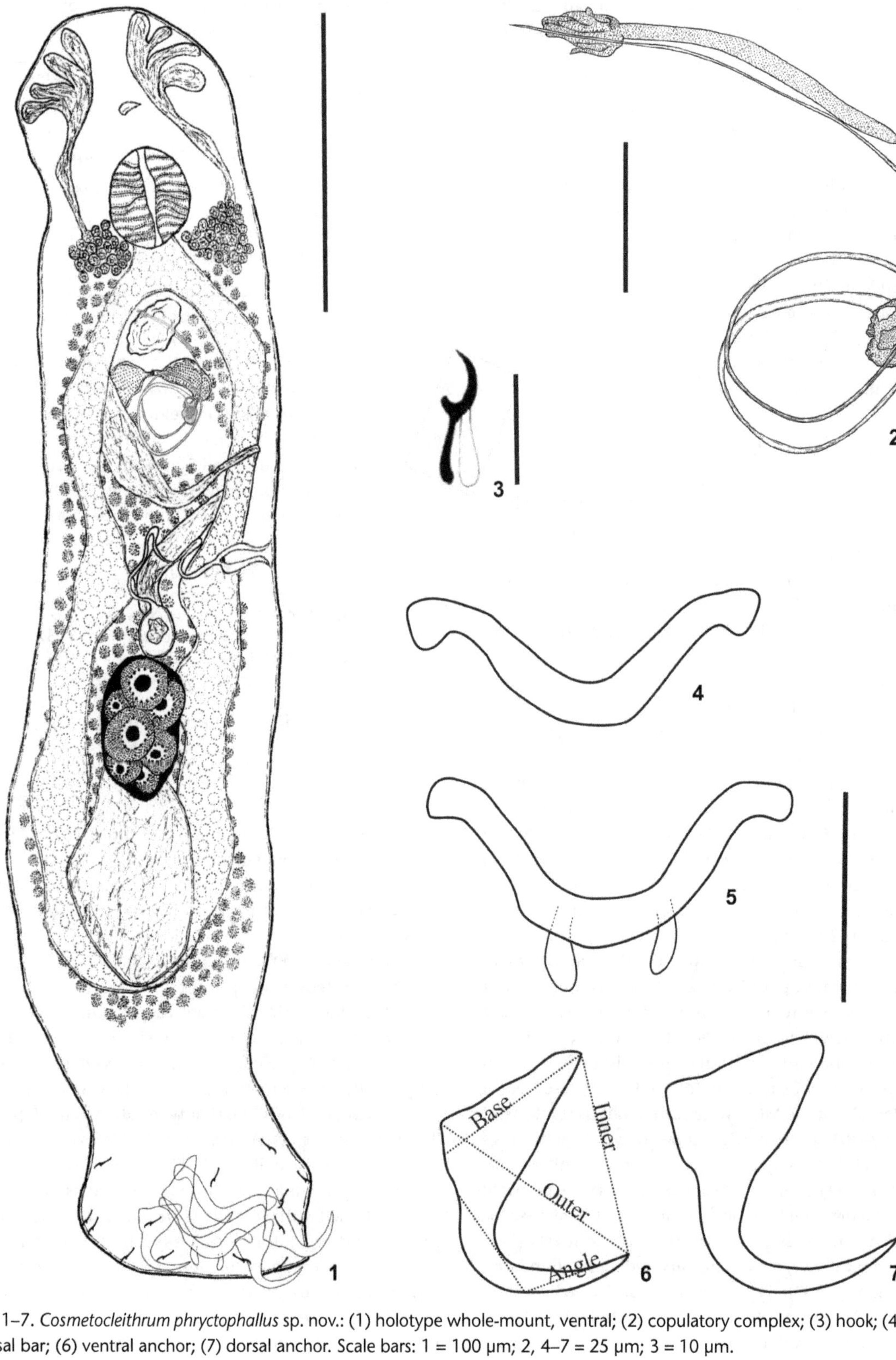

Figures 1–7. *Cosmetocleithrum phryctophallus* sp. nov.: (1) holotype whole-mount, ventral; (2) copulatory complex; (3) hook; (4) ventral bar; (5) dorsal bar; (6) ventral anchor; (7) dorsal anchor. Scale bars: 1 = 100 μm; 2, 4–7 = 25 μm; 3 = 10 μm.

granules (eye-spots), absent. Mouth subterminal; pharynx muscular, sub-spherical, 28 (24–33; n = 9) long, 23 (17–25; n = 9) wide; esophagus short, two intestinal ceca, confluent posteriorly; lacking diverticula. Haptor globose 67 (55–85; n = 9) wide. Anchors similar. Ventral anchor, superficial root, narrow triangular, well developed; deep root inconspicuous; slightly curved shaft and point, forming angle of approximately 90°; point extending just past level of tip of superficial root, outer 27 (24–31; n = 10) long, inner 28 (25–32; n = 10) long, base 18 (11–22; n = 10) (Fig. 6). Dorsal anchor, superficial root triangular, large; deep root inconspicuous; slightly curved shaft and point, forming angle of approximately 75°, point extending just past level of tip of superficial root, outer 28 (21–31; n = 8) long, inner 29 (27–31; n = 8) long, base 19 (16–31; n = 7) (Fig. 7). Ventral bar (Fig. 4) 35 (28–45; n = 10) long, 5 (3–7; n = 10 wide, broadly U-shaped with inflated ends, slightly curved in posterior direction. Dorsal bar (Fig. 5) 36 (28–45; n = 10) long, 4 (3–6; n = 10) wide, U-shaped with rounded ends, slightly curved; two submedial projections posteriorly directed. Hooks similar (Fig. 3), 13 (12–17; n = 10) long, with upright rounded thumb; slightly curved shaft, short; non-dilated shank; filamentous hooklet loop about shank length. Genital pore opening midventral, anterior to copulatory complex; muscular genital atrium. Testis post-germarial, dorsal to germarium, elongated 47 (31–69; n = 6) long, 33 (25–39; n = 6) wide. Vas deferens looping left of intestinal cecum; seminal vesicle a dilatation of vas deferens; long prostatic reservoir with medial constriction. Copulatory complex comprising MCO, accessory piece (Fig. 2). MCO, coiled sclerotized tube 236 (201–299; n = 7) long, with 2 ½ counterclockwise rings, with tapered distal region, base MCO, wide sclerotized margin. Accessory piece 36 (31–50; n = 6) long, non-articulated with MCO, comprising straight rod, distal end with elongate torch-shaped blade, guarding termination of MCO. Germarium 44 (31–69; n = 8) long, 23 (17–31; n = 8) wide, pyriform. Eggs, Mehlis' gland, ootype and uterus not observed. Vagina slightly sclerotized, vaginal pore sinistral, marginal or submarginal, vaginal vestibule cup-shaped, long vaginal canal, sigmoid, slightly sclerotized; seminal receptacle oval. Vitelline follicles dense, coextensive with ceca.

Comparative measurements. Table 3.

Etymology. The specific name (a noun) is from Greek (*phryctos* = burning torch + *phallos* = penis) and refers to the unique shape of the accessory piece.

Remarks. *Cosmetocleithrum phryctophallus* sp. nov. resembles *Cosmetocleithrum akuanduba* sp. nov. by sharing similar morphology of the haptoral structures. However, it can be easily distinguished from *C. akuanduba* sp. nov. due to the presence of a coiled MCO (J-shaped in *C. akuanduba* sp. nov.) and by having a lightly sclerotized vagina with a sigmoid vaginal canal (heavily sclerotized vagina with short, "S"-shaped vaginal canal in *C. akuanduba* sp. nov.). The new species also differs from *C. akuanduba* sp. nov. and all other congeneric species by possessing an accessory piece with an elongate torch-shaped blade.

Table 3. Comparative measurements (µm) of specimens of *Cosmetocleithrum phryctophallus* sp. nov. from the gills of *Hassar orestis* from two locations. MCO = male copulatory organ.

	Belo Monte	N	Vitória do Xingu	N
Body				
Length	356 (250–462)	10	350 (245–510)	6
Width	77(62–95)	10	70(62–90)	6
Haptor				
Wide	67,5(55–85)	9	60,5(50–85)	7
Pharynx				
Length	28(24–33)	9	28(24–33)	6
Width	23(17–25)	9	20(17–23)	7
MCO length	236(201–299)	7	230(201–290)	7
Accessory piece				
Length	36 (31–50)	6	34 (31–45)	5
Ventral Bar				
Length	35(28–45)	10	38(28–45)	8
Width	5(3–7)	10	4(3–7)	10
Dorsal Bar				
Length	46(28–45)	10	35(23–46)	5
Width	4(3–6)	10	4(3–6)	4
Ventral Anchor				
Outer	27(24–31)	10	25(24–30)	5
Inner	28(25–32)	10	27(23–35)	3
Base	18(11–22)	10	17(11–18)	8
Dorsal Anchor				
Outer	29(27–31)	8	25(21–30)	5
Inner	28(21–31)	8	27(27–30)	8
Base	19(16–21)	7	17(16–20)	7
Germarium				
Length	44(31–69)	8	40(31–67)	8
Width	23(17–31)	8	25(17–30)	8
Testis				
Length	47(31–79)	6	45(30–79)	6
Width	33(25–39)	6	30(25–39)	6
Hooks	13 (12–17)	10	15 (12–17)	10

Cosmetocleithrum leandroi sp. nov.

Figs 8–14

Type host. *Hassar gabiru* Birindelli, Fayal & Wosiacki, 2011, Doradidae

Site of infection. Gill filaments.

Type locality. Bacajá River, municipality of Altamira, Pará (03°33′47.1″S, 51°36′50.3″W).

Prevalence. 100% of 15 hosts examined.

Mean intensity. 20 parasites per infected host.

Mean abundance. 20 parasites per host.

Other records. *Hassar gabiru* (Prevalence: 100% of 19 hosts; Mean intensity: 15; Mean abundance: 15), Ilha Grande,

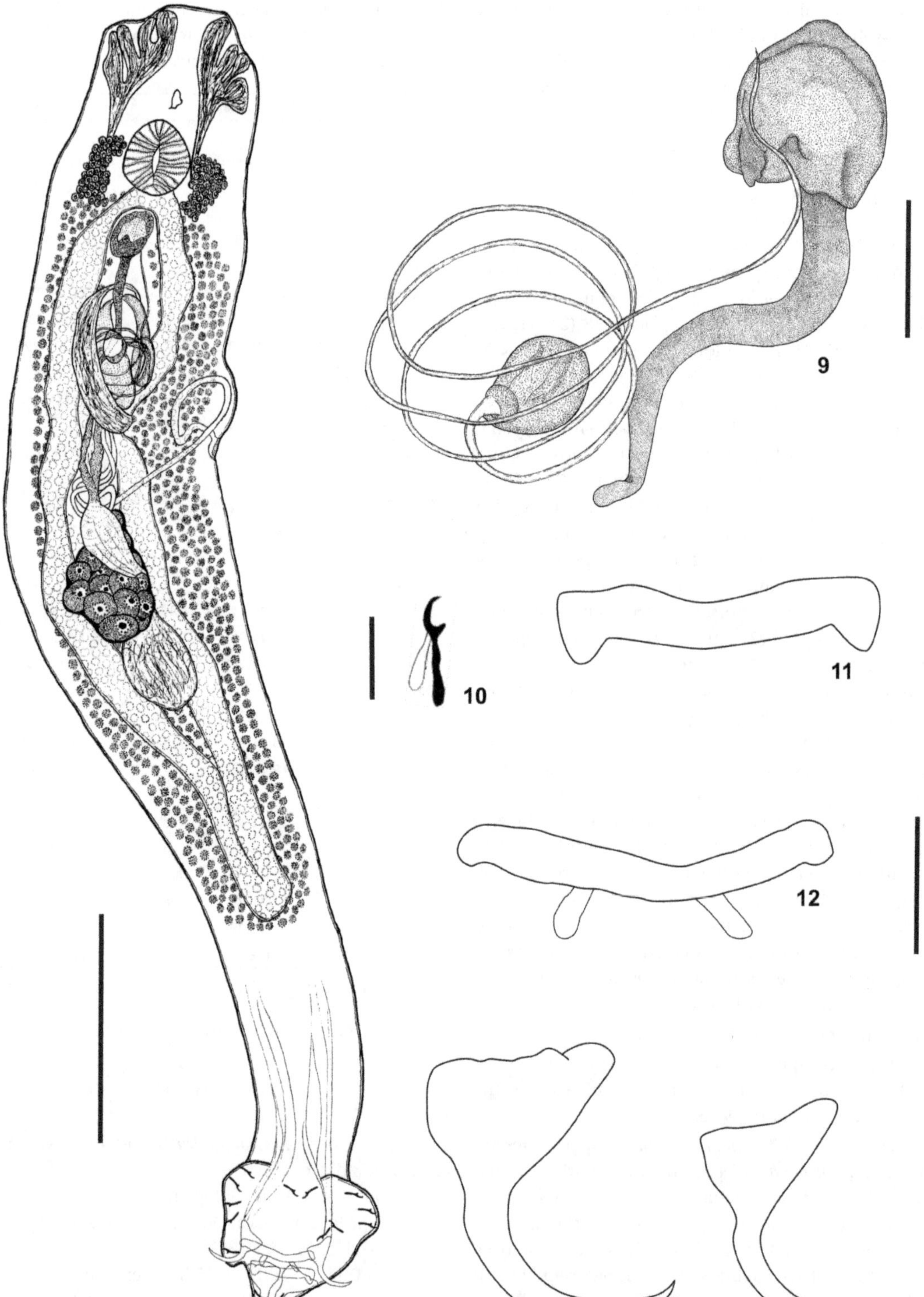

Figures 8–14. *Cosmetocleithrum leandroi* sp. nov.: (8) holotype whole-mount, ventral; (9) copulatory complex; (10) hook; (11) ventral bar; (12) dorsal bar; (13) ventral anchor; (14) dorsal anchor. Scale bars: 8 = 100 μm; 9, 11–14 = 25 μm; 10 = 10 μm.

Xingu River, municipality of Altamira, Pará (03°35′50.2″S, 52°21′22.5″W); *Hassar gabiru* (Prevalence: 100% of 2 hosts; Mean intensity: 2.5; Mean abundance: 2.5), Iriri River, municipality of Altamira, Pará (03°49′06.4″S, 52°41′25.8″W).

Specimens deposited. Holotype, CHIOC 39053 a, and 8 paratypes, CHIOC 39053 b–f, INPA 773, MPEG 0137; 8 vouchers, CHIOC 39054 a–c, INPA 774, MPEG 0138–0140.

Description (based on 10 type specimens, 6 mounted in Hoyer, 4 stained with Gomori's trichrome): Body robust, fusiform, total length including haptor 712 (575–835; n = 8) long, 132 (102–157; n = 8) wide, at level of germarium. Tegument smooth (Fig. 8). Cephalic margin broad; cephalic lobes poorly differentiated; 4 or 5 bilateral pairs of head organs with rod-shaped secretion; cephalic glands unicellular, posterolateral to pharynx. Eyes, pigment granules (eye-spots), absent. Mouth subterminal; pharynx muscular, spherical, 39 (33–46; n = 7) long, 36 (30–42; n = 7) wide; esophagus short; two intestinal ceca, confluent posteriorly, lacking diverticula. Haptor subtriangular 89 (65–110; n = 8) wide. Anchors similar, base wide, superficial and deep root poorly developed. Ventral anchor, curved shaft and point, forming angle of approximately 85°, point extending just past level of tip of superficial root, outer 40 (25–45; n = 10) long, inner 45 (28–52; n = 10) long, base 26 (16–30; n = 10) (Fig. 13). Dorsal anchor, slightly curved shaft and point, forming angle of approximately 110°, point extending just past level of tip of superficial root, outer 35 (30–39; n = 10) long, inner 38 (36–42; n = 10) long, base 20 (15–23; n = 10) (Fig. 14). Ventral bar (Fig. 11) 44 (32–57; n = 9) long, 6 (4–9; n = 8) wide, straight with knobbed ends. Dorsal bar (Fig. 12) 49 (35–65; n = 10) long, 6 (3–8; n = 10) wide, straight with inflated or rounded ends; two submedial projections posteriorly directed. Hooks similar (Fig. 10), 14 (13–15; n = 8) long, with upright rounded thumb; slightly curved shaft, short; non-dilated shank; filamentous hooklet about ½ shank length. Genital pore opening to left of body midline, anterior to copulatory complex; muscular genital atrium. Testis post-germarial, dorsal to germarium, ovoid, 43 (43–44; n = 2) long, 27 (24–30; n = 2) wide. Vas deferens looping left of intestinal cecum; seminal vesicle a dilatation of vas deferens; prostatic reservoir with medial constriction. Copulatory complex comprising MCO, accessory piece (Fig. 9). MCO, coiled sclerotized tube, 572 (550–637; n = 9) long, with 3 ½ counterclockwise rings, with tapered distal region, base of MCO sclerotized. Accessory piece 121 (110–130; n = 5) long, non-articulated with MCO, comprising sigmoid rod, with cup-shaped distal region. Germarium 85 (65–106; n = 5) long, 49 (39–62; n = 5) wide, pyriform. Eggs, Mehlis' gland, ootype and uterus not observed. Vagina heavily sclerotized, vaginal pore sinistral, marginal or submarginal, vaginal canal long, convoluted, heavily sclerotized, proximal region looped, distal region sigmoid; seminal receptacle pyriform. Vitelline follicles dense, coextensive with ceca.

Comparative measurements. Table 4.

Etymology. The specific name is in honor of Dr Leandro Melo de Sousa, Universidade Federal do Pará, Brazil, in recognition of his valuable work on the fish diversity of Xingu River, and also because the specific name of the host species, "gabiru", is the nickname of Dr de Souza (Birindelli et al. 2011). Now, Dr de Sousa has the host and parasite named after him.

Remarks. *Cosmetocleithrum leandroi* sp. nov. resembles *Cosmetocleithrum longivaginatum* Suriano & Incorvaia, 1995 by the general morphology of the bars and by the presence of a long vaginal canal (see Suriano and Incorvaia 1995). However, the new species differs from *C. longivaginatum*, mainly by possessing a MCO comprising a coil of about with 3 ½ rings (6 rings in *C. longivaginatum*) and a sigmoid accessory piece with the distal portion cup-shaped (straight, non-expanded distal accessory piece in *C. longivaginatum*). Also, *C. leandroi* has hooks with similar size, whereas *C. longivaginatum* has 2 different sizes of hooks (pairs 2 to 5 are smaller than pairs 1, 6 and 7). Finally,

Table 4. Comparative measurements (μm) of specimens of *Cosmetocleithrum leandroi* sp. nov. from the gills of *Hassar gabiru* from three locations.

	Bacajá River	N	Xingu River	N	Iriri River	N
Body						
Length	712(575–835)	8	700(535–820)	5	715(580–830)	6
Width	132(102–157)	8	125(100–150)	6	130(102–157)	5
Haptor						
Wide	89(65–110)	8	90(65–100)	5	89(65–110)	8
Pharynx						
Length	39(33–46)	7	40(33–43)	7	39(33–46)	7
Width	36(30–42)	7	30(30–45)	6	36(30–42)	7
MCO length	572(550–637)	9	570(550–640)	5	570(550–637)	9
Accessory piece						
Length	121(110–130)	5	119(100– 135)	6	125(103–120)	5
Ventral Bar						
Length	44(32–57)	9	40(32–50)	5	44(32–57)	9
Width	6(4–9)	8	5(4–8)	7	6(4–9)	8
Dorsal Bar						
Length	49(35–65)	10	50(35–60)	5	49(35–65)	8
Width	6(3–8)	10	5(3–6)	5	5(3–8)	8
Ventral Anchor						
Outer	40(25–45)	10	40(27–45)	4	45(25–45)	7
Inner	45(28–52)	10	45(30–50)	5	45(28–52)	8
Base	26(16–30)	10	27(16–27)	16	27(16–30)	7
Dorsal Anchor						
Outer	35(30–39)	10	34(22–39)	6	34(22–39)	7
Inner	38(36–42)	10	37(21–42)	7	37(21–42)	6
Base	20(15–23)	10	20(15–23)	6	20(15–23)	8
Germarium						
Length	85(65–106)	5	80(65–106)	5	80(65–106)	6
Width	49(39–62)	5	49(39–62)	4	49(39–62)	5
Testis						
Length	43(43–44)	2	42(43–44)	3	43(43–44)	5
Width	27(24–30)	2	27(24–30)	2	27(24–30)	5
Hooks	14(13–15)	8	14(13–15)	6	13(13–15)	6

these species differ in the shape of their anchors, i.e., *C. longivaginatum* has well-defined roots, whereas *C. leandroi* has poorly differentiated roots.

Cosmetocleithrum akuanduba sp. nov.

Figs 15–22

Type host. *Hassar gabiru* Birindelli, Fayal & Wosiacki, 2011, Doradidae

Site of infection. Gill filaments.

Type locality. Ilha grande, Xingu River, municipality of Altamira, Pará (03°35′50.2″S, 52°21′22.5″W).

Prevalence. 52% of 19 hosts examined.

Mean intensity. 3 parasites per infected host.

Mean abundance. 1.6 parasites per host.

Other records. *Hassar gabiru* (Prevalence: 100% of 2 hosts; Mean intensity: 2.5; Mean abundance: 2.5) Iriri River, municipality of Altamira, Pará (03°49′06.4″S, 52°41′25.8″W); *Hassar gabiru* (Prevalence: 60% of 15 hosts; Mean intensity: 3.8; Mean abundance: 2.3) Bacajá River, municipality of Altamira, Pará (03°33′47.1″S, 51°36′50.3″W); *Hassar orestis* (Prevalence: 62% of 24 hosts; Mean intensity: 2.3; Mean abundance: 1.5), Xingu River, Belo Monte Community, municipality of Vitória do Xingu, Pará (03°05′52.5″S, 51°43′18.0″W; 02°47′27.1″S, 51°59′50.0″W).

Specimens deposited. Holotype, CHIOC 39045 a, and 6 paratypes, CHIOC 39045 b-e, INPA 776, MPEG 0141; 8 vouchers, CHIOC 39046 a–b, 39047 a–c, INPA 777, MPEG 0142.

Description (based on 7 type specimens, 1 mounted in Hoyer, 6 stained with Gomori's trichrome): Body fusiform, total length including haptor 451 (360–517; n = 5) long, 110 (92–130; n = 5) wide at level of germarium. Tegument smooth (Fig. 15). Cephalic margin broad; cephalic lobes poorly differentiated; 4 bilateral pairs of head organs with rod-shaped secretion; cephalic glands unicellular, posterolateral to pharynx. Eyes, pigment granules (eye-spots), absent. Mouth subterminal; pharynx muscular, spherical, 32 (30–35; n = 5) long, 28 (25–30; n = 5) wide; esophagus short; two intestinal ceca, confluent posteriorly, lacking diverticula. Haptor globose 90 (75–117; n = 5) wide. Anchors similar. Ventral anchor, superficial root narrow subtriangular, well developed; deep root inconspicuous; slightly curved shaft and point, forming angle of approximately 75°, point extending just past level of tip of superficial root, outer 27(25–29; n = 7) long, inner 30 (25–41; n = 7) long, base 19 (17–20; n = 7) (Fig. 21). Dorsal anchor, superficial root subtriangular; deep root inconspicuous; slightly curved shaft and point forming angle of approximately 70°, point extending just past level of tip of superficial root, outer 25 (22–29; n = 5) long, inner 30 (29–32; n = 5) long, base 19 (17–20; n = 5) (Fig. 22). Ventral bar, variable (Figs 18–19) 41 (30–57; n = 7) long, 6 (4–10; n = 7) wide, broadly U-shaped or straight with inflated or rounded ends. Dorsal bar (Fig. 20) 41 (31–56; n = 7) long, 5 (4–6; n = 7) wide, V-shaped with rounded ends, two submedial projections posteriorly directed. Hooks similar (Fig. 17), 13 (13–14; n = 7) long, with upright truncate thumb; slightly curved shaft, short; non-dilated shank; filamentous hook about shank length. Genital pore opening midventral, anterior to copulatory complex; muscular genital atrium. Testis post-germarial (observed on paratypes), dorsal to germarium, ovoid, 56 (51–64; n = 3) long, 40 (32–47; n = 3) wide. Vas deferens looping left of intestinal cecum; seminal vesicle a dilatation of vas deferens; long prostatic reservoir with medial constriction. Copulatory complex comprising MCO, accessory piece (Fig. 16). MCO with tubular coiled shaft of less than one counterclockwise ring, frequently appearing J shaped, 68 (54–76; n = 4) long, with tapered distal region; base sclerotized, bulbous. Accessory piece 23 (18–30; n = 5) long, non-articulated with MCO, comprising straight rod, with tapered distal region,

Table 5. Comparative measurements (μm) of specimens of *Cosmetocleithrum akuanduba* sp. nov. from the gills of *Hassar gabiru* from three locations.

	Xingu River	N	Iriri River	N	Bacajá River	N
Body						
Length	451(360–517)	5	–	–	421(290–490)	4
Width	110(92–130)	5	–	–	102(72–150)	4
Haptor						
Wide	90(75–117)	5	–	–	87(67-105)	4
Pharynx						
Length	32(30–35)	5	–	–	28(20–35)	4
Width	28(25–30)	5	–	–	26(17–36)	4
MCO length	68(54–76)	4	78(74–86)	5	94(86–98)	5
Accessory piece						
Length	23(18–30)	5	27(20–37)	4	29(22–34)	4
Ventral Bar						
Length	41(30–57)	7	47(45–50)	5	35(32–38)	3
Width	6(4–10)	7	5(4–6)	5	3(3–4)	2
Dorsal Bar						
Length	41(31–56)	7	44(40–47)	4	36(35–38)	2
Width	5(4–6)	7	4(4–5)	4	3(3–4)	3
Ventral Anchor						
Outer	27(25–29)	7	29(28–30)	4	24(23–25)	2
Inner	30(25–41)	7	27(27–29)	4	28(27–29)	2
Base	19(17–20)	7	20(19–21)	4	17(15–20)	2
Dorsal Anchor						
Outer	25(22–29)	5	29(26–31)	5	24(20–29)	2
Inner	30(29–32)	5	28(26–30)	5	25(21–29)	2
Base	19(17–20)	5	19(17–22)	5	16(15–17)	2
Germarium						
Length	51(47–55)	5	–	–	37(31–47)	3
Width	42(35–52)	5	–	–	24(17–31)	3
Testis						
Length	56(51–64)	3	–	–	55	1
Width	40(32–47)	3	–	–	27	1
Hooks	13 (13–14)	7	13(12–14)	16	–	–

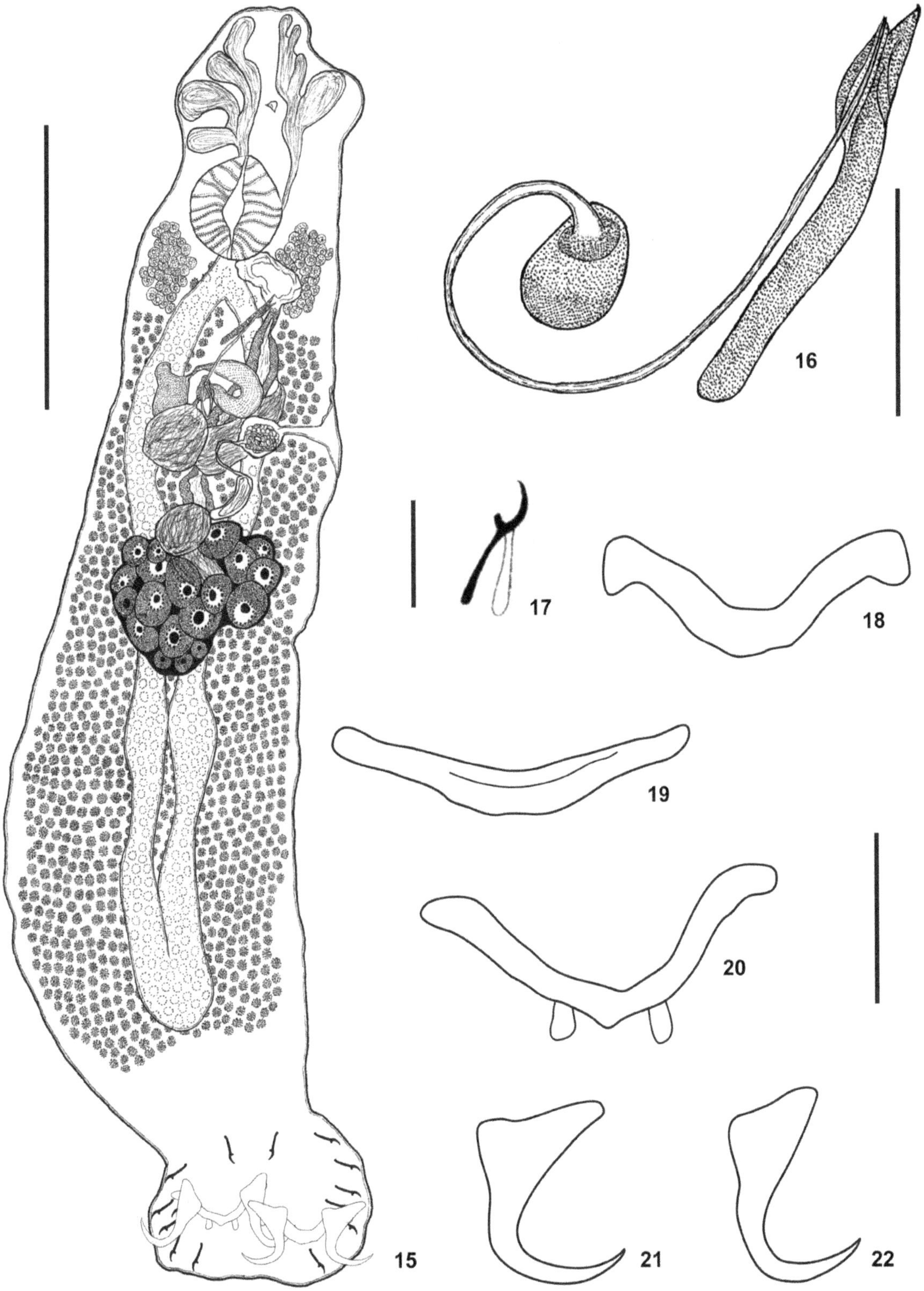

Figures 15–22. *Cosmetocleithrum akuanduba* sp. nov.: (15) holotype whole-mount, ventral; (16) copulatory complex; (17) hook; (18–19) ventral bar; (20) dorsal bar; (21) ventral anchor; (22) dorsal anchor. Scale bars: 15 = 100 µm; 16, 18–22 = 25 µm; 17 = 10 µm.

distal portion with a small gutter guarding distal portion of shaft of MCO. Germarium 51 (47–55; n = 5) long, 42 (35–52; n = 5) wide, oval. Uterus delicate, ventral. Eggs, Mehlis' gland and ootype not observed. Vagina heavily sclerotized, vaginal pore sinistral, marginal or submarginal, vaginal canal slightly sclerotized, proximal region "S"-shaped, distal region expanded; seminal receptacle, spherical. Vitelline follicles dense, coextensive with ceca.

Comparative measurements. Table 5.

Etymology. The name *akuanduba* refers to a divinity called "Akuanduba" of the Arara native people from the Xingu Basin in Pará. According Arara mythic narratives, Akuanduba played his flute to bring order to the world. One day, because of human disobedience, they were thrown into the water. The few survivors had to learn from scratch how to give continuity to life.

Remarks. *Cosmetocleithrum akuanduba* sp. nov. resembles *Cosmetocleithrum parvum* Kritsky, Thatcher & Boeger, 1986 and *Cosmetocleithrum sobrinus* Kritsky, Thatcher & Boeger, 1986 by having a "J" -shaped MCO (see Kritsky et al. 1986). However, the new species differs from these two species by possessing an elongate accessory piece, with sharp distal region, distal tip with a small gutter (accessory piece with proximal arm, hollow bulbous portion distally in *C. parvum*, and accessory piece large, globose, and apparently hollow in *C. sobrinus*). Also *C. akuanduba* has a dorsal anchor with a broad subtriangular superficial root, undeveloped deep root, evenly curved shaft with angle of approximately 70°; (narrow, subrectangular superficial root; developed deep root; evenly curved shaft with angle of approximately 70° and 80° in *C. parvum* and *C. sobrinus*, respectively). The morphological variations in the ventral bar of *C. akuanduba* (Figs 18–19) reported from different hosts and localities (i.e., *H. orestis* collected below the Volta Grande and *H. gabiru* collected above the Volta Grande rapids) were considered intraspecific variations.

Cosmetocleithrum bifurcum Mendoza-Franco, Mendoza-Palmero & Scholz, 2016

Figs 23–29

Syn. *Cosmetocleithrum* sp. 8 Mendoza-Palmero, Blasco-Costa & Scholz, 2015

Cosmetocleithrum bifurcum, Mendoza-Franco et al. (2016): 858–859, figs 6a–h, (descr).

Type host. *Hassar orestis* (Steindachner, 1875), Doradidae.

Type locality. Aquarium Momón River, Iquitos, Peru (03°44′56″ S, 73°15′13″ W).

Site of infection. Gill filaments.

Other records. *Hassar orestis* (Prevalence: 100% of 16 hosts examined; Mean intensity: 9.35 parasites per infected host; Mean abundance: 9.35 parasites per host), Xingu River, Belo Monte Community, municipality of Vitória do Xingu, Pará (03°05′52.5″S, 51°43′18.0″W); *Hassar orestis* (Prevalence: 100% of 8 hosts; Mean intensity: 12.6; Mean abundance: 12.6) Xingu River, municipality of Vitória do Xingu, Pará (02°47′27.1″S, 51°59′50.0″W); *Hassar gabiru* (Prevalence: 100% of 2 hosts; Mean intensity: 1; Mean abundance: 1) Iriri River, municipality of Altamira, Pará (03°49′06.4″S, 52°41′25.8″W); *Hassar gabiru* (Prevalence: 84% of 19 hosts; Mean intensity: 2.6; Mean abundance: 2.2), Ilha Grande, Xingu River, municipality of Altamira, Pará (03°35′50.2″S, 52°21′22.5″W); *Hassar gabiru* (Prevalence: 100% of 15 hosts; Mean intensity: 2.6; Mean abundance: 2.6), Bacajá River, municipality of Altamira, Pará (03°33′47.1″S, 51°36′50.3″W).

Specimens deposited. 17 vouchers, CHIOC 39048 a–b, 39049 a–c, 39050, 39051 a–c, 39052 a–b, INPA 780–781, MPEG 0145–0148.

Comparative measurements. Table 6.

Table 6. Comparative measurements (μm) of specimens of *Cosmetocleithrum bifurcum* from the gills of *Hassar orestis* and *H. gabiru* from Brazil and *H. orestis* from Peru. BR = Brazil, PE = Peru.

	*H. orestis (BR)	N	**H. orestis (PE)	N	*H. gabiru (BR)	N
Body						
Length	292(197–405)	8	(322–420)	17	244(150–325)	8
Width	80(65–107)	8	(51–80)	16	65(52–82)	8
Haptor						
Length	59(52–65)	8	(35–40)	14	43(35–62)	6
Width	44(37–55)	8	(47–59)	14	48(37–55)	6
Pharynx						
Length	22(19–28)	9	(21–25)	17	22(17–27)	8
Width	25(15–31)	9	–	–	20(16–24)	8
MCO length	38(33–44)	7	(31–50)	13	31(30–33)	7
Accessory piece						
Length	21(20–25)	4	(20–23)	15	16(13–20)	4
Ventral Bar						
Length	46(37–55)	6	(33–40)	14	32(26–39)	10
Width	4(3–4)	7	–	–	2(2–3)	11
Dorsal Bar						
Length	31(20–50)	5	(31–38)	14	30(22–40)	8
Width	3(3)	7	–	–	2(2–3)	9
Ventral Anchor						
Outer	29(25–31)	8	27	1	28(25–31)	9
Inner	23(21–25)	8	19	1	21(19–24)	9
Base	14(12–16)	8	(11–12)	25	12(9–17)	9
Dorsal Anchor						
Outer	27(25–33)	7	23	1	24(17–30)	8
Inner	20(18–23)	7	18	1	21(18–29)	8
Base	13(11–17)	7	(10–12)	24	11(9–12)	8
Germarium						
Length	28(46–55)	9	(55–63)	3	42(36–49)	5
Width	22(17–31)	9	(23–40)	3	19(17–24)	5
Testis						
Length	49(41–60)	8	(83–99)	3	41(32–47)	5
Width	24(19–31)	8	(30–47)	3	24(19–30)	5
Hooks	15 (15–16)	6	15	37	14(13–15)	7

* Present study
** Mendoza Franco et al. (2016)

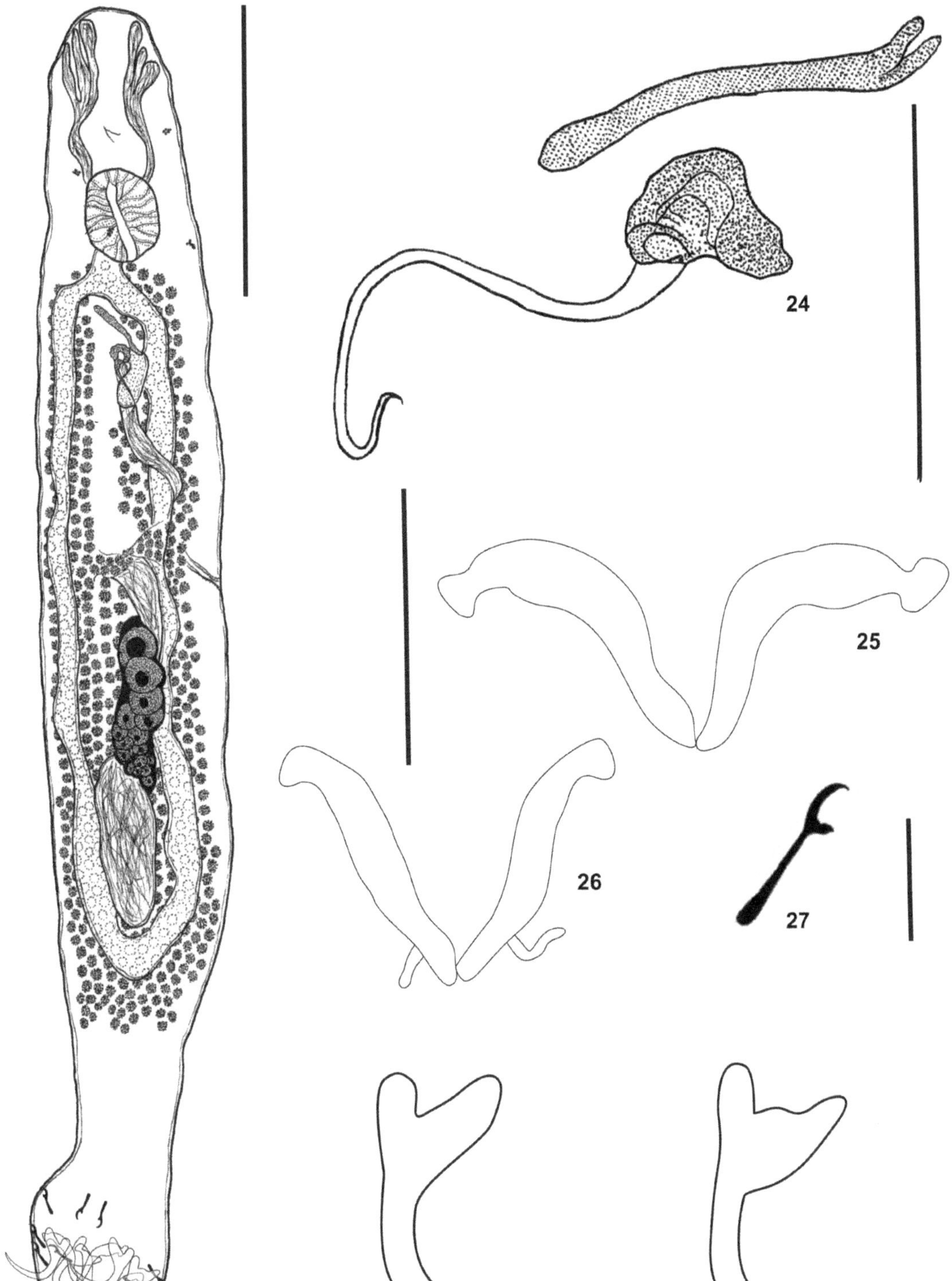

Figures 23–29. *Cosmetocleithrum bifurcum* Mendoza-Franco, Mendoza-Palmero & Scholz, 2016: (23) voucher whole-mount, ventral; (24) copulatory complex; (25) ventral bar; (26) dorsal bar; (27) hook; (28) ventral anchor; (29) dorsal anchor. Scale bars: 23 = 100 μm; 24–26, 28, 29 = 25 μm; 27 = 10 μm.

Remarks. A comparative analysis of *Cosmetocleithrum bifurcum* Mendoza-Franco, Mendoza-Palmero & Scholz, 2016 and specimens of *Cosmetocleithrum* from Xingu River and related tributaries indicated that they are conspecific, mainly because they both share the morphology of the copulatory complex, bars and anchors. The specimens studied here also are similar morphometrically to those specimens from the type host and locality. Mendoza-Franco et al. (2016) recognized differences between hook pair 7 and the other hook pairs (Mendoza-Franco et al. 2016: figs 6G–H), where hook pair 7 has a shaft longer and shank more slender than the other hook pairs. The specimens studied here have hooks similar in shape with erect thumb (thumb directed posteriorly in Mendoza-Franco et al.'s specimens), and evenly curved shaft and point (long shaft, delicate point in Mendoza-Franco et al.'s specimens). *Cosmetocleithrum bifurcum* was primarily characterized by possessing a MCO with about 1–1.5 rings, whereas the specimens studied in the present paper have a sigmoid MCO. However, these variations do not seem to be sufficient enough evidence to propose a new species and we consider them as intraspecific variations.

Vancleaveus klasseni sp. nov.

Figs 30–36

Type host. *Hassar orestis* (Steindachner, 1875), Doradidae

Site of infection. Gill filaments.

Type locality. Xingu River, Belo Monte Community, municipality of Vitória of Xingu, Pará (03°05′52.5″S, 51°43′18.0″W; 02°47′27.1″S, 51°59′50.0″W).

Prevalence. 50% of 24 hosts examined

Mean intensity. 2.3 parasites per infected host.

Mean abundance. 1.6 parasites per host.

Other records. *Hassar gabiru* (Prevalence: 30% of 15 hosts; Mean intensity: 2; Mean abundance: 0.6), Bacajá River, municipality of Altamira, Pará (03°33′47.1″S, 51°36′50.3″W); *Hassar gabiru* (Prevalence: 16% of 12 hosts; Mean intensity: 2; Mean abundance: 0.3), Ilha Grande, Xingu River, municipality of Altamira, Pará (03°35′50.2″S, 52°21′22.5″W).

Specimens deposited. Holotype CHIOC 39058 a, and 9 paratypes, CHIOC 39058 b–h, INPA 778, MPEG 0143; 4 vouchers, CHIOC 39059–39060, INPA 779, MPEG 0144.

Description (based on 10 type specimens, 3 mounted in Hoyer, 7 stained with Gomori's trichrome): Body fusiform, total length including haptor 315 (215–517; n = 9) long, 89 (65–122; n = 9) wide, at level of germarium. Tegument smooth (Fig. 30). Cephalic margin rounded, cephalic lobes inconspicuous; 5 or 6 bilateral pairs of head organs with rod-shaped secretion; cephalic glands unicellular, posterolateral to pharynx. Eyes, pigment granules (eye-spots), absent. Mouth subterminal; pharynx muscular, spherical, 33 (25–36; n = 8) long, 17 (15–20; n = 8) wide; esophagus short; two intestinal ceca, confluent posteriorly, lacking diverticula. Haptor hexagonal 71(52–90; n = 9) wide. Anchors similar. Ventral anchor, divergent roots; superficial root triangular, well developed, with conspicuous fold; deep root developed; slightly curved shaft and point, forming angle of approximately 65°, point extending just past level of tip of superficial root, outer 24 (21–28; n = 6) long, inner 28 (27–32; n = 6) long, base 10 (8–12; n = 6) (Fig. 35). Dorsal anchor, divergent roots; superficial root subtriangular, with conspicuous fold; deep root developed; slightly curved shaft and point, forming angle of approximately 80°, point extending just past level of tip of superficial root, outer 24 (18–27; n = 4) long, inner 29 (29–30; n = 4) long, base 10 (10–11; n = 4) (Fig. 36). Ventral bar (Fig. 33) 35 (27–45; n = 10) long, 4 (3–5; n = 10) wide, broadly U-shaped with inflated ends, posteromedial projection devel-

Table 7. Comparative measurements (µm) of specimens of *Vancleaveus klasseni* sp. nov. from the gills of *Hassar orestis* and *H. gabiru* from Xingu River Basin.

	H. orestis	N	*H. gabiru*	N
Body				
Length	315(215–517)	9	267(215–320)	2
Width	89(65–122)	9	76(65–87)	2
Haptor				
Wide	71(52–90)	9	61(55–67)	2
Pharynx				
Length	33(25–36)	8	29(28–30)	2
Width	17(15–20)	8	18	2
MCO length	96(88–106)	6	98(90–104)	3
Accessory piece				
Length	16 (15–18)	5	23(21–25)	2
Ventral Bar				
Length	35(27–45)	10	38(36–40)	2
Width	4(3–5)	10	4	3
Dorsal Bar				
Length	29(21–38)	10	29(28–30)	3
Width	3(2–3)	10	3(2–4)	3
Ventral Anchor				
Outer	24(21–28)	6	26(25–28)	3
Inner	28(27–32)	6	28(27–30)	3
Base	10(8–12)	6	11(11–12)	3
Dorsal Anchor				
Outer	24(18–27)	4	21(19–23)	4
Inner	29(29–30)	4	25(25–26)	4
Base	10(10–11)	4	10	4
Germarium				
Length	45(29–62)	2	40	1
Width	18(16–20)	2	29	1
Testis				
Length	25(20–30)	2	–	–
Width	19(19–20)	2	–	–
Hooks	21(20–25)	10	22(21–26)	4

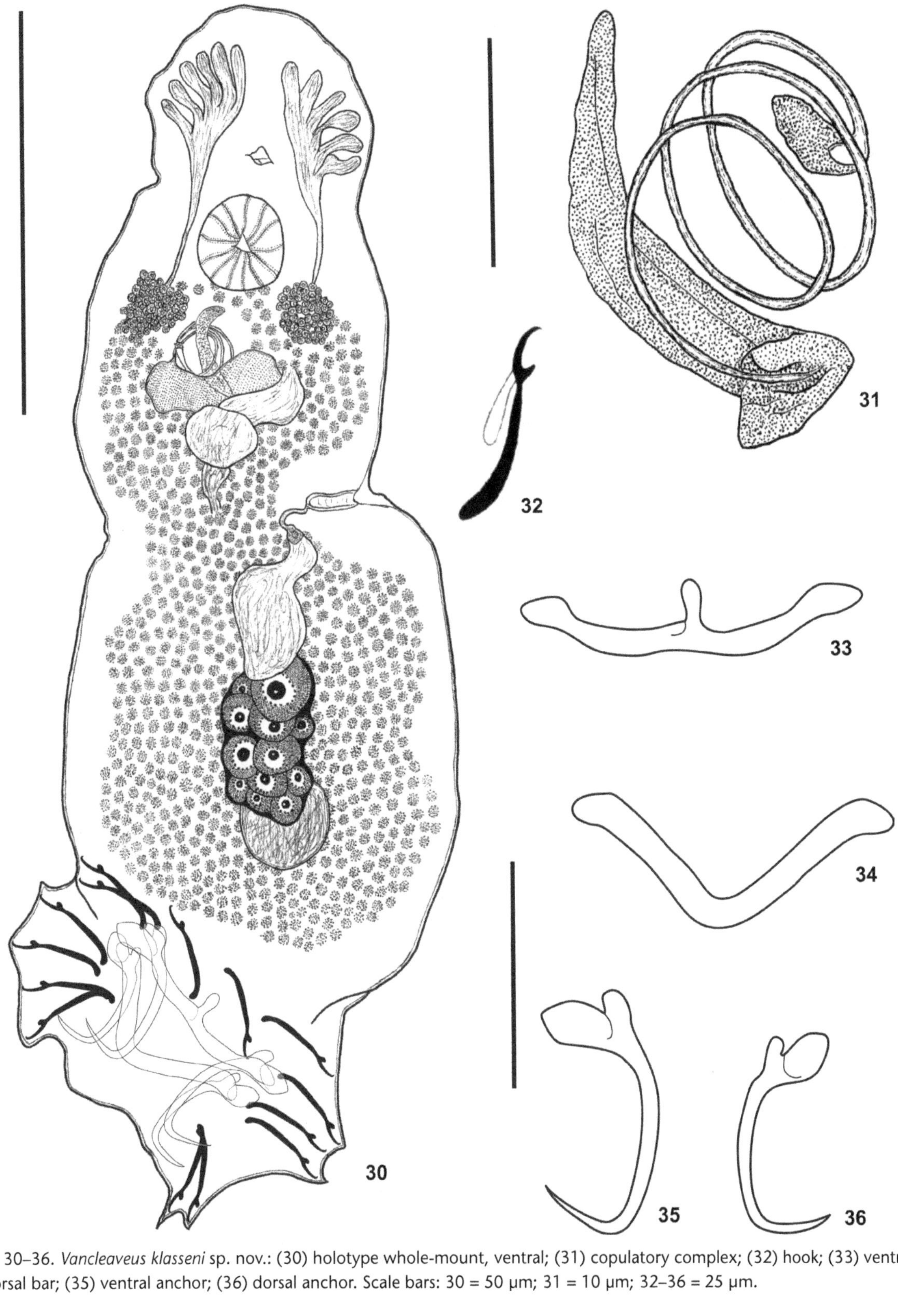

Figures 30–36. *Vancleaveus klasseni* sp. nov.: (30) holotype whole-mount, ventral; (31) copulatory complex; (32) hook; (33) ventral bar; (34) dorsal bar; (35) ventral anchor; (36) dorsal anchor. Scale bars: 30 = 50 µm; 31 = 10 µm; 32–36 = 25 µm.

oped. Dorsal bar (Fig. 34) 29 (21–38; n = 10) long, 3 (2–3; n = 10) wide, V-shaped, with slightly inflated ends. Hooks similar (Fig. 32), 21 (20–25; n = 10) long, with upright slightly acute thumb; slightly curved shaft, short; dilated shank; filamentous hook about ½ shank length. Genital pore opening midventral, anterior to copulatory complex. Testis post-germarial, dorsal to germarium, ovoid 25 (20–30; n = 2) long, 19 (19–20; n = 2) wide. Vas deferens looping left of intestinal cecum; seminal vesicle a dilatation of vas deferens; transversally elongated prostatic reservoir. Copulatory complex comprising MCO, accessory piece (Fig. 31). MCO a coiled sclerotized tube, 96 (88–106; n = 6) long, with 2 ½ counterclockwise rings, with tapered distal region; base a cap with sclerotized margin. Accessory piece 16 (15–18; n = 5) long, elongated, non-articulated with MCO, with tapered proximal region, distal end with small gutter guarding termination of MCO. Germarium 45 (29–62; n = 2) long, 18 (16–20; n = 2) wide, elongate. Eggs, Mehlis' gland, ootype and uterus not observed. Vagina slightly sclerotized, ventral, opening sinistrally; vaginal canal slightly sclerotized, distal region with a bulb serving the connection to the seminal receptacle; seminal receptacle, pyriform. Vitelline follicles dense, coextensive with ceca.

Comparative measurements. Table 7.

Etymology. The species is dedicated to Dr. Gregory J. Klassen in recognition for his work on systematics of monogenoids.

Remarks. *Vancleaveus klasseni* sp. nov. differs from its congeners by (1) possessing anchors with divergent roots; superficial root triangular to subtriangular, well developed, with conspicuous fold; deep root developed; curved shaft and point; (2) MCO a coiled sclerotized tube with 2 ½ rings, base with sclerotized margin; and (3) accessory piece with small gutter in the distal.

DISCUSSION

There are 14 genera of monogenoidean gill parasites associated with six freshwater siluriform families from Neotropics (Braga et al. 2014, Acosta et al. 2018). From this diversity, 11 species of monogenoids from three genera (i.e., *Cosmetocleithrum*, *Paracosmetocleithrum* Acosta, Scholz, Blasco-Costa, Alves & da Silva, 2018 and *Vancleaveus*) are known parasitizing fishes from the Doradidae (Table 1). The richness of monogenoids parasitizing doradids seems to be variable. The doradid, *O. niger* is infected with five different species while other species, for example, bear a single or a few species. Considering the recent evaluation of the number of doradid species, approximately 94 species (Birindelli 2014, Sabaj-Pérez and Hernández 2017), and assuming the presence of two species of monogenoids per host species, a roughly estimated 7% of monogenoidean parasites of Doradidae are known to date.

Species of *Cosmetocleithrum* and *Paracosmetocleithrum* are exclusively found infecting neotropical fishes from the Order Siluriformes. Except for *Cosmetocleithrum striatuli* Abdallah, Azevedo & Luque, 2012, *Cosmetocleithrum laciniatum* Yamada, Yamada, Santos & Anjos, 2017, which occur in different species of Auchenipteridae, and *Cosmetocleithrum longivaginatum*, parasitizing the gills of *Pimelodus albicans* (Valenciennes, 1840) (Pimelodidae), the other species of *Cosmetocleithrum* are restricted to species of Doradidae. *Paracosmetocleithrum trachydorasi* Acosta, Scholz, Blasco-Costa, Alves & da Silva, 2018 is reported from the gills of the doradid, *T. paraguayensis*. On the other hand, species of *Vancleaveus* are primarily parasitic on pimelodid hosts, except for *Vancleaveus klasseni* sp. nov. which is restricted to the doradids; *Vancleaveus cicinnus* Kritsky, Thatcher & Boeger, 1986, shared by species of Doradidae and Pimelodidae; and *Vancleaveus janauacaensis* Kritsky, Thatcher & Boeger, 1986 occurring on species of Doradidae, Pimelodidae and Erythrinidae (Characiformes) (Table 1). According to Graça et al. (2013), the occurrence of *V. janauacaensis* on the erythrinid, *Hoplias malabaricus* (Bloch, 1794) seems to be accidental since the prevalence and mean intensity were low, 4% of 54 hosts and 1 parasite per infected fish, respectively.

Despite the occurrence of *Cosmetocleithrum* and *Vancleaveus* parasitizing the gills of doradids, the most recent cladistic hypothesis for dactylogyrids from catfishes does not support their sister relationship. Mendoza-Palmero et al. (2015) studying partial sequences of the 28S rRNA gene of dactylogyrids, provided a cladistic hypothesis, where monogenoids infecting Neotropical siluriforms are represented by two distinct lineages. In their analysis, species of *Vancleaveus* appears in a clade with species of *Ameloblastella* Kritsky, Mendoza-Franco & Scholz, 2000 and *Unibarra* Suriano & Incorvaia, 1995, closely related to other freshwater dactylogyrids (ie., *Ancyrocephalus* Creplin, 1839, *Onchocleidus* Mueller, 1936, *Ligictaluridus* Beverley-Burton, 1984, *Actinocleidus* Mueller, 1937) parasitizing siluriforms and perciforms with Holartic and Neotropical distributions; while *Cosmetocleithrum* arises as a sister group of *Demidospermus* and unidentified dactylogyrids that occur on pimelodid catfishes. Besides these freshwater dactylogyrid taxa from South America, the clade composed by *Cosmetocleithrum*, *Demidospermus* and Dactylogyridae spp. also includes marine and freshwater dactylogyrids from Eurasia, Africa and South America, supporting the hypothesis of host switching among species of those genera.

Braga et al. (2015) related that the knowledge of phylogenetic relationships of monogenoidean lineages represents the most important information in the task of reconstructing the histories of parasite-host relationships. Understanding patterns in reconstructions of histories in parasite-host systems permits us to manifest possible events of co-speciation, horizontal transference, duplication and extinction. Future studies could elucidate the evolutionary relationships that exist between the lineages on monogenoids and their doradid hosts. The acknowledgement of phylogenetic relationships between the parasite lineages, as well as information about geological events associated with the origin of host-parasite lineages could contribute to a much better comprehension of the evolutionary background involving this host-parasite system.

ACKNOWLEDGMENTS

We would like to thank Tommaso Giarrizzo and Leandro de Sousa from Universidade Federal do Pará, for assistance during the collecting trips; Leandro de Sousa for host identification; Célio Magalhães, Coleção de Invertebrados do Instituto Nacional de Pesquisas da Amazônia allowed access to specimens under his care. This work was supported partially by a Scholarship from Programa de Pós-Graduação em Biodiversidade e Conservação, Universidade Federal do Pará to GBS; and research grants from the Conselho Nacional de Desenvolvimento Científico e Tecnológico (CNPq/PROTAX 001/2015–440526/2015-9) to MVD. The Instituto Chico Mendes de Conservação da Biodiversidade (ICMBio) granted the license for collection of biological material 43381-2. Finally, we thank three anonymous reviewers for their excellent contribution to this manuscript.

LITERATURE CITED

Abdallah VD, Azevedo RK, Luque JL (2012) Three new species of Monogenea (Platyhelminthes) parasites of fish in the Guandu River, southeastern Brazil. Acta Scientiarum, Biological Sciences 34: 483–490. https://doi.org/10.4025/actascibiolsci.v34i4.10466

Acosta AA, Scholz T, Blasco-Costa I, Alves PV, da Silva RJ (2018) A new genus and two new species of dactylogyrid monogeneans from gills of Neotropical catfishes (Siluriformes: Doradidae and Loricariidae). Parasitology International 67(1): 4–12. https://doi.org/10.1016/j.parint.2017.09.012

Birindelli JLO (2014) Phylogenetic relationships of the South American Doradoidea (Ostariophysi: Siluriformes). Neotropical Ichthyology 12(3): 451–564. https://doi.org/10.1590/1982-0224-20120027

Birindelli JLO, Fayal DF, Wosiacki WB (2011) Taxonomic revision of thorny catfish genus *Hassar* (Siluriformes: Doradidae). Neotropical Ichthyology 9(3): 515–542. https://doi.org/10.1590/S1679-62252011000300006

Boeger WA, Vianna RT (2006) Monogenoidea. In: Thatcher VE (Ed.) Aquatic Biodiversity in Latin America – Amazon fish parasites (2nd edn). Pensoft Publishers, Sofia-Moscow, 42–116.

Braga MP, Araújo SBL, Boeger W (2014) Patterns of interaction between Neotropical freshwater fishes and their gill Monogenoidea (Platyhelminthes). Parasitology Research 113: 481–490. https://doi.org/10.1007/s00436-013-3677-8

Braga MP, Razzolini E, Boeger W (2015) Drivers of parasite sharing among Neotropical freshwater fishes. Journal of Animal Ecology 84: 487–497. https://doi.org/10.1111/1365-2656.12298

Buckup PA, Menezes NA, Ghazzi MSA (2007) Catálogo das espécies de peixes de água doce do Brasil. Museu Nacional, Rio de Janeiro, 195 pp.

Bush AO, Lafferty KD, Lotz JM, Shostak W (1997) Parasitology meets ecology on its own terms: Margolis et al. Revisited. Journal of Parasitology 83: 575–583. https://doi.org/10.2307/3284227

Cohen SC, Kohn A, Justo MCN (2013) South American Monogenoidea parasites of fishes, amphibian and reptiles. Oficina de Livros, Rio de Janeiro, 663 pp.

Eschmeyer WN, Fricke R, Van der Laan R (2017) Catalog of Fishes: Genera, Species, References. http://researcharchive.calacademy.org/research/ichthyology/catalog/fishcatmain.asp [Accessed 12/10/ 2017]

Ferreira DO, Tavares-Dias M (2016) Ectoparasites and endoparasites community of *Ageneiosus ucayalensis* (Siluriformes: Auchenipteridae), catfish from Amazon River system in northern Brazil. Journal of Parasitic Diseases. https://doi.org/10.1007/s12639-016-0857-3

Eler ES, Dergam JA, Vênere PC, Paiva LC, Miranda GA, Oliveira AA (2007) The karyotypes of the thorny catfishes *Wertheimeria maculata* Steindachner, 1877 and *Hassar wilderi* Kindle, 1895 (Siluriformes, Doradidae) and their relevance in doradids chromosomal evolution. Genetica 130: 99–103. https://doi.org/10.1007/s10709-006-0023-4

Graça RJ, Ueda BH, Oda FH, Takemoto RM (2013) Monogenea (Platyhelminthes) parasites from the gills of *Hoplias* aff. *malabaricus* (Bloch, 1794) (Pisces: Erythrinidae) in the Upper Paraná River Floodplain, States of Paraná and Mato Grosso do Sul, Brazil. Check List 9: 1484–1487. https://doi.org/10.15560/9.6.1484

Humason GL (1979) Animal Tissue Techniques. W.H. Freeman Co, USA, 661 pp.

Iannacone JÁ, Luque JL (1991) Monogeneos parasitos del "paiche" *Arapaima gigas* (C.) y del "turushuqui" *Oxydoras niger* (V.) en la Amazonia peruana. Boletin de Lima 76: 43–48.

Jeronimo GT, Padua SB, Ventura AS, Gonçalves ELT, Ishikawa MM, Martins ML (2016) Parasitological assessment in the hybrid surubim (*Pseudoplatystoma reticulatum* x *P. corruscans*), with uncommon occurrence of Monogenea parasites. Brazilian Journal of Veterinary Parasitology 25(2): 179-186. https://doi.org/10.1590/S1984-29612016037

Kohn A, Fernandes BMM, Cohen SC (2007) South American Trematodes Parasites of Fishes. Imprinta Express Ltda, Rio de Janeiro, 318 pp.

Kritsky DC, Thatcher VE, Boeger WA (1986) Neotropical Monogenea. 8. Revision of *Urocleidoides* (Dactylogyridae, Ancyrocephalinae). Proceedings of the Helminthological Society of Washington 53: 1–37.

Luque JL, Aguiar JC, Vieira FM, Gibson DI, Santos CP (2011) Checklist of Nematoda associated with the fishes of Brazil. Zootaxa 3082: 1–88.

Mendoza-Franco EF, Mendoza-Palmero CA, Scholz T (2016) New species of *Ameloblastella* Kritsky, Mendoza-Franco & Scholz, 2000 and *Cosmetocleithrum* Kritsky, Thatcher & Boeger, 1986 (Monogenea: Dactylogyridae) infecting the gills of catfishes (Siluriformes) from the Peruvian Amazonia. Systematic Parasitology 93: 847–862. https://doi.org/10.1007/s11230-016-9671-7

Mendoza-Palmero CA, Blasco-Costa I, Scholz T (2015) Molecular phylogeny of Neotropical monogeneans (Platyhelminthes: Monogenea) from catfishes (Siluriformes). Parasites e Vectors 8(164): 2–11. https://doi.org/10.1186/s13071-015-0767-8

Mendoza-Palmero CA, Scholz T, Mendoza-Franco EF, Kuchta R (2012) New species and geographical records of dactylogyrids (Monogenea) of Catfish (Siluriformes) from the Peruvian Amazonia. Journal of Parasitology 98(3): 484–497. https://doi.org/10.1645/GE-2941.1

Mizelle JD, Klucka AR (1953) Studies on monogenetic trematodes. XVI. Dactylogyridae from Wisconsin fishes. American Midland Naturalist 49: 720–733.

Nelson JS, Grande TC, Wilson MVH (2016) Fishes of the World (5th edn). John Wiley, Hoboken, 651 pp. https://doi.org/10.1002/9781119174844

Pantoja WMF, Silva LVF, Tavares-Dias M (2016) Are similar the parasite communities structure of *Trachelyopterus coriaceus* and *Trachelyopterus galeatus* (Siluriformes: Auchenipteridae) in the Amazon basin? Revista Brasileira de Parasitologia Veterinaria 25(1): 46-53. https://doi.org/10.1590/S1984-29612016016

Rasband WS (1997–2016) ImageJ. US National Institutes of Health, Bethesda. http://imagej.nih.gov/ij

Sabaj-Pérez MH, Hernández MA (2017) Taxonomic assessment of the hard-nosed Thornycats (Siluriformes: Doradidae: *Trachydoras* Eigenmann, 1925) with description of *Trachydoras gepharti*, sp. nov. Proceedings of the Academy of Natural Sciences of Philadelphia 166: 1–53. https://doi.org/10.1635/053.166.0102

Santos MD, Brasil-Sato MC (2006) Parasitic community of *Franciscodoras marmoratus* (Reinhardt, 1874) (Pisces: Siluriformes, Doradidae) from the upper São Francisco river, Brazil. Brazilian Journal of Biology 66(3): 931–938. https://doi.org/10.1590/S1519-69842006000500019

Silva Amo, Tavares-Dias M, Jerônimo GT, Martins ML (2011) Parasite diversity in *Oxydoras niger* (Osteichthyes: Doradidae) from the basin of Solimões River, Amazonas state, Brazil, and the relationship between monogenoidean and condition factor. Brazilian Journal of Biology 71(3): 791–796. https://doi.org/10.1590/S1519-69842011000400026

Suriano DM, Incorvaia IS (1995) Ancyrocephalid (Monogenea) parasites from siluriform fishes from the Paranean Platean Icthyogeographical Province in Argentine. Acta Parasitologica 40: 113–124.

Takemoto RM, Pavanelli GC, Lizama MAP, Lacerda ACF, Yamada FH, Moreira LHA, Ceschini TL, Bellay S (2009) Diversity of parasites of fish from the Upper Paraná River floodplain, Brazil. Brazilian Journal of Biology 69: 691–705. https://doi.org/10.1590/S1519-69842009000300023

Tavares-Dias M (2017) Community of protozoans and metazoans parasitizing *Auchenipterus nuchalis* (Auchenipteridae), a catfish from the Brazilian Amazon. Acta Scientiarum 39(1): 123-128. https://doi.org/10.4025/actascibiolsci.v39i1.32836

Thatcher VE (2006) Amazon Fish Parasites (2nd edn). Pensoft Publishers, Sofia–Moscow, 508 pp.

Yamada POF, Yamada FH, Silva RJ, Anjos LAS (2017) A New Species of *Cosmetocleithrum* (Monogenea, Dactylogyridae), a Gill Parasite of *Trachelyopterus galeatus* (Siluriformes, Auchenipteridae) from Brazil, with Notes on the Morphology of *Cosmetocleithrum striatuli*. Comparative Parasitology 84(2): 119–123. https://doi.org/10.1654/1525-2647-84.2.119

Author Contributions: GBS, JFSN and MVD contributed equally to this article.

Competing Interests: The authors have declared that no competing interests exist.

8

In honor of conservation of the Brazilian Atlantic Forest: description of two new damselflies of the genus *Forcepsioneura* discovered in private protected areas (Odonata: Coenagrionidae)

Ângelo Parise Pinto[1], Tom Kompier[2]

[1] *Laboratório de Sistemática de Insetos Aquáticos, Departamento de Zoologia, Universidade Federal do Paraná. Caixa Postal 19020, 81531-980 Curitiba, PR, Brazil.*
[2] *Schoutenstraat 69, 2596SK, Den Haag, The Netherlands.*
Corresponding author: Ângelo Parise Pinto (appinto@ufpr.br)

ABSTRACT. Two new Brazilian Protoneurinae damselflies, *Forcepsioneura regua* **sp. nov.** (holotype male deposited in DZRJ: Brazil, Rio de Janeiro State, Cachoeiras de Macacu municipality, RPPN Reserva Ecológica de Guapiaçu) and *Forcepsioneura serrabonita* **sp. nov.** (holotype male deposited in DZRJ: Brazil, Bahia State, Camacan municipality, RPPNs Serra Bonita) , are described, illustrated and diagnosed based on males and females. The bluish and smaller *F. regua* **sp. nov.** has been confused with at least three previously described species, being very similar to the type species of *Forcepsioneura*, *F. garrisoni* Lencioni, 1999, but lacking a defined tubercle-like process on the posterolateral margin of the median lobe of the prothorax in both sexes, which allows it to be distinguished from all other known species. The shape of the cercus of the male of *F. serrabonita* **sp. nov.** is similar to that of *F. grossiorum* Machado, 2001 and *F. lucia* Machado, 2000, two species with very short ventrobasal process. However, it differs from them mainly by the mediobasal process of the cercus, which is rounded in dorsal view and almost not visible in lateral view. The taxonomic status of *Forcepsioneura* is discussed and a comparison with the other species of the genus is provided. Based on size, habitat and coloration, *Forcepsioneura* can be informally divided into two groups: (1) large, orange-black and montane species, including *F. grossiorum*, *F. itatiaiae* (Santos, 1970), *F. lucia* and *F. serrabonita* **sp. nov.**; (2) small, bluish and lowland species, including *F. garrisoni*, *F. haerteli* Machado, 2001, *F. regua* **sp. nov.** and *F. sancta* (Hagen *in* Selys, 1860). Our findings highlight the urgency in directing collecting efforts to unexplored areas, as well as the importance of private preserves that harbor the type localities as guardians of the threatened and diverse Atlantic Forest diversity. Together these two localities surveyed account for more than 210 species of odonates, representing almost 24% of the number of Brazilian species. Brazil has the greatest number of known species of odonates in the world. This study shows that further research is required in order to fully understand the diversity of *Forcepsioneura*.

KEY WORDS. *Amazoneura*, dragonfly, damselfly, Protoneurinae, Southeastern Brazil, taxonomy.

INTRODUCTION

"...forest cover must be important for adults of the majority of neotropical Odonata. Every effort should be made to preserve forests and forest wetlands in this most biodiverse region of the Earth" (Paulson 2006: 97).

The quotation above illustrates the relevance of forests for the conservation of the insects of the order Odonata, popularly known as dragonflies and damselflies. Life on Earth is facing the sixth mass extinction in the planet's geological history (Barnosky et al. 2011). Distinct from the past, when natural processes caused extinctions, for example, the meteorite impact that ended the era of the giant 'reptiles' and caused the last big extinction event in the Cretaceous Period (Jablonski 2001, Barnosky et al. 2011), the current high rate of biodiversity loss mostly results from human activities. This has led to this epoch being baptized as the Anthropocene (e.g., Crutzen and Stoermer 2000). As a consequence of the high demand for energy, crop production, and encroachment on natural habitats by urbanization, ecosystems are under strong pressure, causing what has been coined of Biodiversity Crisis (e.g., Singh 2002).

Amongst the severely threatened areas of the biosphere, the Atlantic Forest (AF) is recognized as an area of immense species diversity and, for some groups of plants, such as Bromeliaceae, the richest one (Martinelli et al. 2008). Originally, the very complex and heterogeneous AF encompassed about 1,500,000 km^2, mainly along the Brazilian Atlantic Coast (including also small areas in Paraguay and Argentina), and it is estimated that now has been reduced to less than 12% (Ribeiro et al. 2009). The domain, most often referred to as a biome, is composed by herbaceous-arbustive restinga vegetation at sea level, trade off by pluvial forests with tall canopies in the lowland, up to highland fields known as Campos Rupestres of the Southeastern Brazilian Páramos (Safford 1999, Ribeiro et al. 2009). Due to its high levels of biodiversity, endemism, and anthropic pressures, it is considered one of 34 hotspots for conservation (Myers et al. 2000, Mittermeier et al. 2011), even the hottest (Laurance 2009). Its area corresponds to the Parana biogeographical dominion sensu Morrone (2014). As far as Odonates are concerned, the AF holds about 50% of the 878 known Brazilian species (data based on Pinto 2017). However, its original cover corresponds to the most densely populated part of the country, and encompasses, for instance, huge metropolitan areas such as the megacities of São Paulo and Rio de Janeiro, that are among the thirty largest urban centers of the world (UN 2015). The species of *Forcepsioneura* Lencioni, 1999 inhabit this threatened domain. The genus includes small Protoneurinae damselflies that, with the exception of specimens identified as *F. sancta* (Hagen *in* Selys, 1860) from the Cerrado domain of the Brazilian central plateau (see records in Pessacq et al. 2012), are restricted to the AF.

Protoneuridae s.l., as defined by Tillyard (1917), comprises what Tillyard and Fraser (1938: 164) formerly recognized as two parallel series of damselflies with reduced wing venation: the first series encompassing a small homogenous group of genera from the New World, and a second, clearly heterogeneous series from the Old World. Since Tillyard and Fraser (1938), more genera and species have been added to this unnatural assembly, contributing for its original concept representing a polyphyletic group (see Pessacq 2008, Dijkstra et al. 2014). However, the almost exclusively Neotropical representatives from the New World series (Protoneurinae sensu Bridges 1994: IV.7), containing the type genus *Protoneura* Selys *in* Sagra, 1857, and fifteen other genera, were recovered as monophyletic, nested within Coenagrionidae in several analyses (Rehn 2003, O'Grady and May 2003, Bybee et al. 2008, Pessacq 2008, Dijkstra et al. 2014). Restraining true protoneurids to its Western Hemisphere members is an old hypothesis (see Watson 1992), but even its monophyly is not free from controversy (see Pimenta et al. 2015). However, considering that the monophyly of the New World protoneurids is demonstrated as highly supported, its familial status is justified, with other monophyletic groups of genera in Coenagrionidae s.l. each receiving similar level. It is contrary to recent proposals to decrease the rank level of both Pseudostigmatidae (Ingley et al. 2012) and Protoneuridae (Dijkstra et al. 2014) due to these being phylogenetically nested within Coenagrionidae s.l.

South American protoneurids are mostly small sized odonates living near or within forested areas. They are generally weak fliers, frequenting shaded areas near the shoreline of large freshwater bodies and numerous running water environments. Because of very thin general aspect and disruptive colors, protoneurids are difficult to be found in the field, being often overlooked by regular collectors.

Today's Brazilian fauna counts with 70 species of Protoneurinae in 12 genera (Pessacq et al. 2012, Pinto 2017), including the endemic *Forcepsioneura*. This genus was described to include *Phasmoneura ciganae* Santos, 1970 and its type species *Forcepsioneura garrisoni* Lencioni, 1999 (Lencioni 1999). Independently, Angelo B.M. Machado in the same year proposed *P. ciganae* as a junior synonym of *Phasmoneura sancta* Hagen *in* Selys, 1860 (Machado 1999). Indeed, Machado, in a series of seminal works ("Studies on Neotropical Protoneuridae"), was the first author to suggest that *P. sancta* would need to be transferred to its own genus (see Machado 1985) and he (Machado 2000, 2001, 2005) described half of the six currently valid species of *Forcepsioneura*.

Forcepsioneura belongs to a poorly-supported monophyletic group of six genera, along with *Amazoneura* Machado, 2004, *Lamproneura* De Marmels, 2003, *Roppaneura* Santos, 1966, *Phasmoneura* Williamson, 1916, and *Psaironeura* Williamson, 1915 (Pessacq 2008), i.e., the *Roppaneura* clade. All of them have cubitus anterior (CuA) and cubitus posterior plus anal anterior (CuP&AA) veins indistinct (except in *Lamproneura*), antennifer non-carinated dorsally, and male cercus with ventrobasal process and genital ligula with long recurved lateral lobes (Pessacq 2008). The monophyly of this group, as well as that of other formerly erected taxa in Protoneurinae, is uncertain and currently under investigation (Pimenta et al. 2015).

From the six species currently recognized in *Forcepsioneura*, only *F. sancta* was described a long time ago. Most species were discovered recently, thanks to field-works in poor explored areas, and undescribed species are often reported in the literature (e.g., Kompier 2015). At least five undescribed species of *Forcepsioneura*, all from the Brazilian Atlantic Forest, were detected by the first author.

In recognition of conservation efforts to protect AF remnants, we describe here two previously unknown damselflies of "Protoneuridae s.s." (Coenagrionidae: Protoneurinae) recently collected in the states of Bahia and Rio de Janeiro, Brazil. These new damselflies were discovered in protected areas maintained thanks to private associations. The Brazilian AF is one of the most threatened biotic domains of the biosphere, and given its fragmentation and fragility, all conservation actions, including private sector initiatives, must be applauded for their contribution to both the scientific community and society in general.

MATERIAL AND METHODS

Specimens examined are deposited in the following collections. Acronyms, wherever possible, based on the updated version of Arnett et al. (1993) by Evenhuis (2016): DZRJ – Coleção Entomológica "Prof. José Alfredo Pinheiro Dutra", Departamento de Zoologia, Instituto de Biologia, Universidade Federal do Rio do Janeiro, Rio de Janeiro, RJ, Brazil; DZUP – Coleção Entomológica "Pe. Jesus Santiago Moure", Departamento de Zoologia, Setor de Ciências Biológicas, Universidade Federal do Paraná, Curitiba, PR, Brazil; MNRJ – Museu Nacional, Universidade Federal do Rio de Janeiro, Rio de Janeiro, RJ, Brazil; MZSP – Museu de Zoologia, Universidade de São Paulo, SP, Brazil; RWG – Rosser W. Garrison Collection, Sacramento, CA, USA; USNM – National Museum of Natural History, Washington, DC, USA.

General morphology terms follow Pessacq (2008) and Garrison et al. (2010); wing venation is based on the system of Riek and Kukalová-Peck (1984), with additions summarized by Fleck et al. (2003) and Garrison et al. (2010). The terminology proposed by De Marmels (2002) was implemented for the description of the structures of the apical segment of the genital ligula, which we believe corresponds to the second segment following Pfau (1971), while terminology for the ovipositor was based on Matushkina (2004). The genital ligula was extruded using a hook-shaped pin after applying a few drops of 10% ammonia solution on the venter of the second and third abdominal segments (S2–3). In the few instances when genital ligula was fully extracted it was later transferred to a microvial with glycerin and stored with the respective specimen.

All measurements (in mm) and illustrations were made with the aid of a stereomicroscope equipped with a camera lucida. Specimens were photographed with a Leica MZ16 stereomicroscope equipped with a Leica DFC420 camera, and source images combined using Auto-montage software by The Synoptic Group©.

The following abbreviations were used in the text: a.s.l. = above sea level; Ax = antenodal crossveins; Fw = fore wing; GL = genital ligula; Hw = hind wing; MBP = mediobasal process; Px = postnodal crossveins; Pt = pterostigma; S1–10 = abdominal segments; L1–2 = segments of genital ligula; VBP = ventrobasal process.

TAXONOMY

Forcepsioneura regua sp. nov.

Figs 1–6, 17–20, 24–28, 31–32, 35, 37

Forcepsioneura garrisoni nec Lencioni, 1999: — Costa and Oldrini (2005: 200, 208, key, misidentification of specimens from Rio de Janeiro State, Ilha da Marambaia island); — Anjos-Santos and Costa (2006: 41, misidentification based on Costa and Oldrini 2005).

Forcepsioneura haerteli nec Machado, 2001: — Pessacq et al. (2012: 6, misidentification of a male from Rio de Janeiro State, Cachoeiras de Macacu, Japuíba, Santana River).

Forcepsioneura sp.: — Garrison et al. (2010: 358–360, 479, figs 2380, 2383, 2386, 2389, illustrations of wings, middle and posterior lobes of prothorax in laterodorsal view, and caudal appendages in lateral and dorsolateral views of a male from Ilha da Marambaia island, Rio de Janeiro State, Brazil); — Kompier (2015): 112–113, photos of male and female habitus on the field from the type locality, characterization, comparison with *Epipleoneura venezuelensis*, *F. lucia*, and *F. sancta*, notes on habitat and behavior).

Diagnosis. Small, dorsally metallic greenish-blue, lateroventrally light blue and yellowish striped protoneurid; frons angulated; rear of head pale; CuP&AA indistinct; Ax space 1 as long as Ax space 3; GL with long inner fold and distal lateral lobe prolonged into a flagellum; cercus forcipate with mediobasal (MBP) and ventrobasal (VBP) processes, except for ill-defined prothoracic tubercle-like process, all characters agreeing with the definition of *Forcepsioneura* as recently diagnosed (Machado 2009, Garrison et al. 2010).

The comparatively longer VBP, longer than 0.53 of total cercus length (dorsal branch) in lateral view, allows separation of males of *F. regua* sp. nov. from species of the orange-black group (VBP ≤ 0.40 of cercus length in *F. grossiorum*, *F. lucia*, and *F. serrabonita* sp. nov.); anteromesal margin of MBP acute in dorsal view (Fig. 25) separates it from *F. sancta* (largely rounded in dorsal view; sensu Machado 2001); apex of VBP acute and MBP not visible in lateral view (Figs 5, 24) from *F. itatiaiae* (apex of VBP truncated and MBP visible in lateral view); VBP shorter, not reaching ventral margin of S10 in lateral view, and apex strongly curved internally (Figs 5, 24–26) from *F. haerteli* (which has a longer VBP at level of S10, and apex slightly curved internally).

Based on its general color and size, *F. regua* sp. nov. is most similar to *F. garrisoni*. Indeed, these two species are remarkably similar due to the blue stripes on thorax (light blue group). The short MP vein, which reaches distally at most 0.5 of the length of cell distal to the vein descending from subnodus, and the fin-shaped plate of MBP with an acute and strongly anteriorly directed apex in laterodorsal view are unique to these two species and separate them from all other known species in the genus. The short VBP, not reaching ventral margin of S10 ventrally, and posterolateral margin of median lobe of prothorax lacking a defined tubercle-like process distinguish *F. regua* sp. nov. from *F. garrisoni* (VBP long, ventrally reaching the ventral margin of S10, and posterolateral margin of median lobe of prothorax with a strongly defined conical tubercle-like projection). The ill-defined process on prothorax is unique for *F. regua* sp. nov. (Figs 1–3, 17, 27–28).

Despite the fact that the blue coloration and the ill-defined process on the prothorax of *F. regua* sp. nov. appear to be unique to females of this species, females of other *Forcepsioneura* are poorly known. Consequently, females should be identified through comparison with descriptions, figures, and their association with males.

Description of the male holotype. Head (Figs 1–2): Labium, visible parts of maxilla and mandibles (except apex)

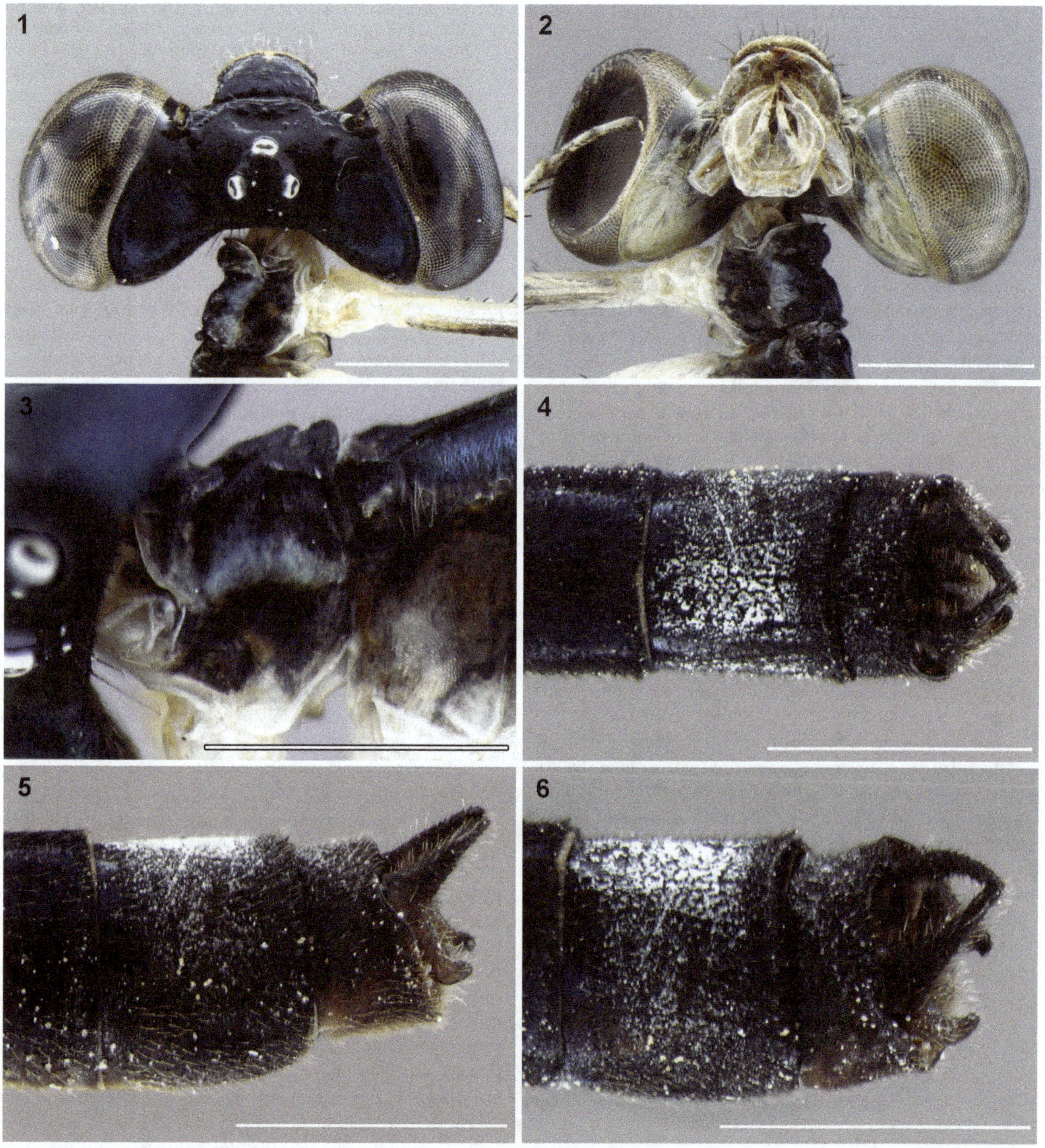

Figures 1–6. *Forcepsioneura regua* sp. nov., holotype male (Brazil. RJ: Cachoeiras de Macacu, DZRJ 2071): (1–2) head in dorsal (1) and ventral (2) views; (3) prothorax in lateral view; (4–6) caudal appendages in dorsal (4), lateral (5) and dorsolateral (6) views. Scale bars = 1 mm.

ivory yellow. Genae yellowish, darkening gradually posteriorly. Labrum dark brown to black with ventral 0.25 yellowish-orange, extended to 0.50 mesally, areas close to clypeolabral suture paler. Anteclypeus with lateral and mesal membranous ventral part yellowish, remainder part forming a rectangular black plate, postclypeus shining black. Antefrons, postfrons and remainder of epicranium black with shining bluish metallic reflections after postfrontal suture over postocular area. Antenna, antennifer and

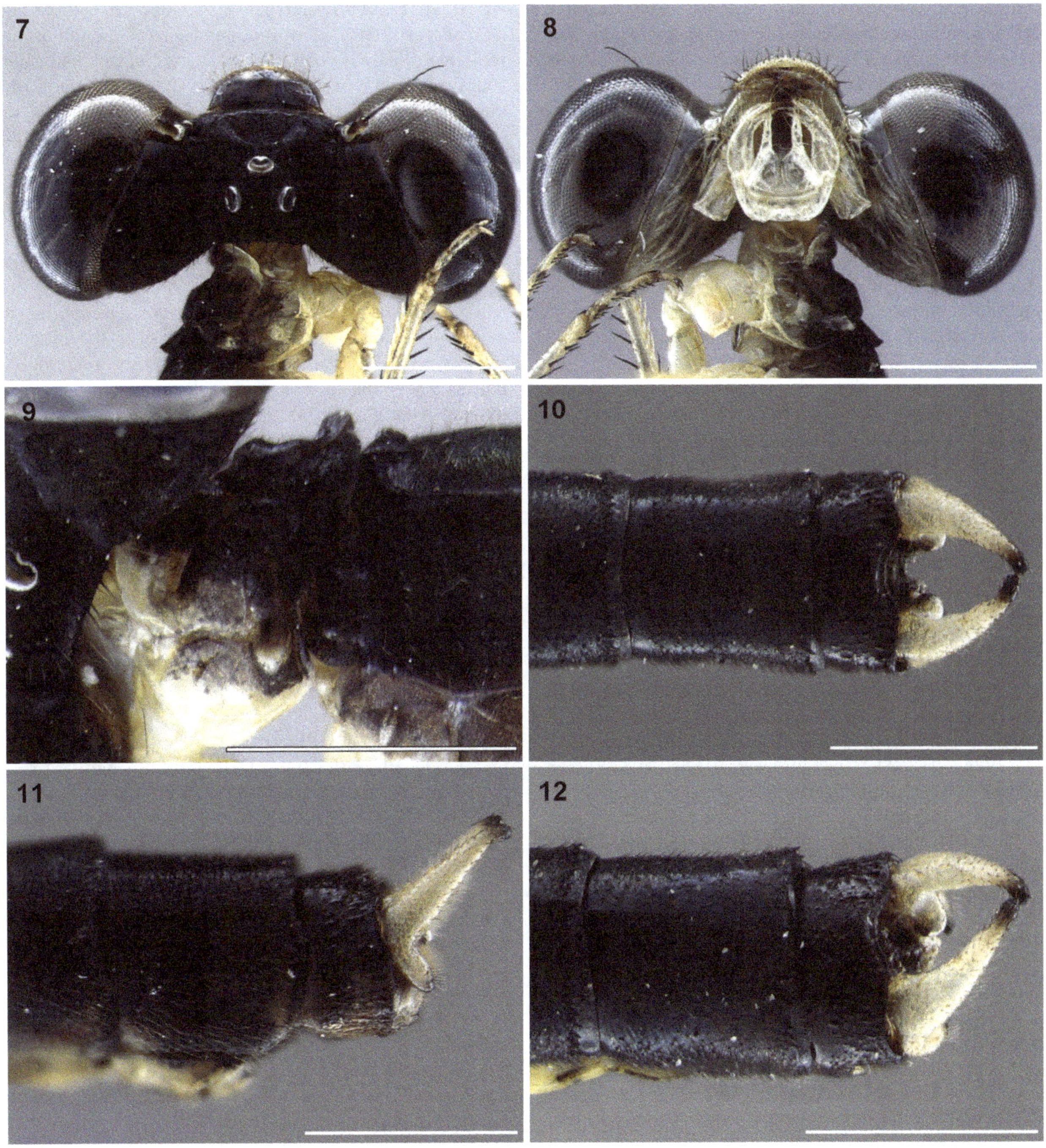

Figures 7–12. *Forcepsioneura serrabonita* sp. nov., holotype male (Brazil. BA: Camacan, DZRJ 0050): (7–8) head in dorsal (7) and ventral (8) views; (9) prothorax in dorsolateral view; (10–12) caudal appendages in dorsal (10), lateral (11) and dorsolateral (12) views. Scale bars = 1 mm.

scape black, posterior surface of pedicel light brown, all parts with laterofrontal surface and distal apex of scape ivory yellow; flagellum lost. Posterior region of the cranium ("postgena" plus "occiput") ivory yellow, 0.33 dorsal brownish-black.

Thorax (Figs 1–3, 17): Prothorax dorsally black, gradually lightening ventrally to brown laterally, ventral 0.3 of pleural sclerites yellow, notum with a ventrolateral whitish-blue stripe; anterior margin of anterior lobe black mesally, yellow laterally, posterolateral margin of median lobe with a very weak projection resembling the tubercle-like process observed in other *Forcepsioneura* (Figs 3, 17); posterior lobe asymmetrically convex, narrower than median lobe in dorsal view, right half slightly deformed (Fig. 17); sternellum (furcasternum sensu Asahina 1954) almost entirely black. Synthorax (Fig. 35) dorsally dark, gradually lightening ventrally; mesepisternum from mesostigmal region to antealar sinus dark brown to black with bluish metallic reflections, with a brown stripe occupying 0.33 of width running longitudinally to mesopleural suture; mesepimeron and metepisternum brown with irregular darker areas and a wide light blue longitudinal stripe running from posterolateral angle of mesinfraepisternum to antealar processes, occupying at maximum 0.33 of mesepimeron to 0.80 of metepisternum width, a mesal dark dot at level of intersegmental suture; metepimeron ivory yellow with a small brown area close to metapleural suture posteriorly; metapostepimeron ivory yellow with a black spot at lateroventral angle. Legs ivory yellow with irregular dark brown to black areas in the articulations of femur-tibia and tarsal segments, dorsal surface of femora, tibiae and apex of pretarsal claws (pretarsus); spurs black, except scale-like ivory yellow two proximal femoral spurs and tibial comb of prothoracic leg; anteroventral surface of femora armed with long and robust spurs, three in pro-, four in meso- and metathoracic femora, posteroventral surface with shorter and thinner spurs, 5 in pro-, 5–6 in meso- and 6–8 in metathoracic femora; anteroventral surface of tibiae armed with 9–10 spurs in pro- (4–5 of tibial comb), 5–6 in meso-, 6 metathoracic, length similar to intervening spaces; pretarsal claws with distinct acute supplementary inferior tooth at ca 0.50 of their length in all legs.

Wings (Fig. 35): Membrane hyaline; venation dark brown to black; Pt black, quadrangular; MP reaches anal margin at the vein descending from subnodus or very slightly distal, covering 2 cells in all wings; Px in Fw 11; Hw 9; RP2 originating at Px 5 in Fw, at Px 4 in Hw.

Abdomen (Figs 4–6, 18–20, 24–26, 35): S1–10 tergites dark brown to black dorsally, light brown to ivory yellow lateroventrally, darker in S8–10, with a very thin line along dorsal carina on S1–8; sternites similar in color to adjacent areas of tergites, yellow ivory in S1, remainder segments yellow to light brown with irregular dark areas, S9 dark brown; pale longitudinal stripe occupying ventral 0.5 of tergites S1–6 laterally, gradually narrowing to ca 0.2 in S8, a narrow line in S9 and ill-defined areas in S10; S3–7 with a narrow anterior pale ring ≤0.1 of segment total length, separated dorsally in S3; S1–7 cylindrical, S8–10 distinctly wider than previous segments, S7 width 0.6 of posterior part of S8 width; posterior margin of S10 with a shallow and wide concavity. Secondary genitalia (Figs 18–20) typical of Coenagrionoidea, anterior lamina with a deep and acute incision, anterior hamule quadrangular, with anteroventral angle acutely projected, posterior hamule almost entirely internalized with a curved thumb-shape; VS longer than wide, maximum width 0.3 of total length in ventral view. Genital ligula (as in Figs 18–20; based on paratype) with L1 smooth, without any kind of special setae; L2 with posterolateral portion of flexure projected distally beyond median region, making its margin concave in ectal view; distal margin of L2 (tip of ligula) with a mesal concavity (Fig. 18), lateral margins prolonged into two curved long flagella, in ectal view basally almost perpendicular, afterwards twisted (Figs 18–20); internal fold proximal to flexure, long with ca 0.4 of L2 total length in lateral view; no sclerotized tubercle at flexure. Epiproct reduced to a membranous-like plate. Cercus (Figs 4–6, 24–26) brown to dark brown, apex of cercus, ventrobasal and mediobasal processes black; cercus in lateral view slightly directed obliquely dorsad, gradually tapering distally; ventrobasal process perpendicular to dorsal branch, length ca 0.66 of cercus, apex stoutly rounded, at distal 0.2 distinctly curved posteriorly, distal edge at 0.7 from distance of ventrobasal process base to margin of S10; mediobasal process not visible in lateral view (Figs 5, 24); tip of cercus blunt; in dorsal view forcipate, proximally wide, slender distally; lateral margin almost straight, internal margin very slightly curved, apexes converging (Figs 4, 25); mediobasal process as a very acute fin-shaped plate, apex strongly directed anteriorly, positioned at basal 0.2 (Fig. 25); in posterior view directed ventrally obliquely and projected from a dilated area of cercus; apex of ventrobasal process broadly acute and abruptly curved in posterior view (Fig. 26). Paraproct orange-yellow, plate-like.

Measurements (mm). Total length (incl. caudal appendages) 33.5; abdomen length (excluding caudal appendages) 28.8; head maximum width 2.9; Fw length 18.5; Hw length 17.4; Fw maximum width 3.3, Hw maximum width 3.3; Pt length 0.47 in Fw, 0.49 in Hw; length of metathoracic femur 1.6; metathoracic tibia 1.7; length of S9+10 in lateral view 1.1; length of cercus (dorsal branch) in lateral view 0.45; length of ventrobasal process in lateral view 0.29.

Female paratype. Similar to holotype but paler, differences described below.

Head (Fig. 27): Labrum greenish-brown to light brown with ventral 0.25 yellowish, extended mesally to 0.50. Antefrons with mesal pale rounded spot. Posterior region of the cranium ("postgena" plus "occiput") ivory yellow, dorsal 0.20 brownish-black.

Thorax (Figs 27–28): Mesal projection of posterior lobe of prothorax with a small concavity divided into two lobes (Fig. 28). Anteroventral surface of femora armed with long and robust spurs, 3 in pro-, 4 in meso- and metathoracic femora, right metafemur with 7 supernumerary spurs, posteroventral surface with shorter and thinner spurs, 5 in pro-, 6 in meso- and metathoracic femora; anteroventral surface of tibiae armed with 5 spurs in meso-, 6 in metathoracic tibiae, length similar to intervening spaces (spurs not visible on the single preserved prothoracic leg). Wings: Pt brown, quadrangular but more rounded than in male; MP reaches anal margin at vein descending from subnodus in Fw and very slightly distal in Hw; Px in Fw 9–10; Hw 11.

Abdomen (Figs 31–32): Tergites of S1–10 dark brown to black dorsally, light brown to ivory yellow lateroventrally, distinctly darker in S8–10; tergites with a very thin pale line along the dorsal carina on S1–8; sternites similar in color to adjacent areas of tergites except 0.5 posterior of S8 dark brown with a transverse posterior yellow stripe contiguous to and with similar width of yellow posterior field of tergite. Ovipositor (Fig. 32) distally reaching level of cercus in lateral view, dark brown, lightening distally, dorsal lobe of third valve (gonoplac of S9) yellow, ventral surface smooth; stylus dark brown, tip yellow. Cercus, epiproct and paraproct dark brown to black.

Measurements (mm). Total length (incl. caudal appendages) 35.2; abdomen length (excluding caudal appendages) 30.5; head maximum width 2.9; Fw length 21.2; Hw length 20.2; Fw maximum width 3.7; Hw maximum width 3.8; Pt length 0.53 in Fw; 0.54 in Hw; length of metathoracic femur 2.0; metathoracic tibia 2.2; length of S9+10 in lateral view 1.0; total length of cercus in lateral view 0.28.

Variation in male paratypes. The nine other males are very similar to holotype except for minor differences described below.

Head: Antefrons ivory yellow with a mesal ill-defined brown line; occipital bar with a yellow spot. Flagellum dark brown. Thorax: Prothorax paler with whitish-blue ventrolateral stripe faint; posterior lobe truncated or with small median concavity. Light blue longitudinal stripe on synthorax occupying at maximum 0.5 of mesepimeron width to almost entire metepisternum. Long and robust spurs on anteroventral surface of femora 3–4 in pro-, 4 in meso- and 4–5 in meta-, posteroventral surface with shorter and thinner spurs, 3–5 in pro-, 4–6 in meso- and 5–8 in metathoracic femora; anteroventral surface of tibiae armed with 9–11 spurs in pro- (4–5 of tibial comb), 5–6 in meso-, 5–6 in metathorax.

Abdomen: Paraproct orange-yellow to dark brown.

Measurements (mm, n = 5). Total length (incl. caudal appendages) 31.7–35.9; abdomen length (excluding caudal appendages) 27.0–30.8; head maximum width 2.8–3.0; Fw length 17.3–19.2; Hw length 16.5–18.0; Fw maximum width 3.0–3.5; Hw maximum width 2.9–3.3; Pt length 0.44–0.58 in Fw; 0.46–0.58 in Hw; length of metathoracic femur 1.7–1.9; metathoracic tibia 1.9–2.1; length of S9+10 in lateral view 1.1–1.3; length of cercus (dorsal branch) in lateral view 0.42–0.60; length of ventrobasal process in lateral view 0.30–0.36.

Larva. Unknown.

Material examined (10 males, 2 females). Holotype male and paratype female (in tandem), BRAZIL. Rio de Janeiro State: Cachoeiras de Macacu municipality, Reserva Ecológica de Guapiaçu (REGUA), forest fragment (22°28′08″S, 42°45′39″W, 42 m a.s.l.), 27.VII.2012, T.M.F. Kompier leg. (DZRJ 2071); 1 male paratype, same data but pond with macrophytes (22°27′10″S, 42°46′13″W, 34 m a.s.l.), 03.XII.2009, A.L. Carvalho & Disciplina Técnicas de Coleta PPGZoo/UFRJ leg. (DZRJ 315, DNA voucher ENT3609); 1 male paratype, same data but stream at small wetland (22°25′51″S, 42°45′39″W, 75 m a.s.l.), 02.III.2013, T.M.F. Kompier leg. (DZRJ 326); 2 male paratypes, same data but 22°28′04″S, 42°45′32″W, 70 m a.s.l., 13.I.2014, (DZRJ 2251, DNA voucher ENT2854; DZRJ 2252, DNA voucher ENT3608); 1 male paratype, same municipality but [District of] Japuíba [22°33′46″S, 42°41′30″W, 26 m a.s.l.], Rio Santana (sic, Rio Macacu?), 08.V.1983, J.M. Costa leg. (MNRJ ODO-0011, DNA voucher ENT2368); 2 male paratypes, Mangaratiba municipality, Ilha de Marambaia, Gruta da Santa (23°03′35″S, 43°57′56″W, 100 m a.s.l.), 02.XII.2000, R.W. Garrison leg. (RWG, DNA voucher ENT3016; DZUP 498845); 1 male and 1 female paratypes, same data but 04.XII.2000 (RWG); 1 male paratype, same data but temporary pool close to areal (MN-RJ-ODO 0208, DNA voucher ENT3607).

Biological and ecological data. Males were observed at seeps from peaty substrate at the forest edge and on very shallow, small streams with abundant leaf litter and limited flow. Even though observations were made on sunny days, they seem to avoid direct sunlight, perching suspended from the tips of leaves and grasses, or on the leaf litter. Streams were either partially or fully covered by trees and bushes. Females were observed in the same habitat. Both males and females were seen flying very slowly, and appeared to float in the air. Copulation lasted up to 10 minutes. Females were observed ovipositing unattended on a muddy substrate. Their habitat was shared with *Heteragrion aurantiacum* Selys, 1862, *H. consors* Hagen in Selys, 1862 and *Acanthagrion gracile* (Rambur, 1842), species that are tolerant to disturbed habitats in typical southeastern Brazilian Atlantic Forest.

Etymology. Noun in apposition in reference to the acronym of the type locality, Reserva Ecológica de Guapiaçu (REGUA), a private reserve maintained for conservation, scientific research and environmental education.

Remarks. This species has been known to specialists for at least eighteen years (e.g., Garrison 2000). Some specimens of the type series of *F. regua* sp. nov. had been previously determined as *F. garrisoni*, *F. haerteli* or *F. sancta* by distinct specialists. For example, Costa and Oldrini (2005) cited this species as *F. garrisoni* from Ilha da Marambaia island in Rio de Janeiro, and provided an illustration of the cercus in lterointernal view (their fig. 30). Even though that illustration agrees with *F. garrisoni*, we think that it was modified from the original description of that species. We believe this because other illustrations in that paper were clearly extracted from previous works, and reproduced with minor modifications. Individuals of the population from Ilha da Marambaia island are distinctly larger than those from the type locality in REGUA.

Forcepsioneura serrabonita sp. nov.

Figs 7–12, 13–16, 21–23, 29–30, 33–34, 36, 38

Diagnosis. A medium, dorsally black and lateroventrally yellowish protoneurid species; frons angulated; rear of head

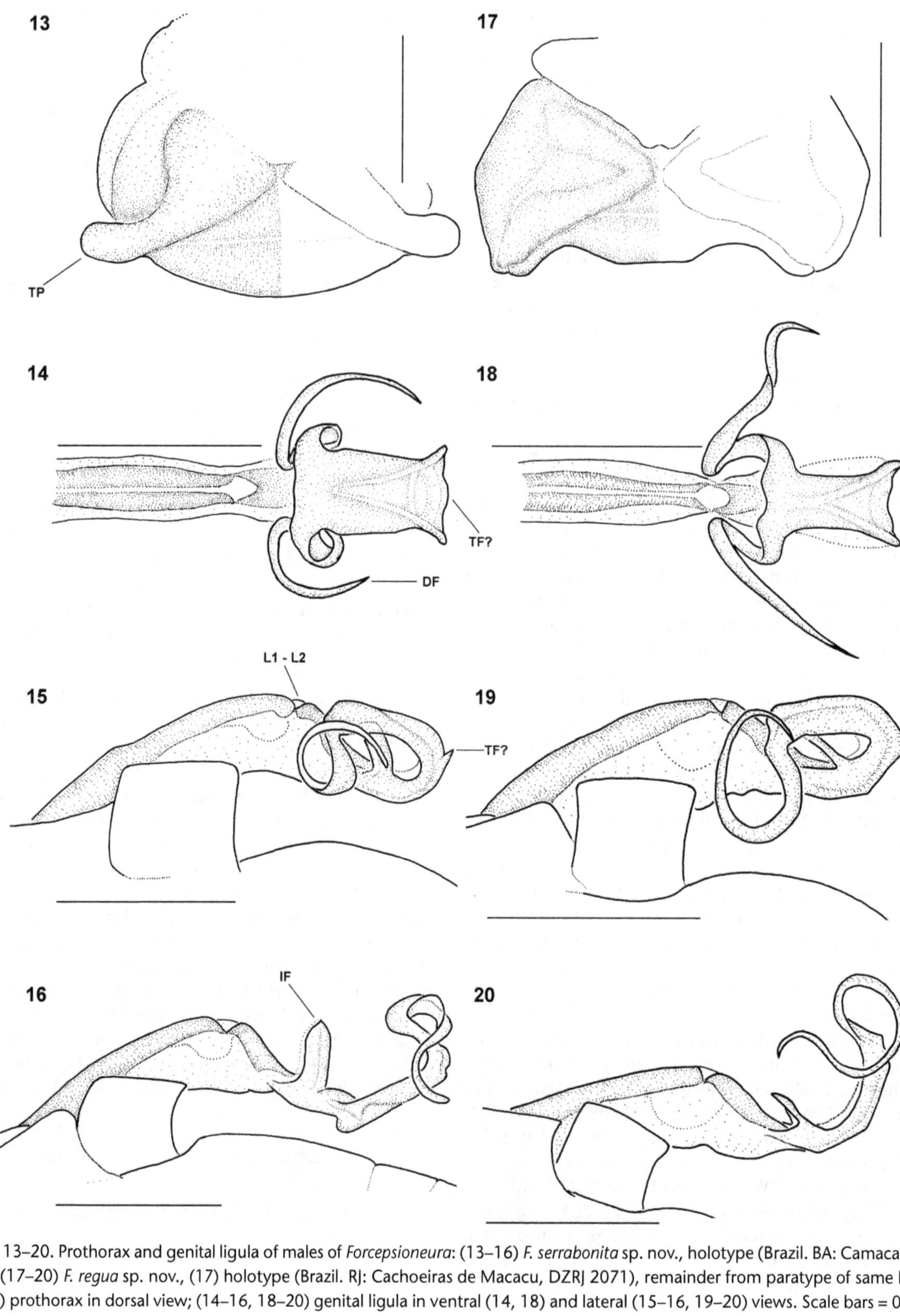

Figures 13–20. Prothorax and genital ligula of males of *Forcepsioneura*: (13–16) *F. serrabonita* sp. nov., holotype (Brazil. BA: Camacan, DZRJ 0050); (17–20) *F. regua* sp. nov., (17) holotype (Brazil. RJ: Cachoeiras de Macacu, DZRJ 2071), remainder from paratype of same locality; (13, 17) prothorax in dorsal view; (14–16, 18–20) genital ligula in ventral (14, 18) and lateral (15–16, 19–20) views. Scale bars = 0.5 mm.

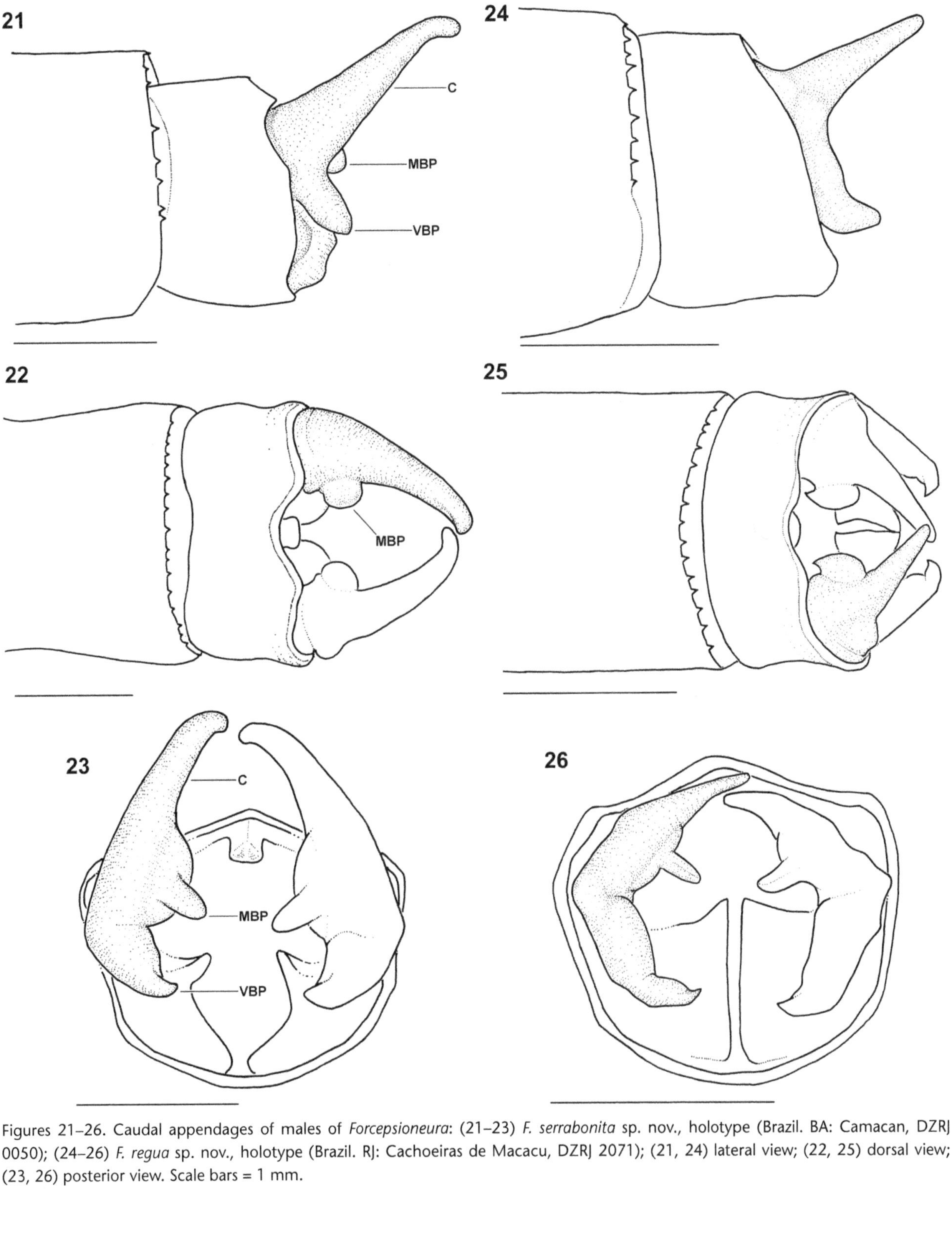

Figures 21–26. Caudal appendages of males of *Forcepsioneura*: (21–23) *F. serrabonita* sp. nov., holotype (Brazil. BA: Camacan, DZRJ 0050); (24–26) *F. regua* sp. nov., holotype (Brazil. RJ: Cachoeiras de Macacu, DZRJ 2071); (21, 24) lateral view; (22, 25) dorsal view; (23, 26) posterior view. Scale bars = 1 mm.

pale; posterolateral margin of median lobe of prothorax with well-defined tubercle-like process; CuP&AA indistinct; Ax space 1 as long as Ax space 3; GL with long inner fold and distal lateral lobe prolonged into a flagellum; cercus forcipate with mediobasal (MBP) and ventrobasal process (VBP), all characters agreeing with the definition of *Forcepsioneura* as recently diagnosed (Machado 2009, Garrison et al. 2010).

Based on general color and shape of male cercus, *F. serrabonita* sp. nov. resembles *F. grossiorum* and *F. lucia*. Males of this orange-black group of large species with robust cercus have a short VBP (as long as ≤ 0.40 of dorsal branch of cercus in lateral view), while it is longer (as long as ≥ 0.55 of dorsal branch) in *F. garrisoni*, *F. haerteli*, *F. itatiaiae*, *F. regua* sp. nov. and *F. sancta*. *Forcepsioneura serrabonita* sp. nov. is distinguished from the first two species by having posterior lobe of prothorax regularly rounded (Figs 9, 13; sinuous in *F. grossiorum*); Pt quadrangular, ratio between costal length and proximal length 0.54 in *F. serrabonita* sp. nov., being similar in size to *F. lucia* (ratio 0.6), but rectangular and distinctly longer in *F. grossiorum* (ratio = 0.8); cercus slender, with anteromesal margin of MBP process rounded in dorsal view (Figs 10, 22; acute in both *F. grossiorum* and *F. lucia*), MBP almost not visible in lateral view (Figs 11, 21; clearly visible in *F. grossiorum* and *F. lucia*).

In view of the large size of specimens of *F. serrabonita* sp. nov., they can be confused with *F. itatiaiae*, from which, in addition to the shorter VBP mentioned above, it can be separated by the shape of the male cercus with strongly cylindrical VBP (Figs 11, 21; blade-like and truncated apex in *F. itatiaiae*), MBP largely rounded dorsally and almost not visible in lateral view (Figs 10–11, 21–22; angulated plate as an axe with concave blade dorsally, and clearly visible in lateral view in *F. itatiaiae*).

Additionally, the distinctive shape of the GL (Figs 14–16), with a large and upright internal fold (obliquely toward anteriorly in the other species), will separate *F. serrabonita* sp. nov. from all species for which the GL has been described.

Females of several *Forcepsioneura* species are poorly known, which hampers a useful diagnosis. Putative females should be identified by comparison with descriptions, figures and association with males.

Description of male holotype. Head (Figs 7–8): Labium, visible parts of maxillae and mandibles (except apex) ivory yellow (Fig. 8). Genae yellowish, darkening gradually posteriorly. Labrum orange-brown (possibly light orange in life) with two lateral and one mesal rounded dark brown spots from clypeolabral suture to 0.5 ventral, pale areas among them darkened. Anteclypeus with lateral and mesal membranous ventral part yellowish, remainder part forming a semicircular dark brown plate, postclypeus almost entirely shining dark brown with ill-defined orange-brown spots. Antefrons dark brown with ill-defined pale areas, postfrons and remainder of epicranium opaque black with a weak copper luster. Antenna, antennifer and scape black, posterior surface of pedicel light dark brown, all parts with laterofrontal surface and distal apex of antennifer and scape ivory yellow; flagellum lost. Posterior region of the cranium ("postgena" plus "occiput") pale brown (Fig. 7), probably light ivory yellow.

Thorax (Figs 9, 36): Prothorax dorsally black, gradually lightening ventrally to yellow at pleural sclerites; anterior margin of anterior lobe ivory yellow; posterolateral margin of median lobe with well-defined tubercle-like process; posterior lobe rounded convex, narrower than median (Figs 9, 13); sternellum (furcasternum sensu Asahina 1954) almost entirely black. Synthorax (Fig. 36) dorsally dark, gradually lightening ventrally; mesepisternum from mesostigmal region to antealar sinus entirely dark brown to black with greenish-copper metallic reflections, except light brown line running longitudinal to mesopleural suture; mesepimeron dorsally light brown, narrowed ventrally by an ivory yellow stripe occupying 0.4 of the lateroventral apex of the sclerite, over mesinfraepisternum it occupies 0.5 ventral; remainder of synthorax ivory yellow, except poorly defined light brown stripe with about 0.3 of width along metapleural suture ending at spiracle level and a small black spot at ventro-distal apex of metapostepimeron. Legs ivory yellow with irregular dark brown to black areas in articulation femur-tibia, dorsal surface of metathoracic femur and apex of pretarsal claws (pretarsus); spurs black except scale-like ivory yellow proximal femoral spurs and tibial comb of prothoracic leg; anteroventral surface of femora armed with long and robust spurs, 5 in pro- and mesothoracic and 6 in metathoracic femora, posteroventral surface with shorter and thinner spurs, 3 in pro- and mesothoracic femora to 6 in metathoracic femur; anteroventral surface of tibiae armed with 9 spurs in pro- (4 of tibial comb), 7 in meso-, 6–7 in metathoracic tibiae, usually longer than intervening spaces; pretarsal claws with distinct acute supplementary inferior tooth at ca 0.50 of their length in all legs.

Wings (Fig. 36): Membrane hyaline; venation dark brown to black; Pt brown, with a narrow internal pale line surrounding the enclosing veins, quadrangular. MP reaches anal margin slightly distal of vein descending from subnodus, covering 2 cells in FW, distal to vein descending from subnodus covering ca 0.5 of a third cell in HW. Px in Fw 12; Hw 11; RP2 originating at or slightly distal to Px 5 in Fw, at or slightly proximal to Px 4 in Hw.

Abdomen (Figs 10–12, 14–16, 21–23, 36): Tergites of S1–10 dark brown to black dorsally, yellow ivory lateroventrally, darker in S8–10; sternites similar in color to adjacent areas of tergite, yellow ivory in S1, remainder segments light yellow to orange-yellow with irregular dark areas; pale longitudinal stripe occupying 0.5 ventral of S1–2 tergites laterally, gradually narrowing to ca 0.3 in S8, S9 with two ill-defined small pale spots, one close to posterior field of S8 and other close to posterior carina, S10 almost entirely black; S3–7 with an narrow anterior pale ring ≤0.1 of total segment length, dorsally separated by a thin line; S1–7 cylindrical, S8–10 distinctly wider than previous segments, S7–8 slightly smashed preventing an accurate measurement of abdominal width in these segments; posterior margin of S10 with a shallow and wide concavity. Secondary genitalia (Figs 14–16) typical of Coenagrionoidea, anterior lamina with a deep and acute incision, anterior hamule quadrangular, with anteroventral angle acutely projected, posterior hamule almost entirely internalized with a curved thumb-shape; VS longer than wide, maximum width 0.4 of total length in ventral view. Genital ligula (as in Figs 14–16) rectangular in ectal view; L1 smooth, without any kind of special setae; L2 with posterolateral portion of flexure projected distally beyond median region, mesally slightly projected in ectal view, in lateral view this projection similar to terminal fold (Fig. 15); distal margin of L2 (tip of ligula) almost straight in ectal view; distal lateral lobe prolonged into a gradually thinner, twisted, flagellum; internal fold proximal to flexure, upright, large, ca 0.4 of L2 total length in lateral view (Fig. 16); no sclerotized tubercle at level of flexure. Epiproct reduced to a membranous-like plate. Cercus (Figs 10–12, 21–23) ivory yellow, with apex, ventrobasal (VBP) and mediobasal (MBP) processes dark brown to black; in lateral view strongly directed obliquely dorsad; VBP in opposite direction, short, apex rounded, distal edge at half distance from ventral margin of S10; MBP visible as a rounded tubercle between cercus and VBP; tip of cercus blunt; in dorsal view forcipate, proximally wide, slender distally, lateral margin gently curved, internal margin slightly more curved, apices converging (Figs 10, 22); MBP as a rounded plate, positioned at 0.33 basal; in posterior view obliquely directed ventrally and projected from a dilated area of cercus; apex of VBP process acute and slightly curved in posterior view (Fig. 23). Paraproct orange-yellow, plate-like.

Measurements (mm). Total length (incl. caudal appendages) 38.7; abdomen length (excluding caudal appendages) 33.0; head maximum width 3.4; Fw length 21.8; Hw length 20.3; Fw maximum width 3.8, 3.7 in Hw; Pt length 0.54 in Fw, 0.58 in Hw; length of metathoracic femur 2.0; metathoracic tibia 2.1; length of S9+10 in lateral view 1.46; total length of cercus (dorsal branch) in lateral view 0.88; length of ventrobasal process in lateral view 0.36.

Female paratype. Similar to holotype. The only differences are described below.

Head (Figs 29–30): Labrum light orange with two lateral and a mesal rounded dark brown spots from clypeolabral suture to 0.5 ventral. Anteclypeus yellowish. Antefrons ivory yellow. Scape articulation and posterior surface of pedicel ivory yellow. Posterior region of the cranium ("postgena" plus "occiput") yellowish, brownish dorsally close to foramen

Thorax (Figs 30, 36): Leg spurs black; anteroventral surface of femora armed with long and robust spurs, 5 in pro-, 4 in meso- and metathoracic femora, posteroventral surface with shorter and thinner spurs, 3 in pro-, 4 in meso- and 6 in metathoracic femora; anteroventral surface of tibiae armed with 10 spurs in pro- (5 of tibial comb), 6 in meso- and metathoracic tibiae, usually longer than intervening spaces.

Wings: MP reaches anal margin slightly distal of the vein descending from subnodus, covering 2 cells and ca 0.25 of a third cell in FW and 0.5 in HW. Px in Fw 13; Hw 11; RP2 originating at or slightly proximal to Px 6 in Fw, at Px 4 in Hw.

Abdomen (Figs 33–34): Coloration very similar to that of male holotype except for the yellow ivory posterior field of S8 and dorsal longitudinal stripe on S9–10, forming an irregular T-spot dorsally on S9. Sternites light yellow to orange-yellow. Ovipositor (Fig. 34) reaching distally level of cercus in lateral view, yellowish-orange, dorsal lobe of third valve (gonoplac of S9) slightly paler, ventral surface smooth; stylus dark brown, tip yellow. Cercus, epiproct and paraproct dark brown to black (Figs 33–34).

Measurements (mm). Total length (incl. caudal appendages) 37.3; abdomen length (excluding caudal appendages) 31.0; head maximum width 3.1; Fw length 21.2; Hw length 19.8; Fw maximum width 3.8; Hw maximum width 3.9; Pt length in Fw 0.54; Pt length in Hw 0.59; length of metathoracic femur 2.2; metathoracic tibia 2.4; length of S9+10 in lateral view 1.0; total length of cercus (dorsal branch) in lateral view 0.32.

Variation in male paratypes. The four other specimens (excluding the teneral one) are very similar to holotype, minor differences are described below.

Head: Labrum brownish to black. Postclypeus shining dark or with ill-defined mesal orange-brown spot. Frons and remainder of epicranium opaque black with a weak copper luster. Antennal flagellum brown to black. Posterior region of cranium ("postgena" plus "occiput") ivory yellow.

Thorax: Anteroventral surface of femora armed with long and robust spurs, 5 in pro-, 3–4 in meso- and 5 in metathoracic femora, posteroventral surface with shorter and thinner spurs, 3 in pro-, 4–5 in meso-, 5–6 in metathoracic femora; anteroventral surface of tibiae armed with 9–10 spurs in pro- (4–5 of tibial comb), 5–6 in meso-, 5–6 in metathoracic tibiae.

Wings: MP reaches anal margin at or slightly distally to vein descending from subnodus, covering from 0.33 to 0.5 of a third cell in HW. Px in Fw 12–15; Hw 10–13; RP2 originating between Px 5–6 in Fw.

Measurements (mm). Total length (incl. caudal appendages) 37.5–40.5; abdomen length (excluding caudal appendages) 31.7–33.7; head maximum width 2.9–3.3; Fw length 20.3–22.3; Hw length 19.5–21.3; Fw maximum width 3.5–3.8; Hw maximum width 3.7–3.8; Pt length 0.56–0.60 in Fw; 0.58–0.64 in Hw; length of metathoracic femur 1.9–2.1; metathoracic tibia 2.1–2.2; length of S9+10 in lateral view 1.3–1.4; total length of cercus (dorsal branch) in lateral view 0.74–0.86; length of ventrobasal process in lateral view 0.12–0.22.

Larva. Unknown.

Material examined (6 males, 1 female). Holotype male, BRAZIL. Bahia State: Camacan municipality, Reserva Particular do Patrimônio Natural (RPPN) Serra Bonita, collecting point "Alojamento" (BA 2012-001; 15°23′16″S, 39°33′58″W, 810 m a.s.l.), 27–29.II.2012, R.A. Carvalho leg. (DZRJ 0050); 1 male paratype, same data but first trickle on road beyond managers house (15°23′19″S, 39°33′57″W, 710 m a.s.l. [822 m]), 5.I.2013, C.M. Flint & O.S. Flint Jr. leg. (USNM); 2 male paratypes, same

Figures 27–30. Head and prothorax in dorsal view of females of *Forcepsioneura*: (27–28) *F. regua* sp. nov., paratype (Brazil. RJ: Cachoeiras de Macacu, DZRJ 2071); (29–30) *F. serrabonita* sp. nov., paratype (Brazil. BA: Camacan, MNRJ ODO-0001). Scale bars = 1 mm.

data but quarry, 20.II.2015 (USNM; MZSP ODO-492, DNA voucher ENT2857); 1 male paratype, same data but 26.II.2015 (USNM); 1 female paratype, same data but spring near telecommunication towers (15°23'02"S, 39°34'19"W; 850 m a.s.l. [880 m]), 25.XII.2013, T.W. Donnelly leg. (MNRJ ODO-0001; ex-TWD collection). Additional specimen (excluded from type series). 1 male (teneral), same data as holotype but trickles at upper water tank (15°23'03", 39°34'19"W, 890 m a.s.l.), 23.XII.2012, C.M. Flint & O.S. Flint, Jr. leg. (USNM).

Etymology. Noun in apposition in reference to the type locality in the complex of RPPNs Serra Bonita, in honor of this protected area and its managers, who have taken positive steps towards the preservation of Atlantic Forest remnants in the southern portion of the biodiversity-rich state of Bahia.

Biological and ecological data. Expected to be a submontane forest species, most likely occurring at seepages similar to the mesohabitat occupied by the large montane species from southeastern Brazil (e.g., *F. itatiaiae* and *F.* cf. *lucia*). The larvae possibly are madicolous and should inhabit shallow streams running over rocky substrates covered by ferns and hornworts/ liverworts forming small waterfalls (Fig. 40). Though we have not observed adult behavior in the field it is plausible to assume that *F. serrabonita* sp. nov. is a forest damselfly like most *Forcepsioneura*.

DISCUSSION

The "Reserva Particular do Patrimônio Natural Serra Bonita" (RPPN, Private Reserve of Natural Heritage) is a complex of protected areas that started from a small fragment bought in the 1990s. Now it covers about 7,500 ha between 300 and 1,080 m above sea level of primary and secondary forests maintained by Instituto Uiraçu (Becker 2017). This complex is located in one of most biodiverse regions of the threatened Atlantic Forest, representing unique fragments with high levels of biodiversity and endemism on montane formations in southern Bahia. Although Serra Bonita remains undersampled, studies have shown

Figures 31–34. Abdominal segments seven to ten (S7–10) of females of *Forcepsioneura*: (31–32) *F. regua* sp. nov., paratype (Brazil. RJ: Cachoeiras de Macacu, DZRJ); (33–34) *F. serrabonita* sp. nov., paratype (Brazil. BA: Camacan, MNRJ ODO-0001). Scale bars for Figs 31–32 = 1 mm.

Figures 35–40. Habitus and habitat of species of *Forcepsioneura*: (35) *F. regua* sp. nov., holotype (Brazil. RJ: Cachoeiras de Macacu, DZRJ 2071); (36) *F. serrabonita* sp. nov., holotype (Brazil. BA: Camacan, DZRJ 0050), (37) *F. regua* sp. nov., live individual at type locality at Reserva Ecológica de Guapiaçu (REGUA); (38) overview of type locality at RPPN Serra Bonita; (39) shallow muddy and slow stream at Itatiaia National Park, typical habitat of small light blue species; (40) seepage covered by hornworts at Três Picos State Park, typical habitat of large orange-black species. Scale bars = 1 mm. Photos: (37) TMFK; (38) RA Carvalho; (39–40) APP.

that it includes at least 810 species of vascular plants, mostly flowering plants (Amorim et al. 2009, Matos et al. 2010), 80 of amphibians (Dias et al. 2014) and a huge number of moths, estimated between five and six thousand species (V. Becker pers. comm.). The species-rich longhorn beetles (Cerambycidae), however, have only been recorded in few numbers, currently only 51 species (Martins and Galileo 2010). In comparison, a preliminary checklist of dragonflies elaborated in collaboration by a few odonatologists already includes almost 90 species.

Like Serra Bonita, also in Southeastern Brazil, the "RPPN Reserva Ecológica de Guapiaçu" originated from private efforts aiming to conserve the AF. This protected area shares a similar history with Serra Bonita, beginning in the 1990s. Originally it was a farm and slowly it became a conservation and restoration portion of the AF. The preserve was founded in the beginning of 2000 and was officially recognized as a RPPN at the state level in 2013, at 302 ha. In 2014 a smaller, 35 ha area was incorporated to it and baptized as REGUA II. Unofficially, the preserve is between 6,500 and 7,500 ha, and 49% to 80% of its area is estimated to be inserted within the Três Picos State Park (Pimentel and Olmos 2011, Soares et al. 2011 for distinct data). The latter is a protected area created by the government of the state of Rio de Janeiro in 2002 and further enlarged in 2009 (INEA 2013, RJ decrees 31,443, 41,990). With elevations from sea level to 2,000 m (Soares et al. 2011), REGUA has a high diversity and is administrated in partnership with the public area of the Três Picos State Park. REGUA also integrates an important watershed for the metropolitan megacity of Rio de Janeiro (Rodríguez Osuna et al. 2014).

Recent studies highlight the relevance of REGUA. For example, 450 species of birds were recorded in REGUA and in neighboring areas (Pimentel and Olmos 2011). As far as we know, 59 species of mosquitoes (Silva et al. 2014) and at least 570 species of butterflies (Soares et al. 2011) have been found there. The most significant aspect of these two protected areas, REGUA and Três Picos State Park, is their status as the richest spots for dragonflies and damselflies in the world, with at least 204 species (see Kompier 2015), surpassing Manu National Park in Madre de Dios province in Peru, the site previously known as the most diverse with 188 species (Paulson 2004; number of species updated by Paulson 2006).

These recent findings, including the two new species described here, bring to attention the urgency in directing collecting efforts at unexplored areas, and also the importance of private preserves that harbor the type localities of described species in the AF. Together, REGUA and Três Picos harbor over 210 species of Odonata, or about 24% of the known Brazilian dragonflies, and represent a major hotspot for the conservation of these insects. As highlighted by Corbet (2006), dragonflies and damselflies depend on forests for survival and these insects are rapidly being deprived of their mesohabitats that allow their existence. Odonatologists should be alert to these processes and focus in knowledge production that promotes mitigation actions to conserve their populations. Charismatic groups of animals, such as insects of the order Odonata, can be used as flagship species for conservation (see Oertli 2008). The high diversity observed in these areas can be used to foster awareness and empathy for the relevance of the conservation of these fragments of AF. Four species of *Forcepsioneura* (out of the eight described) occur in the state of Rio de Janeiro within the National Park of Itatiaia and REGUA and its neighboring areas represent the hotspot for this genus.

When *Forcepsioneura* was erected, Lencioni (1999) designated *F. garrisoni* as the type species, a poorly known species. *Forcepsioneura garrisoni* was only known from the male holotype, until recently when APP collected males in the states of Rio de Janeiro and Paraná that agree fairly well with the original description, and additional specimens from São Paulo, which are deposited in a private collection (F. Lencioni pers. comm.). Besides the type species, the genus also includes the shady forest species *F. grossiorum*, *F. haerteli*, *F. itatiaiae*, *F. lucia*, *F. sancta* and the two new species described here. Of these, *F. grossiorum* is known from a single male from Serra dos Órgãos and *F. haerteli* from two males collected in state of Santa Catarina, the southernmost known record for the genus. All these aspects help to make the generic limits (especially in relation to *Amazoneura*) uncertain because few are known about the included species (e.g., morphological variation, female morphology).

Erected for Amazonian species (Machado 2004), the poorly known *Amazoneura* can be distinguished from *Forcepsioneura* by eight characters as defined by Machado (2009), even though he recognized they were weak to warrant generic delimitation. However, the rounded frons, dark markings on metepimeron and rear of head in *Amazoneura* and the strongly disjunct geographical distribution from *Forcepsioneura* were considered robust enough to support the separation between these two genera. The generic distinction between *Amazoneura* (Amazonia inhabitants) and *Forcepsioneura* (predominantly Atlantic Forest inhabitants) is still unsatisfactory (Machado 2009, Garrison et al. 2010). The inclusion of more species in *Forcepsioneura* and the discovery of undescribed species defy the original generic definition and make a revision necessary.

The two species described here contribute at different levels to challenge the definition of *Forcepsioneura*. In *F. serrabonita* sp. nov. the posterolateral margin of the median lobe is projected into a well-defined tubercle-like process (Figs 9, 13, 30), but it is only poorly projected in *F. regua* sp. nov. (Figs 3, 17, 28; an exception already cited by Machado 2009). The homology of these processes should be investigated in a phylogenetic context because similar projections are observed in a few other genera of Protoneurinae, including *Psaironeura*, a putative representative of the *Roppaneura* clade. The short VBP observed in *F. serrabonita* sp. nov. (Figs 11, 21, 23), though not as short as observed in *Amazoneura* or even in *Roppaneura*, somewhat links species of these genera and probably has a value for generic definition. In addition, *F. lucia* also has very short VBP on the cercus.

The strong separation between AF and Amazonian species, somewhat resembling a vicarious pattern, is questionable. There are records of *F. sancta* (a typical AF species) from the Cerrado domain of the central Brazilian plateau (Pessacq et al. 2012), and an unknown *Forcepsioneura* species was also mentioned from the extreme northern Atlantic Forest of state of Rio Grande do Norte (Irusta and Lencioni 2015). Forested areas in northeastern Brazil have been recently shown to be harbor a mixture between the AF and Amazonian elements (see discussion in Takiya et al. 2016). This puzzle can only be solved, and a more suitable classification proposed, through robust phylogenetic and population level analyses of the genera involved. Indeed, recent investigations using molecular data support the monophyly of *Forcepsioneura* (see Pimenta et al. 2015).

Based on size, general coloration, shape of cercus and habitat, *Forcepsioneura* can be divided into two species groups: (1) includes *F. garrisoni, F. haerteli, F. regua* sp. nov. and *F. sancta*. Small, lowland species that present pale areas predominantly bluish-green, cercus slender with ventrobasal process comparatively long, and distribution almost restricted to coastal areas of the Atlantic Forest (*F. sancta* is an exception), some species even occupying strictly restinga formations, being associated with small shallow streams and muddy pools with slow running water (Fig. 39); (2) includes *F. grossiorum, F. itatiaiae, F. lucia*, and *F. serrabonita* sp. nov. Large, submontane to montane species, pale areas predominately greenish-orange, cercus robust with short ventrobasal process, associated with seepages or small streams to rocky shallow waterfalls (Fig. 40). These groups do not necessarily share common evolutionary histories; however provide a useful grouping to be investigated in phylogenetic analyses.

The light blue *Forcepsioneura* form a complex of strongly similar species including *F. garrisoni, F. regua* sp. nov. and an undescribed species from southern state of Bahia. Although nothing was mentioned about blue hues in the original description of *F. garrisoni* (see Lencioni 1999), we suspect that is also blue. The males recently collected in Rio de Janeiro and Paraná by APP and which have been associated to this species are bluish. Analogous to other small Protoneurinae, such as species of *Drepanoneura* von Ellenrieder & Garrison, 2008, *Epipleoneura* Williamson, 1915, and *Peristicta* Hagen *in* Selys, 1860 (see von Ellenrieder and Garrison 2008, Anjos-Santos and Pessacq 2013, Pessacq 2014), they are strongly similar, and due to their extreme similarity in external morphology and coloration, species identification is only possible through careful examination of the caudal appendages and genital ligula. Therefore, it is possible that this complex of bluish species is a set of cryptic species (remember that even specialists have failed to identify them as distinct entities). Cryptic species might also be involved in distinct populations of *F. lucia*. This species was originally described from Rola Moça State Park, part of the Serra do Espinhaço mountain range in the state of Minas Gerais. However, other populations have recently been identified in Serra da Mantiqueira and Serra do Mar mountain ranges in state of Rio de Janeiro (Serra do Mar populations are mentioned in Kompier 2015) and a third one was found in the state of Espírito Santo. These populations might each represent undescribed species. An ongoing integrative taxonomy research on this group of damselflies has revealed a hidden diversity (Pimenta et al. 2015) and possibly will enlighten key aspects for future investigations about species delimitation in *Forcepsioneura*.

Additionally, Anjos-Santos and Pessacq (2013) considered the mysterious *Peristicta lizeria* Navás, 1920 from the province of Buenos Aires, Argentina, as a potential junior synonym of any of the two southern representatives of *Peristicta*, i.e., *P. aeneoviridis* Calvert, 1909 and *P. forceps* Hagen *in* Selys, 1860. However, there are no characters supporting Navás' (1920) original combination. Generally, Longinos Navás's descriptions of dragonflies are insufficient for identification, and confusing. Most species erected by him still await confirmation, but even considering his inaccuracies the illustration of the caudal appendages is strongly similar to species of the *Roppaneura* clade and not with that observed in *Peristicta*. Based on recent material collected in state of Rio Grande do Sul by APP, that likely represent two unknown species of *Forcepsioneura*, the possibility cannot be dismissed that *P. lizeria* actually belongs to the *Roppaneura* clade, representing a valid species either of *Forcepsioneura*, of *Roppaneura* or even of a new genus.

Finally, this paper is a first step towards an understanding of the diversity and morphology of this remarkable group of damselflies typical of the threatened Atlantic Forest domain. Studies on the morphological and spatial evolution of *Forcepsioneura* are being conducted.

ACKNOWLEDGMENTS

This paper was only possible due to the help of Thomas W. Donnelly, Oliver Flint Jr (USNAM), Rosser W. Garrison (CDFA), and Angelo B.M. Machado (UFMG), who lent us specimens. We also thank ICMBIO/SISBIO for collecting permits; APP is strongly indebted to Rachel A. Carvalho, a colleague and specialist in leafhoppers who collected one of the holotypes and has shared interesting material from her expeditions; the protected area managers and employees, Nicholas and Raquel Locke and Jorge Bizarro (REGUA) and Vitor Becker (Serra Bonita) for their support of scientific research as a way to promote biological conservation. This study was partially supported by the following postdoctoral grants to APP: Conselho Nacional de Desenvolvimento Científico e Tecnológico (CNPq-PDJ proc. 157592/2015-4) and Coordenação de Aperfeiçoamento de Pessoal de Nível Superior (CAPES-PVE proc. A019/2013).

LITERATURE CITED

Amorim AM, Jardim JG, Lopes MMM, Fiaschi P, Borges RAX, Perdiz RDO, Thomas WW (2009) Angiosperms of montane forest areas in southern Bahia, Brazil. Biota Neotropica 9(3): 1–36. https://doi.org/10.1590/S1676-06032009000300028

Anjos-Santos D, Costa J (2006) A revised checklist of Odonata (Insecta) from Marambaia, Rio de Janeiro, Brazil with eight new records. Zootaxa 1300: 37–50.

Anjos-Santos D, Pessacq P (2013) *Peristicta guarellae* sp. nov. from Brazil (Odonata: Protoneuridae). International Journal of Odonatology 16(4): 293–299. https://doi.org/10.1080/13887890.2013.845066

Arnett Jr RH, Samuelson GA, Nishida GM (1993) The Insect and Spider Collections of the World (2nd end). Sandhill Crane Press, Gainesville, 310 pp.

Asahina S (1954) A morphological study of a relic dragonfly *Epiophlebia superstes* Selys (Odonata, Anisozygoptera).The Japan Society for Promotion of Science, Tokyo, 153 pp.

Barnosky AD, Matzke N, Tomiya S, Wogan GO, Swartz B, Quental TB, Marshall C, McGuire JL, Lindsey EL, Maguire KC, Mersey B, Ferrer EA (2011) Has the Earth's sixth mass extinction already arrived? Nature 471(7336): 51–57. https://doi.org/10.1038/nature09678

Becker V (2017) The story of the SBR by Vitor Osmar Becker, PhD. https://www.serrabonita.org/the-story-of-the-serra-bonita [Accessed: 30/09/2017]

Bridges CA (1994) Catalogue of the Family-Group, Genus-Group and Species-Group Names of the Odonata of the World (3rd end). Privately published, Urbana, Illinois, 905 pp. https://doi.org/10.5962/bhl.title.15291

Bybee SM, Ogden TH, Branham MA, Whiting MF (2008) Molecules, morphology and fossils: a comprehensive approach to odonate phylogeny and the evolution of the odonate wing. Cladistics 24(4): 477–514. https://doi.org/10.1111/j.1096-0031.2007.00191.x

Corbet P (2006). Forests as habitats for dragonflies (Odonata). In: Rivera AC (Ed.) Forests and dragonflies: Fourth WDA International Symposium of Odonatology, Pontevedra, Spain, July 2005. Pensoft, Sofia, 13–36.

Costa JM, Oldrini BB (2005) Odonatas da Marambaia. In: Menezes LFT, Peixoto AL, Araujo DSD (Eds) História Natural da Marambaia. Editora da Universidade Rural, Seropédica, 195–230.

Crutzen PJ, Stoermer EF (2000) The "Anthropocene". Global Change Newsletter 41: 17–18. http://www.igbp.net/download/18.316f18321323470177580001401/1376383088452/NL41.pdf

De Marmels J (2002) A study of *Chromagrion* Needham, 1903, *Hesperagrion* Calvert, 1902, and *Zoniagrion* Kennedy, 1917: three monotypic North American damselfly genera with uncertain generic relationships (Zygoptera: Coenagrionidae). Odonatologica 31(2): 139–150. http://natuurtijdschriften.nl/record/592391

Dias IR, Medeiros TT, Nova MFV, Solé M (2014) Amphibians of Serra Bonita, southern Bahia: a new hotpoint within Brazil's Atlantic Forest hotspot. ZooKeys 449: 105–130. https://doi.org/10.3897/zookeys.449.7494

Dijkstra K-DB, Kalkman VJ, Dow RA, Stokvis FR, van Tol J (2014) Redefining the damselfly families: the first comprehensive molecular phylogeny of Zygoptera (Odonata). Systematic Entomology 39(1): 68–96. https://doi.org/10.1111/syen.12035

Evenhuis NL (2016) Abbreviations for insect and spider collections of the world. http://hbs.bishopmuseum.org/codens/codens-inst.html [Accessed: 30/11/2016]

Fleck G, Bechly G, Martínez-Delclòs X, Jarzembowski E, Coram R, Nel A (2003) Phylogeny and classification of the Stenophlebioptera (Odonata: Epiproctophora). Annales de la Société Entomologique de France, Nouvelle Série 39: 55–93. https://doi.org/10.1080/00379271.2003.10697363

Garrison RW (2000) Brazil 2000, or living it up (and collecting) in Rio. Argia 13(1): 12–15.

Garrison RW, von Ellenrieder N, Louton JA (2010) Damselfly genera of the New World. An illustrated and annotated key to the Zygoptera. The John Hopkins University Press, Baltimore, 490 pp.

Ingley SJ, Bybee SM, Tennessen KJ, Whiting MF, Branham MA (2012) Life on the fly: phylogenetics and evolution of the helicopter damselflies (Odonata, Pseudostigmatidae). Zoologica Scripta 41(6): 637–650. https://doi.org/10.1111/j.1463-6409.2012.00555.x

INEA (2013) Parque Estadual dos Três Picos: plano de manejo, resumo executivo. Instituto Estadual do Ambiente (INEA), Rio de Janeiro, 102 pp.

Irusta JB, Lencioni FAA (2015) First record of Pseudostigmatidae (Insecta: Odonata) in the Northeast Region of Brazil. Check List 11(2): 1–3. https://doi.org/10.15560/11.2.1565

Jablonski D (2001) Lessons from the past: evolutionary impacts of mass extinctions. Proceedings of the National Academy of Sciences 98(10): 5393–5398. https://doi.org/10.1073/pnas.101092598

Kompier T (2015). A guide to the dragonflies and damselflies of the Serra dos Orgaos [sic] South-easthern Brazil. Brussels, REGUA Publications, 379 pp.

Laurance WF (2009) Conserving the hottest of the hotspots. Biological Conservation 142(6): 1137. https://doi.org/10.1016/j.biocon.2008.10.011

Lencioni FAA (1999) The genus *Phasmoneura*, with description of *Forcepsioneura* gen. nov. and two new species. Odonatologica 28(2): 127–137. http://natuurtijdschriften.nl/record/592273

Machado ABM (1985) Studies on neotropical Protoneuridae. 5. Redescription of the type of *Phasmoneura ephippigera* (Selys, 1886) (Zygoptera). Odonatologica 14(4): 363–368. http://natuurtijdschriften.nl/record/591723

Machado ABM (1999) Studies on neotropical Protoneuridae 9. *Phasmoneura ciganae* Santos, conspecific with *Phasmoneura sancta* (Hagen) comb. nov. (Zygoptera). Notulae Odonatologica 5(3): 37–38. http://natuurtijdschriften.nl/record/593639

Machado ABM (2000) Studies on neotropical Protoneuridae. 10. *Forcepcioneura lucia* sp. nov. from the Parque Estadual Rola Moça, Minas Gerais, Brazil (Odonata, Zygoptera). Boletim Museu de Biologia Mello Leitão, Nova Série 11/12: 127–134. http://www.boletimmbml.net/pdf/11_07.pdf

Machado ABM (2001) Studies on neotropical Protoneuridae. 11. Two new species of *Forcepsioneura* Lencioni (Odonata-Zygoptera) with a key to the males of the genus. Revista Brasileira de Zoologia 18(3): 845–854. https://doi.org/10.1590/S0101-81752001000300021

Machado ABM (2004) Studies on Neotropical Protoneuridae. 15. *Amazoneura* gen. nov. with description of *A. juruaensis* sp. nov. (Odonata: Zygoptera). Revista Brasileira de Zoologia 21(2): 333–336. https://doi.org/10.1590/S0101-81752004000200029

Machado ABM (2005) *Forcepsioneura grossiorum* spec. nov. from Brazil (Zygoptera: Protoneuridae) [Studies on neotropical Protoneuridae, 17.]. Odonatologica 34(2): 169–172. http://natuurtijdschriften.nl/record/592511

Machado ABM (2008) [2009] Studies on neotropical Protoneuridae. 21. The status of *Amazoneura* Machado, 2004 (Odonata: Protoneuridae). Lundiana 9 (1): 53–56. http://www.icb.ufmg.br/lundiana/full/vol912008/05.pdf

Martinelli G, Vieira CM, Gonzalez M, Leitman P, Piratininga A, da Costa AF, Forzza RC (2008) Bromeliaceae da Mata Atlântica brasileira: lista de espécies, distribuição e conservação. Rodriguésia 59(1): 209–258. http://www.jstor.org/stable/23499386

Martins UR, Galileo MHM (2010) Cerambycidae (Coleoptera) da Serra Bonita. Papéis Avulsos de Zoologia 50: 435–443.

Matos F, Amorim AM, Labiak PH (2010) The ferns and lycophytes of a montane tropical forest in Southern Bahia, Brazil. Journal of the Botanical Research Institute of Texas 4(1): 333–346. http://www.jstor.org/stable/41972015.

Matushkina NA (2004) Comparative morphology of ovipositor in some damselflies (Odonata, Zygoptera). Vestnik Zoologii 38(3): 53–66.

Mittermeier RA, Turner WR, Larsen FW, Brooks TM, Gascon C (2011) Global biodiversity conservation: the critical role of hotspots. In: Zachos F, Habel J (Eds) Biodiversity Hotspots. Springer, Berlin, Heidelberg, 3–22. https://doi.org/10.1007/978-3-642-20992-5_1

Morrone JJ (2014) Biogeographical regionalisation of the Neotropical region. Zootaxa 3782(1): 1–110. https://doi.org/10.11646/zootaxa.3782.1.1

Myers N, Mittermeier RA, Mittermeier CG, Da Fonseca GA, Kent J (2000) Biodiversity hotspots for conservation priorities. Nature 403(6772): 853–858. https://doi.org/10.1038/35002501

Navás L (1920) Insectos Sudamericanos (2ª serie). Anales de la Sociedad Científica Argentina XC: 44–51. http://biodiversitylibrary.org/page/34988004

O'Grady EW, May ML (2003) A phylogenetic reassessment of the subfamilies of Coenagrionidae (Odonata: Zygoptera). Journal of Natural History 37(23): 2807–2834. https://doi.org/10.1080/0022293021000007453

Oertli B (2008) The use of dragonflies in the assessment and monitoring of aquatic habitats. In: Córdoba-Aguilar A (Ed.) Dragonflies and Damselflies: Model organisms for ecological and evolutionary research. Oxford University Press, New York, 79–95. https://doi.org/10.1093/acprof:oso/9780199230693.003.0007

Paulson DR (2004) Critical species of Odonata in the Neotropics. International Journal of Odonatology 7(2): 163–188. https://doi.org/10.1080/13887890.2004.9748208

Paulson DR (2006). The importance of forests to neotropical dragonflies. In: Rivera AC (Ed) Forests and dragonflies: fourth WDA International Symposium of Odonatology, Pontevedra (Spain), July 2005. Pensoft, Sofia, 79–101.

Pessacq P (2008) Phylogeny of Neotropical Protoneuridae (Odonata: Zygoptera) and a preliminary study of their relationship with related families. Systematic Entomology 33(3): 511–528. https://doi.org/10.1111/j.1365-3113.2007.00414.x

Pessacq P (2014). Synopsis of *Epipleoneura* (Zygoptera, Coenagrionidae, "Protoneuridae"), with emphasis on its Brazilian species. Zootaxa 3872(3): 201–234. https://doi.org/10.11646/zootaxa.3872.3.1

Pessacq P, Santos TC, Costa JM (2012) Checklist and updated distribution of Protoneuridae from Brazil. International Journal of Odonatology 15(2): 59–73. https://doi.org/10.1080/13887890.2012.672158

Pfau HK (1971) Struktur und Funktion des sekundären Kopulations apparates der Odonaten (Insecta, Palaeoptera), ihre Wandlung in der Stammesgeschichte und Bedeutung für die adaptive Entfaltung der Ordnung. Zeitschrift für Morphologie und Ökologie der Tiere 70: 281–371. https://doi.org/10.1007/BF00304097

Pimenta AL, Pinto AP, Takiya DM (2015) A first look at the phylogenetic relationships of *Forcepsioneura* and related genera (Coenagrionidae: Protoneurinae). ICO 2015 – International Congress of Odonatology. WDA, Argentina, 15–20.

Pimentel L, Olmos F (2011) The birds of Reserva Ecológica Guapiaçu (REGUA), Rio de Janeiro, Brazil. Cotinga 33: 8–24.

Pinto AP (2017) Odonata. In: Boeger WA, Zaher H, Rafael JA, Valim MP. Taxonomic Catalog of the Brazilian Fauna. PNUD. http://fauna.jbrj.gov.br/fauna/faunadobrasil/171 [Accessed: 26/09/2017]

Rehn AC (2003) Phylogenetic analysis of higher-level relationships of Odonata. Systematic Entomology 28(2): 181–240. https://doi.org/10.1046/j.1365-3113.2003.00210.x

Ribeiro MC, Metzger JP, Martensen AC, Ponzoni FJ, Hirota MM (2009) The Brazilian Atlantic Forest: How much is left, and how is the remaining forest distributed? Implications for conservation. Biological Conservation 142(6): 1141–1153. https://doi.org/10.1016/j.biocon.2009.02.021

Riek EF, Kukalová-Peck J (1984) A new interpretation of dragonfly wing venation based upon early Carboniferous fossils from Argentina (Insecta: Odonatoidea) and basic character states in pterygote wings. Canadian Journal of Zoology 62: 1150–1166. https://doi.org/10.1139/z84-166

Rodríguez Osuna V, Börner J, Nehren U, Prado RB, Gaese H, Heinrich J (2014) Priority areas for watershed service conservation in the Guapi-Macacu region of Rio de Janeiro, Atlantic Forest, Brazil. Ecological Processes 3(16): 1–21. https://doi.org/10.1186/s13717-014-0016-7

Safford HD (1999) Brazilian Páramos I. An introduction to the physical environment and vegetation of the campos de altitude. Journal of Biogeography 26(4): 693–712. https://doi.org/10.1046/j.1365-2699.1999.00313.x

Silva JS, Couri MS, Giupponi APL, Alencar J (2014) Mosquito fauna of the Guapiaçu Ecological Reserve, Cachoeiras de Macacu, Rio de Janeiro, Brazil, collected under the influence of different color CDC light traps. Journal of Vector Ecology 39(2): 384–394. https://doi.org/10.1111/jvec.12114

Singh JS (2002) The biodiversity crisis: a multifaceted review. Current Science 82(6): 638–647. http://www.jstor.org/stable/24106689

Soares A, Bizarro JM, Bastos CB, Tangerini N, Silva NA, da Silva AS, Silva GB (2011) Preliminary analysis of the diurnal Lepidoptera fauna of the Três Picos State Park, Rio de Janeiro, Brazil, with a note on *Parides ascanius* (Cramer, 1775). Tropical Lepidoptera Research 21(2): 66–79. http://troplep.org/TLR/21-2/Tropical-Lepidoptera-Research-21-2-2011-Soarez-et-al.pdf

Takiya DM, Santos APM, Pinto AP, Henriques-Oliveira AL, Carvalho AL, Sampaio BHL, Mattos BC, Moreira FFF, Avelino-Capistrano F, Gonçalves IC, Cordeiro IRS, Câmara JT, Barbosa JF, Souza WRM, Rafael JA (2016) Aquatic insects from the Caatinga: checklist and diversity assessment from Ubajara (Ceará State) and Sete Cidades (Piauí State) National Parks, Northeast Brazil. Biodiversity Data Journal 4: 1–195. https://doi.org/10.3897/BDJ.4.e8354

Tillyard RJ (1917) The biology of dragonflies (Odonata or Paraneuroptera). Cambridge, University Press, 386 pp.

Tillyard RJ, Fraser, FC (1938) A reclassification of the order Odonata. Australian Zoologist 9: 125–169.

UN (2015) World Urbanization Prospects: The 2014 Revision. United Nations (UN), Department of Economic and Social Affairs PD, New York, 493 pp.

von Ellenrieder N, Garrison RW (2008) *Drepanoneura* gen. nov. for *Epipleoneura letitia* and *Protoneura peruviensis*, with descriptions of eight new Protoneuridae from South America (Odonata: Protoneuridae). Zootaxa 1842: 1–34.

Watson JAL (1992) The subfamilies of Protoneuridae (Zygoptera). Odonatologica 21(2): 195–201. http://natuurtijdschriften.nl/download?type=document;docid=591986

Author Contributions: APP analyzed the data and wrote the paper; APP and TK collected the data and reviewed the final draft.

Competing Interests: The authors have declared that no competing interests exist.

9

Description of immature stages and natural history of *Stigmella schinivora* (Lepidoptera: Nepticulidae), a leaf-miner associated with the Brazilian peppertree

Cristiano M. Pereira[1], Denis S. Silva[2], Héctor A. Vargas[3], Gilson R.P. Moreira[2]

[1]*Programa de Pós-graduação em Biologia Animal, Departamento de Zoologia, Instituto de Biociências, Universidade Federal do Rio Grande do Sul. Avenida Bento Gonçalves 9500, 91501-970 Porto Alegre, RS, Brazil.*
[2]*Departamento de Zoologia, Instituto de Biociências, Universidade Federal do Rio Grande do Sul. Avenida Bento Gonçalves 9500, 91501-970 Porto Alegre, RS, Brazil.*
[3]*Departamento de Recursos Ambientales, Facultad de Ciencias Agronómicas, Universidad de Tarapacá. Casilla 6-D, Arica, Chile.*
Corresponding author: Gilson R.P. Moreira (gilson.moreira@ufrgs.br)

ABSTRACT. *Stigmella schinivora* van Nieukerken, 2016 was described from Cataratas de Iguazú, Misiones, Argentina, based on adults reared from *Schinus terebinthifolius* Raddi (Anacardiaceae) leaf mines. The aim of this study is to describe for the first time the external morphology of the immature stages of *S. schinivora* with the aid of light and scanning electron microscopy, based on mines collected on the same host plant, but in Laranjeiras do Sul, Paraná, Brazil. Data on natural history, including histology of the mines, are also provided. The larva passes through four instars, all endophytic, having chewing mouth parts and feeding on the palisade parenchyma. The first three instars are apodous and have a subcylindrical body, bearing only one pair of setae on the tenth abdominal segment; the fourth instar is eruciform, with well-developed ambulatory calli on thorax and abdomen and setae on all tagmata. A serpentine mine is constructed on the adaxial surface, progressively increasing in width during larval development. With the exception of the widened, terminal section, the mine is left filled with larval feces. The fully developed larva of last instar exits through a slit made at the distal end of the mine, building a silk cocoon on the leaf abaxial surface where pupation occurs. This is the first record of *S. schinivora* from Brazil, which was only known from the type locality in Argentina.

KEY WORDS. Atlantic forest, leaf-mining moths, microlepidoptera, nepticulids, *Schinus terebinthifolius*.

INTRODUCTION

The Nepticulidae is one of the most ancient lineages of Lepidoptera, with a global distribution and approximately 884 described species that are divided into 29 genera (van Nieukerken et al. 2016a, van Nieukerken 2018). They are among the smallest extant lepidopterans, having predominantly leaf miner habits and being associated with several plant families (Braun 1917, van Nieukerken et al. 2016a). Their minute size and the scarcity of the material available in collections has led their taxonomy to be mainly based on the morphology of adults, the immature stages rarely being taken into account (e.g., van Nieukerken et al. 2004). Studies that include the general appearance of their mines are not uncommon (e.g., Braun 1917, Stonis et al. 2013, 2014, van Nieukerken et al. 2016b), but specialization of larval feeding on tissues, if any, is largely unknown.

The worldwide distributed genus *Stigmella* Schrank, 1802 currently with ca. 420 species, is the largest genus of Nepticulidae (van Nieukerken 2018). Species identification in this genus can be difficult, since species complexes are common whose adults have similarities in external appearance; in these cases, only subtle differences in the genitalia morphology can be detected among species (Stonis and Remeikis 2016). There are at least 61 species of *Stigmella* recognized for the Neotropical Region (van Nieukerken et al. 2016a); however, this genus is still little studied in this region; thus, this number may not reflect its real diversity in the Neotropics (Puplesis and Robinson 2000, Šimkevičiūtė et al. 2009, Stonis et al. 2014, van Nieukerken et al. 2016a, Stonis and Remeikis 2017). This aspect is even more relevant in Brazil, where there are no records of *Stigmella* yet. This is unexpected since this is a megadiverse country, including biomes such as the Atlantic Forest, known for the great diversity of plants and animals and high endemism indexes (Myers et al. 2000).

Stigmella schinivora van Nieukerken, 2016 was recently described as a leaf miner of the Brazilian peppertree, *Schinus terebinthifolius* Raddi (Anacardiaceae) from the region of Misiones, Argentina (van Nieukerken et al. 2016b). Its description relied on morphology of the male and female genitalia. In the present study, using material collected in southwest Paraná state, Brazil, we present a detailed description of the immature stages of *S. schinivora*, based on light and scanning electron microscopy. We also provide additional information about its natural history, including the histology of its mines on leaves of *S. terebinthifolius*.

MATERIAL AND METHODS

Specimens used in this study came from leaf mines of *S. terebinthifolius* collected in Laranjeiras do Sul municipality, Paraná, Brazil, in 2016 and 2017. They were brought to the Laboratório de Morfologia e Comportamento de Insetos (LMCI), Zoology Department of Federal University of Rio Grande do Sul (UFRGS), Porto Alegre city, and then they were either dissected or kept at room temperature in plastic pots containing moistened cotton for emergence of adults. The adults obtained in the laboratory were identified as *S. schinivora* based on comparison with original descriptions and illustrations of the adult stage, including female and male genitalia, provided by van Nieukerken et al. (2016b).

Adults were pinned and dried. Immature stages were fixed in Dietrich's fluid and preserved in 75% ethanol. For descriptions of the gross morphology, the specimens were cleared in a 10% potassium hydroxide (KOH) solution and slide-mounted in either glycerin jelly or Canada balsam. Observations were performed with the aid of a Leica M125 stereomicroscope, and measurements were performed using an attached ocular micrometer (precision = 0.01 mm). Structures selected to be drawn were previously photographed with a Sony Cyber-shot DSC-H10 digital camera attached to the stereomicroscope, and also by using a Nikon AZ 100M stereomicroscope. Vectorized line drawings were then made with the software Corel Photo-Paint X7, using the corresponding digitalized images as a guide. At least five specimens were used for the descriptions of each morphotype.

For scanning electron microscope analyses, additional specimens were dehydrated in a Bal-tec CPD030 critical-point dryer, mounted with double-sided tape on metal stubs and coated with gold in a Bal-tec SCD050 sputter coater. They were examined and photographed in a JEOL JSM6060 scanning electron microscope at the Centro de Microscopia Eletrônica (CME) of UFRGS.

For plant anatomical descriptions, field-collected leaf parts of *S. terebinthifolius* containing mines of *S. schinivora* were preserved in Dietrich's fluid. Leaf parts containing the different larval instar morphotypes were selected under a stereomicroscope, and freehand cross sections were cut with a razor blade. They were then stained for five seconds with safranin and photographed with a Nikon AZ 100M stereomicroscope.

Vouchers of specimens used in this study were deposited in the insect collection of the Laboratório de Morfologia e Comportamento de Insetos (LMCI), Zoology Department (UFRGS), as follows (all coming from *S. terebinthifolius* leaf-mines collected by the senior author at Laranjeiras do Sul, Paraná, Brazil): 16-23.VII.2016, pinned, dried adults, two females (LMCI 309-10 and 11, with genitalia on slides GRPM 50-151 and 152, respectively), two males (LMCI 309-12 and 13, with genitalia on slides GRPM 50-153 and 154, respectively); 29.XI.2017, immature stages, fixed in Dietrich's fluid, preserved in 70% ethanol, 3 first instar larvae (LMCI 323-2), 3 second instar larvae (LMCI 323-3), 4 third instar larvae (LMCI 323-4), 8 fourth instar larvae (LMCI 323-5) and 3 pupae (LMCI 323-7).

RESULTS

Egg. Flat and oval, firmly adhered to the leaf surface by a glistening substance (Figs 1, 44); average diameter + standard deviation = 0.16 ± 0.002 mm, n = 5. It is covered by a solid, smooth, transparent layer, forming a cap; micropyles and aeropyles were not found.

Larva. Prognathous, with buccal apparatus of chewing type. There are four instars and two morphotypes; the first form corresponds to the first three instars and the second to the last instar. The first morphotype has a subcylindrical, smooth body, without specialized locomotor structures (Fig. 2). The second morphotype has well-developed calli on thorax and abdomen, and setae of variable sizes distributed throughout the body (Fig. 19). We could not find major morphological differences among instars of the first morphotype. However, they can be identified by their size, since corresponding head capsule widths do not overlap (Table 1). The following exponential growth equation was adjusted for the head capsule width: $y = 0.049e^{0.400x}$; n = 37; r = 0.99; $p < 0.0001$.

Penultimate instar. Except for the absence of stemmata, the head of the first morphotype is similar to that of the second one in general color, shape (Figs 3–5), antennae (Fig. 6) and mouth parts (Figs 4, 5, 7, 8), which are described in detail below. The same occurs in relation to thorax and abdomen, including spiracles (Fig. 9). No evident setae were found on the thorax or abdomen of the first morphotype, except for the tenth segment where a pair of conspicuous setae appear dorsolaterally (Figs 10, 11).

Table 1. Variation in size of head capsule width among instars of *Stigmella schinivora* reared on *Schinus terebinthifolius*.

Instar	N	Head capsule width (mm)		
		Mean ± standard error	Range	Growth rate
I	4	0.087 ± 0.003	0.084–0.095	–
II	5	0.116 ± 0.003	0.105–0.126	1.33
III	15	0.154 ± 0.002	0.147–0.168	1.33
IV	15	0.246 ± 0.002	0.231–0.263	1.60

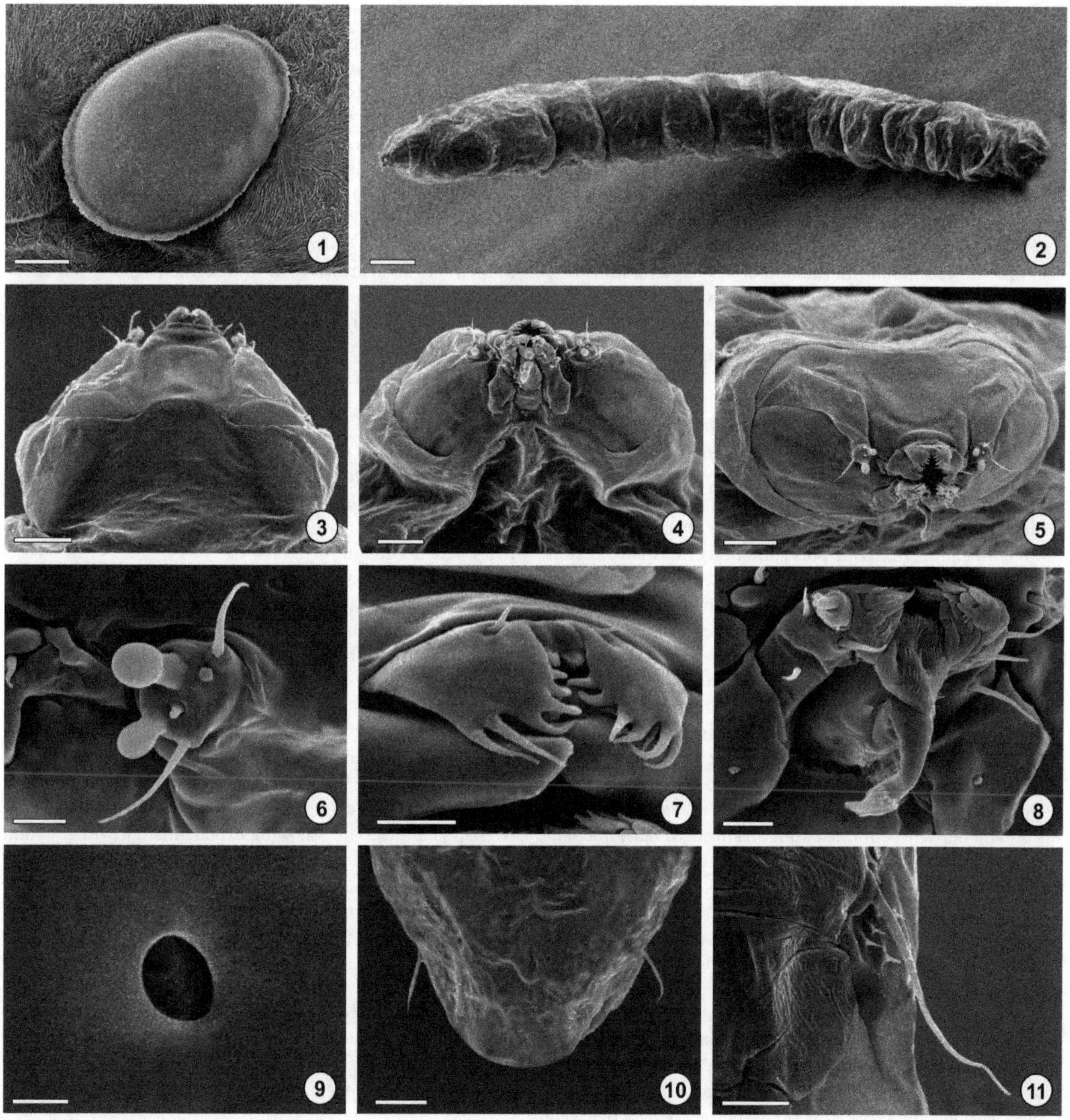

Figures 1–11. Egg and third instar of *Stigmella schinivora* under scanning electron microscopy: (1) egg; (2) general view of larva, lateral; (3–5) head, under dorsal, ventral and anterior views, respectively; (6) antenna, anterior; (7) labrum, anteroventral; (8) labium, showing spinneret in detail, ventral; (9) spiracle of fourth abdominal segment, lateral; (10) last abdominal segment, dorsal; (11) seta of last abdominal segment in detail, dorsal. Scale bars: 50, 100, 20, 20, 20, 20, 5, 5, 1, 25, and 10 μm, respectively.

Last instar. Average length ± standard deviation = 1.42 ± 0.21 mm; n = 5. Head light brown, flattened dorsoventrally, partially concealed within the prothorax, with deep epicranial notch. Frontoclypeus rectangular, longer than wide. Labrum bilobed, with lobes having distal serrated edge, and bearing one pair of short setae mesally (Figs 21, 22); mandibles with well-developed cusps; one long seta on proximal base. Maxilla with well-developed galea and palpi. Labium with tubular spin-

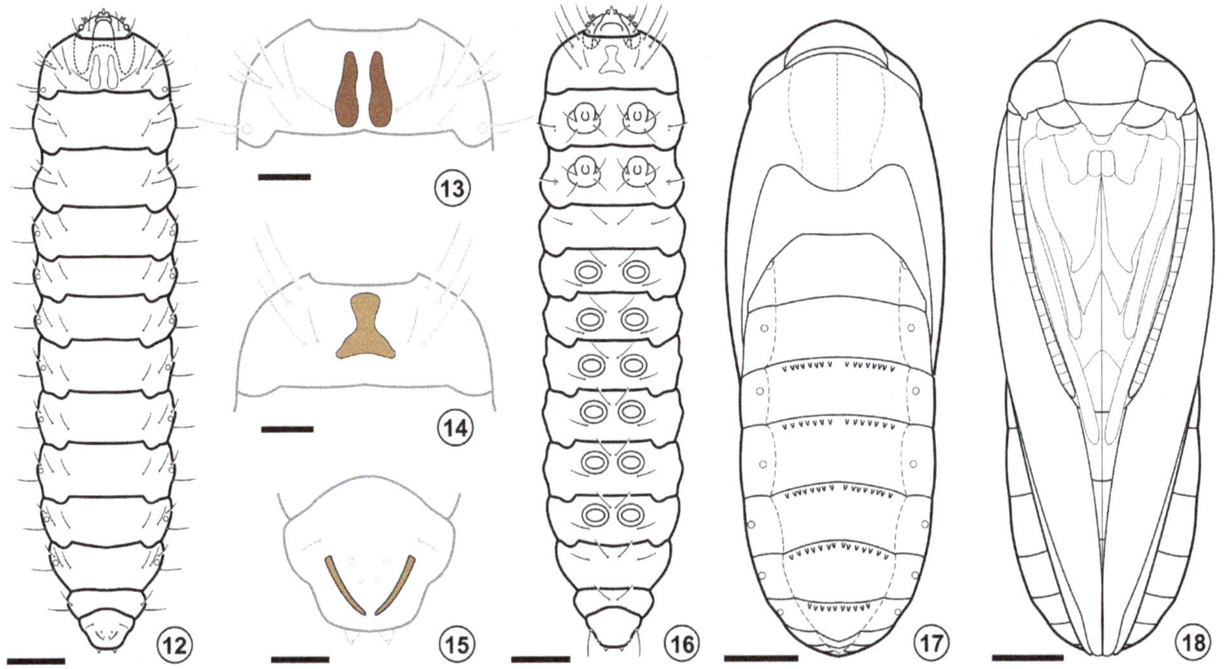

Figures 12–18. Last larval instar and pupal morphology of *Stigmella schinivora* under light microscopy: (12, 16) larva general, dorsal and ventral views, respectively; (13, 14) detail of tergal and sternal prothoracic plates seen through transparency, dorsal and ventral views, respectively; (15) anal rods of last abdominal segment, dorsal. (17, 18) pupa, dorsal and ventral, respectively. Scale bars: 300, 150, 150, 150, 300, 200, and 200 µm, respectively.

neret (Figs 20, 22), with a pair of setae on proximal base; labial palpi unisegmented, bearing a distal seta. Antenna unisegmented, with six apical sensilla; two minute in size, two stout and rounded, and two long and filiform (Fig. 23). A single, circular stemma, posterior to the antenna (Fig. 20). Thorax and abdomen cylindrical, creamy white in preserved material, bearing well-developed filiform setae. Thorax with integument smooth on T1 and sculptured with microtrichia ventrally in T2,3. T1 bearing a light brown shield on the tergum, divided into two elongated, meso-longitudinally arranged plates (Figs 12, 13); a light brown, cup-shaped plate on center of ventral prothoracic sternum (Figs 14, 16). A pair of lateral spiracles without elevated peritreme laterally on prothorax; legs and ambulatory calli absent. T2,3: Dorsal surface smooth; well-developed ambulatory calli ventrally, with the base wider than the transversally rounded apex, bearing an invagination on middle of the posterior wall (Figs 16, 24, 25). Each callus has the base sculptured with microtrichia and the distal edge smooth (Figs 24, 25). Abdominal segments similar in size from A1 to 8; A9 narrower; A10 smaller and subtriangular in shape. Integument mostly smooth on A1 and A10; partially sculptured with microtrichia dorsally on A3-8, laterally on A8-10 and ventrally on A2-9. Spiracles circular, without elevated peritreme (Fig. 27), laterally, from A1 to 8. Pairs of ambulatory calli present on A2-7, differing from the thoracic ones mainly by not having invagination (Fig. 26). A10 smooth, with a pair of light brown, longitudinally arranged, distally converging anal rods that are seen by transparency (Fig. 15); two pairs of triangular-shaped projections, one ventrolaterally (Fig. 31), the other on the distal edge of the segment (Figs 29, 31).

Chaetotaxy of the last instar larva. Head: Most of the setae are absent. Anterior group (A) and stemmatal group (S) unisetose. Substemmatal group (SS) bisetose. Thorax: T1 with thirteen pairs of setae. Dorsal group (D) bisetose; D1 near the lateral margin of the dorsal plate, D2 between D1 and spiracle. Extra dorsal group (XD) bisetose. Subdorsal group (SD) bisetose. Lateral group (L) trisetose; L1 and L3 ventral to spiracle, L3 between spiracle and L1, and L2 anterior to the spiracle, ventral to SD1. Subventral group (SV) trisetose; SV1 near the ventral plate, SV2 near the head capsule, and SV3 between SV1 and SV2. Ventral group (V) unisetose in the ventral plate margin. T2-3 with ten pairs of setae. Dorsal group (D) unisetose. Subdorsal group (SD) and lateral group (L) bisetose; L2 half of the length of L1. Subventral group (SV) trisetose; SV1 and SV3 in the callus, SV2 lateral to the callus. Ventral group (V) bisetose; V1 and V2 on the callus. Abdomen: A1–8 with six pairs of setae. Dorsal group (D) unisetose. Subdorsal group (SD) bisetose; SD1 between D2 and spiracle, SD2 near and anterior to spiracle. Lateral group (L) unisetose; L1 posteroventral to the spiracle. Subventral group (SV) unisetose. Ventral group (V) unisetose, near the ventral medial line. A9 with chaetotaxy similar to the

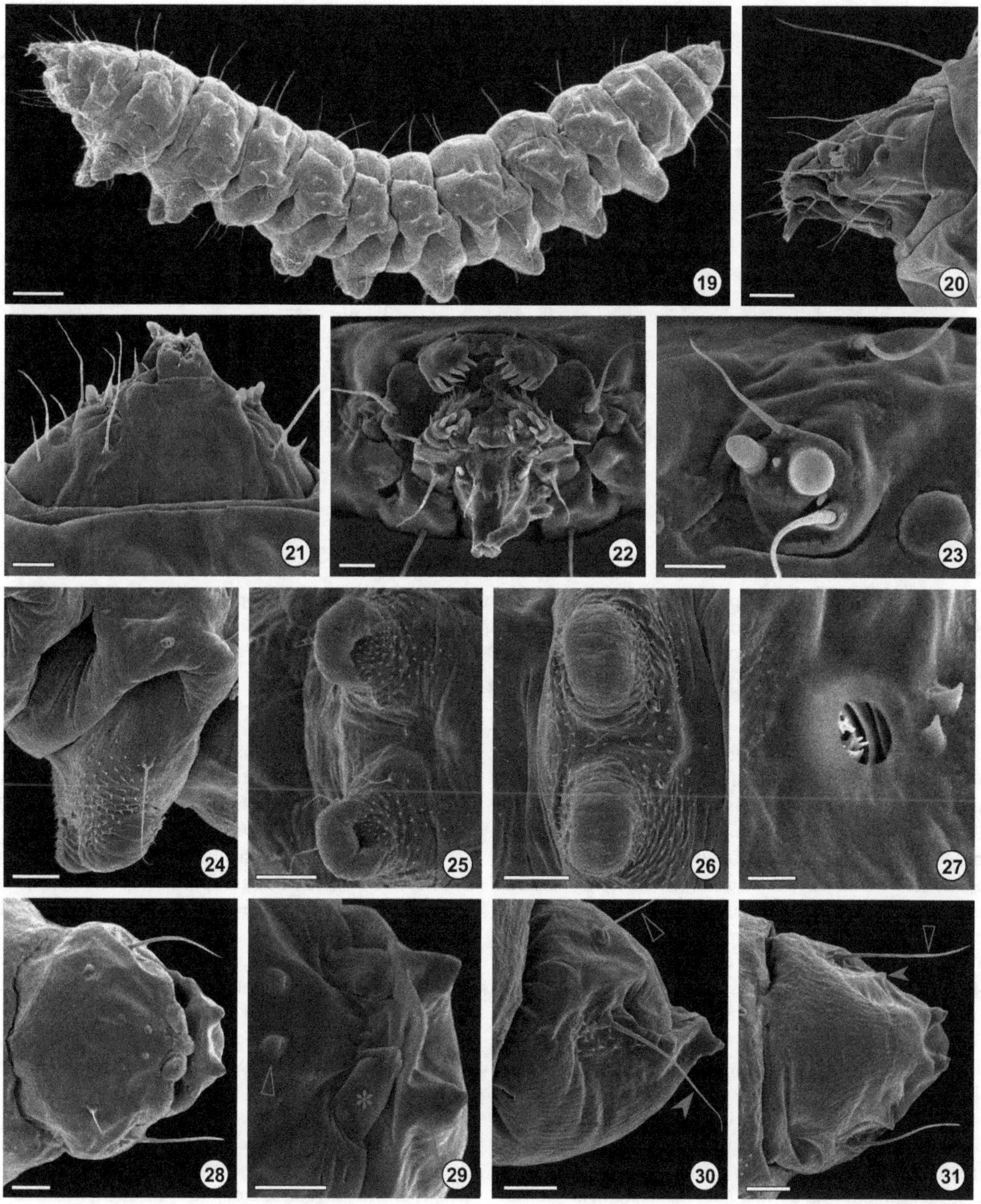

Figures 19–31. Last larval instar of *Stigmella schinivora* under scanning electron microscopy: (19) general view, lateral; (20, 21) head, lateral and dorsal views, respectively; (22) mouthparts, anterior; (23) antenna, anterior; (24, 25) mesothoracic ambulatory calli, lateral and ventral views, respectively; (26) ambulatory calli of sixth abdominal segment, ventral; (27) spiracle of eighth abdominal segment, lateral; (28) last abdominal segment, dorsal; (29) detail of tenth abdominal segment showing distal portion of anal rods (indicated by asterisk), posterior projections (indicated by closed arrow) and vestigial setae (indicated by open arrow), dorsal; (30) D2 and L1 setae, indicated by open and closed arrows, respectively, lateral; (31) last abdominal segment showing lateral projection (indicated by closed arrow) and L1 setae (indicated by open arrow). Scale bars: 200, 20, 150, 10, 10, 50, 50, 50, 5, 20, 20, 20, and 20 μm, respectively.

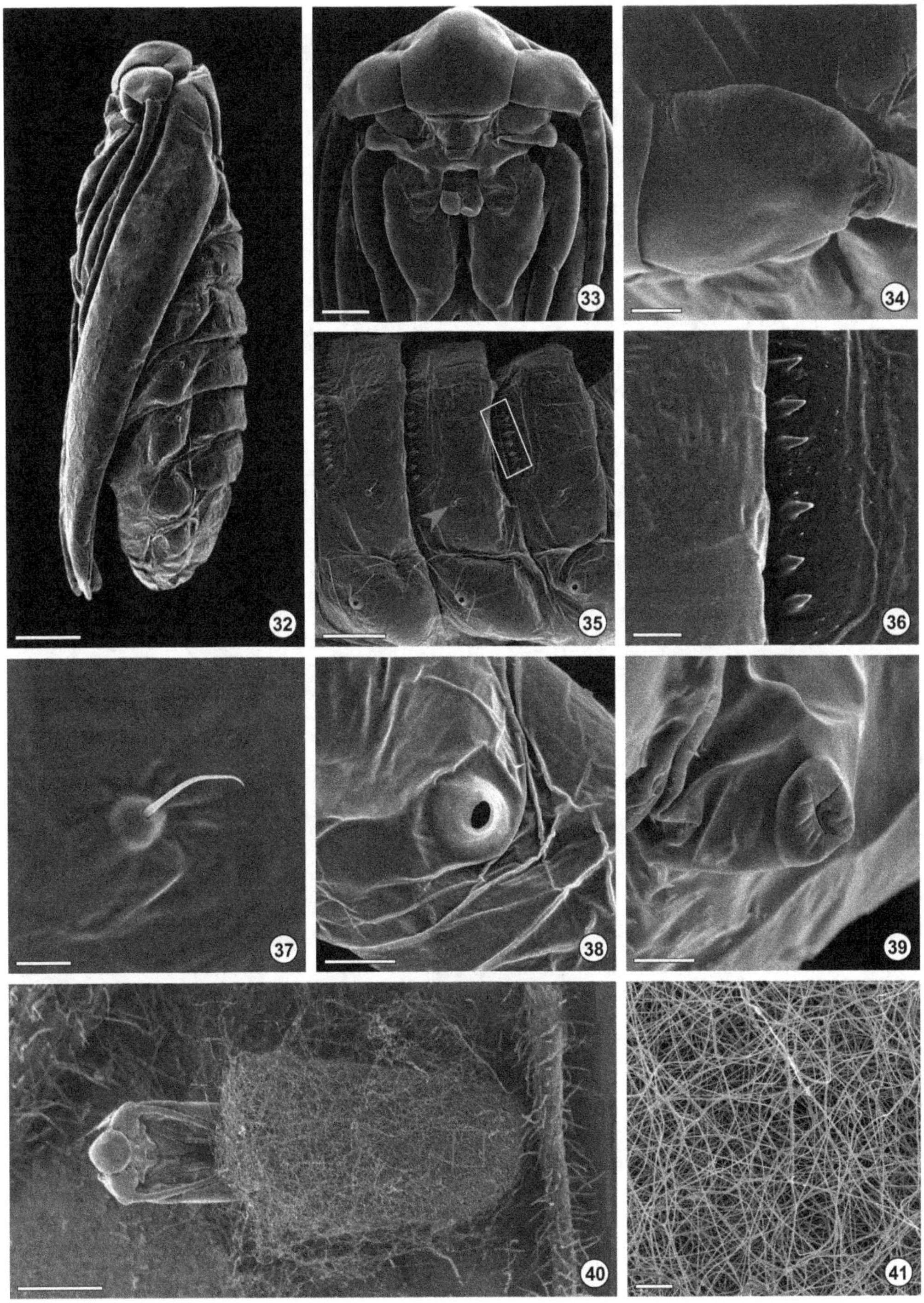

Figures 32–41. Pupa of *Stigmella schinivora* under scanning electron microscopy: (32) general view, lateral; (33) head and mouthparts, ventral; (34) eye-cap in detail, anteroventral; (35) third, fourth and fifth abdominal segments, laterodorsal; (36) detail of dorsal spines of fifth abdominal segment (indicated by rectangular area marked in Fig. 35), laterodorsal; (37) detail of fourth abdominal segment seta (indicated by arrow in Fig. 35), laterodorsal; (38) spiracles of third abdominal segment, laterodorsal; (39) spiracle (apparently closed) of eighth abdominal segment, lateral; (40) cocoon, with pupal exuvium extruded, ventral; (41) weaving pattern of the pupal cocoon surface (enlarged area marked by rectangle in Fig. 40). Scale bars: 200, 100, 40, 100, 30, 10, 20, 10, 500, and 50 μm, respectively.

Figures 42–49. Natural history of *Stigmella schinivora* on *Schinus terebinthifolius*: (42) host-plant leaf bearing leaf mine on the adaxial surface of a foliole (indicated by closed arrow); (43) general view of leaf mine on foliole, showing last instar larva seen through transparence (arrow points to empty egg, and letters indicate position of histological sectioning (treated in Figs 50, 51); (44) empty egg chorium in detail; (45) dissected mine at the final portion, showing last instar larva; (46) exit hole, used by a last instar larva to leave the mine; (47) pupal cocoon adhered to abaxial leaf surface; (48) pupa in detail, after removal of the cocoon, ventral; (49) adult on the leaf, dorsal view. Scale bars: 10, 1, 0.2, 0.5, 0.15, 0.5, 0.5, and 0.5 mm, respectively.

anterior segments, SD and SV group absent. A10 with four pairs of setae. Dorsal group (D) unisetose (represented by D2); two pairs of apparently rudimentary setae (D1 and SD2) (Figs 15, 28, 29). Lateral group (L) unisetose, represented by L1 (Figs 12, 30, 31), longer than D2.

Pupa. Average length ± standard deviation = 1.56 ± 0.03 mm, n = 5. Partially exarate, with distal portion of the wings slightly distant from the abdomen (Fig. 32). Body brownish during early development (darkening later in ontogeny) and flattened dorsoventrally (average maximum width ± standard

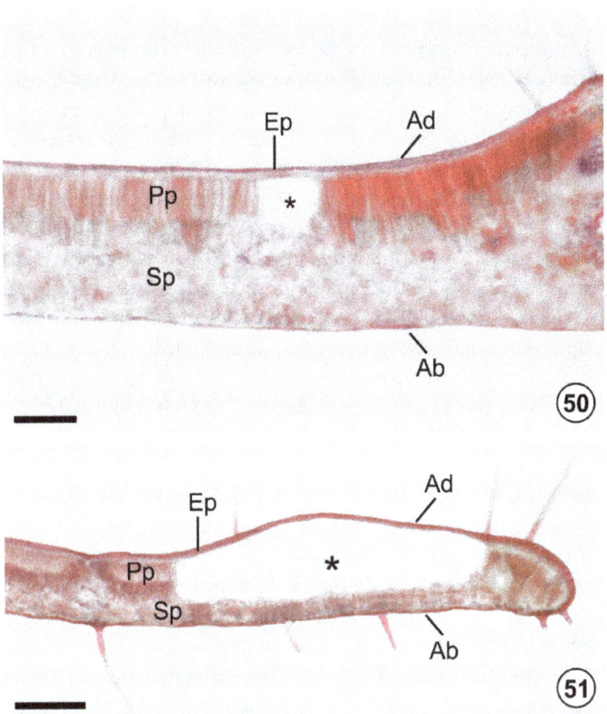

Figures 50–51. Transverse histological sections of *Schinus terebinthifolius* leaf (indicated by dashed lines "a" and "b" in Fig. 43), showing the organization of *Stigmella schinivora* mine during larval ontogeny. (50) First instar, initial linear section of mine (position indicated by letter "a" in Fig. 43); (51) last instar, final section of mine (position indicated by letter "b" in Fig. 43). Asterisks indicate leaf mine. (Ab) Abaxial surface of epidermis, (Ad) adaxial surface of epidermis, (Ep) epidermis, (Pp) palisade parenchyma, (Sp) spongy parenchyma. Scale bars: 0.1 and 0.2 mm, respectively.

deviation = 0.52 ± 0.007 mm, n = 5). Head vertex dome-shaped (Fig. 33), without projections or setae; front and clypeus also smooth, subtrapezoidal; labrum subtriangular and narrow; maxillae well-developed, positioned ventrally to maxillary palpi that are cuneiform and projecting mesally beyond the eyes; labial palpi short, located in between the maxillae. Antenna not reaching the apex of the mesothoracic legs (Fig. 18); scape enlarged, partially covering the eye (Fig. 34). Pronotum as a narrow stripe dorsally (Fig. 17); forewings covering the mesal portion of abdominal segments in ventral view (Fig. 18); hindwings mostly concealed by the forewings; prothoracic and mesothoracic legs visible ventrally, extending respectively to second and fourth abdominal segments; metathoracic legs mostly covered by the wings, extending to distal limit of the abdomen. Abdominal segments A3-7 with rows of posteriorly directed spines, located dorsally on the anterior margin (Figs 17, 35, 36). A pair of dorsal setae, positioned laterally, in the segments T2,3 and A1-7 (Figs 35, 37). Abdominal spiracles with elevated peritreme, opened in A1-7 (Fig. 38), partially closed in A8 (Fig. 39). Cremaster absent.

Life history. The egg is usually laid near a lateral vein on the adaxial leaf surface (Fig. 43). After eclosion the first instar larva bores into the leaf and begins to feed on the parenchyma, filling the empty egg with feces (Fig. 44). The serpentine mine is small and narrow, slowly widening throughout larval ontogeny (Figs 42, 43). From the beginning to the end of the mine, the larva feeds only on the palisade parenchyma (Figs 50, 51), which is formed by two layers of overlapping cells in the leaves of *S. terebinthifolius* (Grisi et al. 2011). The mine is filled fully with feces, which gives it a characteristic blackish appearance (Fig. 43). After completion of development, the last instar larva opens a hole (Figs 45, 46) in distal section of the mine through which it leaves, searching for a pupation site on the abaxial surface either of the same or an adjacent leaf. Cocoon yellowish, cylindrical and silk-made (Figs 40, 41, 47), having flimsy threads, some on the external surface used for attachment to the substrate. It bears a slit on anterior edge through which the pupa projects itself to the outside prior to adult emergence, leaving the exuvium partly protruded from the cocoon (Figs 40, 48, 49).

Densities of *S. schinivora* are generally low in Laranjeiras do Sul populations of *S. terebinthifolius*, and in most cases only one mine occurs per leaf and foliole. Mines with mining larvae of *S. schinivora* were collected mostly during the spring. Apparently, more than one generation occurs per year, which should be further explored.

DISCUSSION

Morphology of nepticulid eggs is still controversial, and has not yet been the subject of any detailed study (Davis 1998). According to Johansson et al. (1990), they are covered by a smooth helmet-shaped egg case, supposedly formed by secretion coming from the female colleterial glands. However, Kobayashi (1996) mentioned the existence of a chorion and presence of micropyle canals on the surface of *Stigmella castanopsiella* (Kuroko, 1978). That of *S. schinivora* is covered by a dome-shaped cap, which can be easily pulled off by physical pressure, thus being detached freely from the remaining egg contents in preserved material. Furthermore, neither micropyles nor aeropyles were found. To better resolve this question, we suggest that oogenesis should be explored in detail for *S. schinivora*, to test whether or not a true chorion is formed in this species.

The four larval instars found here for *S. schinivora* follow the general pattern recorded for nepticulids in general (Johansson et al. 1990). Although barely mentioned in the recent literature, the existence of two larval morphotypes has been known for a long time in other nepticulids, for example *Enteucha acetosae* (Stainton, 1854) (Sich 1908, 1909) and *Trifurcula immundella* (Zeller, 1939) (Sich 1917). The scarcity of setae on early instars was mentioned by van Nieukerken (2007) for *Acalyptris* Meyrick, 1921. The absence of true thoracic legs and also of abdominal prolegs bearing chochets, but prominent thoracic and abdominal ambulatory calli instead follows the general pattern found for the last instar of nepticulids (Davis 1987). We associated the existence of these structures with a need for locomotion outside the mine in search for the place to spin the cocoon and pupate.

The prothoracic dorsal shield found in the last instar of *S. schinivora* is similar to that described for other species of *Stigmella* by Gustafsson (1985); however, the ventral prothoracic plate shows differences compared to those present in other congeneric species described by him. Thus, this structure may provide diagnosable taxonomic characters, and should be explored further regarding its use in identification of *Stigmella* species.

The dorsal sclerotized structures seen by transparency on the last abdominal segment of *S. schinivora* have received different names, such "brace rods" (Stehr 1987), "bar-like" (Gustafsson 1981) and "anal rods" (van Nieukerken 2007, van Nieukerken et al. 2011). We opted for anal rods, since given their position it is very likely that these structures are functionally related to the anus. van Nieukerken et al. (2011) reported that the anal rods may be important in the taxonomy of larvae belonging to *Acalyptris* which should also be further explored in *Stigmella*.

An interesting characteristic of the first morphotype of *S. schinivora* is the presence of a single pair of setae in the tenth abdominal segment. van Nieukerken (2007) reported the presence of similar setae, but in A8 (three pairs) in earlier instars in *Acalyptris*. The large number and size of setae present in the last instar in *S. schinivora* suggest that these structures may be important from a sensorial perspective when outside the mine prior pupation. They have probably not arisen in this instar in particular, but instead were lost in the previous ones in association with their endophytic habit.

The comparison of chaetotaxy in Nepticulidae, particularly in *Stigmella*, showed little variation, suggesting a conserved pattern. Compared to the chaetotaxy described by Gustafsson (1981) for *Stigmella auromarginella* (Richardson, 1890), *S. schinivora* has in T2-3 absence of L3; in A8 presence of L1 and absence of SV2; and in A9 presence of SV2. In comparison to the chaetotaxy described by Gustafsson (1985) for *Stigmella rhomboivora* Gustafsson, 1985, *S. schinivora* has in T2 absence of L3 and in A9 absence of SV1 and presence of L1. Gustafsson (1981) also states that SV3 is absent in *Stigmella plagicolella* (Stainton, 1854) and *Stigmella paradoxa* (Frey, 1858), whereas it is present in *S. schinivora*.

Two setae have been described in the literature for the last abdominal segment of nepticulid larvae (e.g., Gustafsson 1981, 1985), but they have not been named. The designation of D2 and L1 in this study were inferred by comparing locations of setae in previous abdominal segments. We presume the two pairs of vestigial setae found dorsally in the last abdominal segment of *S. schinivora* had not been noticed in previous studies due to their reduced size. They are herein tentatively nominated according to Hinton's system (Stehr 1987), and thus corresponding homologies should be explored further in comparison to other nepticulids.

We are not aware of scanning electron microscopy studies on the pupal morphology of Nepticulidae. The enlarged first antennal segment of *S. schinivora* stands out, associated with the eye cap in the adult, as well as the absence of any trace of a differentiated process on the head dorsum (= cocoon cutter) and a cremaster on the last abdominal segment. These absences are generally found in the family, as there is no need for the cocoon cutter and cremaster, since, as in *S. schinivora*, there is usually a slit anteriorly on nepticulid cocoons through which the pupa projects partially to the outside prior to adult emergence (van Nieukerken et al. 2004). Line drawings and description of the pupa of *S. plagicolella* provided by Patočka and Turčani (2005) are similar to those shown here for *S. schinivora*. However, these authors do not mention the existence of eight closed abdominal spiracles in *S. plagicolella*, which occurs in *S. schinivora*. Compared to other nepticulid genera such as *Trifurcula* Zeller, 1848 (van Nieukerken et al. 2004), *Roscidotoga* Hoare, 2000 (van Nieukerken et al. 2011) and *Acalyptris* (van Nieukerken 2007), differences are found in the arrangement of posteriorly directed abdominal spines in A3-7; they form only one row in the anterior margin in *S. schinivora*, contrary to what is found in these genera in which four to five lines of these spines can be found.

The leaf mine of *S. schinivora* is similar in general shape to congeneric species (e.g. van Nieukerken et al. 2006, Stonis et al. 2013, 2016, Stonis and Remeikis 2017) and to others described for different genera within the Nepticulidae (e.g. van Nieukerken et al. 2011), demonstrating a uniform pattern of the family, even though they may use different host plants. Unfortunately, we did not find other studies addressing histology of mines in Nepticulidae, which precludes comparison with results reported here. Pereira et al. (2017) demonstrated that the damage caused by the gracillariid *Leurocephala chilensis* Vargas & Moreira, 2016 to the leaves of a plant in the same genus, *Schinus molle* L., is different from that caused by *S. schinivora*. Most gracillariids show two different kinds of mandibles during development, which may be used initially for slicing and eating only the adaxial epidermis, as is the case of *L. chilensis*. Chewing mandibles, as in *S. schinivora*, appear only in latter ontogeny for that species, and are also used to eat the palisade parenchyma until the end of larval development. The different adaptations observed in these two species using closely related hostplants reflect different evolutionary patterns of the families in resource usage (Hering 1951, Menken et al. 2010, Doorenweerd et al. 2016).

Finally, it is important to emphasize that morphology of the immature stages in particular has been increasingly taken into account as an aid in species identification among leaf-miner moths, as for example among gracillariids (Davis and Wagner 2011, Kobayashi et al. 2013, Brito et al. 2017). Information on immature stage morphology is also a precondition for understanding interactions of these stages with host plants, particularly when damage on tissues and histology of the mines are explored in conjunction with ontogeny (e.g., Brito et al. 2012, 2013, Vargas et al. 2015, Pereira et al. 2017). Thus, our results not only clarify the morphology of *S. schinivora* immature stages, but also could be used as an integrative framework for characterizing and comparing variation of immature stage morphology and associated host-plant interactions among other nepticulids and beyond.

This is the first report of *S. schinivora* in Brazil, expanding its geographical distribution that was restricted to the type locality in Argentina. *Schinus terebinthifolius* is widely distributed in southern South America (see Davis et al. 2011), and thus the range of *S. schinivora* may be much broader, and should be further explored.

ACKNOWLEDGMENTS

We acknowledge the staff of the Centro de Microscopia Eletrônica/UFRGS for the use of facilities and assistance in scanning electron microscopy analyses. We are grateful to Rosângela Brito for assistance in light microscopy photography, and to José Ricardo Assmann Lemes (UFRGS) for helping with fieldwork. We are especially grateful to Erik J. van Nieukerken (Naturalis Biodiversity Center, The Netherlands) for kindly providing literature and for sharing data on morphology of the egg and first instar of nepticulids. We thank Lafayette Eaton for editing the text. We are also grateful to the reviewers Jonas Rimantas Stonis (Lithuanian University of Educational Sciences, Lithuania) and Germán San Blas (Universidad Nacional de La Pampa, Argentina) for their comments and suggestions. Financial support for this study came in part from project UTA-MAYOR 9718-17 granted to H.A. Vargas by Universidad de Tarapacá, Chile. C.M. Pereira and G.R.P.Moreira were supported, by a doctoral scholarship and research grant from CNPq, Brazil (Process numbers 140496/2015-7 and 16/2551-0000485-4 PRONEX, respectively).

LITERATURE CITED

Brito R, Gonçalves GL, Vargas HA, Moreira GRP (2012) A new species of *Phyllocnistis* Zeller (Lepidoptera: Gracillariidae) from southern Brazil, with life-history description and genetic comparison to congeneric species. Zootaxa 3582: 1–16.

Brito R, Gonçalves GL, Vargas HA, Moreira GRP (2013) A new Brazilian *Passiflora* leafminer: *Spinivalva gaucha* gen. n., sp. nov. (Lepidoptera, Gracillariidae, Gracillariinae), the first gracillariid without a sap-feeding instar. ZooKeys 291: 1–26. https://doi.org/10.3897/zookeys.291.4910

Brito R, Mielke OHH, Gonçalves GL, Moreira GRP (2017) Description of three new species of *Phyllocnistis* Zeller, 1848 (Lepidoptera: Gracillariidae) from the Atlantic Forest, South Brazil, with notes on natural history and phylogeny. Austral Entomology. https://doi.org/10.1111/aen.12298

Braun AF (1917) Nepticulidae of North America. Transactions of the American Entomological Society 43: 155–209.

Davis DR (1987) Nepticulidae. In: Stehr FW (Ed.) Immature Insects. Kendall/Hunt Publishing Company, Dubuque, vol. 1, 350–351.

Davis DR (1998) The Monotrysian Heteroneura. In: Kristensen NP (Ed.) Handbook of zoology, Lepidoptera, moths and butterflies. Walter de Gruyter, Berlin, vol. 1, 65–90.

Davis DR, Mc Kay F, Oleiro M, Vitorino MD, Wheeler GS (2011) Biology and systematics of the leaf mining Gracillariidae of Brazilian Pepper Tree, *Schinus terebinthifolius* Raddi, with descriptions of a new genus and four new species. Journal of Lepidopterists' Society 65: 61–93. https://doi.org/10.18473/lepi.v65i2.a1

Davis DR, Wagner DL (2011) Biology and systematics of the New World *Phyllocnistis* Zeller leafminers of the avocado genus *Persea* (Lepidoptera, Gracillariidae). ZooKeys 97: 39–73. https://doi.org/10.3897/zookeys.97.753

Doorenweerd C, Nieukerken EJ van, Hoare RJB (2016) Phylogeny, classification and divergence times of pygmy leaf-mining moths (Lepidoptera: Nepticulidae): the earliest lepidopteran radiation on Angiosperms? Systematic Entomology 42: 267–287. https://doi.org/10.1111/syen.12212

Grisi FA, Angelo AC, Boeger MRT, Leitão CAE, Galvão SF, Wendling I (2011) Morfoanatomia foliar em mudas de *Schinus terebinthifolius* sob diferentes níveis de saturação hídrica. Floresta 41: 881–894. https://doi.org/10.5380/rf.v41i4.25351

Gustafsson B (1981) Characters of systematic importance in European Nepticulidae larvae (Lepidoptera). Insect Systematics & Evolution 12: 109–117. https://doi.org/10.1163/187631281X00436

Gustafsson B (1985) New species of *Stigmella* from the Gambia (Lepidoptera, Nepticulidae). Tijdschrift voor Entomologie 127: 165–177.

Hering EM (1951) Biology of the leaf miners. Junk, The Hague, 420 pp. https://doi.org/10.1007/978-94-015-7196-8

Johansson R, Nielsen ES, Nieukerken EJ van, Gustafsson B (1990) The Nepticulidae and Opostegidae (Lepidoptera) of north west Europe. Fauna Entomologica Scandinavica 23: 1–739.

Kobayashi S, Huang GH, Nakamura A, Hirowatari T (2013) Four new species of Gracillariidae (Lepidoptera) from China and Japan, and description of the pupal morphology of the genera *Corythoxestis*, *Eumetriochroa*, *Guttigera*, and *Metriochroa*. Zootaxa 3619: 101–129. https://doi.org/10.11646/zootaxa.3619.2.1

Kobayashi Y (1996) Gross embryology of a monotrisian heteroneuran moth, *Stigmella castopsiella* Kuroko (Nepticulidae, Lepidoptera) and its phylogenetic significance. The Lepidopterological Society of Japan 47: 194–200.

Menken SBJ, Boomsma JJ, Nieukerken EJ van (2010) Large-scale evolutionary patterns of host plant associations in the Lepidoptera. Evolution 64: 1098–1119. https://doi.org/10.1111/j.1558-5646.2009.00889.x

Myers N, Mittermeier RA, Mittermeier CG, Fonseca GAB, Kent J (2000). Biodiversity hotspots for conservation priorities. Nature 403: 853–858. https://doi.org/10.1038/35002501

Patočka J, Turčani M (2005) Lepidoptera Pupae: Central European Species. Apollo Books, Stenstrup, vol. 1, 321 pp.

Pereira CM, Silva DS, Gonçalves GL, Vargas HA, Moreira GRP (2017) A new species of *Leurocephala* Davis & Mc Kay (Lepidoptera,

Gracillariidae) from the Azapa Valley, northern Chilean Atacama Desert, with notes on life-history. Revista Brasileira de Entomologia 6: 6–15. https://doi.org/10.1016/j.rbe.2016.11.003

Puplesis R, Robinson GS (2000) A review of the Central and South American Nepticulidae (Lepidoptera) with special reference to Belize. Bulletin of the Natural History Museum 69: 3–114.

Sich A (1908) Notes on the life-history of *Nepticula acetosae* Stt. Entomologist's Record and Journal of Variation 20: 248–252.

Sich A (1909) Notes on the life-history of *Nepticula acetosae* Stt. Entomologist's Record and Journal of Variation 21: 103–106.

Sich A (1917) Notes on the ovum and larva of *Trifurcula immundella*, Zeller. Entomologist's Record and Journal of Variation 29: 117–121.

Šimkevičiūtė A, Stonis JR, Diškus A (2009) Taxonomic checklist of Nepticulidae of Mexico, with the description of three new species from the Pacific Coast (Insecta, Lepidoptera). Acta Zoologica Lituanica 19: 268–277. https://doi.org/10.2478/v10043-009-0037-0

Stehr FW (1987) Order Lepidoptera. In: Stehr FW (Ed.) Immature Insects. Kendall/Hunt Publishing Company, Dubuque, vol. 1, 288–305.

Stonis JR, Remeikis A (2016) Southern Andean *Stigmella sinuosa* complex (Lepidoptera, Nepticulidae): unraveling problematic taxonomy with a pictorial key of adults? Zootaxa 4136: 309-322. https://doi.org/10.11646/zootaxa.4136.2.3

Stonis JR, Remeikis A (2017) Anacardiaceae-feeding Nepticulidae in the Neotropics: description of *Stigmella lilliputica* sp. nov. from Argentina, a pest on Chilean peppertree *Schinus polygama*. Biologija 63: 230–237. https://doi.org/10.6001/biologija.v63i3.3577

Stonis JR, Remeikis A, Diškus A, Noreika R (2013) New Nepticulidae species (Insecta: Lepidoptera) from the Yucatán Peninsula (SE Mexico). Zootaxa 3609: 223–230. https://doi.org/10.11646/zootaxa.3609.2.8

Stonis JR, Remeikis A, Davis DR (2014) Ten new species from the Patagonian Andes (Argentina and Chile), mostly belonging to a newly designated *Stigmella purpurimaculae* group (Lepidoptera: Nepticulidae). Zootaxa 3887: 321–353.

van Nieukerken EJ (2007) *Acalyptris* Meyrick: revision of the *platani* and *staticis* groups in Europe and the Mediterranean (Lepidoptera: Nepticulidae). Zootaxa 1436: 1–48. https://doi.org/10.11646/zootaxa.1436.1.1

van Nieukerken EJ (2018) Nepticulidae and Opostegidae of the world, version 2.0. Scratchpads, biodiversity online. http://nepticuloidea.info [Accessed 09/01/18]

van Nieukerken EJ, Berg C van den, Hoare RJB (2011) A new species of the endemic Australian genus *Roscidotoga* Hoare from rainforests in southern Queensland (Lepidoptera: Nepticulidae). Tijdschrift voor Entomologie 154: 193–201. https://doi.org/10.1163/004074912X13397496981508

van Nieukerken EJ, Doorenweerd C, Hoare RJB, Davis DR (2016a). Revised classification and catalogue of global Nepticulidae and Opostegidae (Lepidoptera, Nepticuloidea). ZooKeys 628: 65–246. https://doi.org/10.3897/zookeys.628.9799

van Nieukerken EJ, Doorenweerd C, Nishida K, Snyers C (2016b) New taxa, including three new genera show uniqueness of Neotropical Nepticulidae (Lepidoptera). ZooKeys 628: 1–63. https://doi.org/10.3897/zookeys.628.9805

van Nieukerken EJ, Mazurkiewicz A, Pałka K (2004) *Trifurcula pallidella* (Duponchel, 1843) (Nepticulidae): distribution, biology and immature stages, particularly in Poland. Nota Lepidopterologica 27: 159–178.

van Nieukerken EJ, Schreurs AEP, Stiphout ML van, Ellis WN (2006) *Stigmella aceris* (Lepidoptera: Nepticulidae), een nieuwe mineermot van esdoorns in Nederland en België. Entomologische Berichten 66: 174–180.

Vargas HA, Brito R, Basilio DS, Moreira GRP (2015) A morphological reappraisal of the immature stages and life history of *Elachista synethes* Meyrick (Lepidoptera, Elachistidae), an Australian leaf miner alien to Chile. Revista Brasileira de Entomologia 59: 265–273. https://doi.org/10.1016/j.rbe.2015.07.002

Author Contributions: GRPM and CMP conceptualized the study. CMP and DSS made the illustrations. CMP, HAV, and GRPM performed the descriptions. CMP, GRPM, HAV, and DSS wrote the first draft. GRPM reviewed and edited the first version of the manuscript.

Competing Interests: The authors have declared that no competing interests exist.

10

Vertical distribution and structure of copepod (Arthropoda: Copepoda) assemblages in two different seasons down to 1,200 m in the tropical Southwestern Atlantic

Cristina de Oliveira Dias[1], Adriana Valente de Araujo[1], Sérgio Luiz Costa Bonecker[1]

[1] *Departamento de Zoologia, Instituto de Biologia, Universidade Federal do Rio de Janeiro. Avenida Prof. Rodolpho Rocco 211, CCS, Bloco A, Laboratório Integrado de Zooplâncton e Ictioplâncton, Sala A0-084, Ilha do Fundão, 21941-590 Rio de Janeiro, RJ, Brazil.*
Corresponding author: Cristina de Oliveira Dias (crcldias@hotmail.com)

ABSTRACT. The vertical distribution of copepod assemblages, ascertained from the surface down to 1,200 m, was investigated during two sampling periods (rainy and dry seasons), at four depths, in the oligotrophic waters of the southwestern Atlantic Ocean. Total density, diversity, and richness differed among sampling depths. Copepod density decreased with depth in the two sampling periods, with a maximum at 1 m and a slight decrease at 800 m. Higher diversities were observed at 250 m and 1,200 m during the rainy season and at 1 m and 1,200 m during the dry season. The maximum number of species was found at 1,200 m during the rainy season and at 1 m during the dry season. Various copepod assemblages were delimited in the water column in the two sampling periods. The deeper copepod assemblages occupied a wider range of depths. Salinity and temperature influenced the structure of copepod assemblages and reflected the hydrographic characteristics of the water masses in the region. *Candacia pachydactyla* (Dana, 1849), *Scolecithrix danae* (Lubbock, 1856), and *Agetus limbatus* (Brady, 1883) were the indicator species found at 1 m. The effects of different environmental factors on the copepod assemblages suggest that these consortia occupy distinct niches in the ocean.

KEY WORDS. Mesozooplanktonic copepod, Southeastern Brazil, Tropical environment.

INTRODUCTION

Oligotrophic regions are more productive and dynamic than previously thought, and have a high degree of biological variability (Dias et al. 2015). Oligotrophic areas constitute more than 75% of oceanic surfaces. They may play a significant role in oceanic biogeochemical budgets and global elemental cycling (Dufour et al. 1999, Corno 2006 and references therein). The southwest Atlantic Ocean is one of the most oligotrophic areas of all oceans worldwide (Andrade et al. 2007).

In zooplankton communities, copepods are known for their diversity and importance in the food web. They are the most abundant primary consumers in the ocean and the main food source for several planktophagous fishes (such as herring, anchovy, sardine, and corvine), fish larvae, and planktonic- and benthic invertebrates (Björnberg 1981).

Unlike other oceanic regions of the southern hemisphere, the southwestern Atlantic Ocean has seldom been qualitatively studied to obtain biogeographical data on zooplankton dynamics, particularly in external slope areas. Most studies in this region have been performed within the uppermost meters of the water column (e.g. Björnberg 1963, Valentin 1984a, b, Campaner 1985, Valentin et al. 1987, Dias 1996, Bassani et al. 1999, Lopes et al. 1999, Bonecker 2006, Lopes et al. 2006, Bonecker et al. 2007, Dias et al. 2015). These studies examined zooplankton (mainly copepod) composition, abundance, distribution, and biomass and have demonstrated the influence of hydrography on spatial variability. The deep sea zooplankton on the Brazilian central coast has not been investigated in detail. Information on the mesopelagic and bathypelagic communities is scarce except for that reported by Bonecker et al. (2014) and Dias et al. (2010). These authors described the compositions of the mesozooplankton and copepod communities, respectively. The sampling range in both cases was from 0 to 2,300 m depth.

The aims of this study were as follows: (a) describe the vertical variation in copepod density and assemblage structure (composition, richness, and diversity) down to 1,200 m over two seasons in the southwest Atlantic Ocean; (b) determine whether copepod population characteristics and environmental parameters are correlated; and (c) investigate copepod assemblages using indicator species. The hypothesis was that copepod assemblages are formed by characteristic indicator species groups arranged along a vertical gradient determined by oceanographic conditions.

MATERIAL AND METHODS

The Campos Basin is located between 20.5°S and 24°S off the central Brazilian coast and covers an area of approximately 100,000 km². In this region, the mean continental shelf width is 100 km and the shelf break is located between the 80-m and 130-m isobaths in the northern- and southern portions, respectively. The slope extends over a width of 40 km and has a mean declivity of 2.5°. Its base is shallower at the northern limit (about 1,500 m) and deeper near the southern limit (about 2,000 m) (Viana et al. 1998). The regional climate is warm and humid with a rainy summer season from November to February and a dry winter season from June to August (Lacerda et al. 2004).

The Brazilian coast is influenced by the warm, oligotrophic Brazil Current, which is a western boundary and flows from the northeast toward the southwest. It forms part of the South Atlantic western boundary current system (Stramma et al. 1990). The Campos Basin is characterized by its water column structure and water mass distributions over the continental shelf and slope. This region has various water masses each with distinct temperatures, salinities, and dissolved oxygen (DO) levels that provide different habitats for pelagic species (Fig. 1). The upper depths include the nutrient-poor subsurface water (Tropical Water, TW) and the relatively cold, nutrient-rich South Atlantic Central Water (SACW, 142–567 m depth). Their temperatures and salinities are below 20 °C and 36.4, respectively (Mémery et al. 2000, Silveira et al. 2000, Pedrosa et al. 2006). Water masses at lower depths include the Antarctic Intermediate Water (AAIW, 567–1,060 m), the Upper Circumpolar Deep Water (UCDW, 1,060–1,300 m), and the North Atlantic Deep Water (NADW, 1,300–3,260) (Viana et al. 1998, Pedrosa et al. 2006, Bonecker et al. 2014).

Mesozooplankton (length >200 μm) samples were collected during oceanographic cruises from February 25-April 13, 2009 (rainy season, RS) and from August 5 to September 17, 2009 (dry season, DS). The sampling stations were distributed along six transects perpendicular to the coast (A, C, D, F, H, and I) in a north-south direction. Each transect contained four sampling stations on the slope between the 400-m and 3,000-m isobaths (400 m, 1,000 m, 1,900 m, and 3,000 m; Fig. 2). Zooplankton samples were not collected in transects B, E, and G. All samples

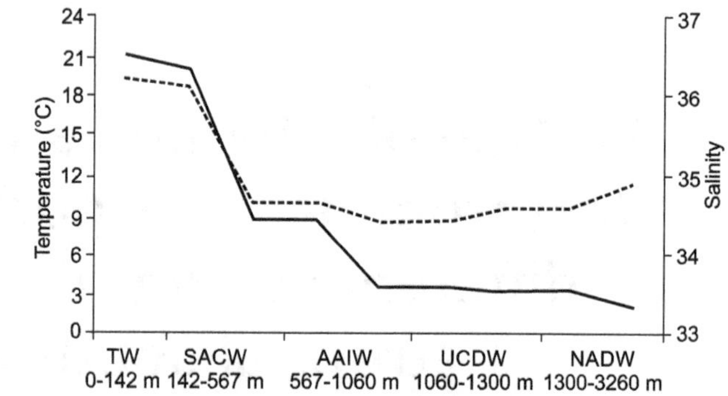

Figure 1. Salinity and temperature of the five water masses (0–3260 m) in the Campos Basin, central Brazilian coast, modified from Bonecker et al. (2014). Solid line = temperature; dashed line = salinity; SS, Subsurface Water; SACW, South Atlantic Central Water; AAIW, Antarctic Intermediate Water; UCDW, Upper Circumpolar Deep Water; NADW, North Atlantic Deep Water.

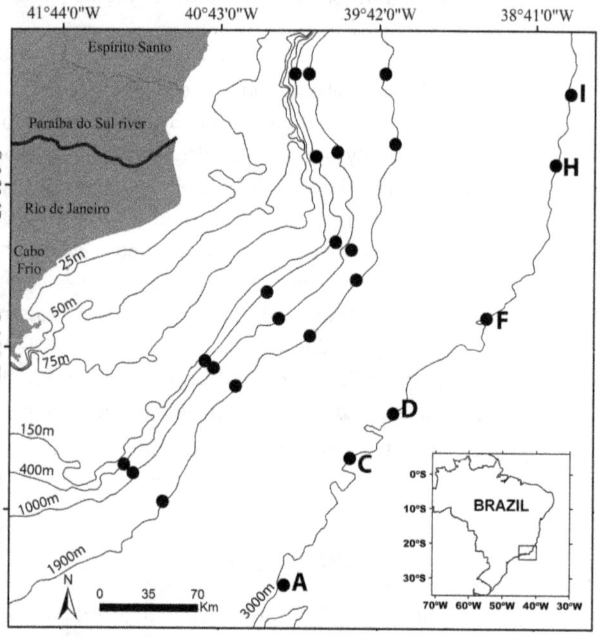

Figure 2. Study area showing sampling stations.

were collected between 6:18 pm and 5:08 am during the rainy season and between 5:57 pm and 5:46 am during the dry season (local time).

Environmental data included (a) water temperature and salinity at all sampling depths, namely, 1 m, 250 m, 800 m, and 1,200 m, by using a CTD (sonde); (b) inorganic nutrients, determined by standard oceanographic methods (Grasshoff et al. 1999); (c) DO, measured continuously in the water column with a sensor coupling in the CTD; and (d) suspended particulate matter (SPM), obtained from water sample filtration. Detailed methodology and discussion of the hydrochemistry of the study area are presented elsewhere (Rodrigues et al. 2014, Dias et al. 2015, Suzuki et al. 2015).

Mesozooplankton samples were collected by horizontal hauls in the previously named water mass nuclei (Tropical Water [TW, 1 m], South Atlantic Central Water [SACW, 250 m], Antarctic Intermediate Water [AAIW, 800 m], and Upper Circumpolar Deep Water [UCDW, 1,200 m]). The horizontal hauls in the slope were taken at 2 knots for 10 min or until the filtered water volume reached 50 m^3. Hauls were made using a MultiNet type midi (Hydro-Bios, Altenholz, Germany, 200 µm white mesh, 50 × 50 frame opening) fitted with a digital flowmeter attached to the inner net mouth and an external meter to assess filtration efficiency. Different nets were used at each depth to prevent sample cross-contamination. The MultiNet was also fitted with a depth gauge. Both the depth and water volume were transmitted to a computer along with the hauls. In the dry season, no samples were collected from the 3,000-m isobath of transects H and I due to logistical problems. A total of 145 samples were analyzed: 78 in the rainy season, and 67 in the dry season.

Samples were fixed and preserved in 4% buffered formalin. In the laboratory, the preserved samples were divided into 1–10 fractions with a Folsom Plankton Splitter (Hydro-Bios) (McEwen et al. 1957) and the resultant subsamples analyzed. All copepods were sorted, identified, and counted under a stereoscopic microscope according to the methods of Bradford-Grieve et al. (1999) and Dias and Araujo (2006). Taxon abundance per cubic meter and copepod species composition were determined for all samples. The total abundance of each species was estimated from both adult and juvenile forms. All specimens were deposited in the copepod collection of the Integrated Zooplankton and Ichthyoplankton Laboratory of the Federal University of Rio de Janeiro (DZUFRJ Copepoda 12622-12631, DZUFRJ Copepoda 12791-12952, DZUFRJ Copepoda 13058-14020, DZUFRJ Copepoda 14059-14189, DZUFRJ Copepoda 14618-14715, DZUFRJ Copepoda 14726-14760, DZUFRJ Copepoda 14834-14975, DZUFRJ Copepoda 15024-15509, DZUFRJ Copepoda 15560-15892, and DZUFRJ Copepoda 16316-16895).

The Shannon-Wiener diversity index (H') and richness were calculated for each station during the sampling period to identify any differences in biodiversity. Pielou's index of evenness was also determined. The Shannon-Wiener diversity index was calculated as follows: $H' = -\sum pi * \ln(pi)$, where pi is the fraction of species i in the total sample. These analyses were run using PRIMER v. 6.1 (Clarke and Gorley 2006). Vertical depth differences in copepod density, richness, diversity, and evenness were tested using generalized linear models (GLM) with gamma family (dispersion = 1). A low additive constant (1) was applied to the copepod density density and evenness data in each sampling station to eliminate zero values in the matrix because the gamma family does not allow for zeros. The analyses were performed using R v. 3.1.3 (R Core 3 Team 2015).

"Station x copepod species" matrices (sampling depths as rows, and copepod species density at each station as columns) were analyzed for each sampling period. Species with occurrences of > 1% and the [log(x + 1)] transformed density data were used to generate a Bray-Curtis similarity matrix. The following multivariate techniques were used to identify various copepod assemblages: (1) an analysis of similarity (ANOSIM) was used to assess whether the copepod assemblage structure varied with depth; (2) a similarity of percentages (SIMPER) analysis was used to identify the species that contributed most to the similarities within groups; and (3) an indicator species analysis (ISA) was used to identify the indicator copepod species at each sampling depth (Dufrêne and Legendre 1997, De Cáceres and Legendre 2009). The ISA index is at a maximum (100%) when all exemplary species are found in a single sample group and the species occurs in all samples from that group. A species was considered an indicator in a particular water mass when its ISA value was significantly higher than that expected by chance alone (compared to one thousand random samples from plots with the same number of species occurrences) and higher than 70%. These indicator values were checked by the Monte Carlo test to establish reliable significance levels (IV,% of perfect indication, $p < 0.05$) The ANOSIM and SIMPER analyses were performed using PRIMER v. 6.1 (Clarke and Gorley 2006). The ISA was carried out using PC-ORD for Windows v. 5.0 (MjM Software, Gleneden Beach, Oregon, USA) (McCune and Mefford 1999).

The relationship between variations in copepod assemblage structure and environmental parameters (temperature, salinity, suspended particulate matter, DO, nitrate, silicate, and inorganic-P or orthophosphate) during the sampling period was analyzed by distance-based linear models (DistLM). The Akaike information criterion for small samples (AICc) was used to select the best model. A distance-based redundancy analysis (dbRDA) was performed to represent the best model selected by DistLM. Both DistLM and dbRDA were performed in PRIMER v. 6.1 (Clarke and Gorley 2006) and PERMANOVA+ for PRIMER (Anderson et al. 2008). All analyses were performed using the matrix for both sampling periods. The results were considered significant when $p < 0.05$.

RESULTS

Hydrography conditions

The environmental conditions during the study period were described in detail by Rodrigues et al. (2014), Dias et al. (2015), and Suzuki et al. (2015). Therefore, they are only briefly summarized here and in Table 1. The lowest temperature and salinity values were obtained during the dry season. At 800 and 1,200 m, both temperature and salinity were relatively uniform. The suspended particulate matter (SPM) ranged from 0.01 mg L^{-1} at 800 m to 4.19 mg L^{-1} at 1 m in the rainy and dry season, respectively. The DO values were homogeneous in both sampling seasons except at 1 m. In both sampling periods, the highest nutrient concentrations were observed in the deepest waters.

Table 1. Minimum and maximum values recorded for temperature (°C), salinity, suspended particulate matter (SPM, mg L^{-1}), dissolved oxygen (DO, mg L^{-1}), nitrate (µmol L^{-1}), silicate (µmol L^{-1}), and orthophosphate (µmol L^{-1}) in samples collected in the Campos Basin water column during the rainy- and dry seasons of 2009. RS, rainy season; DS, dry season.

Depth	Sampling Periods	Temperature		Salinity		SPM		DO		Nitrate		Silicate		Orthophosphate	
		Min	Max	Min	Max	Min	Max	Min	Max	Min	Max	Min	Max	Min	Max
1 m	R.S.	36.68	37.28	25.96	28.50	0.29	1.82	5.78	6.38	0.56	2.19	0.41	9.06	0.01	0.08
	D.S.	36.26	37.12	21.19	24.89	0.15	4.19	4.36	7.14	0.37	4.85	0.39	7.89	0.02	0.44
250 m	R.S.	35.28	35.99	13.64	25.65	0.04	1.93	6.17	6.68	0.58	7.67	0.58	3.53	0.02	0.71
	D.S.	35.12	35.79	12.09	16.75	0.17	0.87	6.22	6.77	1.48	11.41	1.35	7.15	0.26	0.81
800 m	R.S.	34.33	34.40	4.64	5.61	0.01	1.02	6.03	6.84	5.29	32.37	5.33	30.15	0.72	2.00
	D.S.	34.34	34.46	4.39	5.33	0.22	0.67	6.08	6.60	18.18	34.72	19.85	31.31	1.69	1.95
1,200 m	R.S.	34.45	34.57	3.20	3.53	0.33	1.11	5.90	6.06	17.58	31.76	20.93	51.11	1.71	2.08
	D.S.	34.54	34.64	3.21	3.42	0.13	1.05	5.76	5.86	25.42	35.33	37.54	49.88	1.87	2.07

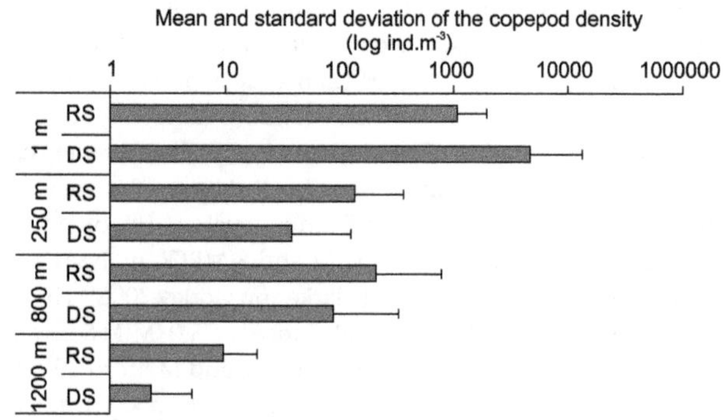

Figure 3. Vertical distribution of the mean and standard deviation of the copepod density (log ind.m^{-3}) during two sampling periods: rainy season, RS; dry season, DS.

Abundance, biodiversity, spatial comparisons, and composition

The total copepod density varied among the sample depths ($p < 0.05$). The highest values were recorded during the rainy season except at 1 m where the copepod density was higher during the dry season than the rainy season (Fig. 3). The vertical pattern showed that copepod density decreased with depth during both sampling periods. The highest densities were observed at 1 m (Mean: RS, 1,102.44 ± 844.03 ind.m^{-3}; DS, 4,593.63 ± 8,743.59 ind.m^{-3}) and at 800 m (Mean: RS, 205.41 ± 585.87 ind.m^{-3}; DS, 86.18 ± 232.34 ind.m^{-3}). The lowest densities were observed at 1,200 m (Mean: RS, 8.52 ± 7.98 ind.m^{-3}; DS, 1.29 ± 1.93 ind.m^{-3}) (Fig. 3).

A total of 139 species were identified in the Campos Basin over both seasons: 111 calanoids, 18 poecilostomatoids, 6 harpacticoids, and 4 cyclopoids. The calanoid order was the most abundant in both sampling periods. It constituted 66–87% of the total copepod density (1 m: 76%; 250 m: 2%; 800 m: 4%; and 1,200 m < 0.05%). Poecilostomatoids (14% of the total copepod density) were the most abundant at 1 m and at 250 m. They were represented primarily by the Oncaeidae and Corycaeidae families, which were relatively abundant at 1,200 m.

In both sampling periods, the cyclopoid copepods identified came exclusively from the Oithonidae family, and the genus *Oithona* Baird, 1843 was more abundant in 250 m. Harpacticoid copepod density peaked at 1 m and was represented mainly by *Macrosetella gracilis* (Dana, 1847). In the mesobathypelagic region (250–800 m), the harpacticoids *Aegisthus* spp. predominated, but their abundance was low compared to that of other genera. Cyclopoid and harpacticoid copepods made up 1–1.5% of the copepod density in both sampling periods (Suppl. materials 1, 2).

Higher diversity values were observed at 250 and 1,200 m in the rainy season and at 1 m and 1,200 m in the dry season. Significant differences in diversity were found across sample depths ($p < 0.05$). Diversity was lower at 800 m (Fig. 4). The maximum number of species was found at 1,200 m during the rainy season (21) and at 1 m during the dry season (17) (Fig. 4). As for total density and diversity, richness differed significantly across water masses ($p < 0.05$). Mean evenness ranged from 0.97 (1 m) to 0.87 (1,200 m) during the rainy season and from 0.98 (1 m) to 0.83 (800 m) during the dry season (Fig. 4). The evenness also differed significantly across sample depths ($p < 0.05$).

Seven copepod species were abundant during the rainy season: *Undinula vulgaris* (Dana, 1849), *Clausocalanus furcatus* (Brady, 1883), *Acrocalanus longicornis* Giesbrecht, 1888, *Paracalanus quasimodo* Bowman, 1971, *Temora stylifera* (Dana, 1849), *Oncaea venusta* Philippi, 1843, and *Farranula gracilis* (Dana, 1849). These constituted 68% of the total copepod density. In the dry season, thirteen species predominated, and represented 86% of the total copepod density: *Calanoides carinatus* (Krøyer, 1849), *C. furcatus*, *Centropages velificatus* (Oliveira, 1947), *Lucicutia flavicornis* (Claus, 1863), *A. longicornis*, *P. quasimodo*, *Rhincalanus cornutus* (Dana, 1849), *Subeucalanus pileatus* (Giesbrecht, 1888), *Temora turbinata* (Dana, 1849), *T. stylifera*, *Oithona plumifera* Baird, 1843, *O. venusta*, and *Onchocorycaeus giesbrechti* (Dahl F., 1894) (Table 2). Contributions of the most representative species varied at the different sampling depths as follows: *C. furcatus*, *F. gracilis*, *P. quasimodo*, and *T. stylifera* (1 m); *O. venusta*, *Scolecithricella minor* (Brady, 1883), *Ctenocalanus vanus* Giesbrecht, 1888, and *Oithona similis* Claus, 1866 (250 m);

Table 2. Vertical distribution of the mean (standard deviation) density (ind.m^{-3}) of the 20 most abundant copepods species found in Campos Basin. RS, rainy season; DS, dry season.

Species	1 m		250 m		800 m		1,200 m	
	RS	DS	RS	DS	RS	DS	RS	DS
Calanoides carinatus	0.27 (1.30)	126.51 (499.99)	1.71 (7.00)	0.72 (3.18)	16.17 (30.56)	31.16 (105.22)	0.13 (0.36)	0.28 (0.04)
Nannocalanus minor	29.99 (76.58)	32.98 (104.26)	0.98 (2.18)	0.06 (0.19)	0.14 (0.31)	0.20 (0.76)	0.16 (0.23)	< 0.01 (< 0.01)
Undinula vulgaris	69.34 (134.22)	39.35 (139,42)	1.51 (3.68)	0.04 (0.13)	0.23 (0.62)	< 0.01 (0.01)	0.17 (0.30)	0.02 (0.01)
Clausocalanus furcatus	132.82 (99.76)	1,139.59 (2,805.30)	4.10 (7.18)	0.38 (1.00)	0.78 (1.81)	0.76 (2.63)	0.72 (1.53)	1.66 (0.41)
Ctenocalanus citer	-	7.76 (34.95)	0.37 (1.83)	< 0.01 (0.02)	42.07 (172.15)	3.26 (12.22)	-	-
Ctenocalanus vanus	-	6.28 (28.47)	0.34 (1.64)	2.51 (10.72)	46.76 (123.43)	0.57 (2.18)	0.12 (0.22)	-
Centropages velificatus	0.13 (0.47)	115.41 (524.23)	-	0.15 (0.72)	-	0.12 (0.38)	-	0.05 (0.01)
Lucicutia flavicornis	219 (5.99)	49.94 (113.34)	0.53 (0.81)	0.32 (0.61)	0.04 (0.10)	< 0.01 (0.02)	0.23 (0.43)	0.02 (0.03)
Acrocalanus longicornis	50.28 (44.69)	62.43 (162,69)	0.59 (1.76)	-	0.27 (0.63)	< 0.01 (0.01)	0.05 (0.13)	-
Calocalanus pavoninus	21.77 (34.57)	19.43 (58,56)	0.39 (1.53)	< 0.01 (0.02)	0.10 (0.25)	< 0.01 (0.01)	0.08 (0.14)	-
Paracalanus quasimodo	68.62 (185.07)	296.29 (610.22)	3.70 (17.54)	0.56 (2.25)	< 0.01 (0.01)	0.09 (0.33)	0.02 (0.06)	-
Rhincalanus cornutus	1.36 (5.06)	47.32 (148.26)	0.14 (0.38)	0.02 (0.06)	0.03 (0.06)	0.51 (1.51)	< 0.01 (0.02)	< 0.01 (< 0.01)
Subeucalanus pileatus	-	45.58 (175.51)	-	0.01 (0.05)	0.10 (0.41)	0.02 (0.05)	-	< 0.01 (0.02)
Temora turbinata	-	111.72 (419.15)	-	0.20 (0.71)	-	0.10 (0.26)	0.02 (0.07)	< 0.01 (0.01)
Temora stylifera	80.97 (109.87)	221.94 (575.14)	3.38 (8.58)	0.05 (0.23)	0.43 (0.80)	0.03 (0.12)	0.10 (0.19)	0.01 (0.02)
Oithona plumifera	2.78 (6.87)	47.91 (115.36)	0.18 (0.57)	1.24 (3.59)	-	0.20 (0.54)	0.05 (0.09)	0.04 (0.07)
Oncaea venusta	80.65 (75,70)	162.50 (278.93)	12.90 (20.78)	5.87 (19.05)	2.34 (3.57)	0.81 (2.25)	1.15 (1.98)	0.24 (0.49)
Onychocorycaeus giesbrechti	24.60 (36.49)	79.38 (219.82)	0.11 (0.27)	0.07 (0.19)	0.005 (0.02)	0.02 (0.06)	0.07 (0.20)	-
Farranula gracilis	204.93 (222.99)	37.12 (51.74)	3.88 (9.51)	0.02 (0.07)	0.60 (0.82)	0.65 (2.44)	0.59 (0.94)	0.03 (0.06)
Macrosetella gracilis	37.31 (99,12)	11.87 (41.39)	3.60 (6.73)	0.27 (0.71)	0.15 (0.25)	0.11 (0.38)	0.32 (0.44)	0.02 (0.01)

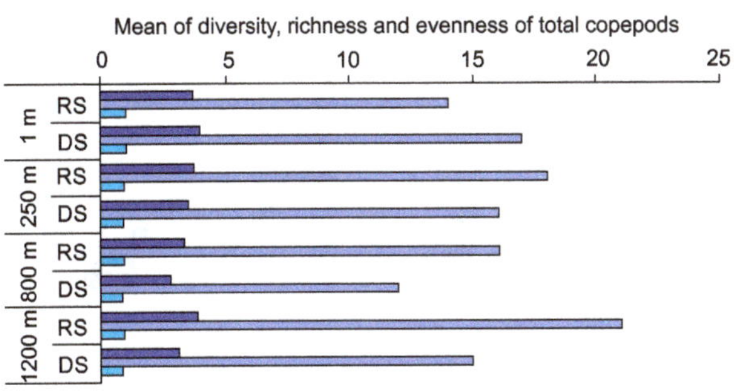

Figure 4. Vertical distribution of the mean diversity, species richness, and evenness of total copepods in the Campos Basin: dark blue bars, diversity; cobalt bars, richness; cyan bar, evenness. RS, rainy season; DS, dry season.

C. carinatus, *Ctenocalanus citer* Heron & Bowman, 1971, and *C. vanus*, (800 m); *O. venusta*, *C. furcatus*, *Conaea rapax* Giesbrecht, 1891, and *F. gracilis* (1,200 m).

Spatial characterization of the copepod assemblages

The similarity analysis revealed that variations in species density were important in determining changes in copepod assemblages along the depths. Pairwise ANOSIM tests during the rainy season demonstrated that sampling depth groups were separated according to their copepod assemblages except for 800 m and 1,200 m (ANOSIM: p = 0.1%; global R = 0.662; Table 3).

The copepod species that contributed most to the similarity of the copepod assemblages at 1 m, 250 m, and 800–1,200 m (SIMPER test) are shown in Table 4. *Oncaea venusta* participated in the copepod assemblages at all depth groups. *Clausocalanus furcatus* and *T. stylifera* occurred in the 1 m, 250 m, and 800–1,200 m assemblages. *F. gracilis* was detected in the 1 m, and 800–1,200 m copepod assemblages. In the 250 m assemblage, *O. similis*, *Triconia* cf. *conifera* (Giesbrecht, 1891), *Lubbockia squillimana* Claus, 1863, *Pleuromamma abdominalis* (Lubbock, 1856), *L. flavicornis*, and *Pleuromamma gracilis* Claus, 1863 were also present. At 800–1,200 m, *C. carinatus*, *T.* cf. *conifera*, and *C. vanus* contributed to the formation of the copepod assemblage.

During the dry season, pairwise ANOSIM tests indicated that the structures of the copepod assemblages at 1 m and at 250 m were distinct from the others (ANOSIM: p = 0.1%; global R = 0.673; Table 3).

The copepod species that contributed most to the similarity of the copepod assemblages at 1 m, 250 m, and 800–1,200 m (SIMPER test) are shown in Table 5. *Clausocalanus furcatus* and *O. venusta* contributed most (>30% similarity) to the formation of the 1 m, 250 m, and 800–1,200 m copepod assemblages. *Farranula gracilis* also contributed to the 1 m assemblage. At 250 m, *T.* cf. *conifera* and *Heterorhabdus spinifrons* (Claus, 1863) contributed the most (along with *O. venusta*) to the formation

Table 3. Analyses of similarity (ANOSIM) between sampling depth groups during the rainy and dry seasons. R, strength of the difference between groups (* significant differences, $p < 0.05$).

Groups of sampling depths	R Statistic	
	Rainy season	Dry season
1 m x 250 m	0.689*	0.759*
1 m x 800 m	0.774*	0.789*
1 m x 1,200	0.951*	0.986*
250 m x 800 m	0.365*	0.359*
250 m x 1,200 m	0.474*	0.547*
800 m x 1,200 m	0.087	0.024

Table 4. Copepod species and their contribution (%) to the similarity of assemblages obtained during the rainy season through the SIMPER analysis.

1 m		250 m		800–1,200 m	
Species	(%)	Species	(%)	Species	(%)
Farranula gracilis	17.48	*Oncaea venusta*	27.45	*Oncaea venusta*	23.02
Clausocalanus furcatus	14.80	*Oithona similis*	14.71	*Calanoides carinatus*	9.93
Oncaea venusta	13.93	*Triconia* cf. *conifera*	7.32	*Clausocalanus furcatus*	9.93
Acrocalanus longicornis	9.69	*Macrosetella gracilis*	6.46	*Triconia* cf. *conifera*	8.42
Undinula vulgaris	9.38	*Oncaea media*	5.47	*Farranula gracilis*	6.68
Temora stylifera	9.10	*Clausocalanus furcatus*	5.20	*Conaea rapax*	4.91
Nannocalanus minor	4.97	*Lubbockia squillimana*	3.47	*Ctenocalanus vanus*	3.69
Others	20.65	*Pleuromamma abdominalis*	2.80	*Undinula vulgaris*	3.37
		Lucicutia flavicornis	2.64	*Temora stylifera*	3.15
		Temora stylifera	2.50	*Macrosetella gracilis*	3.05
		Pleuromamma gracilis	2.26	*Oithona similis*	2.97
		Others	19.72	Others	20.88

of the assemblage *C. rapax* and *C. carinatus* also appeared in the copepod assemblages at 800–1,200 m.

The Indicator Species Analysis identified indicator species only to 1 m depth ($p < 0.05$). They were *Candacia pachydactyla* (Dana, 1849), *Scolecithrix danae* (Lubbock, 1856), and *Agetus limbatus* (Brady, 1883), with indicator values ≥ 88.8%, and all of which occurred in the low-density samples.

Relationship between structure of copepod assemblages and environmental parameters

During the sampling period, the best model (DistLM, AIC = 1,128.4, $R^2 = 0.212$) captured 81.5% of the variability in the fitted model and 16.9% of the variability in the dataset (first two dbRDA axes in Fig. 5). The model included two variables: salinity and temperature, Salinity explained 17% of the variability in copepod assemblage structure. The first dbRDA axis is directly correlated to salinity and temperature. The second dbRDA axis

Table 5. Copepod species and their contribution (%) to the similarity of assemblages obtained during the dry season through the SIMPER analysis.

1 m		250 m		800 m – 1,200 m	
Species	(%)	Species	(%)	Species	(%)
Clausocalanus furcatus	21.03	*Oncaea venusta*	35.81	*Oncaea venusta*	27.63
Oncaea venusta	14.3	*Triconia* cf. *conifera*	17.72	*Conaea rapax*	14.26
Farranula gracilis	8.94	*Heterorhabdus spinifrons*	4.96	*Calanoides carinatus*	13.50
Paracalanus quasimodo	8.20	*Macrosetella gracilis*	4.20	*Clausocalanus furcatus*	12.28
Temora stylifera	6.69	*Oithona setigera*	4.04	*Macrosetella gracilis*	6.58
Undinula vulgaris	4.62	*Oithona plumifera*	3.67	*Triconia* cf. *conifera*	4.31
Lucicutia flavicornis	4.37	*Oncaea media*	3.64	Others	21.44
Onchocorycaeus giesbrechti	3.86	*Scolecithricella minor*	2.92		
Acrocalanus longicornis	2.71	*Clausocalanus furcatus*	2.50		
Nannocalanus minor	2.63	Others	20.54		
Triconia cf. *conifera*	2.54				
Macrosetella gracilis	2.32				
Others	17.79				

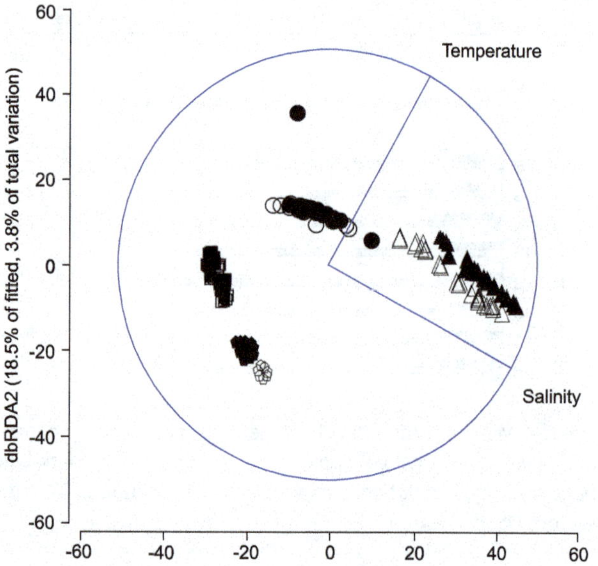

Figure 5. DbRDA ordination derived from DistLM analysis, including the main environmental parameters explaining the variability in copepod assemblage structures during the sampling period. Temp, temperature; Sal, salinity. Black triangle, rainy season 1 m; open triangle, dry season 1 m; black circle, rainy season 250 m; open circle, dry season 250 m; black square, rainy season 800 m; open square, dry season 800 m; black polygon, rainy season 1,200 m; open polygon, dry season 1,200 m.

is directly correlated to temperature and inversely correlated to salinity. The first dbRDA axis separated the assemblages sampled at 1 m and 250 m, and the deep assemblages (800 m and 1,200 m). The second axis discriminated the assemblages sampled at the lower depths. The 800 m and 1,200 m assemblages are separated in the plot below.

DISCUSSION

Hydrography conditions

The environmental parameters described along the water column reflect the hydrological characteristics of the water masses in this region (Niencheski et al. 1999, Rezende et al. 2007). Temperature and salinity values recorded in the area decreased from the subsurface (1 m) down to deep water (1,200 m). The inorganic nutrient concentrations had a vertical distribution pattern typical of oceanic regions, namely, increasing from the surface to deeper waters. The nutrient concentrations recorded in the surface layer of the Campos Basin are characteristic of oligotrophic waters found in the Brazil Current. They are indicative of the low primary production in the oceanic region (Metzler et al. 1997). Deep waters are more enriched especially along the continental margin of southeastern Brazil. Nutrients are added when the surrounding currents remobilize sediment (Silveira et al. 2000, Suzuki et al. 2015).

Abundance, biodiversity, spatial comparisons, and copepod assemblage composition

Copepod density and diversity distinguished the shallowest (1 m) and deepest (250–1,200 m) waters during both sampling periods. In both seasons, the total abundance decreased with depth except for a slight increase at 800 m. Richness and diversity increased in the mesopelagic and bathypelagic regions. Dusk-to-dawn sampling in the present study may have introduced bias in the data due to the diel vertical zooplankton migration. Nevertheless, other studies showed patterns similar to those observed here. Vinogradov et al. (1998) found copepod richness maxima at mesobathypelagic depths in both daytime- and nighttime vertical hauls (0–3,774 m) in the frontal zone between the Gulf Stream and the Labrador Current. In another study in the subtropical and subarctic North Pacific Ocean, Steinberg et al. (2008) noted density peaks at the surface and in the mesopelagic zone in both day and night vertical hauls (0–1,000 m). The authors attributed the density trend to the presence of a very high number of taxa feeding on suspended or sinking detritus in the mesopelagic zone and the occurrence of a large population of ontogenetically migrating calanoid copepods. In the present study, the abundance of large copepods – *C. carinatus*, *R. cornutus*, *Undeuchaeta plumosa* (Lubbock, 1856), *Heterorhabdus papilliger* (Claus, 1863), *Pleuromamma* spp. Giesbrecht & Schmeil, 1898, and *Lophothrix latipes* (Scott T., 1894) – and cold water species (*C. citer*) in deep waters may account for the observed vertical abundance and diversity patterns.

Although vertical hauls were used at the greatest depths, the very low density, richness, and diversity values found at 1,200 m follow the same trend reported in vertical studies of the southwestern Atlantic Ocean (Dias et al. 2010, Bonecker et al. 2014). In the oligotrophic Levantine Sea of the Eastern Mediterranean, Weikert et al. (2001) also observed this pattern for deep-sea plankton communities sampled in daytime- and nighttime hauls (0–4,175 m).

Most of the copepods species observed in this study were previously reported for the oceanic waters of Brazil (Gusmão et al. 1997, Lopes et al. 1999, Cavalcanti and Larrazábal 2004, Dias et al. 2010), the west-central Gulf of Mexico (López-Salgado et al. 2000), the eastern tropical Pacific (northwestern Mexican coast) (Suárez-Morales et al. 2000), the Mediterranean Sea and the North Atlantic (Andersen et al. 2001, 2004), and the eastern tropical Atlantic Ocean from 6°S to 4°N and at 6°W, 12°W, 18°W, and 24°W (Piontkovski et al. 2003). Small calanoids were dominant at 1 m (smaller calanoids) and large calanoids (like *C. carinatus*) at 800 m during both sampling periods. Poecilostomatoid copepods (Oncaeidae and Corycaeidae families) were more abundant than the calanoids at 1 m, 250 m, and 1,200 m. The same results were reported by Roman et al. (1993) who found the concentrations of calanoids usually decreased with depth in the tropical Pacific. Poecilostomatoids dominated the deepest stratum off Bermuda (Paffenhöfer and Mazzocchi 2003) and in the Red Sea, the Arabian Sea, and the Eastern Mediterranean Sea down to a maximum depth of 1,850 m (Böttger-Schnack 1994). Vertical- and horizontal zooplankton distributions depend on food availability and predation pressure (Fernandez-Alamo and Farber-Lorda 2006). In the present study, an abundance of calanoid suspension-feeders was reported in the chlorophyll-rich (upper) layers, whereas the detritivorous poecilostomatoids had a slightly deeper but wider distribution. Paffenhöfer and Mazzocchi (2003) hypothesize that feeding behavior determines the depth ranges that are occupied by specific genera. According to these authors, young forms and adults of a certain species occupy a flexible vertical range that meets their main needs. The major variables include temperature, food quality, food quantity, and predation.

During the rainy season, the five most abundantly identified calanoid copepod species were mainly tropical and oceanic. During the previous dry season, an abundance of neritic calanoids was noted (*C. velificatus*, *S. pileatus*, *P. quasimodo*, and *T. turbinata*) and cold water species (*C. carinatus*). These observations can be attributed to lower salinity and temperature values, respectively. Lopes et al. (1999, 2006) reported similar findings for the Rio de Janeiro coast, on the eastern continental shelf of Rio de Janeiro State and the southern Brazilian shelf, respectively, and, Dias et al. (2010) for the continental shelf of the Campos Basin.

The vertical distribution of the copepod assemblages varied due to differences in the environmental conditions along the water column. The copepod assemblages occupied different niches at each sampling depth. Variations in the vertical distribution of the species must be considered to understand their relationships at each depth. Species-specific depth preferences and ranges created statistically distinct communities at each depth. The copepod assemblages detected at 1 m consisted mainly of tropical- and subtropical species indigenous to the tropical oceanic waters (TW) of Brazil (Cavalcanti and Larrazábal 2004, Lopes et al. 2006, Neumann-Leitão et al. 2008, Nunes and Resgalla Jr 2012). *Candacia pachydactyla, S. danae,* and *A. limbatus* are considered indicators at this depth. These species are epi- and mesopelagic in oceanic waters and have been detected in the tropical- and subtropical waters of the Atlantic, Pacific, and Indian Oceans (Bradford 1973, Bradford-Grieve et al. 1999, Razouls et al. 2000, 2005–2018).

In a study of the Campos Basin, Dias et al. (2010) suggested that oceanic waters influence the differences in copepod composition and abundance. Deep-water species contribute to the upper water masses. The vertical distribution of *C. carinatus* is limited to the deeper waters below the thermocline and the oxygen minimum layer (Verheye et al. 2005, Gaard et al. 2008, Teuber et al. 2013). The vertical distribution of plankton assemblages helps explain the organization and dynamics of pelagic communities and the vertical flux of organic matter throughout the water column (Nogueira-Júnior et al. 2015). Large-scale spatial patterns of certain zooplankton groups coincide with surface water masses, circulation, and upwelling regions. This coincidence aligns with the ecological and dynamic partitioning of the pelagic ecosystem (Fernandez-Alamo and Farber-Lorda 2006). Sournia (1969) indicated that such obvious seasonal changes are not always evident in tropical oceans. What does occur is a succession of small changes in plankton stocks determined by local weather conditions and water mass movements.

The parameters responsible for the signatures of the water masses sampled at each depth correlate with the separation of the copepod assemblages along the water column. The water masses at the four depths have their own hydrographic characteristics. The physicochemical parameters (salinity and temperature) were the main factors influencing copepod assemblage structure and distinction. McManus and Woodson (2012) proposed that plankton interacts with their environment across a range of scales and theorized that the spatial distributions of oceanic organisms (from the smallest plankton to the largest predators) are determined by the interactions between organism behavior and physical oceanographic structure and processes. It has been reported that in marine ecosystems, vertical stratification along the water column influences the distribution of fish larvae and zooplankton groups (Pilar Olivar et al. 2010, Nogueira-Júnior et al. 2015, Doxa et al. 2016). In this study, the model generated for the sampling period influenced the compositions of the copepod assemblages at each sampling depth.

FINAL CONSIDERATIONS

This study shows that various environmental parameters influenced the spatial patterns of copepod assemblages. This observation implies that these consortia occupy distinct niches throughout the Campos Basin. The sample grid included in this study is more extensive than that used in any previous study of the southwest Atlantic Ocean. The findings of this study helped explain the vertical distribution of the copepod species in this region.

ACKNOWLEDGMENTS

This study is part of the Habitats Project–Campos Basin Environmental Heterogeneity by CENPES/PETROBRAS. The authors thank the team of the Zooplankton and Ichthyoplankton Integrated Laboratory of the Federal University of Rio de Janeiro. We thank Cláudio de S. Ressur and José R.S. Silva for sorting the samples. We also wish to thank Luiz Fernando L. Fernandes and Suzanna Vianna for help with sample identification, and Rodolfo Paranhos and Marina S. Suzuki for providing the environmental parameters. Finally, we wish to thank PETROBRAS who assisted with the sampling and the analyses.

LITERATURE CITED

Andersen V, Gubanova A, Nival P, Ruellet T (2001) Zooplankton community during the transition from spring bloom to oligotrophy in the open NW Mediterranean and effects of wind events. 2. Vertical distributions and migrations. Journal of Plankton Research 23: 241–263. https://doi.org/10.1093/plankt/23.3.243

Andersen V, Devey C, Gubanova A, Picheral M, Melnikov V, Tsarin S, Prieur L (2004) Vertical distributions of zooplankton across the Almeria-Oran frontal zone (Mediterranean Sea). Journal of Plankton Research 26: 275–293. https://doi.org/10.1093/plankt/fbh036

Anderson MJ, Gorley RN, Clarke KR (2008) PERMANOVA+ for PRIMER: Guide to Software and Statistical Methods. Plymouth, PRIMER-E, 274 pp.

Andrade L, Gonzalez AM, Rezende CA, Suzuki M, Valentin JL, Paranhos R (2007) Distribution of HNA and LNA bacterial groups in the southwest Atlantic Ocean. Brazilian Journal of Microbiology 38: 330–336. https://doi.org/10.1590/S1517-83822007000200028

Bassani C, Bonecker ACT, Bonecker SLC, Nogueira CR, Reis JML, Nascimento LR (1999) Plâncton do Litoral Norte do Estado do Rio de Janeiro (21°00′ a 23°30′S) – Análise e Síntese do Conhecimento. Oecologia Brasiliensis 7: 99–120. https://doi.org/10.4257/oeco.1999.0701.05

Björnberg TSK (1963) On the marine free-living copepods off Brazil. Boletim do Instituto Oceanográfico de São Paulo 13: 3–142. https://doi.org/10.1590/S0373-55241963000100001

Björnberg TSK (1981) Copepoda. In: Boltovskoy E (Ed.) Atlas del zooplancton del Atlántico Sudoccidental. Mar del Plata, Instituto Nacional de Investigación y Desarrollo Pesquero, 587–679.

Bonecker SLC (2006) Atlas de Zooplâncton da Região Central da Zona Econômica Exclusiva Brasileira. Rio de Janeiro, Museu Nacional, 232 pp.

Bonecker SLC, Dias CO, Fernandes LDA, Avila LR (2007) Zooplâncton. In: Valentin JL (Ed.) Características Hidrobiológicas da Região Central da Zona Econômica Exclusiva Brasileira (Salvador, BA, ao Cabo de São Tomé, RJ). Brasília, Ideal Gráfica e Editora, 125–140.

Bonecker SLC, Araujo AV de, Carvalho PF de, Dias CO, Fernandes LFL, Migotto AE, de Oliveira OMP (2014) Horizontal and vertical distribution of mesozooplankton species richness and composition down to 2,300 m in the southwest Atlantic Ocean. Zoologia 31: 445–462. https://doi.org/10.1590/S1984-46702014000500005

Böttger-Schnack R (1994) The microcopepod fauna in the eastern Mediterranean and Arabian Sea: a comparison with the Red Sea fauna. Hydrobiologia 292/293: 271–282. https://doi.org/10.1007/978-94-017-1347-4_36

Bradford JM (1973) Revision of family and some generic definitions in the Phaennidae and Scolecithricidae (Copepoda: Calanoida). New Zealand Journal of Marine and Freshwater Research 7: 133–152. https://doi.org/10.1080/00288330.1973.9515460

Bradford-Grieve JM, Markhaseva EI, Rocha CEF, Abiahy B (1999) Copepoda. In: Boltovskoy D (Ed.) South Atlantic zooplankton. Leiden, Backhuys Publishers, 869–1098.

Campaner AF (1985) Occurrence and distribution of copepods (Crustacea) in the epipelagial off southern Brazil. Boletim do Instituto Oceanográfico de São Paulo 33: 5–27. https://doi.org/10.1590/S0373-55241985000100002

Cavalcanti EAH, Larrazábal MEL de (2004) Macrozoozooplâncton da Zona Econômica Exclusiva do Nordeste do Brasil (segunda expedição oceanográfica– REVIZEE/NE II) com ênfase em Copepoda (Crustacea). Revista Brasileira de Zoologia 21(3): 467–475. https://doi.org/10.1590/S0101-81752004000300008

Clarke KR, Gorley RN (2006) PRIMER v6: User Manual/Tutorial. Plymouth, PRIMER-E, 190 pp.

Corno G (2006) Primary Production Dynamics in the North Pacific Subtropical Gyre. PhD thesis. Oregon, USA, Oregon State University. Available online at: https://ir.library.oregonstate.edu/concern/graduate_thesis_or_dissertations/zc77sv465

De Cáceres M, Legendre P (2009) Associations between species and groups of sites: indices and statistical inference. Ecology 90: 3566–3574. https://doi.org/10.1890/08-1823.1

Dias CO (1996) Copépodes da costa leste do Brasil. Brazilian Archives of Biology and Technology 39: 113–122.

Dias CO, Araujo AV (2006) Copepoda. In: Bonecker SLC (Ed.) Atlas do zooplâncton da região central da Zona Econômica Exclusiva. Rio de Janeiro, Museu Nacional, Série Livros/Documentos REVIZEE Score Central 21: 21–99.

Dias CO, Araujo AV, Paranhos R, Bonecker SLC (2010) Vertical copepod assemblages (0–2300 m) off southern Brazil. Zoological Studies 49: 230–242. http://zoolstud.sinica.edu.tw/Journals/49.2/230.pdf

Dias CO, Araujo AV, Vianna SC, Loureiro Fernandes LF, Paranhos R, Suzuki M.S, Bonecker SLC (2015). Spatial and temporal changes in biomass, production and assemblage structure of mesozooplanktonic copepods in the tropical south-west Atlantic Ocean. Journal of the Marine Biological Association of the United Kingdom 95: 483–496. https://doi.org/10.1017/S0025315414001866

Doxa A, Holon F, Deter J, Villéger S, Boissery P, Mouquet N (2016) Mapping biodiversity in three-dimensions challenges marine conservation strategies: The example of coralligenous assemblages in North-Western Mediterranean Sea. Ecological Indicators 61: 1042–1054. https://doi.org/10.1016/j.ecolind.2015.10.062

Dufour P, Charpy L, Bonnet S, Garcia N (1999) Phytoplankton nutrient control in the oligotrophic South Pacific subtropical gyre (Tuamotu Archipelago). Marine Ecology Progress Series 179: 285–290. https://doi.org/10.3354/meps179285

Dufrêne M, Legendre P (1997) Species assemblages and indicator species: the need for a flexible asymmetrical approach. Ecological Monographs 67: 345–366. https://doi.org/10.1890/0012-9615(1997)067[0345:SAAIST]2.0.CO;2

Fernandez-Alamo MA, Farber-Lorda J (2006) Zooplankton and the oceanography of the eastern tropical Pacific: A review. Progress in Oceanography 69: 318–359. https://doi.org/10.1016/j.pocean.2006.03.003

Gaard E, Gislason A, Falkenhaug T, Søiland H, Musaeva E, Vereshchaka A, Vinogradov G (2008) Horizontal and vertical copepod distribution and abundance on the Mid-Atlantic Ridge in June 2004. Deep-Sea Research II 55: 59–71. https://doi.org/10.1016/j.dsr2.2007.09.012

Grasshoff K, Ehrhardt M, Kremling K (1999) Methods of seawater analysis. Weinheim, Wiley-VCH, 600 pp. https://doi.org/10.1002/9783527613984

Gusmão LMO, Neumann-Leitão S, Nascimento-Vieira DA, Silva TA, Ilva APS, Porto-Neto FF, Moura MCO (1997) Zooplâncton oceânico entre os Estados do Ceará e Pernambuco-Brasil. Trabalhos do Instituto de Oceanografia da Universidade Federal de Pernambuco 25: 17–30. Available online at: https://www.researchgate.net/publication/316358751_Zooplancton_Oceanico_entre_os_Estados_do_Ceara_e_Pernambuco_Brasil

Lacerda LD, Rezende CE, Ovalle AR, Carvalho CE (2004) Mercury distribution in continental shelf sediments from two offshore oil fields in southeastern Brazil. Bulletin of Environmental Contamination and Toxicology 72: 178–185. https://doi.org/10.1007/s00128-003-0257-0

Lopes RM, Brandini FP, Gaeta SA (1999) Distribution patterns of epipelagic copepods off Rio de Janeiro (SE Brazil) in summer 1991/1992 and winter 1992. Hydrobiologia 411: 161–174. https://doi.org/10.1023/A:1003859107041

Lopes RM, Katsuragawa M, Montú MA, Muelbert JH, Dias JF, Gorri C, Brandini FP (2006) Zooplankton and ichthyoplankton distribution on the southern Brazilian shelf: an overview. Scientia Marina 70: 189–202. https://doi.org/10.3989/scimar.2006.70n2189

López-Salgado I, Gasca R, Suárez-Morales E (2000) La comunidad de copépodos (Crustacea) en los giros a mesoescala en el occidente del Golfo de México (julio,1995). Revista de Biologia Tropical 48: 169–179.

McCune B, Mefford MJ (1999) PC-ORD. Multivariate analysis of ecological data. Oregon, Gleneden Beach, MjM Software, v. 5.0.

McEwen GF, Johnson MW, Folsom TR (1957) A statistical analysis of the performance of the Folsom plankton sample splitter, based upon test observations. Archiv für Meteorologie, Geophysik und Bioklimatologie 7: 502–527. https://doi.org/10.1007/BF02277939

McManus MA, Woodson CB (2012) Plankton distribution and ocean dispersal. The Journal of Experimental Biology 215: 1008-1016. https://doi.org/10.1242/jeb.059014

Mémery LM, Arhan XA, Alvarez-Salgado MJ, Messia H, Mercier CG, Castro A, Rios F (2000) The water masses along the western boundary of the south and equatorial Atlantic. Progress in Oceanography 47: 69–98. https://doi.org/10.1016/S0079-6611(00)00032-X

Metzler PM, Gilbert PM, Gaeta AS, Lublan J (1997) New and regenerated production in the South Atlantic off Brazil. Deep-Sea Research 44: 363–384. https://doi.org/10.1016/S0967-0637(96)00129-X

Neumann-Leitão S, Eskinazi Sant'anna EM, Gusmão LMO, Nascimento-Vieira DA, Paranaguá MN, Schwamborn R (2008) Diversity and distribution of the mesozooplankton in the tropical Southwestern Atlantic. Journal of Plankton Research 30: 795–805. https://doi.org/10.1093/plankt/fbn040

Niencheski LF, Baumgarten MGZ, Roso RH, Bastos LAP (1999). Oceanografia química – levantamento bibliográfico e identificação do estado atual do conhecimento. Rio de Janeiro, MMA, CIRM, FEMAR, 171 pp.

Nogueira-Júnior M, Brandini FP, Codina JCU (2015) Diel Vertical Dynamics of Gelatinous Zooplankton (Cnidaria, Ctenophora and Thaliacea) in a Subtropical Stratified Ecosystem (South Brazilian Bight). PLoS ONE 10: e0144161. https://doi.org/10.1371/journal.pone.0144161

Nunes RD, Resgalla C Jr (2012) The zooplankton of Santa Catarina continental shelf in southern Brazil with emphasis on Copepoda and Cladocera and their relationship with physical coastal processes. Latin American Journal of Aquatic Research 40: 893–913. https://doi.org/10.3856/vol40-issue4-fulltext-7

Paffenhöfer J, Mazzocchi MG (2003) Vertical distribution of subtropical epiplanktonic copepods. Journal of Plankton Research 25: 1139–1156. https://doi.org/10.1093/plankt/25.9.1139

Pedrosa P, Paranhos R, Suzuki MS, Andrade L, Silveira ICA, Schmidt ACK, Falcão APC, Lavrado HP, Rezende CE (2006) Hidroquímica de massas de água oceânicas em regiões da margem continental brasileira Bacia de Campos. Estado do Rio de Janeiro, Brasil. Geochimica Brasiliensis 20: 104–122. https://doi.org/10.21715/gb.v20i1.238

Pilar Olivar M, Emelianov M, Villate F, Uriarte I, Maynou F, Álvarez I, Morote E (2010) The role of oceanographic conditions and plankton availability in larval fish assemblages off the Catalan coast (NW Mediterranean). Fisheries Oceanography 19: 209–229. https://doi.org/10.1111/j.1365-2419.2010.00538.x

Piontkovski S, Williams R, Ignatyev S, Bolyachev A, Chesalin M (2003) Structural-functional relationships in the pelagic community of the eastern tropical Atlantic Ocean. Journal of Plankton Research 25: 1021–1034. https://doi.org/10.1093/plankt/25.9.1021

R Core Team (2015) R: A Language and Environment for Statistical Computing. Vienna, R Foundation for Statistical Computing, v. 17 3.1.3.

Razouls S, Razouls C, de Bovée F (2000) Biodiversity and biogeography of Antarctic copepods. Antarctic Science 12: 343–362. https://doi.org/10.1017/S0954102000000407

Razouls C, de Bovée F, Kouwenberg J, Desreumaux N (2005–2018) Diversity and geographic distribution of marine planktonic copepods. Available on line at: http://copepodes.obs-banyuls.fr/en [Accessed 27/03/2018]

Rezende CE, Andrade L, Suzuki MS, Faro BCMT, Gonzalez ASM, Paranhos R (2007) Hidroquímica. In: Valentin JL (Ed.) Características hidrobiológicas da região central da Zona Econômica Exclusiva brasileira (Salvador, BA, ao Cabo de São Tomé, RJ). Brasília, Ideal gráfica, Série Documentos REVIZEE/SCORE Central, 31-60.

Rodrigues SV, Marinho MM, Jonck CCC, Gonçalves ES, Brant VF, Paranhos R, Curbelo MP, Falcão AP (2014) Phytoplankton community structures in shelf and oceanic waters off southeast Brazil (20°–25°S), as determined by pigment signatures. Deep-Sea Research I 88: 47–62. https://doi.org/10.1016/j.dsr.2014.03.006

Roman MR, Dam HG, Gauzens AL, Napp JM (1993) Zooplankton biomass and grazing at the JGOFS Sargasso Sea time series station. Deep-Sea Research II 40: 883–901. https://doi.org/10.1016/0967-0637(93)90079-I

Silveira ICA, Schmidt ACK, Campos EJS, Godoi SS, Ikeda YA (2000) Corrente do Brasil ao largo da costa leste Brasileira. Brazilian Journal of Oceanography 48: 171–183. https://doi.org/10.1590/S1679-87592000000200008

Sournia A (1969) Cycle annuel du phytoplankton et de la production primaire dans les mers tropicales. Marine Biology 3: 287–303. https://doi.org/10.1007/BF00698859

Steinberg DK, Cope JS, Wilson SE, Kobari T (2008) A comparison of mesopelagic mesozooplankton community structure in the subtropical and subarctic North Pacific Ocean. Deep-Sea Research II 55: 1615–1635. https://doi.org/10.1016/j.dsr2.2008.04.025

Stramma L, Ikeda Y, Peterson RG (1990) Geostrophic transport in the Brazil Current region north of 20°S. Deep-Sea Research 37: 1875–1886. https://doi.org/10.1016/0198-0149(90)90083-8

Suárez-Morales E, Franco-Gordo C, Saucedo-Lozan M (2000) On the pelagic copepod community of the central Mexican tropical Pacific (Autumn, 1990). Crustaceana 73: 751–761. https://doi.org/10.1163/156854000504778

Suzuki MS, Rezende CE, Paranhos R, Falcão AP (2015) Spatial distribution (vertical and horizontal) and partitioning of dissolved and particulate nutrients (C, N and P) in the Campos Basin, Southern Brazil. Estuarine, Coastal and Shelf Science 166: 4–12. https://doi.org/10.1016/j.ecss.2014.12.031

Teuber L, Schukat A, Hagen W, Auel H (2013) Distribution and Ecophysiology of Calanoid Copepods in Relation to the Oxygen Minimum Zone in the Eastern Tropical Atlantic. PLoS ONE 8: e77590. https://doi.org/10.1371/journal.pone.0077590

Valentin JL (1984a). Analyse des paramétres hydrobiologiques dans la remontée de Cabo Frio (Brésil). Marine Biology 82: 259–276. https://doi.org/10.1007/BF00392407

Valentin JL (1984b) Spatial structure of the zooplankton community in the Cabo Frio region (Brazil) influenced by coastal upwelling. Hydrobiologia 113: 183–199. https://doi.org/10.1007/BF00026607

Valentin JL, Monteiro-Ribas MA, Mureb E (1987) Sur quelques zooplanctontes abondants dans l'upwelling de Cabo Frio (Brésil). Journal of Plankton Research 9: 1195–1226. https://doi.org/10.1093/plankt/9.6.1195

Verheye HM, Hagen W, Auel H, Ekau W, Loick N, Rheenen I, Wencke P, Jones S (2005) Life strategies, energetics and growth characteristics of *Calanoides carinatus* (Copepoda) in the Angola-Benguela frontal region. African Journal of Marine Science 27: 641–651. https://doi.org/10.2989/18142320509504124

Viana AR, Faugeères JC, Kowsmann RO, Lima JAM, Caddaah LFG, Rizzo JG (1998) Hydrology, morphology and sedimentology of the Campos continental margin, offshore Brazil. Sedimentary Geology 115: 133–157. https://doi.org/10.1016/S0037-0738(97)00090-0

Vinogradov ME, Shushkina EA, Nezlin NP, Arnautov GN (1998) Vertical distribution of zooplankton in the frontal zone of the Gulf Stream and Labrador Current. Journal of Plankton Research 20: 85–103. https://doi.org/10.1093/plankt/20.1.85

Weikert H, Koppelmann R, Wiegratz S (2001) Evidence of episodic changes in deep-sea mesozooplankton abundance and composition in the Levantine Sea Eastern Mediterranean. Journal of Marine System 30: 221–239. https://doi.org/10.1016/S0924-7963(01)00060-4

Author Contributions: SLCB designed the study, participated in fieldwork and revised the paper; COD and AVA analyzed the data and wrote the paper.

Competing Interests: The authors have declared that no competing interests exist.

11

Relating flight initiation distance in birds to tropical dry forest restoration

Marcela Osorio-Beristain[1], Ángeles Rodríguez[1], Cristina Martínez-Garza[1], Raúl Ernesto Alcalá[1]

[1]*Centro de Investigación en Biodiversidad y Conservación, Universidad Autónoma del Estado de Morelos. 62209, Cuernavaca, Morelos, México.*
Corresponding author: Marcela Osorio Beristain (mosorio@uaem.mx)

ABSTRACT. Flight initiation distance (FID), defined as the distance at which an individual flees from an approaching predator, might depend on how the individual perceives the risk of being predated. Using a standardized walking approach method on focal bird individuals, we investigated whether different levels of vegetation cover (habitat) influence the perception of predation risk. To do this, we worked in an area of tropical dry forest in central Mexico that is currently part of a restoration ecology experiment. We hypothesized that restoration decreases individual's predation risk perception by increasing the complexity of the vegetation cover. The escape responses of three tropical birds with different diets and foraging strategies were also contrasted. There was no effect of habitat on FID, suggesting that birds in both habitats perceived predation risk in a similar manner. There was, however, a difference in FID among species: the Golden-cheeked Woodpecker tolerated closer human presence before flight than the Inca Dove and Streak-backed Oriole. This difference is likely due to the use of an alternative avoidance strategy of this species, which uses trunks for hiding. To decrease birds' perceived predation risk, restoration intervention plans should include a mosaic of larger excluded plots located near relatively well-conserved sites to increase the area covered by vegetation.

KEY WORDS. Escape from predators, foraging strategies, habitat quality, predation risk, tropical birds.

INTRODUCTION

Predation risk is one of the main selective forces shaping animal behavior (Lima and Dill 1990). In particular, behavioral strategies that reduce individuals' exposure or likelihood of attack by predators should be favored (Lima 1998, Valeix et al. 2009). Predation risk results from a combination of factors affecting both attack and capture probabilities that include the structure of the environment, vegetative cover, social factors, and distribution and abundance of predators (Frid and Dill 2002), besides intrinsic species characteristics (Lima and Zollner 1996, reviewed in Weston et al. 2012). Because antipredator behavior responds to changes in these factors rather than the predation rate per se, it is expected that animals' responses to nonlethal human disturbance follow the same decision rules (Frid and Dill 2002). Riskier human approaches could reduce the opportunities for foraging and habitat use in general if the level of human disturbance is high (Miller et al. 2001, Fernández-Juricic 2000). This effect could be stronger in open habitats (Finney et al. 2005), where high visibility could increase the distance at which threats are detected (Whittingham and Karl 2004); consequently, individual perception of security should improve as vegetation cover increases (see meta-analysis in Stankowich and Blumstein 2005). On the other hand, some species may feel more secure in open areas because they can escape faster (see Ripple et al. 2001) or because areas with higher vegetation cover may allow predators to hide (Hopcraft et al. 2005). In the Nuthatch *Sitta europaea* (Linneaus, 1758), Sittidae, lower predation risk, measured as the time investment in vigilance while foraging, was associated positively with closeness to vegetation (Mozetich and Carrascal 1995). In ground-foraging birds, denser habitat structure increases visual obstruction (Butler and Gillings 2004) and may increase predation risk; alternatively, denser structure may reduce predation risk if birds are cryptic when under cover (review in Whittingham and Karl 2004 for examples in both alternatives).

Fleeing is a main predator avoidance strategy. However, it involves large potential costs due to the energetic cost of flight itself and the many costs associated with abandoning fundamental activities such as courtship, breeding, and foraging (Lima and Dill 1990). Individuals should therefore adjust their sensitivity to an approaching predator based on this balance of costs and benefits. This decision may vary within the same population, depending, for instance, on the time of day and the condition of satiation (Piratelli et al. 2015).

Flight initiation distance (FID), or the distance at which potential prey initiates flight when approached by a predator (Ydenberg and Dill 1986) is a useful measure of this predation risk sensitivity. Measuring how behavior changes depending on environmental characteristics allows the comparison of derived fitness consequences among habitats with contrasting characteristics. For example, behavior measurements enable the identification of critical resources that determine whether or not a site is suitable for a particular species (Lindell 2008).

The study of escape behavior has an important potential role in evaluating habitat quality within the framework of restoration ecology. Restoration ecology not only aims for the recovery of biodiversity per se, but more broadly aims to re-establish natural processes reduced or eliminated by human-induced disturbance. Several studies have shown that after eliminating anthropogenic disturbance in an area, the structural complexity of vegetation (i.e. density, biomass, vertical strata, percent cover) increases by natural succession, from secondary vegetation to mature forest (Lamb and Erskinepd Parrotta 2005). Heterogeneous landscapes resulting from an experimental restoration approach often include a gradient of successional stages (Howe and Martínez-Garza 2014), making them ideal systems for testing the effectiveness of using escape behavior (FID) as an indicator of the restoration of ecological processes.

Animals respond to approaching humans similarly to terrestrial predators (Frid and Dill 2002). Therefore, a standardized walking approach to focal individuals mimics the effect of an approaching predator and allows the quantification of FID. We applied this standardized approach to compare FID in three bird species with different diet and foraging strategies to evaluate the success of one of the few long-term restoration ecology experiments performed within a tropical dry forest. This biome is particularly important for conservation and restoration goals due to its high biological value in terms of endemic species as well as human habitation and resource use (Quesada et al. 2009). In this experiment, we compared FID in two habitats: areas under continued human disturbance and plots where anthropogenic disturbance has been eliminated since 2006, when cattle were removed as part of a larger ongoing restoration experiment (Martínez-Garza et al. 2011). Since restoration intervention has increased vegetation complexity (for example, De la O-Toris et al. 2012), this may lead to decreased predation risk perception, and therefore, we would expect lower FID in the three focal bird species in restored areas than in areas under disturbance. On the other hand, if higher vegetation complexity increase predation risk, FID will be higher in the restored areas. In addition, we examined between-species differences in escape behavior among two canopy-foraging birds *Icterus pustulatus* (Wagler, 1829), Icteridae, and *Centurus chrysogenys* (Vigors, 1839), Picidae and one ground-forager *Columbina inca* (Lesson, 1847), Columbidae. We expected lower FID in *I. pustulatus* and *C. chrysogenys* than in *C. inca*, since canopy-foraging species should be less vulnerable to terrestrial predators.

MATERIAL AND METHODS

Study site

The Sierra de Huautla Biosphere Reserve is a federally protected area (ca. 59 030 ha) located in the southern part of the state of Morelos, Mexico. It borders the states of Guerrero and Puebla (18°20′10″, 18°34′20″ N and 98°51′20″, 98°08′15″ W). The topography is variable and elevation ranges from 900 to 1,300 m. The mean annual temperature is 24.5 °C and the average total rainfall is 850 mm, with peak rainfall occurring between June and October. Climate is classified as tropical subhumid with summer rains (CONAGUA, Gerencia Regional Balsas). The original vegetation is tropical dry forest (Trejo and Dirzo 2000), and nearly 60 % of the total area is relatively well-conserved, particularly in steeply sloped areas (CONANP 2005). However, around 40 % of the total surface has been degraded due to deforestation for agricultural fields, wood extraction, and cattle ranching.

A large portion of tropical dry forest near the town of El Limón de Cuauchichinola (ca. 300 inhabitants, 1220 m asl) was deforested in the early 1990s, used for maize cultivation for six years, and subsequently abandoned. Despite abandonment, the area is still degraded, since the secondary forest has been used mainly to sustain cattle ranching (~ 7 head/ha, De la O-Toris et al. 2012).

Study species

All observations were conducted on medium-sized species (Howell and Webb 1995) of resident birds. Trials were carried out when each focal bird was perched alone, and during the non-breeding season, since parental care could increase risky behavior (Frid and Dill 2002). Bird species were selected because they constitute a representative and resident avifauna in the Mexican tropical dry forest (Urbina-Torres 2000), and also for their contrasting foraging behavior. The Streak-backed Oriole (*I. pustulatus*, 19–23 cm) is a frugivorous-insectivorous species that feeds in the canopy; both sexes are ornately colored with carotenoid-based yellow-orange body plumage (Jaramillo and Burke 1999, Murphy et al. 2009). The Golden-cheeked Woodpecker (*C. chrysogenys*, 19–21.5 cm) is a bark insectivore that feeds while perching on branches. Males are barred black and white with bright yellow-orange colors on the nape whereas females have a grayish crown (Howell and Webb 1995). The Inca Dove (*C. inca*, 20.5–23 cm) is a granivorous ground feeder

(Howell and Webb 1995). This dove is a secretive species that hides in low, dense vegetation, but it also occurs in human settlements (Singhal et al. 2016).

Restoration context

In 2006, 14 experimental plots (50 × 50 m) with similar vegetation and history of use were located and marked (details in Martínez-Garza et al. 2011, 2016). The plots were situated within an area of some 50 ha near the biological station "El Limón". Distance among experimental plots ranged from 0.08 to 1.59 km (0.72 ± 0.46; mean ± standard deviation). The most abundant trees species in the disturbed areas were *Acacia cochliacantha* (Humbolt & Bonpland ex Willd, 1806), *Mimosa benthamii* (Macbr, 1919), Fabaceae and *Ipomoea pauciflora* (Martens & Galeotti, 1845), Convolvulaceae, these species dominate the post-agricultural landscape, and the experimental plots (Martínez-Garza et al. 2011). There were also a few *Bursera* spp. trees, specifically maintained by local people due to their ethnobotanical value (Maldonado et al. 2013).

Eight experimental plots were enclosed with electric fences to avoid the intrusion of small and large domestic animals, and the remaining plots were marked but remained open to domestic animals. By the sixth year after exclusion, marked differences in the vegetation matrix were detected. For example, herbaceous vegetation (Poaceae and Asteraceae) cover was 20 times greater in the excluded than in non-excluded plots (Hernández-Flores et al. 2016), and tree vegetation cover was 60-80 % in the excluded plots and less than 10% in the non-excluded plots (Santana-Huicochea et al. in review). Meanwhile, in non-excluded plots the soil remained bare and degraded, sustaining poor herb biomass due to constant grazing and soil compaction by cattle (De la O-Toris et al. 2012).

Experimental design

The observer walked slowly in the plot searching for a focal individual. Once a bird was identified, the distance between the observer and the individual was recorded as escape behavior is influenced by the intruder's starting distance (Blumstein 2003). The observer then scored FID by walking directly toward the bird at a constant pace (ca.1 m/s) while maintaining eye contact, recording both the perching height and the distance at which focal individual initiated flight. Distance measurements were done with a laser rangefinder (Leica DISO D5).

Direct starting distance calculated using the Pythagorean Theorem was used as a measurement of FID, since escape behavior is often influenced by the height at which the bird was perched at in the tree (Geist et al. 2005). Measurements of FID were carried out in restored and perturbed areas (n = 8 plots per habitat). The study was conducted during April and May 2012 between 6:30 and 8:30 am. Because birds were not banded, only one trial per species was carried out per day in each plot to minimize the possibility of pseudoreplication due to observing the same individual. In total, we performed 54 trials on *I. pustulatus*, 18 trials on *C. inca*, and 50 trials on *C. chrysogenys*.

Statistical analyses

An analysis of covariance (ANCOVA) with independent slopes was performed to evaluate the effect of habitat, species, and intruder starting distance on FID. The effect of habitat was included as a nominal variable with two levels (restored and perturbed plots) with fixed effects. Species identity was included as a three-level nominal variable with fixed effects, and intruder's starting distance was included as the covariate (Blumstein 2003). To understand the effect of habitat on the expected relationship between starting distance and FID, the interaction between starting distance and habitat was included as was the interaction between species and starting distance. Raw FID was log-transformed to fulfill the assumption of homoscedasticity of residuals. All statistical tests were performed using JMP statistical software (version 4.0.3, SAS Institute, Cary, NC).

RESULTS

The ANCOVA utilized to evaluate the effect of habitat, identity of birds on FID explained 22 % of the total variance. According to the effect tests of the analysis of covariance, the species identity and starting distance (the covariate) showed a marginal and a highly significant effect, respectively, on the variation in FID (Table 1). There was a positive relationship between FID and starting distance. There was no significant effect of habitat or the interaction terms on FID (Table 1).

Due to the fact that, in this ANCOVA, the effect of habitat and both interaction terms were non-significant, a model that included only the effect of species identity, the starting distance,

Table 1. ANCOVA results for species identity, habitat, the covariate and the interaction terms on FID.

Source	DF	Sum of Squares	F	p
Species (S)	2	2.99	2.51	> 0.05
Habitat (H)	1	0.36	0.61	> 0.05
Covariate (C)	1	6.75	11.31	< 0.005
S×C	2	0.72	0.61	> 0.050
H×C	1	0.31	0.52	> 0.050

Table 2. ANCOVA results for species identity, the covariate and the interaction term on FID.

Source	DF	Sum of Squares	F	p
Species (S)	2	3.66	3.08	< 0.05
Covariate (C)	1	6.38	10.75	< 0.005
S×C	2	0.77	0.65	> 0.05

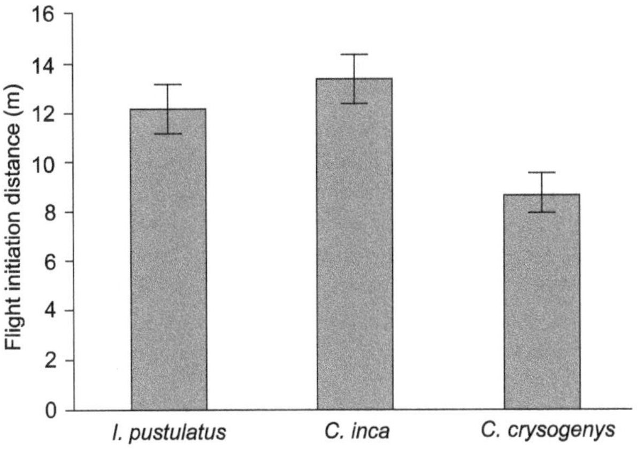

Figure 1. Mean (± SE) flight initiation distance scored in the three avian species evaluated.

and their interaction was also tested. The effect test of this ANCOVA showed statistical significance only for the effect of species identity and the covariate were statistically significant (Table 2). The post hoc mean comparison test (Tukey HSD) indicated that there was no difference in FID between *C. inca* and *I. pustulatus*, whereas the FID of *C. crysogenys* was significantly lower (they allowed a closer approach) than in the other two species (Fig. 1).

DISCUSSION

The significant effect of starting distance (the covariate) indicates that birds in our study, were on alert to potential predators. This is consistent with the hypothesis that animals should not pay the attentional cost for continued monitoring of an approaching predator and should flush at a greater distance as starting distance increases (Blumstein 2003). However, our results show no association between the escape behavior and habitat, demonstrating no effect yet of restoration intervention on FID for these three species after six years. In contrast, there was a significant between-species difference; the Golden-cheeked Woodpecker was more tolerant of human approach than Inca Doves or Streak-backed Orioles in both habitats. We discuss the likely explanation for the lack of effect of restoration intervention and the implication of our results on the design of restoration projects in tropical dry forests. We also discuss the causes explaining inter-specific differences in the perception of security independent of restoration intervention.

We propose three possible non-exclusive explanations for the lack of differences in FID between habitats. First, while restoration actions clearly increased cover of both herbaceous (Hernández-Flores et al. 2016) and tree vegetation, it is possible that the vertical complexity in the restored areas may have been insufficient to increase birds' perception of security. This is because although planted tree increased cover their vertical growth was still limited after six years.

Second, it is also possible that the lack of effect of habitat could result from the geographic scale at which restored plots are present within the landscape. The 50 × 50 m plots could represent very small vegetated patches in comparison with the overall landscape (~50 ha), which is mainly composed of open areas and a broken canopy of trees. Birds may have been unable to perceive the decreased risk at this scale, and the landscape's fragmented physiognomy may have even produced edge effects with higher predation risk. Small patches were perceived as risky sites by birds (Batary and Baldi 2004) and rodents (Baker and Brown 2010) in forest restoration plots in Costa Rica, whereas increased patch area and canopy cover were associated with longer duration of bird visits (Fink et al. 2009). Also in Costa Rica, restoration patches of tens to a few hundred square meters offered less cover from predation risk for insectivorous and omnivorous birds, including Cherriés tanager, *Rampochelus costaricensis* (Lesson, 1831), Thraupidae; Ruffous-capped warbler, *Basileuterus rufifrons* (Swainson, 1838), Parulidae; Common tody-fly-catcher, *Todirostrum cinereum* (Linnaeus, 1766), Tyrannidae; and Plain wren, *Thryothorus modestus* (Cabanis, 1861), Troglodytidae (Morrison et al. 2010). The authors therefore recommended that patches of at least a few thousand square meters be planted to reduce the predation risk perception of birds.

Third, the study site is within an ecologically fragmented matrix, surrounded by secondary forest, maize cultivation plots, open areas with limited vegetation and eroded soil. Individuals could have been sensitized by frequent proximity to humans or cows (i.e. perceive greater risk from repeated exposure to the predator type, see Stankowich and Blumstein 2005). Indeed, a meta-analysis found that experience with predators increased the prey's perception of predation risk by 38 % (Stankowich and Blumstein 2005), and the same study reported a greater FID in bird species during hunting seasons, when species face higher exposure to human predators than at other times. In the case of our study site, birds are frequently exposed to human and cow presence, since the plots are surrounded by paths where people frequently walk, ride on horseback, or travel in small trucks; these areas are also used by cattle during the rainy season. It is possible that habituation to repetitive human and cow stimuli resulted in similar levels of wariness both in open sites and those with more vegetative cover. On the other hand, it is possible that local birds have been selected for tolerance to humans and cows (see Dongen et al. 2015). It would be interesting to evaluate this hypothesis in our study site.

The decreased FID in the Golden-cheeked Woodpecker compared to the other two species was contrary to our prediction. This was also contrary to the general expectation that FID increases with body size (see Blumstein 2006, and Piratelli et al. 2015), since this was the largest of the three species studied. We suspect that the decrease in FID was due to specialized feeding habit. Golden-cheeked Woodpeckers perch on small branches and move around the trunks while gleaning prey from the bark. This behavior allows individuals to hide when a predator approaches, essentially providing an alternative strategy to flight. In this study, woodpeckers were observed moving around the trunk (i.e. maintaining the trunk between themselves and the "predator") and observing the "predator" before taking flight. This hiding behavior could account for the higher tolerance in terms of FID to an approaching predator than the other species studied. Streak-backed Orioles, on the other hand, are likely unable to apply this hiding strategy due to their conspicuous coloration. In this part of the species' geographic range, both sexes are ornately colored with carotenoid-based yellow-orange body plumage and vivid crimson-orange feathers on the head and breast (Murphy et al. 2009). In species where camouflage/hiding is not a predator avoidance option, individuals are likely to be less tolerant to an approaching predator in any habitat. It would be interesting to test this prediction in species where one sex has a more conspicuous ornamental trait to verify whether FID is larger in the more conspicuous sex (see Møller et al. 2015).

Interestingly, the Inca Dove, a species that occurs in urbanized sites, was less tolerant to human approach than Golden-cheeked Woodpeckers. The Inca Dove is a ground forager with low aspect ratio. Individuals find small seeds by picking at the soil surface in open sites. While foraging, individuals maintain their heads down and have limited predator detection capacity. Birds with a lower aspect ratio have longer FID as they have limited potential to escape from predation (Møller 2014). The Columbiforme clade has developed a greater mass-specific power output due to relatively shorter wings with higher beat frequencies (Jackson 2009), which enables an energetic vertical flight to escape from predation. Inca Dove behavior may be subject to a "flush early and avoid the rush" trade-off (Blumstein 2010), where the higher energetic investment of fleeing early is compensated by a reduction of the attention cost of monitoring an approaching predator (Samia et al. 2013).

Animals' behavior and decisions can be useful indicators to evaluate the success of habitat management actions for birds in different context as restoration ecology, urban parks and protected areas. In this study, we found that for medium sized bird species, small (ca. 2500 m^2) and isolated restored plots may not decrease perceived predation risk. Therefore, to decrease birds' perceived predation risk, restoration plans should include a mosaic of larger excluded plots located near relatively well-conserved sites to increase the area covered by vegetation. Wildlife managers should use FID in zones that are far removed from human impact on wild species to avoid possible sensitization or habituation effects. So, in conjunction with restoration plans it is important to identify buffer zones in order to avoid disturbance induced by human activity.

ACKNOWLEDGEMENTS

This study was funded by the Programa para el Desarrollo Profesional Docente en Educación Superior UAEMOR-CA 56. MOB was supported by an Individual Project from the Universidad Autónoma del Estado de Morelos and CMG by the Consejo Nacional de Ciencia y Tecnología Project #80027. The Centro de Investigación en Biodiversidad y Conservación supported researchers at the El Limón Biological Station. MOB is grateful to ECOLAB from University Paul Sabatier for its support during her sabbatiacal stay. Gregory Charre made useful comments on the manuscript and Lynna Kiere reviewed the English. We are grateful to Juan Camacho for his participation in the fieldwork and to Gerardo Pacheco from El Limón, Luz María Ayestarán, and Alondra Nicolás for their contributions to the maintenance of the experimental pots. This research was carried out in accordance with Mexican laws and ethical guidelines.

LITERATURE CITED

Baker MAA, Brown JS (2010) Islands of fear: effect of wooded patches suitability of the stipped mouse in a South Africa grassland. Functional Ecology 24: 1311–1322. https://doi.org/10.1111/j.1365-2435.2010.01757.x

Batary P, Baldi A (2004) Evidence of an edge effect on avian nest success. Conservation Biology 18: 384–400. https://doi.org/10.1111/j.1523-1739.2004.00184.x

Blumstein DT (2003) Flight initiation distance in birds is dependent on intruder starting distance. Journal of Wildlife Management 67: 858–857. https://doi.org/10.2307/3802692

Blumstein DT (2006) Developing an evolutionary ecology of fear: how life history and natural history traits affect disturbance tolerance in birds. Animal Behavior 71: 383–399. https://doi.org/10.1016/j.anbehav.2005.05.010

Blumstein DT (2010) Flush early and avoid the rush: a general rule of antipredator behavior?. Behavioral Ecology 21: 444–442. https://doi.org/10.1093/beheco/arq030

Butler SJ, Gillings S (2004) Quantifying the effects of habitat structure on prey detectability and accessibility to farmland birds. Ibis 146(Suppl. 2): 121–130. https://doi.org/10.1111/j.1474-919X.2004.00352.x

CONANP (2005) Programa de Conservación y Manejo de la Reserva de la Biosfera Sierra Huautla. Comisión de Areas Naturales Protegidas, México, 210 pp.

De la O-Toris J, Maldonado B, Martínez-Garza C (2012) Efecto de la perturbación en la comunidad de herbáceas nativas y ruderales de una Selva estacional Mexicana. Botanical Sciences 90: 464–480. https://doi.org/10.17129/botsci.475

Dongen WFD, Robinson RW, Weston MA, Mulder RA, Guay PJ (2015) Variation at the DR4D4 locus is associated with wariness and local site selection in urban back swans. BMC Evolutionary Biology 15: 253. https://doi.org/10.1186/s12862-015-0533-8

Fernández-Juricic E (2000) Local and regional effects of pedestrians on forest birds in a fragmented landscape. Condor 102: 242–255. https://doi.org/10.1650/0010-5422(2000)102[0247:LAREOP]2.0.CO;2

Fink RB, Lindell CA, Morrison EB, Zahawi RA, Holl KD (2009) Patch size and tree species influence the number and duration of bird visits in forest restoration plots in southern Costa Rica. Restoration Ecology 17: 474–486. https://doi.org/10.1111/j.1526-100X.2008.00383.x

Finney SK, Pearce-Higgins JW, Yalden DW (2005) The effect of recreational disturbance on an upland breeding bird, the golden plover Pluvialis apricaria. Biological Conservation 121: 56–63. https://doi.org/10.1111/j.1526-100X.2008.00383.x

Frid A, Dill LM (2002) Human caused disturbance stimuli as a form of predation risk. Conservation Ecology 6: 11. http://www.consecol.org/vol6/iss1/art11, https://doi.org/10.5751/ES-00404-060111

Geist C, Liao J, Libby S, Blumstein DT (2005) Does intruder group size and orientation affect flight initiation distance in birds? Animal Biodiversity and Conservation 28: 67–73.

Hernández-Flores J, Osorio-Beristain M, Martínez-Garza C (2016) Ant foraging as an indicator of tropical dry forest restoration. Environmental Entomology 45: 999–994. https://doi.org/10.1093/ee/nvw054

Hopcraft, JGC, Sinclair ARE, Packer C (2005) Planning for success: Serengeti lions seek prey accessibility rather than abundance. Journal of Animal Ecology 74: 555–566. https://doi.org/10.1111/j.1365-2656.2005.00955.x

Howe HF, Martínez-Garza C (2014) Restoration as experiment. Botanical Sciences 92: 1–10.

Howell SNG, Webb S (1995) A guide to the birds of México and Northern Central America. Oxford University Press, 1010 pp.

Jackson BE (2009) The allometry of bird flight performance. PhD thesis, Montana, United States, University of Montana. https://scholarworks.umt.edu/etd/960

Jaramillo A, Burke P (1999) New World Blackbirds. University Press, Princeton, 427 pp.

Javůrková V, Šizling AL, Kreisinger J, Albrecht T (2012) An alternative theoretical approach to escape decision making: the role of visual cues. PLoSOne 7: e32522. https://doi.org/10.1371/journal.pone.0032522

Lamb D, Erskinepd, Parrotta JA (2005) Restoration of degraded tropical forest landscapes. Science 310: 1621–1632. https://doi.org/10.1126/science.1111773

Lindell CA (2008) The value of animal behavior in evaluations of restoration success. Restoration Ecology16: 192–203. https://doi.org/10.1111/j.1526-100X.2008.00389.x

Lima SL (1998) Stress and decision making under the risk of predation: recent developments from behavioral, reproductive, and ecological perspectives. Advances in the Study of Behavior 27: 212–290. https://doi.org/10.1016/S0065-3454(08)60366-6

Lima SL, Dill LM (1990) Behavioural decisions made under the risk of predation. Canadian Journal of Zoology 68: 616–640. https://doi.org/10.1139/z90-092

Lima SL, Zollner PA (1996) Towards a behavioral ecology of ecological landscapes. Trends in Ecology and Evolution 11: 131–135. https://doi.org/10.1016/0169-5347(96)81094-9

Maldonado B, Caballero J, Delgado-Salinas A, Lira R (2013) Relationship between use value and ecological importance of floristic resources of seasonally dry tropical forest in the Balsas river Basin, México. Economic Botany 67: 12–29. https://doi.org/10.1007/s12231-013-9222-y

Martínez-Garza C, Osorio-Beristain M, Valenzuela-Galván D, Nicolás-Medina A (2011) Inter and intra-annual variation in seed rain in a secondary tropical forest excluded from chronic perturbation. Forest Ecology and Management 262: 2202–2218. https://doi.org/10.1016/j.foreco.2011.08.013

Martínez-Garza C, Osorio-Beristain M, Alcalá-Martínez RE, Valenzuela-Galván D, Mariano N (2016) Ocho años de restauración experimental en las selvas estacionales de México. In: Ceccon E, Martínez-Garza C (Eds) Experiencias Mexicanas en la Restauración de los Ecosistemas. CRIM-UNAM, CONABIO, UAEM, México, 333–354.

Miller SG, Knigh RL, Miller CK (2001) Wildlife responses to pedestrian and dogs. Wildlife Society Bulletin 29: 121–132.

Møller AP (2014).Life history, predation and fligth initiation distance in a migratory bird. Journal of Evolutionary Biology 27: 1101–1113. https://doi.org/10.1111/jeb.12399

Møller AP, Samia DSM, Weston MA, Guay PJ, Blumstein DT (2014) American Exceptionalism: Population Trends and Flight Initiation Distances in Birds from Three Continents PloS One 9: e107883.

Møller AP, Samia DSM, Weston MA, Guay PJ, Blumstein DT (2015) Flight initiation distance in relation to sexual dichromatism and body size in birds from three continents. Biological Journal of Linnean Society 117: 828–831. https://doi.org/10.1111/bij.12706

Morrison EB, Lindell CA, Holl KD, Zahawi RA (2010) Patch size effect on avian foraging behaviour: implications for tropical forest restoration design. Journal of Applied Ecology 47: 131–138. https://doi.org/10.1111/j.1365-2664.2009.01743.x

Mozetich IM, Carrascal LM (1995) The influence of the foraging niche on the behavioural response to increased predation risk in three-gleaning birds. Ardeola 42: 45–56.

Murphy TG, Hernández D, Osorio-Beristain M, Montgomerie RM, Omland K (2009) Female status signaling and seasonal variation in territoriality in the elaborate monomorphic streak-backed oriole. Behavioral Ecology 20: 1001–1006. https://doi.org/10.1093/beheco/arp089

Piratelli AJ, Favoretto GR, De Almeida A, Maximiano MF (2015) Factors affecting escape distance in birds. Zoologia 32: 434–444. https://doi.org/10.1590/s1984-46702015000600002

Quesada M, Sánchez-Azofeifa GA, Alvarez-Añorve M, Stoner KE, Avila-Cabadilla L, Calvo-Alvarado J, Castillo A, Espirito-Santo MM, Fagundes M, Fernándes GE, Gamon J, Lopezaraiza-Mikel M, Lawrence DL, Cordeira-Morellato LP, Powers JS, De S Neves F, Rosas-Guerrero V, Sayago R, Sánchez-Montoya G (2009) Succession and management of tropical dry forests in the Americas: Review and new perspectives. Forest Ecology and Management 258: 1011–1024. https://doi.org/10.1016/j.foreco.2009.06.023

Ripple WJ, Larsen EJ, Renkin RA, Smith DW (2001) Trophic cascades among wolves, elk, and aspen on Yellowstone National Park's northern range. Biological Conservation 102: 227–234. https://doi.org/10.1016/S0006-3207(01)00107-0

Samia DSM, Nomura F, Blumstein DT (2013) Do animals generally flush early and avoid the rush? A meta-analyses. Biology Letters 9: 20130016. https://doi.org/10.1098/rsbl.2013.0016

Singhal A, Ingty A, Sinha P, Ghoshal K (2016) American Museum of Natural History Birds of North America. Penguin Random House DK Publishing, New York.

Stankowich T, Blumstein DT (2005) Fear in animals: a meta-analyses and review of risk assessment. Proceedings of the Royal Society of London B 272: 2622–2634. https://doi.org/10.1098/rspb.2005.3251

Trejo I, Dirzo R (2000) Deforestation of seasonally dry tropical forest: a national and local analysis in Mexico. Biological Conservation 94: 133–142. https://doi.org/10.1016/S0006-3207(99)00188-3

Urbina-Torres F (2000) New distributional information of birds from the state of Morelos, Mexico. Bulletin of the British Ornithologists' Club 120: 1–15.

Valeix M, Loveridge AJ, Chamaille-James S, Davidson Z, Murindagomo F, Fritz H, MacDonald DW (2009) Behavioral adjustments of African herbivores to predation risk by lions: spatiotemporal variations influence habitat use. Ecology 90: 23–30. https://doi.org/10.1890/08-0606.1

Weston MA, McLeod M, Blumstein DT, Guay PJ (2012) A review of flight-initiation distances and their application to managing disturbance to Australian birds. Emu 112: 262–286. https://doi.org/10.1071/MU12026

Whittingham MJ, Karl LE (2004) The effects of habitat structure on predation risk of birds in agricultural landscapes. Ibis 146: 2012–220. https://doi.org/10.1111/j.1474-919X.2004.00370.x

Ydenberg RC, Dill LM (1986) The economics of fleeing from predators. Advances in the Study of Behavior 16: 222–249. https://doi.org/10.1016/S0065-3454(08)60192-8

Author Contributions: MOB conceived, designed the experiment and took the data in the field with AR. MOB wrote the paper, REA analyzed the data. CMG, REA commented on the paper.

Competing Interests: The authors have declared that no competing interests exist.

12

Breeding biology of *Tyrannus melancholicus* (Aves: Tyrannidae) in a restinga reserve of southeastern Brazil

Hermes Daros[1], William Barbosa Dutra[1], Charles Duca[2]

[1] *Programa de Pós-graduação em Ecologia de Ecossistemas, Universidade Vila Velha. Campus Boa Vista, 29102-920 Vila Velha, ES, Brazil.*
[2] *Laboratório de Ecologia de Populações e Conservação, Universidade Vila Velha. Campus Nossa Senhora da Penha, 29102-623 Vila Velha, ES, Brazil.*
Corresponding author: Charles Duca (cduca@uvv.br)

ABSTRACT. Existing knowledge of the breeding success and life history characteristics of most Neotropical bird species is scarce. Here, we help fill this gap by analyzing aspects of the breeding biology of the Tropical Kingbird *Tyrannus melancholicus* (Vieillot, 1819), which is a good model for this kind of study as it is a common species occurring in various environments, including urban areas, but little is known about its life history. We provide results concerning the breeding period, clutch size, incubation and nestling periods, description of nests, eggs and nestlings, and the plants used for nest sites by this species. Fifty-four nests were monitored over two seasons (2012–2014) in a protected area in southeastern Brazil. Nesting began at the end of the dry season and the beginning of the rainy season. The frequency of active nests varied according to variations in rainfall for each breeding season analyzed. The means and standard deviations of the incubation period (14.2 ± 1.9 days), nestling period (15.1 ± 0.8 days) and clutch size (2.5 ± 0.7 eggs) were similar to values reported for other Neotropical passerines. Twenty-one plant species used as nest trees and for the construction of the nests were identified. The results show that *T. melancholicus* is not highly selective when choosing plant species used for nest construction.

KEY WORDS. Atlantic Forest, breeding season, clutch size, life history, Tropical Kingbird.

INTRODUCTION

The breeding biology and life history characteristics of most Neotropical bird species are poorly known (Martin 1996, Stutchbury and Morton 2008), particularly when it comes to endemic species (Mason 1985, Robinson et al. 2000). Although several studies have investigated the breeding biology of Brazilian Neotropical birds in the Cerrado (Alves and Cavalcanti 1990, Medeiros and Marini 2007, Marini et al. 2009a, b, Duca and Marini 2011), Pantanal (Pinho et al. 2006, Nobrega and Pinho 2010), Atlantic Forest (Alves et al. 2002, Duca and Marini 2004) and grasslands of southern Brazil (Chiarani and Fontana 2015, Repenning and Fontana 2016), the data they generated are limited compared to what is known about birds from temperate regions. The nesting behavior of many tyrannids (New World tyrant flycatchers), for example, is completely unknown, and detailed information on the breeding biology of this group should be expanded (Fitzpatrick 2004).

Tyrannus Lacepede, 1799 includes migratory birds that live in semi-open areas bordering dense vegetation, frequently in cities and often near freshwater bodies (Sigrist 2014). This genus is composed of 13 species, six of which are found in South America (Clements 2007), including the Tropical Kingbird *Tyrannus melancholicus* (Vieillot, 1819), which feeds on fruits and insects (Sick 1997, Höfling and Camargo 2002). Although it is widely distributed in diverse environments, little is known about this species, including its reproductive biology (Legal 2007). In contrast, numerous studies have been conducted on the breeding biology of some of its congeners, such as the Fork-tailed Flycatcher *Tyrannus savana* (Vieillot, 1808) (Teul et al. 2007, Marini et al. 2009a), Western Kingbird *Tyrannus verticalis* (Say, 1823) (Murphy 1988, Bergin 1997) and Eastern Kingbird *Tyrannus tyrannus* (Linnaeus, 1766) (Murphy 2007, Cooper et al. 2009).

The objective of this study was to evaluate the breeding biology of *T. melancholicus* in a restinga area on the southeastern coast of Brazil and to determine the following nesting attributes: i) mean number of breeding attempts per season; ii) duration of the breeding period; iii) characterization of the nests and eggs; iv) clutch size; v) duration of incubation and nestling periods; vi) offspring development and; vii) identification of the plant species in which nests are placed.

MATERIAL AND METHODS

The study was conducted in the Setiba Environmental Protection Area (Área de Proteção Ambiental de Setiba, hereafter APA-Setiba), which includes the Paulo César Vinha State Park (hereafter PEPCV). The APA-Setiba has an area of 12,960 ha. Inside the APA-Setiba is the PEPCV, which consists of a sand-coastal plain of approximately 1,500 ha, with a length of approximately 12 km and a mean width of 2 km. This park is located in the city of Guarapari, state of Espírito Santo, Brazil, between the coordinates 20°33'–20°38'S and 40°23'–40°26'W (Martins et al. 1999). According to the Köppen classification, the region has a monsoon climate (Am) (Alvares et al. 2013), with hot and rainy summers and dry winters, a mean annual temperature of 23.3 °C, a mean annual rainfall of 1,307 mm, and a mean annual relative humidity of 80 % (Fabris and Cesar 1996). This region has 225 recorded bird species (Venturini et al. 1996), which represent 22.1 % of the total number of bird species described for the Atlantic Forest (MMA 2000).

The PEPCV and APA-Setiba are located in the Atlantic Forest biome, specifically in a restinga ecosystem (sand-coastal plain), where the characteristic vegetation is dominated by non-flooded open shrubland formations composed of herbaceous plants, shrubs and even trees (Pereira 2003). This non-flooded open shrubland formation is characterized by islands of vegetation interspersed with areas of white sand formed by marine quartz, with elevated temperatures and salinity and poor soil (Pereira 2003). The shrubs primarily consist of *Clusia hilariana* Schlechtendal (Clusiaceae) and *Protium icicariba* (DC.) Marchand (Burseraceae) organized into thickets and interthickets. Forest edges are found along with open vegetation formations. The forest formations of the restinga have an upper-canopy strata between 6 and 10 m, a lower-canopy strata between 3 and 4 m, and a few emergent trees reaching 18 to 20 m, giving the trees a mean height of 8.4 m (Pereira 2003, Assis et al. 2004a, b).

The data were collected weekly from August 2012 to March 2014, which included two breeding seasons (2012/2013 and 2013/2014). Birds were captured using mist nets and marked with metal bands provided by the National Center for Bird Research and Conservation/Chico Mendes Institute for Biodiversity Conservation Centro Nacional de Pesquisa e Conservação de Aves Silvestres/Instituto Chico Mendes de Conservação da Biodiversidade – CEMAVE/ICMBIO, license number 3138/7 and with unique combinations of color-bands. We obtained monthly records of captured and/or recaptured birds, except for August 2012 and June and July 2013. As a cloacal protuberance is a morphological characteristic present only in males (Wolfson 1952, Nakamura 1990) and no captured male with a cloacal protuberance was recorded as also having a brood patch, we used these characteristics to identify the sex of captured individuals: individuals captured during the breeding season were sexed based on the existence or lack of a brood patch and/or cloacal protuberance. The sex of captured birds that had no cloacal protuberance or brood patch was not classified. Estimates of the breeding period were based on the dates in which the studied nests were active, the mating behaviors, and the presence of brood patches and cloacal protuberances in the captured individuals.

The search for nests was actively conducted starting in August of 2012 and extending through February of 2014. The nests encountered were monitored at intervals of 2–3 days and assessed for their contents (empty, eggs, or nestlings). For each nest, the dates of egg laying and hatching and the date that the nestlings left the nest were recorded. The final outcome was also recorded (success, preyed upon, or abandoned). This monitoring provided information for estimating the clutch size, incubation and nestling periods, and nestling development. Data about nestling development were collected via the visual inspection of nestlings, noting skin, bill and feather color and developmental progression.

Only nests that were found before the first egg was laid and where at least one egg survived to hatching were considered for the incubation period estimation. The nestling period was estimated using only successful nests for which the hatching date was known. After the nests became inactive, parts of the nest plants were collected for their identification.

Nest measurements were preferentially performed in the early stages of incubation to avoid the natural deformation resulting from use. The external diameter (mm), internal diameter (mm), height (mm) and depth of the brood chamber (mm) of the nests were measured. Brood chamber depth represents the distance between the center of the chamber to the plane of the top of the nest. The height of the nest above ground was also measured (m). Due to their fragility, the eggs were weighed and measured (length and width) only once between the third and fifth day of the incubation period. The morphological appearance of the eggs (color and shape) and the final clutch size per nest were also recorded.

Monthly rainfall data, collected at the weather station nearest (≈ 45 km) to the study site, were obtained from the Capixaba Institute of Research, Technical Assistance and Rural Extension (Instituto Capixaba de Pesquisa, Assistência Técnica e Extensão Rural – INCAPER) to establish the relationship between rainfall and the number of active nests present at the study site.

A Kolmogorov-Smirnov test was used to assess whether the data were normally distributed. The means are reported with standard deviations. A one-sample goodness-of-fit test (Williams G-test) was performed to test for variations in clutch size in each season. Simple linear regression was used to analyze relationships between nest plant species and the number of nests, and phytosociological studies were used to obtain the frequency of the occurrence of plant species at the study site (CEPEMAR 2007a, b). All statistical analyses were performed according to Zar (2010), assuming a 5 % significance level. The PAST statistical packages (Hammer et al. 2001) and BioEstat 5.0 (Ayres and Ayres-Jr 2000) were used.

RESULTS

A total of 47 breeding groups were monitored during the study. Fifteen adults were banded from 12 different breeding groups. Among the 47 breeding groups monitored, only two were composed of more than one pair; the other 45 consisted of one breeding pair. A total of 54 nests were monitored (28 nests from the 2012/2013 breeding season and 26 nests from the 2013/2014 breeding season). *Tyrannus melancholicus* breeding pairs performed one to three consecutive breeding attempts within the same breeding season. Only one breeding pair made three attempts, and five breeding pairs made two attempts. The mean number of breeding attempts was 1.15 ± 0.42.

Nesting period

Tyrannus melancholicus performed breeding activities from September to February. The first evidence of reproduction (mating behavior) was observed within the first weeks of September. During the 2012/2013 breeding season, the first nest was found on October 26, 2012; this nest contained nestlings in an advanced stage of growth. Active nests were recorded between October of 2012 and February of 2013. During the 2013/2014 breeding season, the first nest was found on October 2, 2013, and it contained two eggs, which agreed with the onset of breeding activity in September. Active nests were recorded between October of 2013 and February of 2014. During the first year of study, there was a single peak in active nests in December of 2012 (Fig. 1). In the second year of study, the nesting peaks were different from those of the previous year. There were two peaks as follows: a smaller peak between late November and early December and a larger peak between January and early February (Fig. 2). Based on both breeding seasons studied, the nesting period lasted approximately five months (141.5 ± 2.1 days).

In both years, nesting activity began in October and peaked in December in the first year and January/February in the second year. November was the rainiest month in the first breeding season (2012/2013) (Fig. 3), and the nesting peak occurred thereafter. During the second breeding season (2013/2014), December was the rainiest month (Fig. 3); the first nesting peak occurred between November and December, and there was a subsequent drop in nest establishment in December, followed by an increase in the number of active nests from the beginning of January through the first week of February.

Nest construction

The time taken to construct nests was observed for two breeding pairs. One pair completed nest construction in six days and another completed construction in eight days. All other nests were found in advanced stages of construction. Nest building was observed throughout the breeding season, with building peaks coinciding with the peaks in nesting activity. On two occasions, monitored nests were abandoned before eggs were laid. Based on the observations of banded pairs for which sex was identified (n = 3), only the females were seen building nests. The reuse of nests from other breeding seasons was also observed, where the pairs remodeled their old nest for their own use (n = 4; 7.4%). Some pairs reused the same site to construct

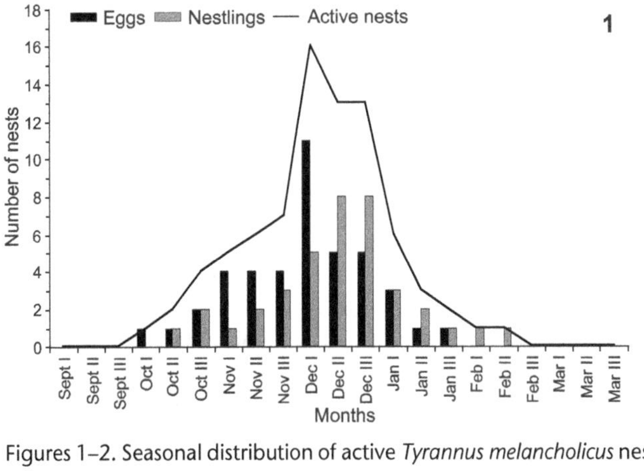

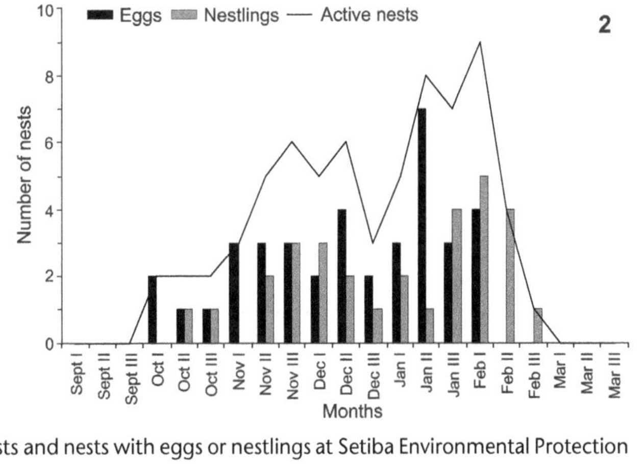

Figures 1–2. Seasonal distribution of active *Tyrannus melancholicus* nests and nests with eggs or nestlings at Setiba Environmental Protection Area, Brazil, during the breeding seasons of 2012/2013 (**1**) and 2013/2014 (**2**). Roman letters indicate 10-day periods of each month: I = from day 1 to 10; II = from day 11 to 20; III = from day 21 to 30 (or 31).

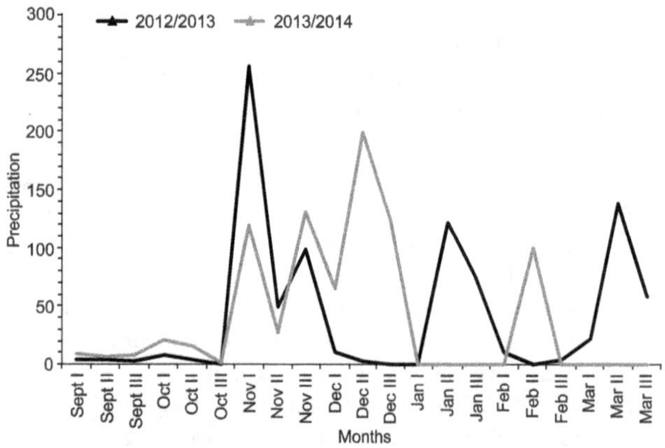

Figure 3. Mean 10-day rainfall in millimeters (mm) from 2012 to 2014 in the Municipality of Guarapari, southeastern Brazil. Roman letters indicate 10-day periods of each month: I = from day 1 to 10; II = from day 11 to 20; III = from day 21 to 30 (or 31). (Source: Instituto Capixaba de Pesquisa, Assistência Técnica e Extensão Rural – INCAPER).

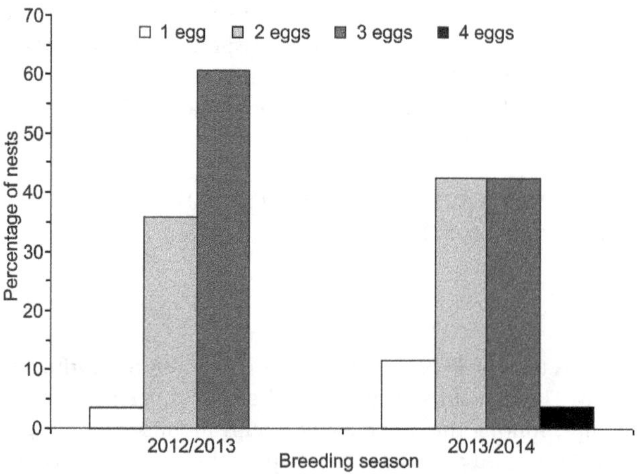

Figure 4. Clutch size of *Tyrannus melancholicus* at Setiba Environmental Protection Area, Brazil, during the breeding seasons of 2012/2013 and 2013/2014.

a new nest (n = 2; 3.7%) after the old nest had disappeared. In addition to reusing nests between breeding seasons, the same nest in the same season was reused for a new breeding attempt (n = 4; 7.4 %). Nest repairs were not observed after eggs were laid.

Characterization of nests and eggs

Nests produced by *T. melancholicus* were always open and cup shaped, with two different types of lining. The external part, which provided protection and support for the nest, had a thicker lining, which included sticks. The internal part of the nest, which held the eggs and nestlings, had a thinner lining composed of small roots, tendrils and frayed fibers from dried leaves. Nests were constructed on emerging branches with little foliage present for protection, allowing almost full exposure to sunlight, and they were always built in the bifurcation of branches. Although the nests were found both close to the ground (0.73 m) and higher up in the trees (3.62 m), nests were, on average, 1.98 ± 0.59 meters (n = 54) from the ground. No nests were found inside bushes or in the tree canopies. The nests measured, on average, 122.35 ± 10.71 mm in external diameter, 77.72 ± 7.42 mm in internal diameter, 73.44 ± 11.08 mm in height and 55.28 ± 5.19 mm in depth (n = 25).

The eggs were ovoid, with colors ranging from white to cream to salmon and with spots and marks ranging between burgundy and brown in color. On average, the eggs were 24.30 ± 1.06 mm in length and 17.23 ± 0.53 mm in width and weighed 2.53 ± 0.19 g (n = 30).

Clutch size

Clutch sizes varied from between one and four eggs, with a mean of 2.5 ± 0.7 eggs (n = 54). Although there was a large number of nests with three eggs in the first year of study, in the second year, there was the same number of nests with two or three eggs, including one nest with four eggs (Fig. 4). No significant differences were found in clutch sizes between the two years evaluated (G = 8.038, df = 27, p = 0.999). The mean clutch size in the first year was 2.6 ± 0.6 (n = 28), and it was 2.4 ± 0.8 (n = 26) in the second year.

Incubation time and feeding of chicks

Egg laying occurred on consecutive days, and nestlings hatched asynchronously since incubation started after the first egg was laid. The mean incubation period was 14.2 ± 1.9 days (n = 10), ranging between 12 and 17 days. The mean nestling period was 15.1 ± 0.8 days (n = 8), ranging between 14 and 16 days.

We observed that only the females incubated the eggs and kept the nestlings warm. The male acted as a sentinel during incubation. The male was always alert and occupied a high perch and alerted the female when there was a potential threat. During the nestling period, both the male and female took turns monitoring the nest and searching for and delivering food to the nestlings. After leaving the nest, fledglings hid in dense groups of bushes, making it impossible to observe them. Despite this, adult behavior indicated that parental care continued for several days after the fledglings left the nest, but it was not possible to define for how long it occurred.

Tyrannus melancholicus showed aggressive, defensive behavior during both the incubation and nestling periods when potential predators approached the nest. Agonistic behaviors were observed by *T. melancholicus* towards the Southern Caracara *Caracara plancus* (Miller, 1777) and Yellow-headed Caracara *Milvago chimachima* (Vieillot, 1816) (both Falconidae), Roadside Hawk *Rupornis magnirostris* (Gmelin, 1788) (Acciptridae), Guira Cuckoo *Guira guira* (Gmelin, 1788) and Smooth-billed Ani *Crotophaga ani* Linnaeus, 1758 (both Cuculidae).

Offspring development

The data on the morphological characteristics and development of offspring were taken from observations of 31 nests that reached the nestling phase. At hatching, the nestlings had their eyes closed. Light beige plumage was observed throughout the body interspersed with areas of bare skin, with a higher concentration of plumage on the head and back. The skin was pink, and the bill was bright yellow. Starting on the fifth day, the skin began to darken and the first signs of the emergence of feather shafts appeared on the back and wings. The eyes began to partially open at this time. On the eighth day, the shafts already exhibited half of the length of the feathers, which began to hide the skin. The eyes were completely open, and the bill was darker. On the thirteenth day, covert feathers had completely emerged, although the rectrices and remiges were not fully developed. On the day that the fledglings left the nest, they had an adult-like appearance but with a yellow gape and shorter rectrices than adults.

Nest plants

Nest plants were identified for 61 *T. melancholicus* nests. The plant species in which nests were most frequently placed was *Kielmeyera albopunctata* Sadddi followed by *Byrsonima sericea* (Table 1). One exotic species, *Acacia mangium* (Fabaceae), was also used. Only one plant species could not be identified. There was no relationship between the frequency of nest plant species at the study site and the number of nests constructed in each plant species ($r^2 = 0.121$, df = 9, $p = 0.738$) (Table 1).

Table 1. Nest plant species used, number and percentage of *Tyrannus melancholicus* nests per plant species at Setiba Environmental Protection Area, Guarapari, Brazil.

Nest plant species	Number of nests	Percentage of nests (%)
Kielmeyera albopunctata	11	18.0
Byrsonima sericea	9	14.8
Agarista revoluta	6	9.8
Erythroxylum nitidum	6	9.8
Clusia hilariana	4	6.6
Guapira opposita	3	4.9
Manilkara subsericea	3	4.9
Myrciaria floribunda	3	4.9
Humiria balsamanifera	2	3.3
Marlierea neuwiedeana	2	3.3
Neomitranthes obtusa	2	3.3
Acacia mangium	1	1.6
Gomidesia martiana	1	1.6
Guapira pernambucensis	1	1.6
Ilex integerrima	1	1.6
Myrsine umbellata	1	1.6
Ocotea notata	1	1.6
Ouratea cuspidata	1	1.6
Protium icicariba	1	1.6
Salzmania nitida	1	1.6
Unidentified	1	1.6

DISCUSSION

Tyrannus melancholicus attempted successive breeding attempts within the same season, with a maximum of three attempts per season. The behavior of performing multiple breeding attempts in the same season is commonly reported among tyrannids (e.g., Haverschmidt 1971, Skutch 1997, Lopes and Marini 2005, Hoffmann and Rodrigues 2011). In the same restinga area studied here, up to four breeding attempts in one season were observed for the Tropical Mockingbird *Mimus gilvus* (Vieillot, 1807) (Mimidae) (Luiz Carlos de Araújo and Charles Duca 2016, unpublished data). The number of attempts made by some Mimidae species is thought to be influenced by predation and breeding success (Laskey 1962). In a transition area between the Atlantic Forest and the Cerrado, up to three attempts were observed for the Gray-backed Tachuri *Polystictus superciliaris* (Wied, 1831) (Tyrannidae) (Hoffmann and Rodrigues 2011), whereas up to four attempts were observed for the Campo Suiriri *Suiriri affinis* (Vieillot, 1818) and up to five for the Chapada Flycatcher *Suiriri islerorum* (Zimmer, Whittaker & Oren, 2001) in an area of the Cerrado in central Brazil (Lopes and Marini 2005).

New attempts at reproduction within the same season can occur for various reasons, the most common of which is predation pressure (Stutchbury and Morton 2001). The number of breeding attempts can be positively influenced by the overabundance of food (Simons and Martin 1990, Hoi et al. 2004), as well by the presence of helpers (Cockburn 1998, Cafrey 1999). According to Roper (2005), in some situations, only a high number of breeding attempts are able to ensure breeding success for birds in tropical regions, which are known for their high predation pressure. It is possible that some bird species begin their breeding period early to increase the number of attempts in case of failure (Lopes and Marini 2005, Hoffmann and Rodrigues 2011). Moreover, the number of breeding attempts in a breeding period may vary based on experience, with older females performing more breeding attempts than inexperienced females (Roper 2005).

Nesting period

Tyrannus melancholicus had a relatively long nesting period of approximately five months. In the Cerrado of central Brazil, *T. savana* was found to have a breeding period varying from three to four months (Thaís Martins Pimentel unpublished data) and from two to three months (Marini et al. 2009a), which were shorter than the breeding period found in this study. This pattern found for *T. savana* is supported by other studies of tyrannids (Aguilar and Marini 2007, Marini et al. 2009a) and other Passeriformes from the south-central region of Brazil (Pinho et al. 2006,

Rodrigues 2009, Duca and Marini 2011). Moreover, nesting periods equally long as that of *T. melancholicus* were observed in the same south-central region (Marini and Durães 2001) and in southeastern Brazil (Aguilar et al. 1999, Duca and Marini 2004), particularly for *M. gilvus* (six months), which was studied in the same area examined in the present study (Luiz Carlos de Araújo and Charles Duca, unpublished data).

The breeding period of *T. melancholicus* began at the end of the dry season and the beginning of the rainy season in southeastern Brazil, coinciding with the first rains. Mating activities were initiated in September, and the nesting period began in the first week of October. Up to this point, the two breeding seasons considered in this study were similar, but they began to differ at the end of November. In the first breeding season (2012/2013), peak rainfall was observed in November (403.3 mm), followed by a relatively dry December (13.6 mm). After the rainy November, the highest peak in nesting activity occurred over the entire month of December. In the following breeding season (2013/2014), November also had high rainfall (276.1 mm), but peak rainfall did not occur until December (387.6 mm). With the rainfall pattern differing from the previous year, the peak in nesting activity also differed, occurring in the second half of January and the first half of February. This correlation between breeding and rainfall over two breeding seasons suggests that the timing of breeding was driven by the timing of rainfall. A relationship between the nesting season and rainfall was also observed for a species of tanager in the Cerrado of central Brazil, with variation in the start of breeding activities related to changes in mean monthly rainfall (Duca and Marini 2011).

Several other studies have shown a relationship between rainfall and the nesting period of birds in tropical biomes (Aguilar et al. 2000, Mezquida 2002, Auer et al. 2007). This relationship may be linked to the increase in the abundance of food due to rain (Boag and Grant, 1984, Sick 1997). Birds may regulate the timing of their breeding period to coincide with periods of food abundance, which would allow sufficient nutrition for the production of eggs and sufficient food for nestling development (Perrins 1970, Ramo and Busto 1984). According to some studies (Tanaka and Tanaka 1982, Ramo and Busto 1984), an increase in rainfall directly increases insect abundance, thereby favoring insectivorous birds (Oniki and Willis 1983, Jahn et al. 2010). Thus, flexibility in the start of the nesting period allows the birds to address climate unpredictability and, consequently, adjust reproduction to avoid periods with a low abundance of food (Duca and Marini 2011).

Nest construction

Nests were built only by females, which is a characteristic found in other tyrannids such as *S. affinis* and *S. islerorum* (Lopes and Marini 2005), the Lesser Elaenia *Elaenia chiriquensis* Lawrence, 1865 (Tyrannidae) (Medeiros and Marini 2007) and *P. superciliaris* (Hoffmann and Rodrigues 2011). The participation of the male is restricted to a vigilance role while the female prepares the nest. The shape and material used for nest construction were similar to those described for *T. melancholicus* in Santa Catarina (Legal 2007) and for *T. savana* in the Cerrado of central Brazil (Marini et al. 2009a).

The reuse of nests within the breeding season (7.4 %) and the reuse of old nests (7.4 %) and old nest sites between breeding seasons (3.7 %) were observed. Legal (2007) observed the reutilization of nest sites by *T. melancholicus* during two consecutive years. The reuse of nests by species that make open nests is relatively rare (Bergin 1997) but has been recorded for tyrannids with elaborate nests and specialized nest sites, with nests being reused within the same season and between seasons (Aguilar and Marini 2007).

The reuse of nests may confer advantages, such as reduced energy expenditure in the building of nests, and disadvantages, such as the presence of parasites and weak nest structures (Bergin 1997, Friesen et al. 1999). The low availability of nesting sites and the construction of elaborate nests that resist the effects of time may be directly related to the reuse of nesting sites (Aguilar and Marini 2007). Among the advantages of reusing nesting sites is foreknowledge of local predators, food availability, the best nesting sites and competitors (Lindberg and Sedinger 1997, Reed et al. 1998). Fidelity to nesting sites is also a known explanation for this behavior (Bergin 1997, Friesen et al. 1999) and is frequently described for migratory birds with philopatric tendencies, such as *T. melancholicus* (Harvey et al. 1979). Studies also show that breeding success repeated over several years at the same site favors the return of the bird to that site (Harvey et al. 1979, Hepp and Kennamer 1992).

Characterization of nests and eggs

The dimensions listed by Legal (2007) in the description of a *T. melancholicus* nest showed both similarities and differences to the dimensions found in our study. The external and internal lengths were close to the means found in this study (see Legal 2007), whereas the distance from the ground and nest height and depth differed from the means found in this study. The height of the nest from the ground found for the congener *T. savana* in Cerrado (Marini et al. 2009a) was similar to that found for *T. melancholicus* in this study. *Tyrannus melancholicus* lives in different habitats, and it is expected that it can adjust some nest parameters according to habitat characteristics.

The eggs were similar in appearance to those found by Legal (2007) (white with spots ranging from brown to burgundy). However, in the present study, variations in eggshell color were noted: cream and salmon eggs were also found. The mean length and width of the eggs were similar to those found by Legal (2007). Similarities in egg parameters were found when comparing the attributes of *T. melancholicus* eggs with those of its congener *T. savana* (e.g., Mezquida 2002, Di Giacomo 2005, Marini et al. 2009a). In comparison with the eggs of three congeners, *T. tyrannus*, *T. verticalis* and the Scissor-tailed Flycatcher

Tyrannus forficatus (Gmelin, 1789), found in temperate regions, similarities were seen in length and width, and differences were observed in weight (Murphy 1988).

Clutch size

The most common clutch size found in the present study (three eggs) was the same as that of one nest found in southern Brazil (Legal 2007). For the congener *T. savana* in central Brazil, the majority of the nests contained three eggs (Marini et al. 2009a).

The clutch size for *T. melancholicus* was similar to that of *M. gilvus* at the same study site (Luiz Carlos de Araújo and Charles Duca, unpublished data) and was similar to that of several other species from the Brazilian Cerrado (Lopes and Marini 2005, Medeiros and Marini 2007). Small clutch sizes are consistently observed for Neotropical birds, which have smaller clutch sizes compared to birds from temperate regions (Stutchbury and Morton 2008). This difference between temperate and tropical regions can often be observed in the clutch sizes of congeners of *T. melancholicus* in temperate regions, which have mean clutch sizes of 3.4 eggs for *T. tyrannus*, 4.4 for *T. verticalis* and 4.7 for *T. forficatus*, with even larger clutch sizes of up to five and six eggs for *T. verticalis* and *T. forficatus* (Murphy 1988, Jahn et al. 2014).

The main hypotheses explaining the variation in clutch size between these two regions relate to the cost of producing eggs and feeding offspring, as well as variations in temperature, relative humidity, and photoperiods (Ricklefs 2000, Stutchbury and Morton 2008). According to Martin (1987), the ability of parents to feed their offspring can limit the clutch size because larger clutches are costlier to feed. The high predation rate in Neotropical regions may also favor smaller clutch sizes because adults invest in smaller numbers of eggs and larger numbers of breeding attempts to ensure breeding success (Slagsvold 1982).

Incubation and nestling periods

The mean incubation period for *T. melancholicus* (14.2 days) was shorter than that reported for one nest of the same species monitored in southern Brazil (17 days; Legal 2007). For the congener *T. savana*, the incubation period has been reported to vary from 13.6 (Marini et al. 2009a) to 16 days (Thaís Martins Pimentel, unpublished data). Variations in the incubation period may occur depending on environmental conditions and food availability (Murphy 1986, Rotenberry and Wiens 1989).

In forest environments in the Atlantic Forest, the length of the incubation period for *T. melancholicus* was longer than that of the Black-cheeked Gnateater *Conopophaga melanops* (Vieillot, 1818) (Conopophagidae) (12 days) (Alves et al. 2002) and shorter than that of the Red-rumped Cacique *Cacicus haemorrhous* (Linnaeus, 1766) (Icteridae) (18.2) (Duca and Marini 2004). The incubation period of the studied species was similar to that of *M. gilvus* (14 days) in a restinga within the same area were *T. melancholicus* was investigated for the present study (Luiza Carlos de Araújo and Charles Duca, unpublished data). In studies on tyrannids in the Cerrado, similar incubation periods were reported for *S. affinis* and *S. islerorum* (15.2 days; Lopes and Marini 2005) and the Plain-crested Elaenia *Elaenia cristata* (Pelzeln, 1868) (Tyrannidae) (15.2 days; Marini et al. 2009b), but periods were longer for *P. superciliaris* (17.7 days; Hoffmann and Rodrigues 2011). According to Sick (1997), the mean incubation period of most passerine species in the Atlantic Forest is approximately 15 days, and the present study corroborates this trend.

The mean nestling period of *T. melancholicus* (15.1 days) was shorter than the mean time reported by Legal (2007) (18 days) for this species in southern Brazil but was similar to that of other Neotropical tyrannids with open nests (e.g., Medeiros and Marini 2007, Marini et al. 2009a, b, Mezquida 2002, Hoffmann and Rodrigues 2011). The nestling period in this study was shorter than that reported for other Passeriformes, such as *S. affinis* (18.9 days), *S. islerorum* (18.3 days) (Lopes and Marini 2005), *C. melanops* (18 days) (Alves et al. 2002) and *C. haemorrhous* (23.4 days) (Duca and Marini 2004). Shorter incubation and nestling periods may be desirable to reduce the time of exposure to predators (Martin 1987), but early abandonment of the nest can put the fledglings at risk and compromise the development of flight ability and, consequently, the ability to escape from predators (Marini et al. 2009a).

Nest plants

The species of plants in which *T. melancholicus* placed nests suggest that those plant species may be used more often than plant species that are more readily available at the study site. Parameters can be considered selective when applied out of proportion to their level of availability in the habitat, thus showing a process of selection by the animal (Johnson 1980). Indeed, such selection may occur because some birds have a close relationship with the plant species in which they nest (Petersen and Best 1985). Additionally, there is evidence that birds are able to identify nesting sites that reduce the risk of predation due to the difficulty of accessing the site or the amount of vegetation cover above the nests (Stauffer and Best 1986, Martin and Roper 1988).

Tyrannus melancholicus did not appear to be particular about the plant species in which it builds its nest; nests were found in 21 different plant species, most of which were common in the study area. However, the results show that two plant species together accounted for over 30 % of the nest sites [*Kielmeyera albopunctata* (18.0 %) and *Byrsonima sericea* (14.8 %)]. In a vegetation structure survey conducted in the same study area (CEPEMAR 2007b), a relative frequency of occurrence of 4.0 % was found for *K. albopunctata* and of 1.5 % for *B. sericea*, which were seventh and twenty-first in frequency of occurrence, respectively. In this same study, the species most frequently found were *Clusia hilariana* (6.1 %), *Protium icicariba* (4.7 %), *Guapira pernambucensis* (Casar.) Lundell (Nyctaginaceae) (4.7 %) and *Ocotea notata* (Nees and Mart.) (Lauraceae) (4.7 %).

All of these species were used as nest plants by *T. melancholicus*, albeit with a low frequency (6.6 % for *C. hilariana* and 1.6 % for each of the others).

In a study conducted at the same study site, it was found a similar pattern for *M. gilvus*, with the use of 30 plants species for nesting, which were also the most common plants in the area (Rodrigo Morais Pessoa and Charles Duca, unpublished data). However, the plant species containing the highest number of *M. gilvus* nests (15.7 %) was *P. icicariba*. The use of the more abundant plant species (e.g., *K. albopunctata* and *C. hilariana*) in the study area as nest sites is supported by the hypothesis known as the "potential-prey hypothesis" (Martin 1993). This hypothesis suggests that nests located in shrub species that are abundant in the area are safer from predation due to the increased availability of potential sites for nest construction. Thus, there is a reduction in the efficiency of predators in locating nests, and the predation pressure is consequently lessened.

ACKNOWLEDGMENTS

We thank Instituto Estadual do Meio Ambiente (IEMA) for a research permit. H. Daros and W.B. Dutra received scholarships from Fundação de Amparo à Pesquisa e Inovação do Espírito Santo (FAPES), processes 58033084 and 59672021. Fundação Nacional de Desenvolvimento do Ensino Superior Particular (FUNADESP) funded this study (process 38/2012). We thank A. Jahn and the anonymous reviewer, who kindly made suggestions on the manuscript.

LITERATURE CITED

Aguilar TM, Leite LO, Marini MÂ (1999) Biologia de nidificação de *Lathrotricus euleri* (Cabanis 1968) (Tyrannidae) em fragmento de mata em Minas Gerais. Ararajuba 7: 125–133.

Aguilar TM, Maldonado-Coelho M, Marini MÂ (2000) Nesting biology of the Gray-hooded Flycatcher (*Mionectes rufiventris*). Ornitologia Neotropical 11: 223–230.

Aguilar TM, Marini MÂ (2007) Nest and nest-site reuse within and between breeding seasons by three neotropical flycatchers (Tyrannidae). Brazilian Journal of Biology 67: 537–540. https://doi.org/10.1590/S1519-69842007000300020

Alvares CA, Stape JL, Sentelhas PC, Gonçalves JLM, Sparovek G (2013) Köppen's climate classification map for Brazil. Meteorologische Zeitschrift 22: 711–728. https://doi.org/10.1127/0941-2948/2013/0507

Alves MAS, Cavalcanti RB (1990) Ninhos, ovos e crescimento de filhotes de *Neothraupis fasciata*. Ararajuba 1: 91–94.

Alves MAS, Rocha CFD, Sluys MV, Vecchi MB (2002) Nest, eggs and effort partitioning in incubation and rearing by a pair of the Black-cheecked gnateater, *Conopophaga melanops* (Passeriformes, Conopophagidae), in an Atlantic Rainforest area of Rio de Janeiro, Brazil. Ararajuba 10: 67–71.

Assis AM, Pereira OJ, Thomaz LD (2004a) Fitossociologia de uma floresta de restinga no Parque Estadual Paulo César Vinha, Setiba, município de Guarapari (ES). Revista Brasileira de Botânica 27: 349–361. https://doi.org/10.1590/S0100-84042004000200014

Assis AM, Thomaz LD, Pereira OJ (2004b) Florística de um trecho de floresta de restinga no município de Guarapari, Espírito Santo, Brasil. Acta Botânica Brasileira 18: 191–201. https://doi.org/10.1590/S0102-33062004000100016

Auer SK, Bassar RD, Fontaine JJ, Martin TE (2007) Breeding biology of passerines in a subtropical montane Forest in Northwestern Argentina. The Condor 109: 321–333. https://doi.org/10.1650/0010-5422(2007)109[321:BBOPIA]2.0.CO;2

Ayres M, Ayres-Jr M (2000) BioStat 3.0: aplicações estatísticas nas áreas das ciências biológicas e médicas. Belém, Sociedade Civil de Mamirauá.

Bergin TM (1997) Nest reuse by Western Kingbird. The Wilson Bulletin 109: 735–737.

Boag PT, Grant PR (1984) Darwin's Finches (Geospiza) on Isla Daphne Major, Galapagos: breeding and feeding ecology in a climatically environment. Ecological Monographs 54: 463–489. https://doi.org/10.2307/1942596

Cafrey C (1999) Feeding rates and individual contributions to feeding at nests in cooperatively breeding Western American Crows. The Auk 116: 836–841. https://doi.org/10.2307/4089347

CEPEMAR (2007a) Plano de manejo da Área de Proteção Ambiental de Setiba. Vitória, Technical report, 1004 pp. Available online at: https://iema.es.gov.br/APA_Setiba [Accessed: 25/01/2014]

CEPEMAR (2007b) Plano de manejo do Parque Estadual Paulo César Vinha. Vitória, Technical report, 1162 pp. Available online at: https://iema.es.gov.br/PEPCV [Accessed: 25/01/2014]

Chiarani E, Fontana CS (2015) Breeding biology of the lesser grassfinch (*Emberizoides ypiranganus*) in southern Brazilian upland grasslands. The Wilson Journal of Ornithology 127: 441–456. https://doi.org/10.1676/14-137.1

Clements JF (2007) The Clements checklist of the birds of the World. Cornell University Press, New York,.

Cockburn A (1998) Evolution of helping behavior in cooperatively breeding birds. Annual Review of Ecology and Systematics 29:141–177. https://doi.org/10.1146/annurev.ecolsys.29.1.141

Cooper NW, Murphy MT, Redmond LJ, Dolan AC (2009) Density-dependent at the first reproduction in the eastern kingbird. Oikos 118: 413–419. https://doi.org/10.1111/j.1600-0706.2008.16997

Di Giacomo AG (2005) Aves de la Reserva El Bagual. In: Di Giacomo AG, Rapovickas SF (Ed.) Historia natural y paisaje de la reserva el bagual, provincia de Formosa. Asociación Ornitológica del Plata, Buenos Aires, 201–465.

Duca C, Marini MÂ (2004) Aspectos da nidificação de *Cacicus haemorrhous* (Passeriformes, Icterinae) no sudeste do Brasil. Ararajuba 12: 25–32.

Duca C, Marini MÂ (2011) Variation in breeding of the Shrike-Like Tanager in Central Brazil. The Wilson Journal of Ornithology 123: 259–265. https://doi.org/10.1676/10-116.1

Fabris LC, Cesar O (1996) Estudos florísticos em uma mata litorânea no sul do estado do Espírito Santo, Brasil. Boletim do Museu de Biologia Mello Leitão 5: 15–46.

Fitzpatrick JW (2004) Family Tyrannidae (Tyrant-flycatchers). In: Del Hoyo J, Elliott A, Christie D (Eds) Handbook of the Birds of the World. Lynx Edicions, Barcelona, 170–461.

Friesen LE, Wyatt VE, Cadman MD (1999) Nest reuse by Wood Thrushes and Rose-breasted Grosbeaks. The Wilson Bulletin 111: 132–133.

Harvey PH, Greenwood PJ, Perrins CM (1979) Breeding area fidelity of Great Tits (*Parus major*). Journal of Animal Ecology 48: 305–313. https://doi.org/10.2307/4115

Hammer O, Harper DAT, Ryan PD (2001) PAST: Paleontological Statistics software package for education and data analysis. Paleontologia Eletronica 4: 9.

Haverschmidt F (1971) Notes on the life history of the Rusty margined Flycatcher in Surinam. The Wilson Bulletin 83: 124–128.

Hepp GR, Kennamer RA (1992) Characteristics and consequences of nest-site fidelity in Wood Ducks. The Auk 109: 812–818. https://doi.org/10.2307/4088155

Hoffmann D, Rodrigues M (2011) Breeding biology and reproductive success of *Polystictus superciliaris* (Aves: Tyrannidae), an uncommon tyrant-flycatcher endemic to the highlands of eastern Brazil. Zoologia 28: 305–311. https://doi.org/10.1590/S1984-46702011000300004

Höfling E, Camargo HFA (2002) Aves no campus. EDUSP e IB-USP, São Paulo, 168 pp.

Hoi H, Kristin A, Valera F, Hoi C (2004) Clutch enlargement in Lesser Gray Shrikes (*Lanius minor*) in Slovakia when food is superabundant: a maladaptive response. The Auk 121: 557–564. https://doi.org/10.1642/0004-8038(2004)121[0557:CEILGS]2.0.CO;2

Jahn EA, Levey DJ, Mamani AM, Saldias M, Alcoba A, Ledezma MJ, Flores B, Vidoz JQ, Hilarion F (2010) Seasonal differences in rainfall, food availability, and the foraging behavior of Tropical Kingbirds in the southern Amazon Basin. Journal of Field Ornithology 81: 340–348. https://doi.org/10.1111/j.1557-9263.2010.00290.x

Jahn E, Tuero DT, Mamani AM, Bejarano V, Masson DA, Aguilar E (2014) Drivers of clutch-size in Fork-tailed Flycatchers (*Tyrannus savana*) at temperate and tropical latitudes in South America. Emu 114: 337–342. https://doi.org/10.1071/MU13084

Johnson DH (1980) The comparison of usage and availability measurements for evaluating resource preference. Ecology 61: 65–71. https://doi.org/10.2307/1937156

Laskey AR (1962) Breeding biology of Mockingbirds. The Auk 79: 596–606. https://doi.org/10.2307/4082641

Legal E (2007) Aspectos da nidificação do Suiriri, *Tyrannus melancholicus* (Vieillot, 1819), (Aves: Tyrannidae) em Santa Catarina. Atualidades Ornitológicas On-Line 140: 51–52. http://www.ao.com.br/download/ao140_51.pdf [Accessed: 21/01/2014]

Lindberg MS, Sedinger JS (1997) Ecological significance of brood-site fidelity in Black Brant: spatial, annual, and age-related variation. The Auk 115: 436–446. https://doi.org/10.2307/4089202

Lopes LE, Marini MÂ (2005) Biologia reprodutiva de *Suiriri affinis* e *S. islerorum* no Cerrado do Brasil Central. Papéis Avulsos de Zoologia 45: 127–141. https://doi.org/10.1590/S0031-10492005001200001

Marini MÂ, Durães R (2001) Annual patterns of molt and reproductive activity of passerines in South-central Brazil. The Condor 103: 767–775. https://doi.org/10.1650/0010-5422(2001)103[0767:APOMAR]2.0.CO

Marini MÂ, Lobo Y, Lopes LE, França LF, Paiva LV (2009a) Biologia reprodutiva de *Tyrannus savana* (Aves, Tyrannidae) em cerrado do Brasil Central. Biota Neotropica 9: 55–63. https://doi.org/10.1590/S1676-06032009000100007

Marini MÂ, Silveira MB, Sousa NM, Borges FJ (2009b) Biologia reprodutiva de *Elaenia cristata* (Tyrannidae) no Cerrado do Brasil Central. Neotropical Biology and Conservation 4: 3–12. https://doi.org/10.4013/nbc.2009.41.01

Martin TE (1987) Food as a limit on breeding birds: a life-history perspective. Annual Review of Ecology and Systematics 18: 453–487. https://doi.org/10.1146/annurev.es.18.110187.002321

Martin TE (1993) Nest predation among vegetation layers and habitat types: revising the dogmas. American Naturalist 141: 897–913. https://doi.org/10.1086/285515

Martin TE (1996) Life history evolution in tropical and south temperate birds: what do we really know? Journal of Avian Biology 27: 263–271. https://doi.org/10.2307/3677257

Martin TE, Roper JJ (1988) Nest predation and nest-site selection of a western population of the Hermit Thrush. The Condor 90: 51–57. https://doi.org/10.2307/1368432

Martins MLL, Carvalho-Okano RM, Luceno M (1999) Cyperaceae do Parque Estadual Paulo César Vinha, Guarapari, Espírito Santo, Brasil. Acta Botânica 13: 1999. https://doi.org/10.1590/S0102-33061999000200008

Mason P (1985) The nesting biology of some passerines of Buenos Aires, Argentina. In: Buckley PA, Foster MS, Morton ES, Ridgely RS, Buckley FG (Eds) Neotropical Ornithology. American Ornithologist's Union, Ornithological Monographs 36: 954–972. https://doi.org/10.2307/40168328

Medeiros RCS, Marini MÂ (2007) Biologia reprodutiva de *Elaenia chiriquensis* (Lawrence) (Aves, Tyrannidae) em Cerrado do Brasil Central. Revista Brasileira de Zoologia 24: 12–20. https://doi.org/10.1590/S0101-81752007000100002

Mezquida ET (2002) Nidificacion de ocho especies de Tyrannidae em la Reserva de Ñacuñán, Mendoza, Argentina. Hornero 17: 31–40.

MMA (2000) Avaliação e ações prioritárias para a conservação da biodiversidade da Mata Atlântica e Campos sulinos. Secretaria de Biodiversidade e Florestas, Ministério do Meio Ambiente (MMA),Brasília.

Murphy MT (1986) Temporal components of reproductive variability in Eastern Kingbirds (*Tyrannus tyrannus*). Ecology 67: 1483–1492. https://doi.org/10.2307/1939079

Murphy MT (1988) Comparative reproductive biology of Kingbirds (Tyrannus spp.) in eastern Kansas. The Wilson Bulletin 100: 357–376.

Murphy MT (2007) Lifetime reproductive success of female eastern kingbirds (*Tyrannus tyrannus*): influence of lifespan, nest predation and body size. The Auk 124: 1010–1022. https://doi.org/10.1642/0004-8038(2007)124[1010:LRSOFE]2.0.CO;2

Nakamura M (1990) Cloacal protuberance and copulatory behavior ot he Alpine Accentor (*Prunella collaris*). The Auk 107: 284–295. https://doi.org/10.2307/4087611

Nobrega PFA, Pinho JB (2010) Biologia reprodutiva e uso de habitat por *Cantorchilus leucotis* (Lafresnaye, 1845) (Aves, Troglodytidae) no pantanal, Mato Grosso, Brasil. Papeis Avulsos de Zoologia 50: 511–516. https://doi.org/10.1590/S0031-10492010003100001

Oniki Y, Willis EO (1983) Breeding records of birds from Manaus, Brazil; IV. Tyrannidae to Vireonidae. Revista Brasileira de Biologia 43: 45–54.

Pereira OJ (2003) Restinga: origem, estrutura e diversidade. In: Jardim MAG, Bastos NNC, Santos JUM (Eds) Desafios da botânica brasileira no novo milênio: inventário, sistematização e conservação da diversidade vegetal. Museu Paraense Emilio Goeldi, Sociedade Botânica do Brasil, Belém, 177–179.

Perrins CM (1970) The timing of bird's breeding seasons. Ibis 112: 242–255. https://doi.org/10.1111/j.1474-919X.1970.tb00096.x

Petersen KL, Best LB (1985) Nest-site selection by Sage Sparrows. The Condor 87: 217–221. https://doi.org/10.2307/1366885

Pinho JB, Lopes LE, Morais DH, Fernandes AM (2006) Life history of the Mato Grosso Antbird *Cercomacra melanaria* in the Brazilian Pantanal. Ibis 148: 321–329. https://doi.org/10.1111/j.1474-919X.2006.00542.x

Ramo C, Busto B (1984) Nidificacion de los Passeriformes em los Llanos de Apure (Venezuela). Biotropica 16: 59–68. https://doi.org/10.2307/2387896

Reed ET, Cooch EG, Goudie RI, Cooke F (1998) Site fidelity of Black Brant wintering and spring in the strait of Georgia, British Columbia. The Condor 100: 426–437. https://doi.org/10.2307/1369708

Repenning M, Fontana CS (2016) Breeding biology of the Tropeiro Seedeater (*Sporophila beltoni*).The Auk 133: 484–496. https://doi.org/10.1642/AUK-15-226.1

Ricklefs RE (2000) Density dependence, evolutionary optimization, and the diversification of avian life histories. The Condor 102: 9–22. https://doi.org/10.1650/0010-5422(2000)102[0009:DDEOAT]

Robinson WD, Robinson TR, Robinson SK, Brawn JD (2000) Nesting success of understory forest birds in central Panama. Journal of Avian Biology 31: 151–164. https://doi.org/10.1034/j.1600-048X.2000.310207.x

Rodrigues SS (2009) Biologia e sucesso reprodutivo de *Mimus saturninus* (Aves: Mimidae) no cerrado. Master's Dissertation, Universidade de Brasília, Brasília, DF. Available online at: http://repositorio.unb.br/handle/10482/4585 [Accessed: 10/02/2014]

Roper JJ (2005) Try and try again: Nest predation favors persistence in a neotropical birds. Ornitologia Neotropical 16: 253–262.

Rotenberry JT, Wiens JA (1989) Reproductive biology of shrubsteppe passerine birds: geographical and temporal variation in cluth size, brood size, and fledging success. The Condor 91: 1–14. https://doi.org/10.2307/1368142

Sick H (1997) Ornitologia Brasileira. Editora Nova Fronteira, Rio de Janeiro, 862 pp.

Sigrist T (2014) Guia de campo: avifauna brasileira. Editora Avis Brasilis, São Paulo, 608 pp.

Simons LS, Martin TE (1990) Food limitation of avian reproduction: an experiment with the cactus wren. Ecology 71: 869–876. https://doi.org/10.2307/1937358

Skutch AF (1997) Life of the Flycatcher. University of Oklahoma Press, Norman, 162 pp.

Slagsvold T (1982) Clutch size variation in passerine birds: the nest predation hypothesis. Oecologia 54: 159–169. https://doi.org/10.1007/BF00378388

Stauffer DF, Best LB (1986) Nest site characteristics of open-nest birds in riparian hábitat in Iowa. The Wilson Bulletin 98: 231–242.

Stutchbury BJM, Morton ES (2001) Behavioral Ecology of Tropical Birds. Academic Press, San Diego, 165 pp.

Stutchbury BJM, Morton ES (2008) Recent advances in the behavioral ecology of tropical birds. The Wilson Journal of Ornithology 120: 26–37. https://doi.org/10.1676/07-018.1

Tanaka LK, Tanaka LK (1982) Rainfall and seasonal changes in arthropod abundance on a tropical oceanic island. Biotropica 14: 114–123. https://doi.org/10.2307/2387740

Teul M, Piaskowski VD, Williams KM (2007) The breeding biology of the fork-tailed flycatcher (*Tyrannus savana*) in a lowland pine savanna habitats in Belize. Ornitologia Neotropical 18: 47–59.

Venturini AC, Ofranti AMS, Varejão JBM, Paz PR (1996) Aves e mamíferos na restinga: Parque Estadual Paulo César Vinha, Setiba, Guarapari, Espírito Santo. Vitória, Secretaria de Estado de Desenvolvimento Sustentável, 68 pp.

Wolfson A (1952) The cloacal protuberance: a means for determining breeding condition in live male passerines. Bird-Banding 23: 159–165. https://doi.org/10.2307/4510381

Zar JH (2010) Biostatistical analysis. Pearson Prentice-Hall, London, 944 pp.

Author Contributions: HD and CD designed the experiments; HD and WBD conducted the experiments; HD and CD analyzed the data and wrote the paper.

Competing Interests: The authors have declared that no competing interests exist.

13

Redescription of *Ventania avellanedae* (Stylommatophora: Odontostomidae), a land snail endemic to the Ventania Mountain System, Argentina

Julia Pizá[1,2], Néstor J. Cazzaniga[1], Natalia S. Ghezzi[1]

[1] *Departamento de Biología, Bioquímica and Farmacia, Universidad Nacional del Sur. San Juan 670, B8000ICN Bahía Blanca, Buenos Aires, Argentina.*
[2] *Instituto de Ciencias Biológicas y Biomédicas del Sur, Universidad Nacional del Sur-CONICET. B8000ICL Bahía Blanca, Buenos Aires, Argentina.*
Corresponding author: Julia Pizá (jpiza@uns.edu.ar)

ABSTRACT. Although the presence of apertural folds and lamellae is the most recognizable character of the Odontostomidae, some species lack them, mostly in *Anctus* Martens, 1860, *Bahiensis* Jousseaume, 1877 and *Moricandia* Pilsbry & Vanatta, 1898. *Eudioptus avellanedae* Doering, 1881 – a slender odontostomid species that lacks even the slightest trace of folds or lamellae in its shell aperture – was however transferred to *Odontostomus* by Pilsbry in 1902 on the basis of its building forward of the aperture-margins. It is currently placed in its own monotypic subgenus, *Cyclodontina* (*Ventania*) Parodiz, 1940, on the basis of about the same argument. In this paper we redescribe its shell morphology and, for the first time, describe the internal anatomy of the pallial complex and the reproductive and digestive systems. The presence of a spongy gland in the pallial complex; of a short penis sheath with no retractor muscle; of a bursa copulatrix duct longer than spermoviduct, and of an epiphallic gland strongly support the inclusion of this unusual species in Odontostomidae. The species is diagnosable by the sculpture of the protoconch, which is not smooth as previously described, but has waved axial ribs crossed by spiral lines in young specimens; the distinctive external and internal shape of the bursa copulatrix duct; the internal penis wall divided in three regions of different sculpture; the smooth inner wall of the vagina; the long and cylindrical epiphallus with a distal widening indicating the presence of an epiphallic gland, and the penis retractor muscle inserted in the distal end of a short flagellum. These characters support the validity of *Ventania* Parodiz, 1940, different from *Cyclodontina* Beck, 1837.

KEY WORDS. Anatomy, conservation, endemisms, Gastropoda, Orthalicoidea.

INTRODUCTION

Ventania Parodiz, 1940 is a genus of terrestrial snails including a single species from the southern Buenos Aires province, Argentina. *Ventania avellanedae* (Doering, 1881) exclusively inhabits the orographic system of Ventania, of Paleozoic origin (Ventana and Curamalal Sierras). Although a specimen from Golfo Nuevo (northern Patagonia) was cited by Pilsbry (1902), the presence of this species outside the Ventania Mountain System was never confirmed by further collections.

Up to now, this snail was classified exclusively by its shell morphology. Originally described as a member of Bulimulidae, its slightly expanded aperture with a faintly reflected outer lip was the basis for Pilsbry (1902) to include this species in Odontostomidae. However, due to its slender aspect and lack of even the slightest trace of folds or lamellae in the shell aperture, its generic assignment was uncertain up to the establishment of the monotypic subgenus *Cyclodontina* (*Ventania*) by Parodiz, 1940, who also described the radula of this species. Nevertheless, radular features did not add any diagnostic character to further support that generic position.

This same species was described by Pilsbry (1896) from Sierra de la Ventana position as *Anctus* (?) stearnsianus. Soon after, Pilsbry (1902) acknowledged that this species was, in fact, a junior synonym of *Eudioptus avellanedae* Doering, 1881 (Bulimulidae) but he stated that there was an "absolute" (though not specified) difference with other species of *Eudioptus* Albers, 1860, and concluded that *E. avellanedae* was similar to some species of *Odontostomus* (*Spixia*) Pilsbry & Vanatta, 1898, hypothesizing that it had totally lost the apertural folds and lamellae, as it seems to have partially occurred in *Plagiodontes patagonicus* (d'Orbigny, 1835), with which *E. avellanedae* concurs in the same habitat. Parodiz (1940) realized that *E. avellanedae* was not a *Spixia* and compared it with species in *Bahiensis* Jousseaume, 1877 (at that time, considered another section of *Odontostomus* Beck, 1837), in which some species have remnants of apertural dentition while others lack any fold in the aperture. Besides the apertural differences, a main argument to split the monotypic *Odontostomus* (*Ventania*) from *Bahiensis* and *Spixia* was the completely smooth protoconch of the former. Parodiz (1940) also stressed that *Ventania* has a fully convex outer surface on the last whorl because, having no folds occluding the aperture, there are no depressions behind the lip, which are characteristic of *Spixia* and other forms in *Odontostomus* Beck, 1837. According to Parodiz (1940) *Ventania* was part of *Odontostomus* because its radular structure was fully similar to that of some species in the group of *O.* (*Spixia*) *alvarezi* (d'Orbigny, 1835).

Parodiz (1944) modified the classificatory scheme by moving the subgenera *Spixia* Pilsbry & Vanatta, 1898, *Scalarinella* Dohrn, 1875, *Plagiodontes* Doering, 1876, *Ventania* Parodiz, 1940, *Bahiensis* Jousseaume, 1877 and *Moricandia* Pilsbry & Vanatta, 1898 from *Odontostomus* to *Cyclodontina* Beck, 1837. All of them were afterwards treated as full genera (Fernández 1973, Campos-Salgado and Santos-Coelho 2003, Scarabino 2003, Simone 2006, Cabrera and Martínez 2012), with the only exception of *Ventania*, since no author has discussed this taxonomic uncertainty on new characters. Two papers on conservation of land snails (Burela et al. 2003, Delhey et al. 2005) cited this species as *Ventania avellanedae*, with no mention to the taxonomic criterion to *Ventania* being raised to the genus level.

Some morphological features described in the literature do not fit well with the actual, fresh material (e.g., protoconch sculpture of young specimens, radular teeth morphology), and nothing is currently known on the soft anatomy of this snail species. Our new information leads to consider *Ventania* a different monotypic genus, different from *Cyclodontina*. We follow Breure et al. (2010) and Breure and Romero (2012) in considering Odontostomidae at the family level and Simone (2006) and Colley (2013) in considering *Bahiensis* and *Cyclodontina* as different genera.

MATERIAL AND METHODS

We studied shells and living snails collected in several sites of the Ventania Mountain System (Cerro Curamalal Chico, 37°41'S, 62°18'W, 10 specimens, Natalia Ghezzi leg., 1999; Cerro Curamalal Grande 37°43'S, 62°13'W, 5 specimens, Julia Pizá leg. 2014; Cerro Ventana, 38°09'S, 61°48'W, 12 specimens, Natalia Ghezzi and Julia Pizá leg. 1999; Sierra de la Ventana, 38°07'S, 61°47'W, 5 specimens, Julia Pizá leg. 2011). The analysis was complemented with shell specimens of the malacological collection at the Museo Argentino de Ciencias Naturales "Bernardino Rivadavia" (MACN, Buenos Aires): MACN 9487: Sierra de la Ventana, 6 shells, Hans Seckt leg., 1918. MACN 11270: Sierra de la Ventana, 1 shell, Martín Doello-Jurado leg., 1920. MACN 21263: Sierra de la Ventana, 3 shells, Horacio J. Harrington leg., 1923. MACN 23090: Sierra de la Ventana, 4 shells, Alberto Castellanos leg., 1937. MACN 1644: Sierra de la Ventana, 5 shells, Héctor S. Gavio leg., 1939. MACN 24395: Sierra de la Ventana, 1 shell, Oscar Kühneman leg., 1940. MACN 18596: Sierra de Curamalal, 2 shells, Augusto C. Scala leg., 1929. MACN 3228: Sierra de la Ventana, 1 shell, María de las Mercedes Richard leg., 1943. MACN 11085: Sierra de Curamalal, 2 shells, Lucas Kraglievich leg., 1920. MACN 34037: Cerro Ventana, 21 living specimens, Natalia Ghezzi and Julia Pizá leg., 1999.

Morphometric variables were obtained on scaled sketches made with a drawing device on a stereoscopic microscope (variables defined as Pizá and Cazzaniga 2003): shell length (SL), shell width (SW), last whorl length (LWL), aperture length (AL), aperture width (AW), major angle (MA) and spiral angle (SA). Total whorl number (WN) was determined by direct observation under a stereoscopic microscope. Measurements are quoted as: mean ± standard deviation (coefficient of variation) [minimum–maximum]. Cephalopodium data were measured with a scaled eyepiece: foot length, foot width, head length, length of the ocular tentacles, and length of the fore tentacles. Teleoconch and protoconch sculpture were studied with a LEO EVO 40-XVP scanning microscope, on shells of young and adult specimens.

Dissections were made on freshly dead specimens. Radulae and mandibles were obtained and prepared according to Ploeger and Breure (1977).

TAXONOMY

Orthalicoidea Albers, 1860
Odontostomidae Pilsbry & Vanatta, 1898

Ventania Parodiz, 1940

Cyclodontina (*Ventania*) Parodiz, 1940.

Type and only species. *Ventania avellanedae* (Doering, 1881).

Diagnosis. Shell elongate-ovate, rather thin, glossy, tawny to brown in color, whorls slightly convex, protoconch with axial ribs crossed by spiral lines, teleoconch with fine, irregular, radial wrinkles and wavy spiral striae; elongate-ovate aperture occupying ⅓ of the total shell length and lacking any fold or lamella; peristome slightly expanded and reflected. Short flagellum with the thin penis retractor muscle attached at its distal end. Epiphallus long and distally widened indicating the presence of an internal epiphallic gland. Penial papilla absent.

Ventania avellanedae (Doering, 1881)

Figs 1–20

Eudioptus avellanedae Doering, 1881: 64, pl. 1, figs 2, 3; Breure 1974: 112.
Anctus (?) *stearnsianus* Pilsbry, 1896: 41; Baker 1963: 230; Breure 1974:124.
Odontostomus (*Spixia*) *avellanedae*: Pilsbry 1902: 91, pl. 11, figs 56–59.
Odontostomus (*Ventania*) *avellanedae*: Parodiz 1940: 229, fig. 5; Parodiz 1942: 334, pl. 3, fig. 27; Richardson 1993: 44.
Cyclodontina (*Ventania*) *avellanedae*: Parodiz 1944: 5; Parodiz 1957: 28; Fernández and Castellanos 1973: 281; Fernández 1973: 121; Cuezzo et al. 2013.
[?] *Spixia avellanedae* [sic]: Breure 1974: 125.
Ventania avellanedae: Burela et al. 2003: 176, Delhey et al. 2005: 11.

Type locality. Doering stated that this snail inhabit on "rocas cuarcíticas de la Sierra de Currumalan" [Quartzite rocks in the Curamalal sierra], and also in "cerros o promontorios de la Sierra de la Ventana, cerca del Fuerte Argentino" [mountains or promontories in Sierra de la Ventana, near Fuerte Argentino], i.e., 19 km west of Tornquist City (38°05'S–62°13'W), on the road joining the National routes 35 and 33, on the banks of the Sauce Chico stream.

Diagnosis. Protoconch with waved axial ribs crossed by spiral lines. Penis retractor muscle inserted in the distal end of a short flagellum. Epiphallus long and cylindrical with a distal widening indicating the presence of an internal epiphallic gland. Internal wall of the penis with three areas of different sculpture. Internal wall of the vagina smooth. Middle portion of the bursa copulatrix duct distinctly enlarged, with an internal sculpture of lengthy longitudinal lamellae in the wide region proximal to the bursa copulatrix, and a thick wall with perpendicular zig-zag grooves in the distal one.

Redescription. Shell (Figs 1–10). Elongate-ovate, glossy, tawny to brown in color, rather thin; sometimes with some dark brown bands irregularly spread, and marked growth lines; 6 to 7.5 moderately convex whorls with shallow but well defined sutures.

Elongate-ovate aperture occupying ⅓ of the total shell length and lacking any fold or lamella (Fig. 1). Peristome slightly expanded and reflected (Figs 1, 2); inner lip turning on the columellar region although with a sharp edge. Parietal callus thin but well-defined. Columellar axis slightly inclined rightwards. Umbilicus narrow (Fig. 2).

Protoconch (Fig. 3) with a finely costulated surface, with marked waved axial ribs that often branch, crossed by a weaker spiral striation. Protoconch of adult snails smooth as a result of erosion (Figs 2, 4, 8). Protoconch-teleoconch boundary well-defined (Figs 3, 8). Teleoconch with wrinkle-like growth lines (Figs 1, 2, 8–10).

Shell measurements (in mm): SL: 21.85 ± 1.76 (0.080) [18.33–26.07]; SW: 7.86 ± 0.61 (0.078) [6.55–9.05]; LWL: 14.13 ± 0.92 (0.065) [12.26–16.07]; AL: 8.60 ± 0.78 (0.091) [6.90–10.36]; AW: 5.01 ± 0.45 (0.091) [3.90–6.00]; MA: 132.07 ± 3.19 (0.024) [125–138]; SA: 31.07 ± 2.76 (0.089) [26–39].

Cephalopodium (Figs 2, 11). Greenish brown. Head shorter than the tail, which does not surpass the length of the spire. Optic tentacles about four times longer than fore tentacles. Buccal opening large, round, with a pair of well-developed rounded lobes or lower lip appendices, which do not extend over the mandible (Fig. 11). Upper lip conspicuous but lacking a third buccal lobe or upper lip appendix.

Digestive system (Figs 12–15). Buccal mass spheroidal. Mandible arquated (Fig. 12), formed by 12 imbricated plates directed towards the central zone. Central plates rectangular, more or less narrow, of variable width; lateral plates from rect-

Figure 1. Shell morphology of *Ventania avellanedae*: general view showing shell variability and teleoconch striation. Scale bar: 5 mm.

Figure 2. Living specimen of *Ventania avellanedae* showing erosion in adult shells, umbilicus and peristome slightly reflexed.

angular to triangular, of uniform size; the more external plates are much larger, almost twice as long as the central ones. The surface of the plates is smooth; their borders are projected outwards, generating an irregular outline.

Radular teeth transversally arranged on a straight line (Fig. 13). Central tooth tricuspid and smaller than the lateral ones (Fig. 14); central cusp or mesocone of the central tooth romboid, with a free border that usually reach the basal part of the tooth; lateral cusps or ectocones much shorter than the central one. Lateral teeth bicuspid, with a long mesocone and a small ectocone located in an opposite position to the central tooth (Fig. 14). Marginal teeth bicuspid, with a thin sharpen mesocone and a smaller ectocone, similar each other; inserted between the normal bicuspid teeth, some teeth have a bifurcated ectocone giving the tooth a tricuspid aspect (Fig. 15). The form of the teeth gradually varies towards the radula edges, the cusps becoming less definite. Despite the gradualness of the transition from lateral to marginal teeth, they are usually distinguishable because marginal teeth are more elongated, have a more developed free grasping edge and slender mesocones.

Pallial complex (Fig. 16). Narrow and elongated, 18.55 mm long on average (16.40–22.45) including the mantle collar, extends along about one and a half whorl, i.e. roughly ¼ of the total length of the snail body. The pinkish triangular kidney occupies ¼ to ⅓ of the lung length. Internal structure of the kidney with longitudinal, ondulated lamellae, in close contact among them. On the upper zone of the kidney, lamellae are widened and more globose; however they do not demarcate different regions, as their inner structure is homogeneous.

The primary ureter runs along the rectal side of the kidney up to the top of the lung cavity; it then turns down along the rectum and forms the secondary ureter, which opens proximally in the ureteric pore at the level of the middle point of the kidney. From this point on, the secondary ureter is open and ends at the pneumostome.

The pericardium, located in the upper columellar side of the pallial system, is 1.75–2.55 mm long. It is continuous with the prominent pulmonary vein that runs parallel to the rectum and reaches the mantle collar. The afferent marginal vein branches out approximately from the distal third of the pulmonary vein, equalling about 45% of its length. Adrectal area between the rectum and the pulmonary vein and between the pulmonary vein and the marginal afferent vein with a moderate vascularization. A marginal vein of weak development branches out from the last portion of the pulmonary vein and runs along the mantle collar border. The interramus area is deeply excavated and of variable shape, either triangular or rectangular. The mantle collar includes a whitish spongy pallial gland.

Genital system (Figs 17–20). Ovotestis composed by groups of acini extending over the inner, excavated face of the third whorl of the digestive gland. Collecting channels going out the ovotestis lead into the hermaphrodite duct, a twisted duct with a swollen central portion forming the vesicula seminalis. It ends at the white fertilization pouch-spermathecal complex composed by a proximal swollen portion and a thin, long blind sac (Figs 17, 20). This complex is visible on the basal side of the albumen gland. The hyaline-yellow albumen gland is elongated, folded over its basis, and lies against the fore concave surface of the digestive gland over the digestive pouch.

Spermoviduct long and very curvilineal formed by the twisted, hyaline orange uterus, ending at the free oviduct and the white and glandular prostate, and continuous with the vas deferens.

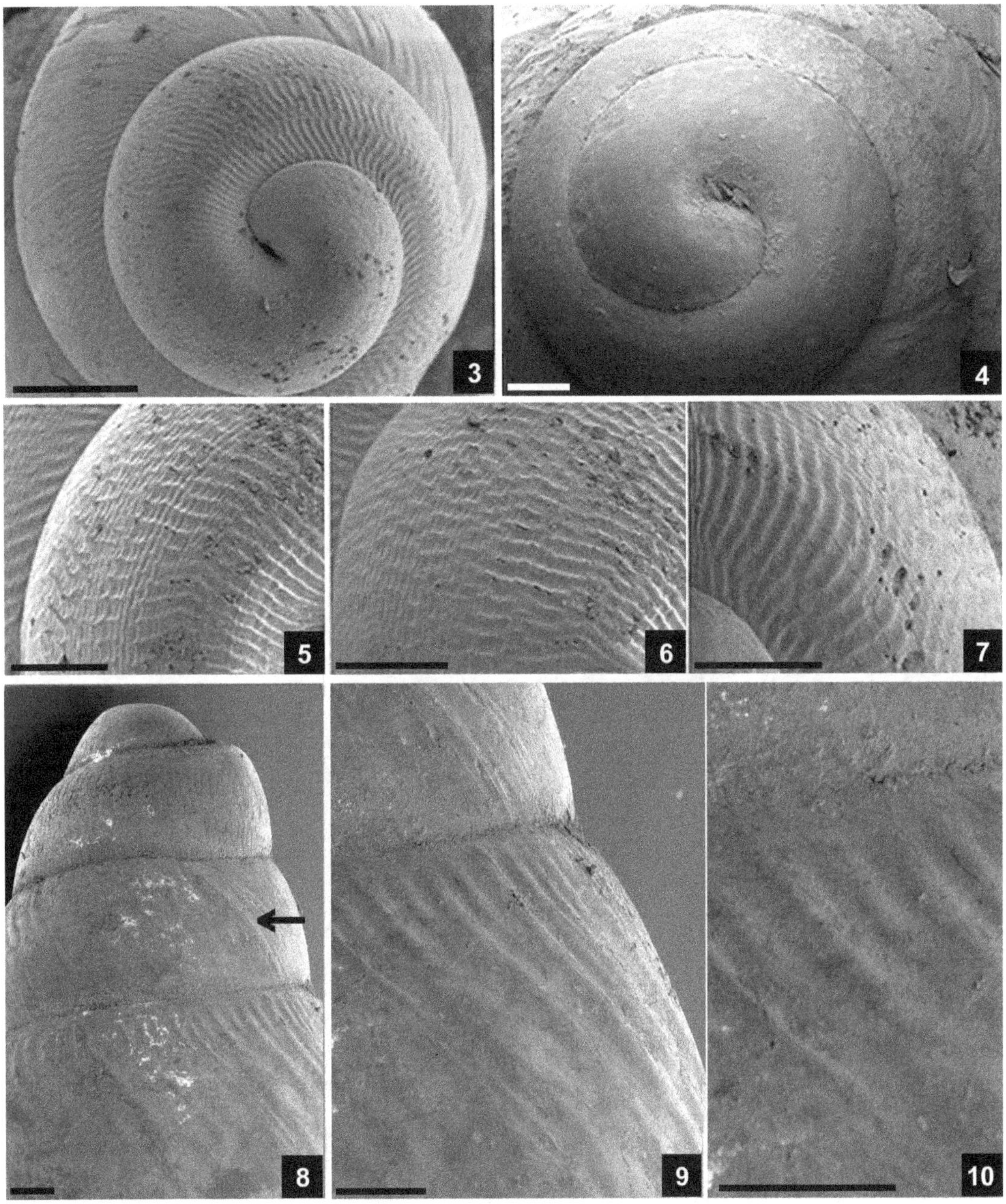

Figures 3–10. Scanning electron micrographs of the protoconch and teleoconch of *Ventania avellanedae*: (3) protoconch of a juvenile shell; (4) protoconch of an adult shell; (5–7) details of the axial and spiral sculpture of the protoconch with increasing degrees of erosion; (8–10) transition from protoconch to teleoconch, and details of the teleoconch of an adult specimen. Arrow indicates transition protoconch-teleoconch. Scale bar: 3 = 0.5 mm; 4–10 = 0.2 mm.

The bursa copulatrix is spherical, and its duct is longer than the spermoviduct, which has two portions of different diameter. The part proximal to the bursa copulatrix slightly broadens towards the transition point at the distal third, where its diameter decreases by half and slightly widens again towards the end. The inner wall presents a characteristic sculpture. The proximal wide region has lenghty longitudinal lamellae, while the distal one has a thick wall with perpendicular zig-zag grooves

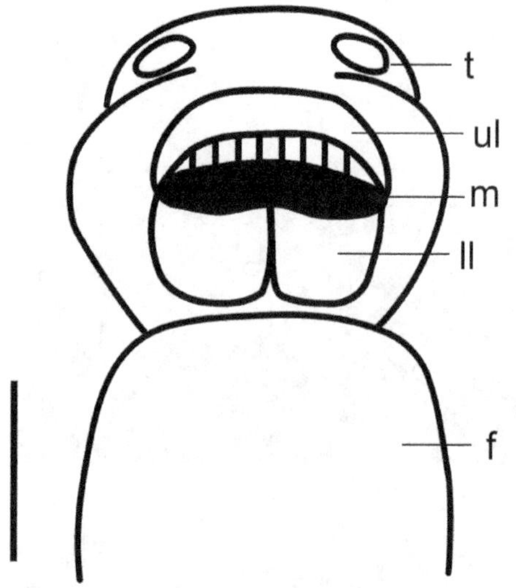

Figure 11. Buccal morphology of *Ventania avellanedae*. (f) foot, (ll) lower lip, (m) mandible, (t) tentacle, (ul) upper lip. Scale bar: 1 mm.

(Fig. 19). The duct of the bursa copulatrix and the free oviduct empty into the short and cylindrical vagina with spongy but smooth inner wall.

The vas deferens is a tubule of constant diameter that emerges just above the bifurcation of the vagina to the free oviduct and the bursa copulatrix duct. It runs attached to the vagina and the penial complex surface, passing underneath the penis sheath, and finally gets into the epiphallus–flagellum boundary. It is clearly visible and of thick diameter.

The male terminal genitalia or phallic complex is composed by a well delimited flagellum, epiphallus and penis. Flagellum, short and thin, occupies 1/9 of the phallic complex; the small and thin retractor muscle inserts on its distal end (Fig. 17). The limit between flagellum and epiphallus is marked by the calibre shift and the entering of the vas deferens. The epiphallus is longer than flagellum and penis, occupies about 2/3 of the phallic complex, presents a distal widening indicating the presence of an internal epiphallic gland, and its inner wall presents four well defined straight folds. The limit between epiphallus and penis is demarcated by a constriction. Penis occupies about 1/5–1/4 of the

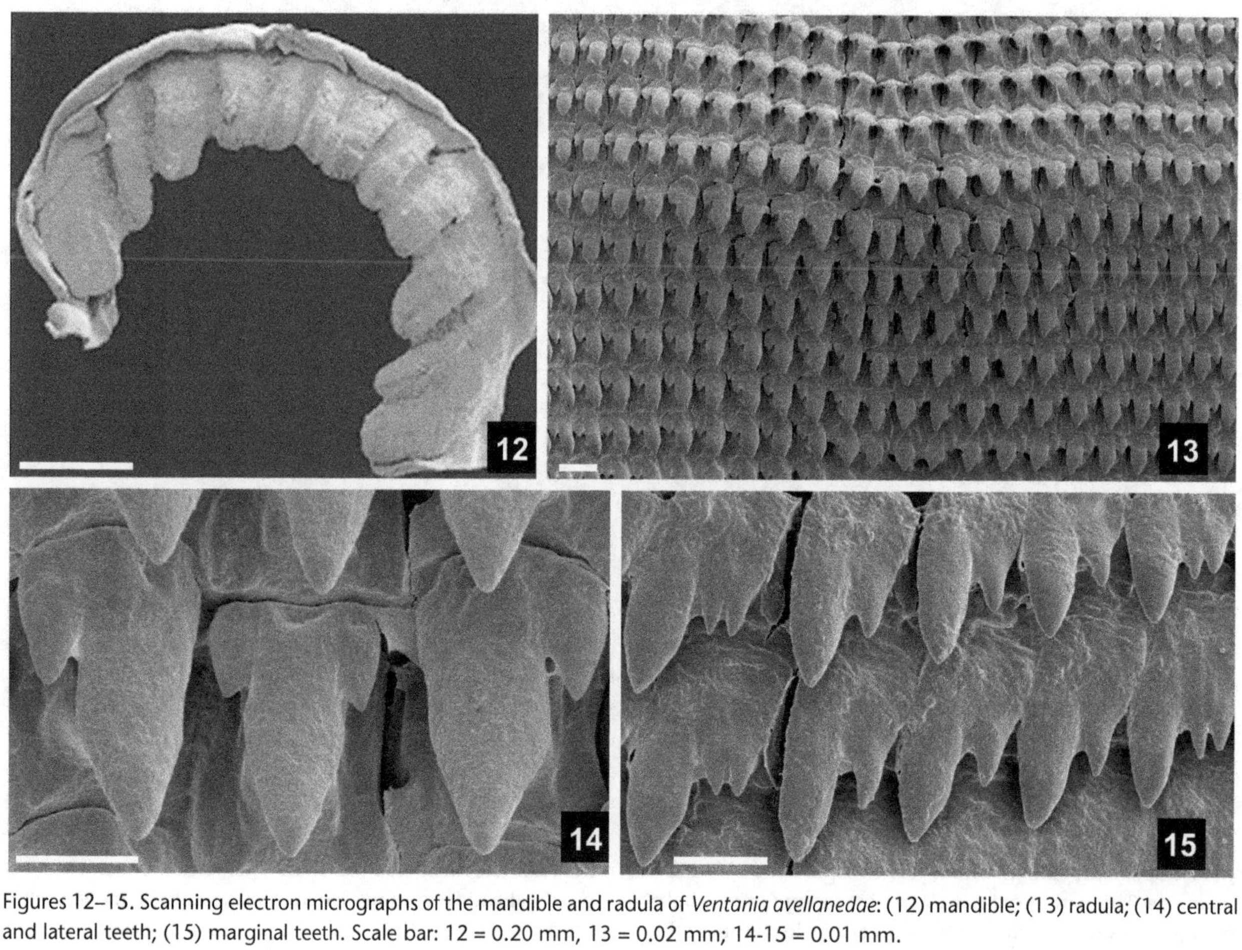

Figures 12–15. Scanning electron micrographs of the mandible and radula of *Ventania avellanedae*: (12) mandible; (13) radula; (14) central and lateral teeth; (15) marginal teeth. Scale bar: 12 = 0.20 mm, 13 = 0.02 mm; 14-15 = 0.01 mm.

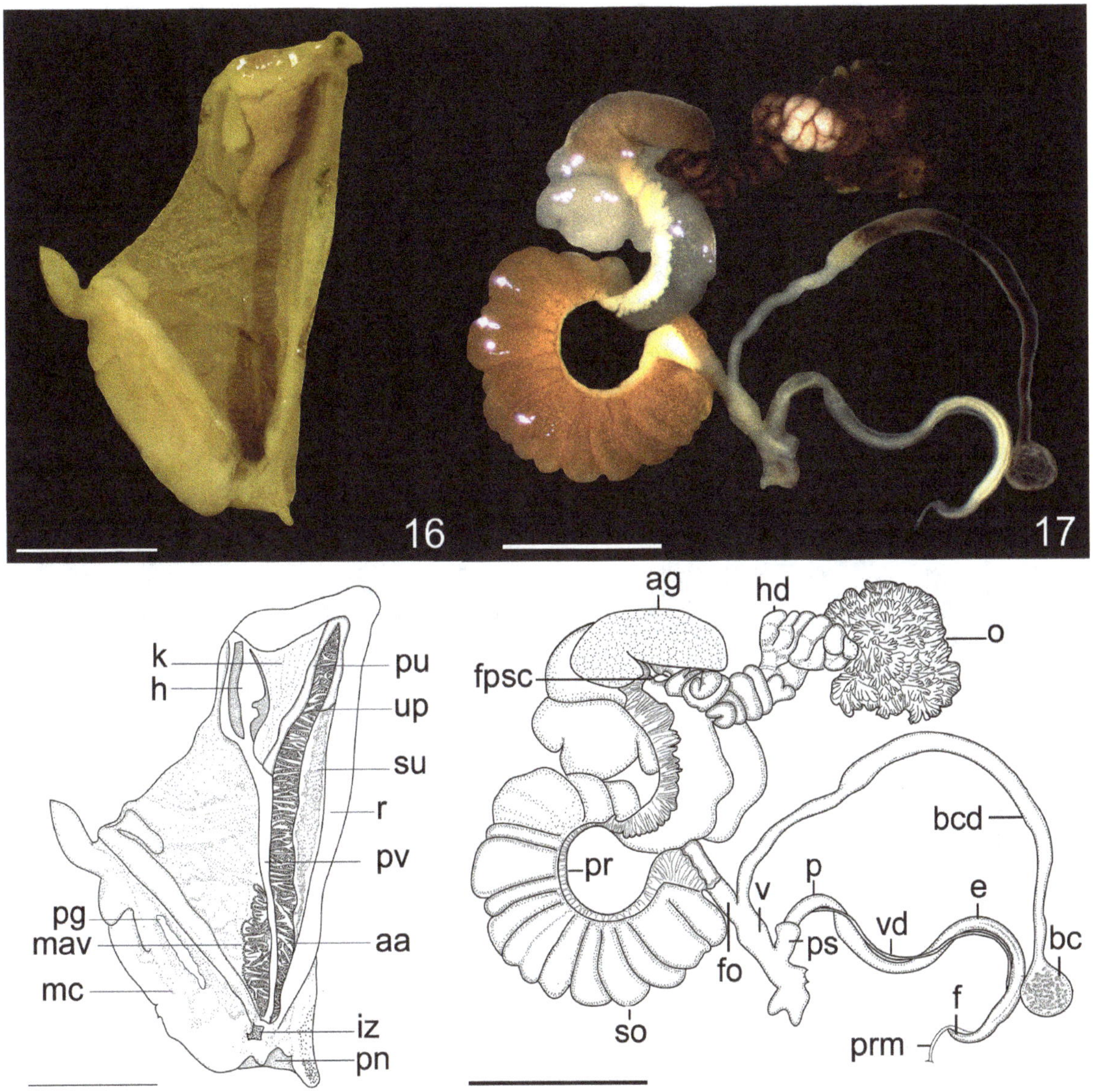

Figures 16–17. Internal anatomy of *Ventania avellanedae*: (16) pallial complex; (17) genital system. (aa) adrectal area, (ag) albumen gland, (bc) bursa copulatrix, (bcd) bursa copulatrix duct, (e) epiphallus, (f) flagellum, (fo) free oviduct, (fpsc) fertilization pouch-spermathecal complex, (h) heart, (hd) hermaphroditic duct, (iz) interramus zone, (k) kidney, (mav) marginal afferent vein, (mc) mantle collar, (o) ovotestis, (p) penis, (pg) pallial gland, (pn) pneumostome, (pr) prostate, (prm) penis retractor muscle, (ps) penial sheath, (pu) primary ureter, (pv) pulmonary vein, (r) rectum, (so) spermoviduct, (su) secondary ureter, (up) ureteric pore, (v) vagina, (vd) vas deferens. Scale bar: 5 mm.

phallic complex. It is cylindrical and relatively slender, with a strong muscular wall. A short muscular sheath surrounds its distal end (Fig. 17). Inner wall of the penis with folds and pilasters delimiting three different regions (Fig. 18): the proximal region, with longitudinal irregularly festooned pilasters and folds in the direction of penis length; the medium region, with transversal lamellae; and a distal, variable portion that has thin irregular longitudinal folds in some specimens and a spongy surface in others.

Distribution and habitat. *Ventania avellanedae* is endemic to the Ventania Mountain System, in southern Pampas, Argentina. This area was included in the biogeographic Chacoan Subregion, Pampa province by Morrone (2006). The distribution of the species is restricted to mountainous environments, mainly living in rock crevices, but it can be also found under rocks or associated to plants in rocky microhabitats (Delhey et al. 2005), so it is a rock dweller species as defined by Heller (1987).

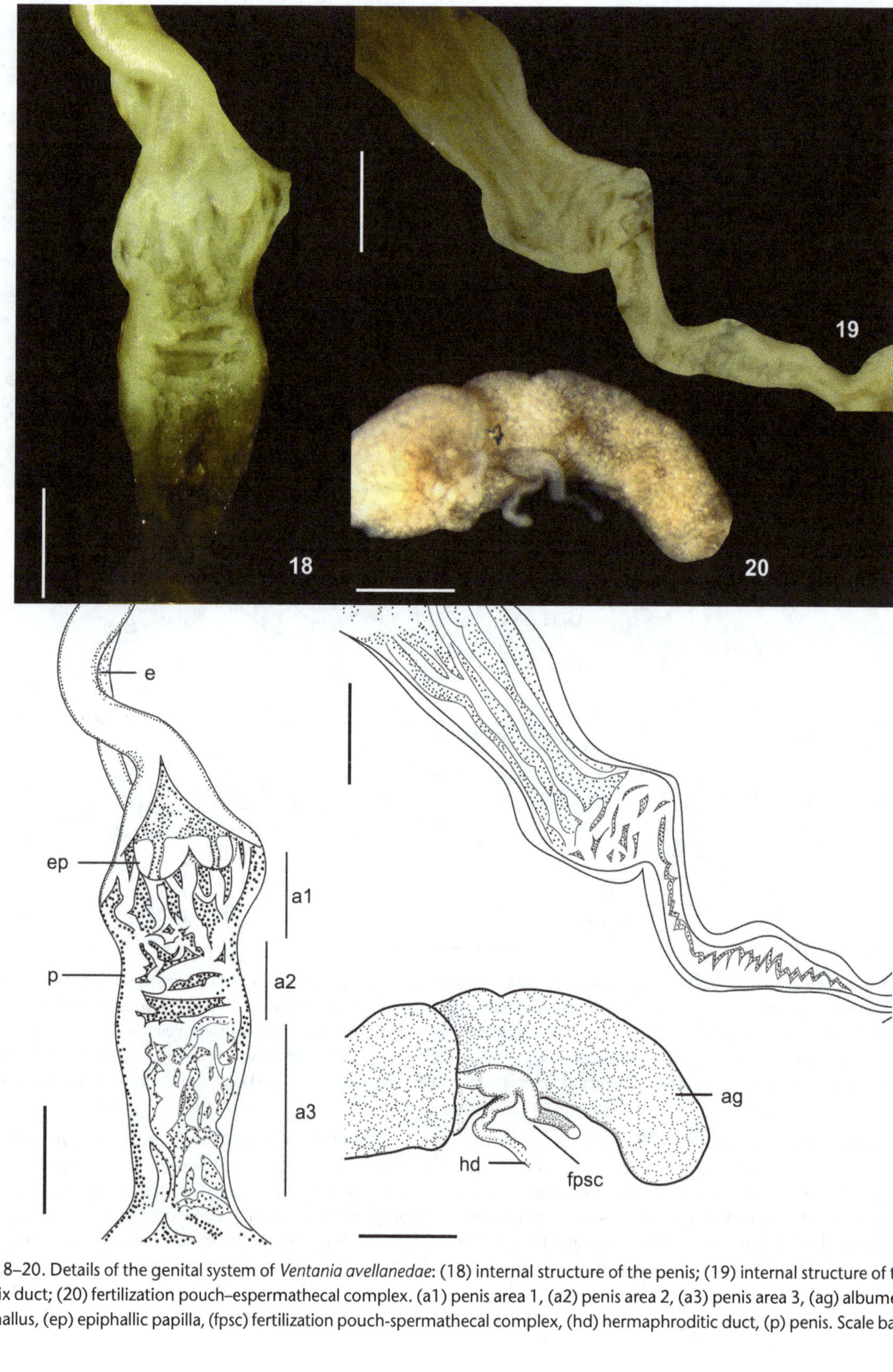

Figures 18–20. Details of the genital system of *Ventania avellanedae*: (18) internal structure of the penis; (19) internal structure of the bursa copulatrix duct; (20) fertilization pouch–espermathecal complex. (a1) penis area 1, (a2) penis area 2, (a3) penis area 3, (ag) albumen gland. (e) epiphallus, (ep) epiphallic papilla, (fpsc) fertilization pouch-spermathecal complex, (hd) hermaphroditic duct, (p) penis. Scale bar: 1 mm.

DISCUSSION

The absence of apertural folds and lamellae, together with the scarcity of taxonomic arguments, prompted us to discuss the position of this endemic species at the family and genus levels. Odontostomidae Pilsbry & Vanatta, 1898 was formally defined mostly by having several folds and lamellae ("apertural teeth") partially occluding the aperture. Breure (1979) stated that species of Odontostomidae usually have low number of mandibular plates (less than 20) and that the ovotestis size is larger in the Bulimulidae than in the Odontostomidae. However, both characters are greatly variable and show many exceptions.

Cuezzo (2011) performed a phylogenetic analysis considering sperm ultrastructure and other morphological characters of some species of Bulimulidae and Odontostomidae. As a result, Odontostomidae, represented by *Plagiodontes daedaleus* (Deshayes, 1851), *Clessinia pagoda* Hylton Scott, 1967 and *Spixia tucumanensis* (Parodiz, 1941), is considered a monophyletic clade with some morphological apomorphies (pallial gland present; penis sheath basal, short, cuff-like; bursa copulatrix duct longer than the spermoviduct; presence of epiphallic papilla; protoconch sculptured) and some sperm apomorphies (not considered here).

Miranda (2016) presented a cladogram in the context of a phylogenetic study of genus *Bostryx* Troschel, 1847, and proposed the following discrete apomorphies for Odontostomidae (represented by *Spixia tucumanensis* and *Plagiodontes daedaleus*): presence of apertural lamellae (1), wide umbilicus (2), secondary ureter open (3), pallial gland present (4), triangular interramus area (5), vagina wider in its proximal portion (6), with inner wall with pilasters (7); bursa copulatrix duct longer than spermoviduct (8), with a middle portion approximately equal to the distal portion (9); penis sheath basal (10); vas deferens passing under the penis sheath (11) and adhered to penis retractor muscle (12); penis retractor muscle thick (13) and inserted in the flagellum in distal portion (14).

Ventania presents all the diagnostic characters proposed for Odontostomidae by Breure (1979) and Cuezzo (2011) (excluding the apertural lamellae and folds) and most of the character states hypothesized as synapomorphies of the family by Miranda (2016). Unlike as stated by Miranda (2016), *Ventania* has a narrow umbilicus, a short and cylindrical vagina with a smooth inner wall, a bursa copulatrix duct with two different areas, and the penis retractor muscle thin and the vas deferens not adhered to it. The interramus area is variable, triangular or rectangular.

Taxonomic value of the "apertural teeth" is heterogeneous among land snail families. For example, Endodontidae, Punctidae, Pupillidae, Eniidae and Camaenidae include species with and without apertural folds within a single taxon (Solem 1976, 1983, Cuezzo 2003, Nekola and Coles 2010, Gümüs and Neubert 2012); however, most Orthalicoidea having well developed apertural folds are included in a special family, usually with no information on internal anatomy. It is to be noted that some Bulimulidae and Orthalicidae (two taxa that normally lack any apertural folds) develop some columellar structures and, sometimes, also basal or parietal ones that should be equivalent to some "teeth" of the Odontostomidae, as is the case of *Drymaeus coarctatus* (Pfeiffer, 1845) or *Eudolichotis distorta* (Bruguière, 1789). On the other hand, several species of Odontostomidae are deprived of apertural folds, e.g. *Clessinia oblita* (Reeve, 1848), *C. neglecta* (Pfeiffer, 1847) and several species in *Bahiensis* Jousseaume, 1877 and *Moricandia* Pilsbry & Vanatta, 1898.

Regarding the apomorphies stated in Miranda (2016) that are absent in *Ventania avellanedae*, it is to be noted that *Cyclodontina tudiculata* and *Bahiensis punctata* also present a cylindrical vagina. The penis retractor muscle is thin and attached to the flagellum in all *Cyclodontina* and *Bahiensis* species of known anatomy. The interramus area is variable in shape in *Ventania avellanedae*. Besides, *Plagiodontes* and *Spixia* species present either triangular or rectangular interramus zones (Pizá and Cazzaniga 2009, 2010, 2012, 2016, Salas Oroño 2007, 2010).

The presence of a spongy gland, and an opened secondary ureter in the pallial complex, the penis sheath muscular and short without retractor muscle, the vas deferens passing under penis sheath, the presence of an epiphallic gland, and the duct of the bursa copulatrix longer than the spermoviduct support the hypothesis that *Ventania* should be included in the Odontostomidae.

The protoconch of *Ventania* was described as smooth (Pilsbry 1902, Parodiz 1940, 1942, Fernández and Castellanos 1973), as also did Pilsbry (1896) for *Anctus stearnsianus*, a junior synonym of *Ventania avellanedae*. However, the protoconch of young specimens of *Ventania avellanedae* is actually sculptured with waved axial ribs crossed by spiral lines, similar to that of *Plagiodontes dentatus* (cf. Pizá & Cazzaniga, 2003). Species in *Cyclodontina* and *Bahiensis* may have different protoconch sculptures; e.g., Colley (2013) mentioned that the protoconch has axial ribs in *B. guarani*; a faint punctuation in *B. punctatissimus* (Lesson, 1830); axial lines of stippling aspect in *B. ringens* (Dunker, 1847), and weak axial lines with stippling spiral lines in *C. catharinae* (Pfeiffer, 1856). Sculpture vanishing in adult shells is frequent in species living in mountain areas with high proportion of quartz stones, as it was described by Cazzaniga et al. (2005) for *Plagiodontes patagonicus* (d'Orbigny, 1835), and by Pizá et al. (2006) for *P. rocae* Doering, 1881, two Odontostomidae species sympatric to *Ventania avellanedae*.

The radula of *V. avellanedae* does not fit the description and drawings by Parodiz (1940, 1943) since the radula of the analyzed fresh specimens had bi– and tricuspid teeth, not a pentacuspid central tooth; lateral teeth were very variable, and the marginal ones were very elongated. This radula is similar to that of some species of Odontostomidae, e.g. *Cyclodontina tudiculata* (Martens, 1868) (Colley, personal communication) and *Spixia cuezzae* Salas Oroño, 2010. So, the view of Parodiz (1940, 1943) that the radula of *Ventania* should not have any generic difference with other *Odontostomus* is here reaffirmed.

The morphology of the genital system was compared with an alcohol preserved specimen of *Bahiensis guarani* (d'Orbigny, 1835) – an Odontostomidae from Brazil and northern Argentina deprived of apertural folds –, and the available information on *B. punctatissimus*, *C. tudiculata* (Martens, 1868) and *C. rhodinostoma* (d'Orbigny, 1835) (Colley, personal communication).

Ventania is similar to *Cyclodontina* and *Bahiensis* in the shape and position of the thin penis retractor muscle, located at the distal end of the flagellum. *Ventania avellanedae* differs from *C. tudiculata* (Martens, 1868), *C. rhodinostoma* (d'Orbigny, 1835), and *B. punctatissimus* (which lacks epiphallus) by the bursa copulatrix duct shape and the distal widening of the epiphallus. It also differs from *Bahiensis guarani* (d'Orbigny, 1835) because the latter has an inner penial papilla and the inner wall of the vagina has a sculpture of longitudinal lamellae. Sculpture of the inner wall of penis and vagina is unknown in the remaining species of *Bahiensis* and *Cylclodontina*. *Ventania avellanedae* differs from the known species in *Cyclodontina* and *Bahiensis* by its penis inner wall divided in three different areas defined by different sculptures; the long and cylindrical epiphallus widening in the transition to the penis corresponding with the inner epiphallic gland, and differences in the shape and inner sculpture of the bursa copulatrix duct.

Ventania avellanedae is a rare rock dweller species restricted to rocky habitats of the Ventania Mountain System. As a strict endemic species, it deserves a high conservation priority. The distribution area of *V. avellanedae* includes the "Ernesto Tornquist" Provincial Park, a reserve protecting the natural highland grasslands in southern Pampas. However, native flora and fauna of the park is impacted by feral horses and exotic woody plants. The effects of these disturbances on natural vegetation and bird community dynamics were already studied in this area. Delhey et al. (2005) studied the effect of some environmental conditions and the impact of horses and exotic pines on the distribution and abundance of four autochthonous macrosnail species, including *V. avellanedae*. Macrosnails were absent from sites with exotic pines, but presented different responses to feral horses. *Ventania avellanedae* did not seem to be very affected by horses because the preferred resting sites are rock fissures, where the snails are somewhat safe from that impact.

ACKNOWLEDGMENTS

To Alejandro Tablado, curator at the Invertebrate Collection of the Museo Argentino de Ciencias Naturales "Bernardino Rivadavia" (MACN), for the material loans and kind facilities. To Gabriela Cuezzo for loaning the alcohol preserved specimen of *Bahiensis guarani* and kind advice. To Eduardo Colley for sending us valuable unpublished material. To Martín Carrizo for helping during sampling and photographs. To Cecilia Gutierrez Ayesta (Centro Científico-Tecnológico CONICET Bahía Blanca) for her help in obtaining the micrographs. To Juan Francisco Araya and an anonymous reviewer who helped to improve the manuscript. JP is a staff researcher of the Consejo Nacional de Investigaciones Científicas y Técnicas (CONICET); NJC is a staff researcher of the Comisión de Investigaciones Científica de la Provincia de Buenos Aires.

LITERATURE CITED

Baker HB (1963) Type land shells in the Academy of Natural Sciences of Philadelphia, II. Land Pulmonata, exclusive of North America north of Mexico. Proceedings of the Academy of Natural Sciences of Philadelphia 115: 191–259.

Breure ASH (1974) Catalogue of Bulimulidae (Gastropoda, Euthyneura), II. Odontostominae. Basteria 38: 109–127.

Breure ASH (1979) Systematics, phylogeny and zoogeography of Bulimulinae (Mollusca). Zoologische Verhandelingen 168: 1–215.

Breure ASH, Groenenberg DSJ, Shilthuizen M (2010) New insights in the phylogenetic relationships within the Orthalicoidea (Mollusca, Stylommatophora) based on 28S sequence data. Basteria 74(1/3): 25–31.

Breure ASH, Romero PE (2012) Support and surprises: molecular phylogeny of the land snail superfamily Orthalicoidea using a three-locus gene analysis with a divergence time analysis and ancestral area reconstruction (Gastropoda: Stylommatophora). Archiv für Molluskenkunde 141(1): 1–20. https://doi.org/10.1127/arch.moll/1869-0963/141/001-020

Burela S, Ghezzi NS, Delhey VK, Pizá J, Cazzaniga NJ (2003) Caracoles terrestres endémicos y lombrices del área protegida del sistema de Ventania. In: Vaquero MC, Cernadas de Bulnes MN (Eds) II Jornadas Interdisciplinarias del Sudoeste Bonaerense, Bahía Blanca, Argentina, June 6-8, 2002, vol. 3, 169–182.

Cabrera F, Martínez S (2012) The oldest Odontostomidae (Mollusca: Gastropoda): *Bahiensis priscus* n.sp. (Paleocene, Uruguay). Palaontologische Zeitschrift 86: 451–456. https://doi.org/10.1007/s12542-012-0145-1

Campos-Salgado N, Santos-Coelho AC (2003) Moluscos terrestres do Brasil (Gastrópodes operculados ou não, exclusive Veronicellidae, Milacidae e Limacidae). Revista de Biología Tropical 51(Suppl. 3): 149–189.

Cazzaniga NJ, Pizá J, Ghezzi NS (2005) Intraspecific clinal variation in *Plagiodontes patagonicus* (Gastropoda: Orthalicidae, Odontostominae), an endemic species from Argentina. Journal of Natural History 39(24): 2203–2216. https://doi.org/10.1080/00222930400004578

Colley E (2013) Taxonomia, Macroecologia e Ecologia de Gastropoda terrestre (Mollusca, Orthogastropoda) do Estado do Paraná, Brasil. PhD thesis, Curitiba, Brazil: Universidade Federal do Paraná. https://acervodigital.ufpr.br/handle/1884/32096

Cuezzo MG (2003) Phylogenetic analysis of the Camaenidae (Mollusca: Stylommatophora) with special emphasis on the American taxa. Zoological Journal of the Linnean Society 138: 449–476. https://doi.org/10.1046/j.1096-3642.2003.00061.x

Cuezzo MG (2011) Ultrastructure of spermatozoa in Orthalicidae (Mollusca, Gastropoda, Stylommatophora) and its systematic implications. Zoomorphology 130(3): 147–166. https://doi.org/10.1007/s00435-011-0128-x

Cuezzo MG, Miranda MJ, Ovando XMC (2013) Species catalogue of Orthalicoidea in Argentina (Gastropoda: Stylommatophora). Malacologia 56(1/2): 135–191. https://doi.org/10.4002/040.056.0210

Delhey V, Burela S, Pizá J, Ghezzi N, Cazzaniga NJ (2005) Conservation of land snails in the mountain grasslandsof the Argentinian pampas. Tentacle, the Newsletter of the IUCN/SSC Mollusc Specialist Group 13: 11–13.

Doering A (1881) Moluscos: Informe oficial de la comisión científica de la expedición al Río Negro bajo las órdenes del General D. Julio Argentino Roca (Buenos Aires). Buenos Aires, Ostwald y Martínez, 64 pp.

Fernández D (1973) Catálogo de la malacofauna terrestre argentina. Comisión de Investigaciones Científicas de la Provincia de Buenos Aires, La Plata, 197 pp.

Fernández D, Castellanos ZA (1973) Clave genérica de la malacofauna terrestre argentina. Revista del Museo Argentino de Ciencias Naturales 11: 263–285.

Gümüş BA, Neubert E (2012) New taxa of terrestrial molluscs from Turkey (Gastropoda, Pristilomatidae, Enidae, Hygromiidae, Helicidae). ZooKeys 171: 17–37. https://doi.org/10.3897/zookeys.171.2273

Heller J (1987) Shell shape and land-snail habitat in a Mediterranean and desert fauna. Biological Journal of the Linnean Society 31: 257–272. https://doi.org/10.1111/j.1095-8312.1987.tb01992.x

Miranda MJ (2016) Cladistic analysis of the Argentinian species of the genus *Bostryx* (Gastropoda, Stylommatophora) based on morphological evidence. Papeis Avulsos de Zoologia 56(10):109–133. https://doi.org/10.11606/0031-1049.2016.56.10

Morrone JJ (2006) Biogeographic areas and transition zones of Latin America and the Caribbean islands based on panbiogeografic and cladistic analyses of the entomofauna. Annual Review of Entomology 51: 467–494. https://doi.org/10.1146/annurev.ento.50.071803.130447

Nekola JC, Coles BF (2010) Pupillid land snails of Eastern North America. American Malacological Bulletin 28(2): 29–57. https://doi.org/10.4003/006.028.0221

Parodiz JJ (1940) *Ventania* nuevo subgénero de *Odontostomus*. Notas del Museo de la Plata. Zoología 5(42): 227–234.

Parodiz JJ (1942) Los Odontostominos de la Argentina, segunda parte. Physis 19(53): 319–343.

Parodiz JJ (1944) Contribuciones al conocimiento de los moluscos terrestres sudamericanos, II. Comunicaciones Zoológicas del Museo de Historia Natural de Montevideo 1(11): 1–6.

Parodiz JJ (1957) Catalogue of the land Mollusca of Argentina (cont.). Nautilus 71(1): 22–30.

Pilsbry HA (1896) A new species of *Bulimus*. Nautilus 10(4): 41.

Pilsbry HA (1902) Manual of Conchology. Second Series: Pulmonata. Conchological Section of the Academy of Natural Sciences, vol. 14, Philadelphia, 302 pp.

Pizá J, Cazzaniga NJ (2003) Redescription, shell variability and geographical distribution of *Plagiodontes dentatus* (Wood, 1828) (Gastropoda: Odontostomidae) from Uruguay and Argentina. Zootaxa 154: 1–23.https://doi.org/10.11646/zootaxa.154.1.1

Pizá J, Ghezzi NS, Cazzaniga NJ (2006) A rare land snail endemic from Argentina: *Plagiodontes rocae* Doering1881 (Gastropoda: Orthalicidae, Odontostominae). Archiv für Molluskenkunde 135(1): 91–99. https://doi.org/10.1127/arch.moll/0003-9284/135/091-099

Pizá J, Cazzaniga NJ (2009) A new species of *Plagiodontes* from Argentina, and new data on the anatomy of four other species in the genus (Gastropoda: Orthalicidae, Odontostominae). Journal of Natural History 43(23/24): 1437–1471. https://doi.org/10.1080/00222930902903244

Pizá J, Cazzaniga NJ (2010) Allopatry and anatomical distinctiveness of two puzzling land snails in genus *Plagiodontes*, from Argentina (Gastropoda: Orthalicidae, Odontostominae). Malacologia 53(1): 1–24. https://doi.org/10.4002/040.053.0101

Pizá J, Cazzaniga NJ (2012) Barrels and pagodas: clarifying the identity and variability of two nominal taxa of carinated land snail in the genus *Plagiodontes* (Gastropoda: Orthalicidae, Odontostominae). Journal of Natural History 46 (7–8): 383–410. https://doi.org/10.1080/00222933.2011.640440

Pizá J, Cazzaniga NJ (2016) *Plagiodontes parodizi*, a new species from Argentina (Gastropoda: Odontostomidae). Journal of Conchology 42(3): 1–10.

Ploeger S, Breure ASH (1977) A rapid procedure for preparation of radulae for routine research with the scanning electron microscope. Basteria 41: 47–52.

Richardson CL (1993) Bulimulacea: Catalog of species. Amphibulimidae, Anadromidae, Gangrellidae, Odontostomidae, Orthalicidae. Tryonia 27: 1–164.

Salas Oroño E (2007) Taxonomic review of the *Spixia pyriformis* species complex (Gastropoda: Pulmonata: Odontostominae). Zootaxa 1498: 1–25. https://doi.org/10.5281/zenodo.177060

Salas Oroño E (2010) A new species of *Spixia* from Argentina (Gastropoda, Stylommatophora, Odontostominae). Journal of Conchology 40(3): 305–313.

Scarabino F (2003) Lista sistemática de los Gastropoda terrestres vivientes de Uruguay. Comunicaciones de la Sociedad Malacológica del Uruguay 8(79): 203–214.

Simone LRL (2006) Land and freshwater mollusks of Brazil. EGB, São Paulo, 390 pp.

Solem A (1976) Endodontoid land snails from Pacific Islands (Mollusca: Pulmonata: Sigmurethra) Part I. Family Endodontidae. Field Museum of Natural History, Chicago, 508 pp.

Solem A (1983) Endodontoid land snails from Pacific Islands (Mollusca: Pulmonata: Sigmurethra). Part II. Families Punctidae and Charopidae, Zoogeography. Field Museum of Natural History, Chicago, 336 pp.

Author Contributions: JP, NJC and NSG contributed equally to this article.

Competing Interests: The authors have declared that no competing interests exist.

14

What are the main local drivers determining richness and fishery yields in tropical coastal fish assemblages?

Cynthia D. Souza[1], Vandick S. Batista[1], Nidia N. Fabré[1]

[1]*Instituto de Ciências Biológicas e da Saúde, Universidade Federal de Alagoas. Avenida Lourival Melo Mota, Tabuleiro do Martins, 57072-900 Maceió, AL, Brazil.*
Corresponding author: Vandick S. Batista (vandickbatista@gmail.com)

ABSTRACT. Seasonal ecological effects caused by temperature and photoperiod are typically considered minimal in the tropics. Nevertheless, annual climate cycles may still influence the distribution and abundance of tropical species. Here, we investigate whether seasonal patterns of precipitation and wind speed influence the structure of coastal fish assemblages and fishing yields in northeast Brazil. Research trips were conducted during the rainy and dry seasons using commercial boats and gear to sample the fish community. Diversity was analyzed using abundance Whittaker curves, diversity profiles and the Shannon index. Principal Component Analysis (PCA) was used to analyze associations between the abundance of species and various environmental variables related to seasonality. A total of 2,373 fish were collected, representing 73 species from 34 families – 20 of which were classified as both frequent and abundant. Species richness was greater and more equitable during the rainy season than the dry season – driven by changes in the precipitation rather than to wind speed. Species diversity profiles were slightly greater during the rainy season than the dry season, but this difference was not statistically significant. Using PCA was identified three groups of species: the first associated with wind speed, the second with precipitation, and the third with a wide range of sampling environments. This latter group was the largest and most ecologically heterogeneous. We conclude that tropical coastal fish assemblages are largely influenced by local variables, and seasonally mediated by annual changes related to precipitation intensity and wind speed, which in turn influences fishery yields.

KEY WORDS. artisanal fishing, fishes, wind, precipitation, seasonality, gillnet.

INTRODUCTION

Water quality and nutrient availability in coastal waters are influenced by environmental factors such as the spatial and temporal patterns of precipitation, land drainage, and wind patterns (Kennedy et al. 2002). Seasonal changes in biophysical factors can modify the composition of communities, affecting the occurrence and distribution of species within and between trophic levels (Blaber et al. 1995, Brown et al. 1997, Walther et al. 2002). In coastal ecosystems, these effects are usually mediated through temporal changes in temperature profile, the extent and magnitude of precipitation and continental runoff, wind patterns and storm frequency (Bernal-Ramírez et al. 2003, Jury 2011, Kennedy et al. 2002). Studies on the effects of changes in climate and biological productivity in the eastern Atlantic have documented a range of seasonal effects on trophic levels, coral reefs, juvenile dispersal, fishery yields and fish species abundance (e.g. Muehe and Garcez 2005, Ciotti et al. 2010, Costa et al. 2010, Leão et al. 2010, Schroeder and Castello 2010). Even so, the generalization of these findings is not possible as climate and local landscape characteristics may drive communities to different profiles.

In the tropics, where annual changes in temperature and photoperiod are minimal, seasonality is heavily determined by precipitation (Lowe-McConnell 1987), which generates dry and rainy seasons (Figueroa and Nobre 1990). Such seasonal variation, though less apparent than that observed in temperate areas, has the capacity to influence fish behavior (Wootton 1990) and, consequently, may influence the distribution and abundance of species within a given area (Laevastu and Hayes 1981) with a knock-on effect on fishery yields (Hilborn and Walters 1992). The consequences of these variability includes the estuarization process, when coastal waters resembles estuarine conditions (Longhurst and Pauly 1987b), even affecting the functional diversity of fish assemblages (Passos et al. 2016).

Even facing those large-scale patterns, local effects are detectable after higher scale effects being known and considered.

In this study, the objectives were to determine how seasonal patterns of precipitation and wind influence the temporal structure of tropical fish assemblages and the yields of the coastal gillnet fishery.

MATERIAL AND METHODS

Samples were collected in the central coastal zone of the state of Alagoas, north-eastern Brazil, near the main regional fisheries harbor (Fig. 1).

The seasonal patterns are typically tropical and consist of a rainy season from March to August and a dry season from September to February (Macêdo et al. 2004). Precipitation is high on the coastal plain; with an annual average of 1,800 mm. Coastal currents are conditioned by winds and tides and, in the rainy season, trade winds. During the rainy season, the southeast quadrant (SE) experiences more frequent and intense winds, while during the dry season, such winds flow from the northeast quadrant (NE) (Araújo et al. 2006).

Trips were made on crescent moon days in an 8 m wooden boat with an ice capacity of 500 kg and a one-cylinder, diesel-powered B18 engine. Six sampling trips were conducted between October 2010 to August 2011, with three occurring during the rainy season (June/July/August) and three during the dry season (October/December/February). Each trip lasted three days, totaling 30 launches of effective fishing.

Sets were established between two fishing areas with mud and gravel substrates, known locally as "Lama Grande" and "Tira da Pedra" (Fig. 1). The first set was approximately 11 km from Jaraguá Harbor at a depth of approximately 12 m; the second was located approximately 14 km from the port of Jaraguá in a depth of approximately 20 m. The distance between the centroids of the sites was approximately 6.5 km.

A commercial gillnet 1,330 m long and 1.5 m high was used, with a mesh size of 40 mm between opposite knots and a nylon thread size of 50 mm. Gillnets were set near the surface parallel to the seabed and was fixed at both ends with anchors. The mean ± standard deviation of the duration of each set was 3.51 ± 1.01 hours.

For each launch, the set position and duration were recorded, wind direction and intensity, the type and number of each species caught and the total length (cm) of each fish caught. All fishes were physically anesthetized by hypothermia on board and killed freezing on ice.

Additional data on precipitation, wind speed and direction were obtained from INMET/SEMARH, Brazilian government. To facilitate the communication of our results to environmental managers and members of local communities, a scale of wind intensity was created based on the reports of the fishermen on board and the daily wind speed recorded by INMET/SEMARH as follows: null = 0.0 m/s; weak ≤ 3.0 m/s; strong > 3.0 m/s. Similarly, the precipitation values recorded by INMET/SEMARH were categorized as very low = 0–31 mm; low = 32–84 mm; moderate = 85–137 mm; moderate-high = 138–196 mm; high = 197–200 mm and very high ≥ 201 mm.

From these data, there was calculated: the number of species and specimens per set, the mean length per species per

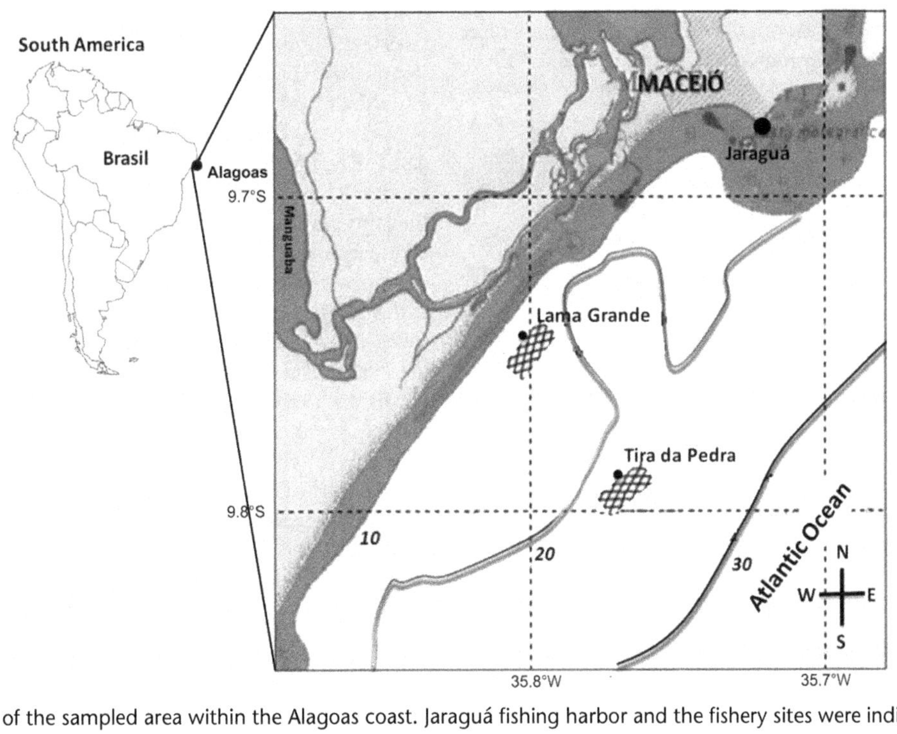

Figure 1. Location of the sampled area within the Alagoas coast. Jaraguá fishing harbor and the fishery sites were indicated.

set, the catch per set (kg), the CPUE (catch per unit effort) with standardized effort (1,330 m gillnet * set hours), the mean wind speed in m/s (mean of 6 hours per day: 3 hours before + 2 hours during the set + 1 hour after) and the total monthly precipitation (mm). All data were compared between the dry and rainy seasons.

Data analysis

Samples of all species were taken to the laboratory (SISBIO 1837810) after each fishing trip and identified using a variety of keys (Lessa and Nóbrega 2000, Menezes and Figueiredo 1980). One or two type specimens of each species were placed in a standard collection as reference material (see Suppl. material 1).

Analyses of diversity were conducted using three different methods: Whittaker abundance curves (Whittaker 1960, Whittaker et al. 2001), diversity profiles (Melo 2008) using Rényi series – Eq. (B.1) (Magurran 2004) and the Shannon-Wiener diversity index (H') – Eq. (B.2), including a bias correction term (Hammer et al. 2001). A modified t-test was used to analyze the Shannon index, estimating the variance by Eq. (B.3) (Magurran 2004). Diversity analyses were conducted using the program PAST (Hammer et al. 2001).

Variables were analyzed by univariate and factorial analyses of variance (ANOVA). The frequency of wind direction during the surveys was analyzed using the chi-square test (Legendre and Legendre 1998). Univariate and factorial analyses were performed in Statistica v.8.

The classification of abundant and frequent species by season was based on numerical abundance (PN %) and the frequency of occurrence (FO %), and the classification included four categories (Garcia and Sobrinho 2001): 1) Abundant and common (PN % > 100/S and BF % ≥ 50 %), where S is the total number of species per season; 2) Abundant but uncommon (PN % > 100 I/O and FO % < 50 %); 3) Sparse but common (PN % < 100/S and BF % ≥ 50 %); 4) Sparse and uncommon or occasional (PN % < 100/S and FO % < 50 %).

Because the variables used in the exploratory analysis were all quantitative and linear, a Principal Component Analysis (PCA) was performed (Legendre and Legendre 1998). To facilitate the PCA and focus on species of high commercial value, there were selected 16 species from the most abundant and frequent species and included precipitation and wind speed data. The multivariate model was tested by analysis of similarity (ANOSIM), and groups were identified by the greatest influence test SIMPER (Similarity Percentage) using Bray Curtis distance index (Clarke 1993). Multivariate analysis was performed using Statistica v. 8 and tests were done in PAST.

RESULTS

Diversity

There were collected 2,373 fishes representing 73 species and 34 families (Suppl. material 1). From these species, there were 39 categorized as abundant but uncommon, 51 as occasional species, and 51 and 20 as abundant and common species, respectively (the same species can belong to more than one category depending on the weather station) (Suppl. material 1). The frequent and abundant species, from most to least abundant, were as follows: Carangidae: *Caranx crysos* (Mitchill, 1815) (blue runner) (N = 360); Sciaenidae: *Larimus breviceps* Cuvier, 1830 (shorthead drum) (N = 321); Scombridae: *Euthynnus alletteratus* (Rafinesque, 1810) (little tunny) (N = 301); Scombridae: *Scomberomorus brasiliensis* Collette Russo & Zavala-Camin, 1978 (serra spanish mackerel) (N = 254); Haemulidae: *Conodon nobilis* (Linnaeus, 1758) (barred grunt) (N = 168); Clupeidae: *Opisthonema oglinum* (Lesueur, 1818) (Atlantic thread herring) (N = 108); Carangidae: *Chloroscombrus chrysurus* (Linnaeus, 1766) (Atlantic bumper) (N = 98), Sciaenidae: *Menticirrhus littoralis* (Holbrook, 1847) (gulf kingcroaker) (N = 72); Ariidae: *Bagre bagre* (Linnaeus, 1766) (coconut sea catfish) (N = 57); Lutjanidae: *Lutjanus synagris* (Linnaeus, 1758) (lane snapper) (N = 56); Ariidae: *Cathorops spixii* (Agassiz, 1829) (madamango sea catfish) (N = 51); Carangidae: *Caranx hippos* (Linnaeus, 1766) (jack crevalle) (N = 50); Sciaenidae: *Macrodon ancylodon* (Bloch & Schneider, 1801) (king weakfish) (N = 42); Centropomidae: *Centropomus parallelus* Poey, 1860 (fat snook) (N = 41); Haemulidae: *Haemulopsis corvinaeformis* (Steindachner, 1868) (roughneck grunt) (N = 40) and Engraulidae: *Cetengraulis edentulus* (Cuvier, 1829) (atlantic anchoveta) (N = 35). Among these 20 species, seven were prevalent during the rainy season and three in the dry season (Suppl. material 1). Rarefaction curves estimated for the samples caught during dry and rain seasons using the bootstrap richness estimator yield different patterns of relative diversity for both seasons for all sample sizes (Fig. 2). Although the dry season richness is closer at high sample size, the asymptotic level was reached for both seasons and the difference remained.

The size and slope of the curves in the Whittaker abundance diagram indicated that the rainy season (Fig. 3) had greater species richness and equitability than the dry season (Fig. 4). In the rainy season, the predominant species in order of decreasing abundance were *C. crysos*, *E. alletteratus*, *S. brasiliensis* and *L. breviceps*. Apart from these species, abundance decreased gradually with increasing richness. In the dry season, the predominant species were *L. breviceps*, *C. nobilis*, *C. chrysurus* and *C. crysos*.

The diversity profiles indicated that diversity was marginally higher in the rainy season than in the dry season. The relevant alpha values are close to one indicating greater abundance. Diversity varied more in richness than in abundance between seasons (Fig. 5). However, the Shannon diversity index did not differ between seasons (t–test, p > 0.05).

Interaction of physical and biological variables

The univariate ANOVA showed significant differences between season in the number of species, number of fish, CPUE and mean species length (Tab. 1). The highest average observed values were observed in the rainy season (Fig. 6).

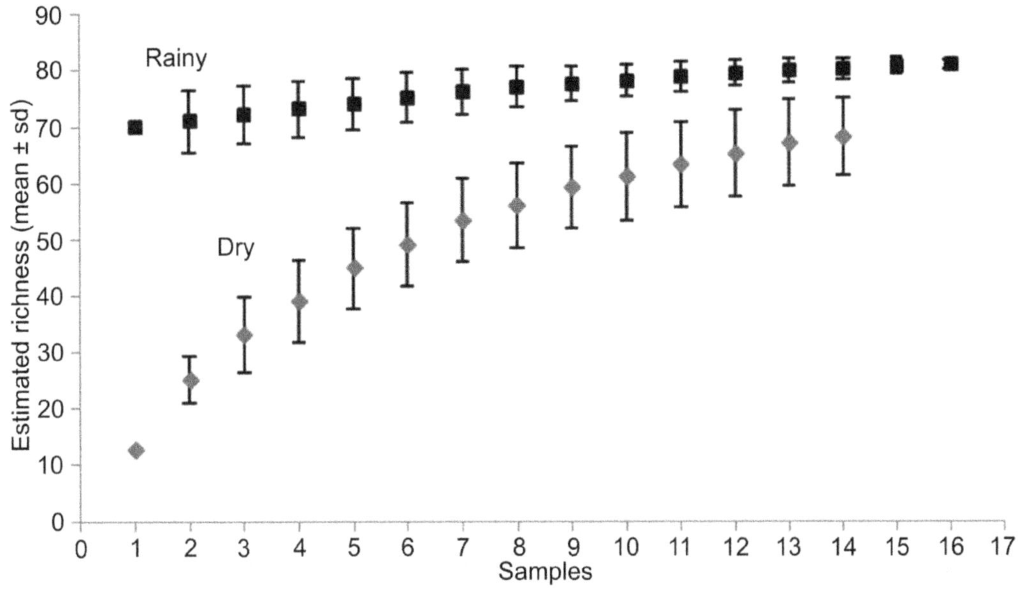

Figure 2. Rarefaction curves for the samples caught during dry and rain seasons using the bootstrap richness estimator.

Table 1. Analysis of variance tables to compare seasonal effects on species number, number of fishes, CPUE and mean length for fishes caught at the coast of Alagoas.

Effect	df	SS	MS	F	p
Species number					
Intercept	1	4392.30	4392.30	276.58	<0.01
Season	1	208.03	208.03	13.10	<0.01
Error	28	444.67	15.88		
Total	29	652.70			
Number of fishes					
Intercept	1	187704.30	187704.30	73.28	<0.01
Season	1	28644.30	28644.30	11.18	<0.01
Error	28	71716.40	2561.30		
Total	29	100360.70			
CPUE (Kg)/(m*h)					
Intercept	1	0.00	0.00	51.53	<0.01
Season	1	0.00	0.00	8.02	0.01
Error	28	0.00	0.00		
Total	29	0.00			
Mean length (cm)					
Intercept	1	27394.55	27394.55	1171.77	<0.01
Season	1	369.39	369.39	15.80	<0.01
Error	28	654.60	23.38		
Total	29	1024.00			

To identify those variables responsible for these seasonal differences, the influences of precipitation and wind speed were tested using a factorial ANOVA. Neither wind speed nor the interaction between precipitation and wind speed had significant seasonal effects (Tab. 2). However, there were significant effects of precipitation on the number of species and the number and average size of fish (but not the CPUE) (Tab. 2). The changes were greatest with respect to the number and species of fish, which exhibited linear increasing trends with increasing precipitation. Changes in average length and CPUE, while significant, did not follow the same trend (Fig. 8 and 9). Although the CPUE did not vary seasonally, possibly due to the selectivity of the net, the average catch was higher in the rainy season ($p < 0.05$; Fig. 7), indicating a possible influence of precipitation on the catch. Although the variance analysis revealed significant differences in average size between seasons, the size range (20 to 44 cm) remained consistent across seasons.

The chi-square ($\chi 2$) indicated that the winds of the northeast quadrant (NE) predominated during the dry season, and the winds of the southeast quadrant (SE) predominated during the rainy season. No quadrants predominated with respect to source winds in December (i.e. during the rainy season) (Fig. 10).

The PCA clustered species into three groups (Fig. 11): Group one was associated with wind speed and consisted of *C. chrysurus*, *L. breviceps*, *C. parallelus* and *H. corvinaeformis*, with the former two species being dominant in the dry season. The second group was formed by two species: *C. crysos*, *S. brasiliensis*, *E. alletteratus*, *O. oglinum* and *L. synagris* (Fig. 11), the first three of which were dominant in the rainy season. Species of Group three were associated with areas of high precipitation and wind speed or from high turbidity waters: *C. nobilis*, *M. littoralis*, *C.*

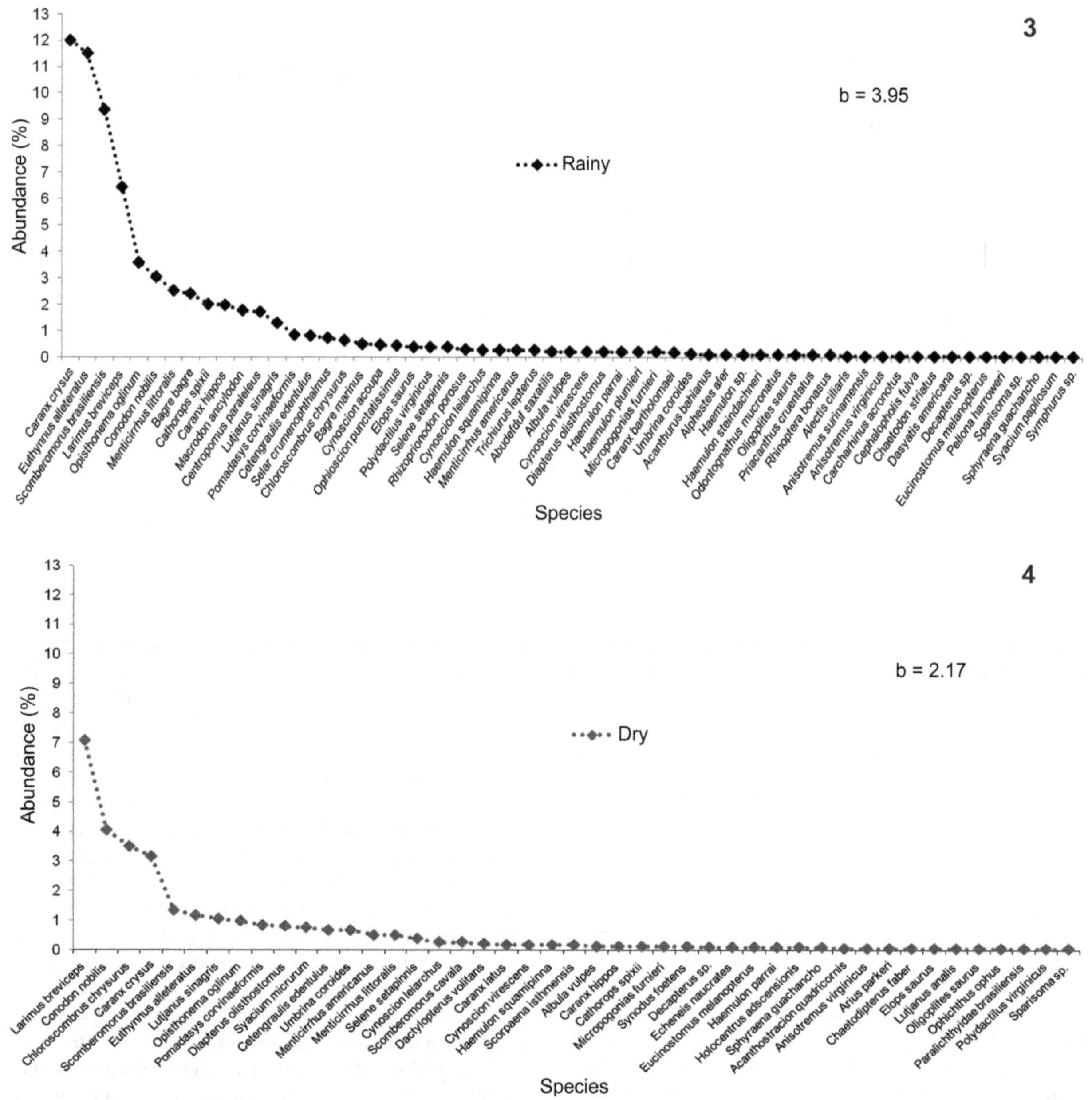

Figures 3-4. Whittaker plot of fishes from Alagoas coast sampled during the rainy season (**3**) and the dry season (**4**).

spixii, *C. hippos*, *C. edentulus*, *B. bagre*, *M. ancylodon*; these latter two species were only recorded during the rainy season. The first two PCA axes explained 57.32 % of the total variation, with factor 1 primarily explained by precipitation (40.43 %) and factor 2 by wind speed (16.89 %) (Fig. 11). The similarity analysis revealed a significant interaction between precipitation and wind speed (ANOSIM: $r = 0.6$, $p < 0.01$). The SIMPER test identified precipitation as contributing the most to the interaction (55.67 %) and wind speed the least (0.48 %) (Tab. 3).

DISCUSSION

The results support the hypothesis that precipitation and winds are significant drivers of fish species richness and fishing yields in the coastal tropical waters. The temporal dynamics and quality of water and nutrients are ultimately affected by climate variation, especially in precipitation and wind, generating environmental conditions that can affect the structure of estuarine and marine systems (Kennedy et al. 2002) and seasonal

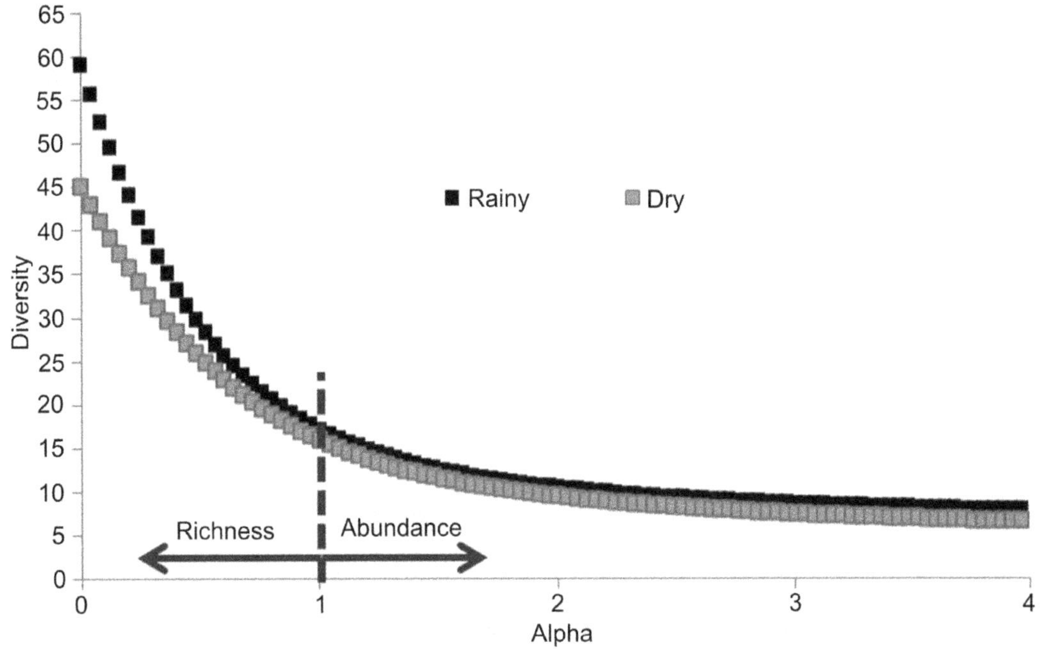

Figure 5. Diversity profile of the dry and rain seasons on the coast of Alagoas. Alpha = 1 Emphasizes the separation of richness and diversity profile.

shifts in biotic responses (Lowe-McConnell 1987). In the current study, seasonal differences were observed in the number of species caught, number of individuals, CPUE and average species length. Moreover, seasonal variation in precipitation appears to influence all measures except for CPUE.

More generally, there was found that the communities sampled with a bottom gillnet off the coast of Alagoas were characterized by relatively high species richness (n = 73), corresponding to 72 % of richness as estimated by the bootstrap method. The 16 most frequent species (Suppl. material 1) belong to nine families common in several pelagic, reef and estuary regions of the eastern coast of Brazil (e.g. Araújo et al. 2002, Costa et al. 2003, Tubino et al. 2007, Lessa et al. 2009, Rangely et al. 2010, Carneiro and Salles 2011). However, the communities were dominated by few species (n = 20) that were both abundant and frequent as observed in other marine areas (e.g. Godefroid et al. 2003, Lira and Teixeira 2008).

Mean yield was higher during the rainy season than during the dry season. This is not straightforward to interpret and the results of other seasonality studies vary widely with respect to periods of higher yield depending on the type of gear used, the amount of effort (Béné and Tewfik 2001, Pet-Soede et al. 2001); and the types of environments and species sampled (Jury 2011, Tubino et al. 2007). In contrast to the current study, yields were found to be higher in the dry season in the bay and estuarine environments of Rio de Janeiro, Brazil (Tubino et al. 2007) and Paranaguá Bay, southern Brazil (Vendel et al. 2003). In these regions, both primary and secondary yields and fish concentrations were higher during the dry season (Allen 1982). Despite such regional differences, Tubino et al. (2007) also reported a predominance of *C. crysos* during the dry season in tropical estuaries, when high temperatures and biological production makes these environments more attractive to enhance reproduction of marine species (Araújo et al. 1998). With respect to our data, increases in abundance and the rainy season catch could be explained by the increase in drainage (Day et al. 2012, Dittmar et al. 2001) originating from the large number of mangroves, lagoons and rivers along the Brazilian coast (Lara 2003, Araújo et al. 2006). The nutrient-rich sediments of these environments are carried to the marine environment by the rains, increasing the biological productivity of coastal waters.

The pattern of wind direction affected the seasonality of assemblage structure, with a predominance of trade winds from the northeast quadrant (NE) in the dry season and a predominance of those from the southeast quadrant (SE) in the rainy season – this is typical of regional atmospheric patterns (Servain and Legler 1986). However, as precipitation modulates seasonality in the tropics (Lowe-McConnell 1987, Macêdo et al. 2004), its effect was stronger than that of wind speed (wind speed was the least influential variable in the multivariate analysis). A weak effect of wind speed on fish assemblages were also reported for Paranaguá Bay (Vendel et al. 2003), where the highest wind speeds occurred in summer. North-eastern and south-eastern trade winds in the study region, which are associated with precipitation in the rainy season, should affect coastal currents (Hazin 2009). These currents then carry nutrients from

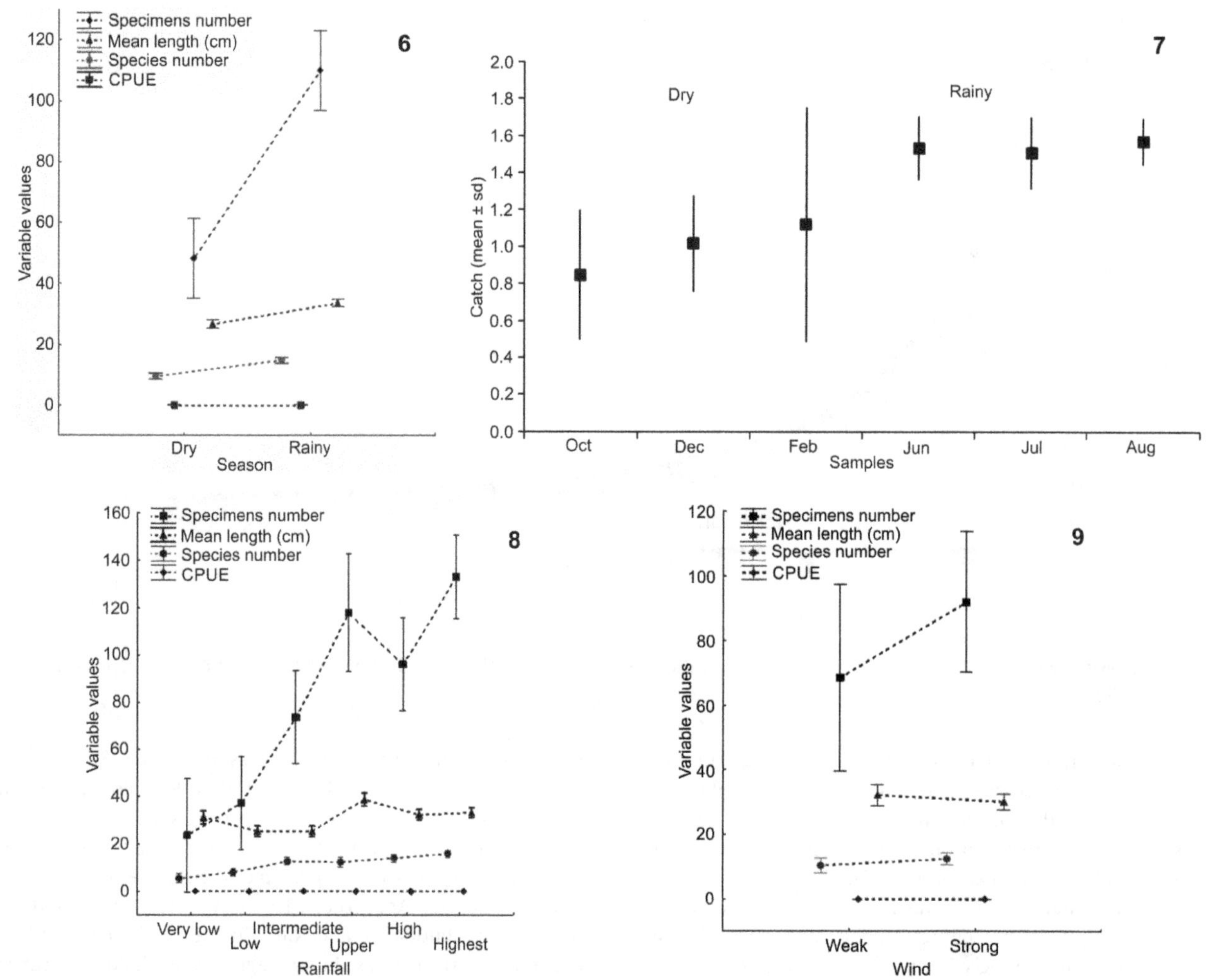

Figures 6–9. Mean ± sd of the number of fishes, fish total length, species richness, and CPUE by climatic seasons (**6**); CPUE (mean ± sd) by month (**7**); precipitation by month (**8**) and wind strength by month (**9**).

mangroves, lagoons and rivers to the surface and column waters of marine environments (Dittmar et al. 2001, Day et al. 2012), thereby increasing fishing yields.

Our multivariate analysis separated species into three groups. All species that were both abundant and frequent in the dry season where associated with wind. An example of this group is *L. breviceps*, a demersal species that is one of the most abundant fish on the Brazilian coast (Lira and Teixeira 2008, Souza et al. 2008). Individuals and species of the family Sciaenidae, with representatives found in all three groups were very abundant in the present work. Species of this family are very abundant along the northeastern coast (Lessa et al. 2009) and include marine, estuarine and freshwater species throughout the world (Nelson 2006).

Group two contained no estuarine resident species, only pelagic or reef species. Group two species are more flexible to environmental changes related to seasonality and its effects on marine dynamics (Longhurst and Pauly 1987a). Carangidae, Clupeidae, Scombridae, and Lutjanidae were represented in this group. Larvae of the former three families have been recorded along the north-eastern coast of Brazil (Mafalda Jr et al. 2006). The families Carangidae and Clupeidae contain many surface pelagic coastal species (Zavala-Camin 1983). Scombridae contains pelagic ocean species that use coastal waters only as nurseries (Moyle and Cech 1988). However, species of Scombridae, such as *E. alleteratus* and *S. brasiliensis* (group 2), are captured at various stages of growth in coastal areas of north-eastern Brazil (Lessa et al. 2009) due to the narrow continental shelf region. Group three contained the greatest number of species that inhabit the widest variety of habitats. It contained primarily estuarine or estuarine-dependent species, with estuarine, pelagic and reef species also present. Here, the association of the estuarine

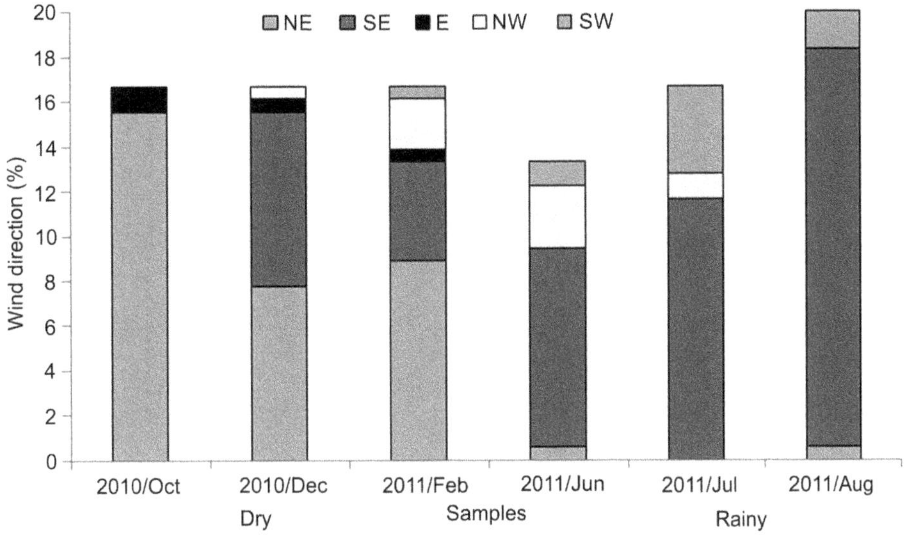

Figure 10. Frequency and direction of winds in each sampled month on the coast of Alagoas.

Table 2. Analysis of variance tables comparing effects of rainfall and winds on species number, number of fishes, CPUE and mean length for fishes caught at the coast of Alagoas.

Effect	df	SS	MS	F	p
Species number					
Intercept	1	3280.08	3280.08	281.60	<0.01
Rainfall	5	331.79	66.36	5.70	<0.01
Wind	1	26.45	26.45	2.27	0.15
Rainfall * Wind	5	84.59	16.92	1.45	0.25
Error	18	209.67	11.65		
Total	29	652.70			
Number of fishes					
Intercept	1	161309.48	161309.48	86.87	<0.01
Rainfall	5	41261.62	8252.32	4.44	0.01
Wind	1	3445.32	3445.32	1.86	0.19
Rainfall * Wind	5	17838.55	3567.71	1.92	0.14
Error	18	33422.58	1856.81		
Total	29	100360.70			
CPUE (kg)/(m*h)					
Intercept	1	0.00	0.00	38.50	<0.01
Rainfall	5	0.00	0.00	2.27	0.09
Wind	1	0.00	0.00	0.03	0.87
Rainfall * Wind	5	0.00	0.00	0.67	0.65
Error	18	0.00	0.00		
Total	29	0.00			
Mean length (cm)					
Effect	1	24276.52	24276.52	1016.97	<0.01
Intercept	5	522.17	104.43	4.37	0.01
Rainfall	1	27.11	27.11	1.14	0.30
Wind	5	144.26	28.85	1.21	0.34
Rainfall * Wind	18	429.68	23.87		
Error	29	1024.00			

Table 3. Similarity percentage analysis (SIMPER) tested for differences in the main species composition and abiotic variables (rainfall and wind strength) at the coast of Alagoas.

Variable	Contribution	%
Rainfall	34.62	55.67
Caranx crysos	3.72	61.65
Euthynnus alletteratus	3.43	67.17
Larimus breviceps	3.43	72.68
Scomberomorus brasiliensis	2.92	77.38
Conodon nobilis	1.72	80.15
Opisthonema oglinum	1.30	82.24
Chloroscombrus chrysurus	1.09	84.00
Menticirrhus littoralis	0.88	85.42
Bagre bagre	0.82	86.74
Cathorops spixii	0.71	87.88
Caranx hippos	0.70	89.01
Macrodon ancylodon	0.62	90.01
Lutjanus synagris	0.59	90.96
Centropomus parallelus	0.56	91.86
Cetengraulis edentulus	0.43	92.56
Haemulopsis corvinaeformis	0.39	93.18
Wind strength	0.30	93.66

environment with precipitation patterns and the life cycles of Neotropical species is more evident, confirming the importance of precipitation levels in influencing fish species richness and abundance in the coastal tropics.

Gillnets were chosen to collect data because they are the favored fishing method of artisanal fishers in tropical regions (Castello 2010, Godınez-Domınguez et al. 2000, Hovgård and Lassen 2000), including northeastern Brazil (Lessa et al. 2009). It also can capture several species of varying sizes(Nielsen and

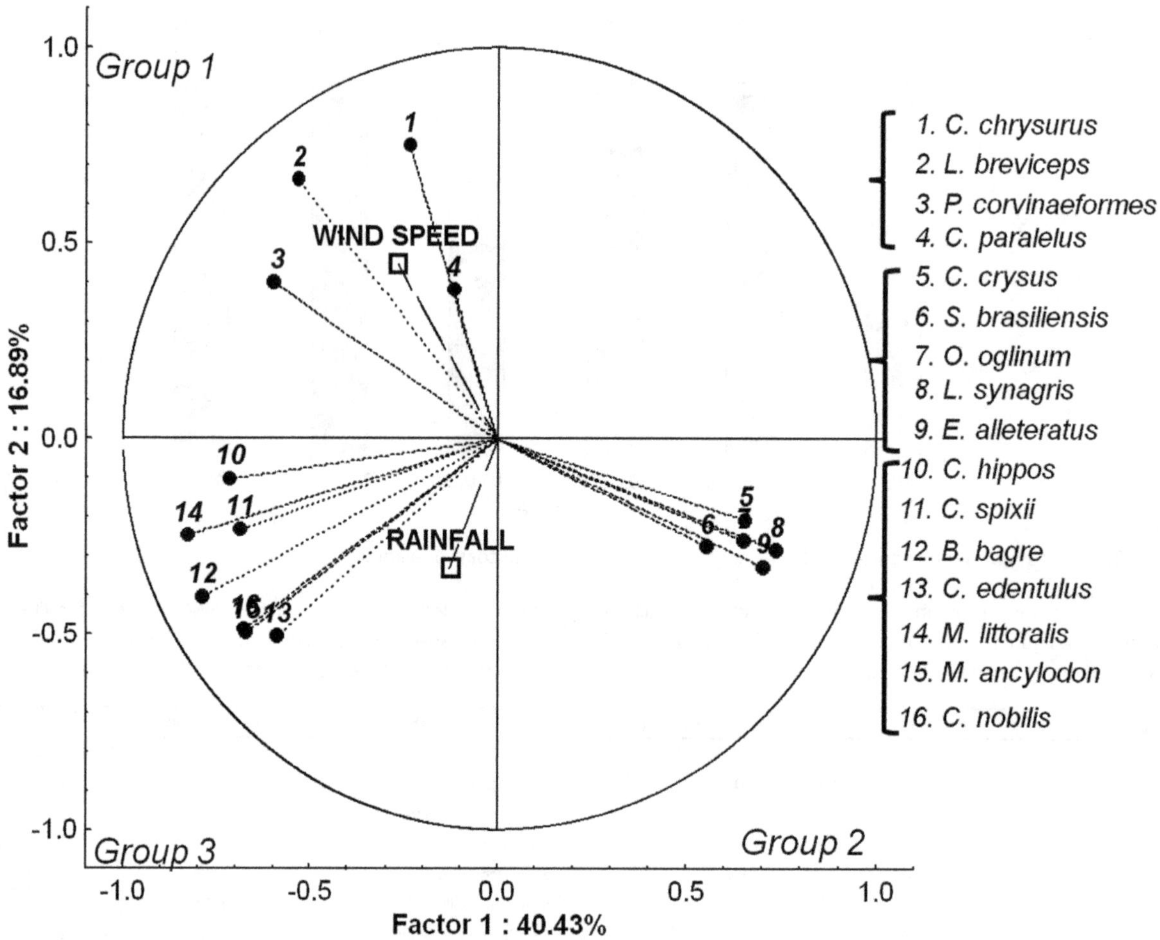

Figure 11. PCA of species abundance associated to precipitation and wind speed.

Johnson 1983), and tend to be less harmful to the aquatic environment than other methods (Hovgård and Lassen 2000). Moreover, they are also inexpensive, easy to repair and technologically simple (Hovgård and Lassen 2000).

Species co-occurrence is complex and poorly understood along neotropical waters (Andrade-Tubino et al. 2008, Azevedo et al. 2006, Barletta et al. 2010). Hidden behind precipitation and wind effects may appear the influence of trophic relationships, reproductive cycles (Keddy and Weiher 1999) and may also a stochastic component (Grossman et al. 1982). Moreover, this influence may be modulated by specific variables correlated to the rivers input into the coastal areas, including salinity, river flow (Gillson et al. 2012, Mitchell et al. 1999), turbidity (Castillo-Rivera et al. 2002, Cyrus and Blaber 1987, 1992, Johnston et al. 2007, Whitfield 1999) or pollution (Lekve et al. 2002). Nevertheless, evidences provided here indicated for consistent seasonal changes determined by precipitation and wind direction/intensity in the distribution and abundance of species within coastal fish assemblages. These ecological changes have inevitable knock-on effects on fisheries yields and the composition of the catch and should be carefully considered when developing coastal conservation, as well as for fisheries policy and regulations.

ACKNOWLEDGEMENTS

This study was financially supported by the Conselho Nacional de Desenvolvimento Científico e Tecnológico (CNPq) – process 560385/2008-0 and 482201/2010-0, Coordenação de Aperfeiçoamento de Pessoal de Nível Superior (CAPES) and the Fundação de Amparo à Pesquisa do Estado de Alagoas (FAPEAL) – process 20080531775-5. We are particularly grateful to the "Jupirá" and its crew. We are also thankful for the assistance of the LAEPP team, the comments on earlier versions of this manuscript by A.R. Carvalho and C.L. Sampaio. Thanks also to R.J. Ladle for his comments and English revision.

LITERATURE CITED

Allen LG (1982) Seasonal abundance, composition and productivity of the littoral fish assemblage in upper Newport Bay, California. Fishery Bulletin 80: 769–790.

Andrade-Tubino MF, Ribeiro ALR, Vianna M (2008) Organização espaço-temporal das ictiocenoses demersais nos ecossistemas estuarinos brasileiros: uma síntese. Oecologia Brasiliensis 12: 5. https://doi.org/10.4257/oeco.2008.1204.05

Araújo FG, Gomes da Cruz-Filho A, Costa de Azevedo MC, Clistenes de Alcantara Santos A (1998) Estrutura da comunidade de peixes demersais da Baía de Sepetiba, RJ. Revista Brasileira de Biologia 58: 417–430. https://doi.org/10.1590/S0034-71081998000300007

Araújo JN, Martins AS, Costa KG (2002) Idades e crescimento da cioba, *Ocyururs chrysurus*, da Costa Central do Brasil. Brazilian Journal of Oceanography 50: 47–57. https://doi.org/10.1590/S1679-87592002000100004

Araújo T, Santos R, Seoan J, Manso V (2006) Erosão e progradação no litoral brasileiro – Alagoas. In: Muehe D (Ed.) Erosão e progradação no litoral brasileiro, Brasília, MMA, 197–212. https://doi.org/10.1016/j.ecss.2005.09.006

Azevedo MCC, Araújo FG, Pessanha ALM, Araújo Silva M (2006) Co-occurrence of demersal fishes in a tropical bay in southeastern Brazil: A null model analysis. Estuarine, Coastal and Shelf Science 66: 315–322.

Barletta M, Jaureguizar AJ, Baigun C, Fontoura NF, Agostinho AA, Almeida-Val VMF, Corrêa MFM (2010) Fish and aquatic habitat conservation in South America: a continental overview with emphasis on neotropical systems. Journal of Fish Biology 76: 2118–2176. https://doi.org/10.1111/j.1095-8649.2010.02684.x

Béné C, Tewfik A (2001) Fishing effort allocation and fishermen's decision making process in a multi-species small-scale fishery: analysis of the conch and lobster fishery in Turks and Caicos Islands. Human Ecology 29: 157–186. https://doi.org/10.1023/A:1011059830170

Bernal-Ramírez JH, Adcock GJ, Hauser L, Carvalho GR, Smith PJ (2003) Temporal stability of genetic population structure in the New Zealand snapper, *Pagrus auratus*, and relationship to coastal currents Marine Biology 142: 567–574. https://doi.org/10.1007/s00227-002-0972-9

Blaber SJM, Brewer DT, Salini JP (1995) Fish communities and the nursery role of the shallow inshore waters of a tropical bay in the gulf of Carpentaria, Australia. Estuarine, Coastal and Shelf Science 40: 177–193. https://doi.org/10.1016/S0272-7714(05)80004-6

Brown JH, Valone TJ, Curtin CG (1997) Reorganization of an arid ecosystem in response to recent climate change. Proceedings of the National Academy of Sciences 94: 9729–9733. https://doi.org/10.1073/pnas.94.18.9729

Carneiro PB de M, Salles R de (2011) Caracterização da pescaria com rede de emalhar derivante realizada no município de Fortaleza, Estado do Ceará. Arquivos de Ciências do Mar 44: 69–80.

Castello J (2010) O futuro da pesca da aquicultura marinha no Brasil: a pesca costeira. Ciência e Cultura 62(3): 32–35.

Castillo-Rivera M, Zavala-Hurtado JA, Zárate R (2002) Exploration of spatial and temporal patterns of fish diversity and composition in a tropical estuarine system of Mexico. Reviews in Fish Biology and Fisheries 12: 167–177. https://doi.org/10.1023/A:1025051027676

Ciotti A, Garcia CAE, Jorge DSF (2010) Temporal and meridional variability of Satellite-estimates of surface chlorophyll concentration over the Brazilian continental shelf Pan-American. Journal of Aquatic Sciences 5: 236–253.

Clarke KR (1993) Non-parametric multivariate analyses of changes in community structure. Austral Ecology 18: 117–143. https://doi.org/10.1111/j.1442-9993.1993.tb00438.x

Costa MBSF, Mallmann DLB, Pontes PM, Araujo M (2010) Vulnerability and impacts related to the rising sea level in the Metropolitan Center of Recife, Northeast Brazil. Pan-American Journal of Aquatic Sciences 5: 341–349.

Costa PAS, Braga C, Otávio L (2003) Reef fisheries in Porto Seguro, eastern Brazilian coast. Fisheries Research 60: 577–583. https://doi.org/10.1016/S0165-7836(02)00145-5

Cyrus DP, Blaber SJM (1987) The influence of turbidity on juvenile marine fish in the estuaries of Natal, South Africa. Continental Shelf Research 7: 1411–1416. https://doi.org/10.1016/0278-4343(87)90046-X

Cyrus DP, Blaber SJM (1992) Turbidity and salinity in a tropical northern Australian estuary and their influence on fish distribution. Estuarine, Coastal and Shelf Science 35: 545–563. https://doi.org/10.1016/S0272-7714(05)80038-1

Day JW, Kemp WM, Alejandro Y, Crump BC (2012) Estuarine Ecology. Hoboken, New Jersey, 550 pp. https://doi.org/10.1002/9781118412787

Dittmar T, Lara RJ, Kattner G (2001) River or mangrove? Tracing major organic matter sources in tropical Brazilian coastal waters. Marine Chemistry 73: 253–271. https://doi.org/10.1016/S0304-4203(00)00110-9

Figueroa SN, Nobre CA (1990) Precipitation distribution over central and western tropical South America. Climanalise 5: 36–45.

Garcia AM, Sobrinho JPV (2001) O aumento da diversidade de peixes no estuário da Lagoa dos Patos durante o episódio El Niño 1997–1998. Atlântica 23: 133–152.

Gillson J, Suthers I, Scandol J (2012) Effects of flood and drought events on multi-species, multi-method estuarine and coastal fisheries in eastern. Australia Fisheries Management and Ecology 19: 54–68. https://doi.org/10.1111/j.1365-2400.2011.00816.x

Godefroid RS, Spach HL, Schwarz R Jr, Queiroz G, Oliveira Neto JF (2003) Efeito da lua e da maré na captura de peixes em uma planície de maré da Baía de Paranaguá, Paraná, Brasil. Boletim do Instituto de Pesca 29: 47–55.

Godınez-Domınguez E, Rojo-Vázquez J, Galvan-Pina V, Aguilar-Palomino B (2000) Changes in the structure of a coastal fish assemblage exploited by a small-scale gillnet fishery during an El

Nino-La Nina event. Estuarine, Coastal and Shelf Science 51: 773–787. https://doi.org/10.1006/ecss.2000.0724

Grossman GD, Moyle PB, Whitaker JO Jr (1982) Stochasticity in structural and functional characteristics of an Indiana stream fish assemblage: a test of community theory. The American Naturalist 120: 423–454. https://doi.org/10.1086/284004

Hammer Ø, Harper DAT, Ryan PD (2001) PAST: paleontological statistics software package for education and data analysis. Palaeontologia Electronica 4: 1–9. Available online at: http://palaeo-electronica.org/2001_1/past/issue1_01.htm [Accessed: 05/11/2011]

Hazin FHV (2009) Meteorologia e sensoriamento remoto, oceanografia física, oceanografia química, oceanografia geológica. Programa REVIZEE Score Nordeste, Fortaleza, 248 pp. Available online at: https://www.mar.mil.br/secirm/documentos/revizee/score-nordeste-vol1.pdf [Accessed: 29/11/2011]

Hilborn R, Walters CJ (1992) Quantitative fisheries stock assessment: choice, dynamics and uncertainty. Springer, Boston, 570 pp. https://doi.org/10.1007/978-1-4615-3598-0

Hovgård H, Lassen H (2000) Manual on estimation of selectivity for gillnet and longline gears in abundance surveys. FAO, Rome, 84 pp.

Johnston R, Sheaves M, Molony B (2007) Are distributions of fishes in tropical estuaries influenced by turbidity over small spatial scales? Journal of Fish Biology 71: 657–671. doi: org/10.1111/j.1095-8649.2007.01536.x

Jury MR (2011) Environmental influences on Caribbean fish catch. International Journal of Oceanography 2011: 1–11. https://doi.org/10.1155/2011/174729

Keddy P, Weiher E (1999) Introduction: the scope and goals of research on assembly rules. In: Weiher E, Keddy P (Eds) Ecological assembly rules: perspectives, advances, retreats. Cambridge University Press, Cambridge, 1–20. https://doi.org/10.1017/CBO9780511542237.001

Kennedy VS, Twilley RR, Kleypas JA, Cowan JH Jr, Hare SR (2002) Coastal and Marine Ecosystems & Global Climate Change-Potential Effects on U.S. Resources. Pew Center on Global Climate Change, Arlington, VA, 64 pp. Available on line at: https://www.c2es.org/site/assets/uploads/2002/08/marine_ecosystems.pdf [Accessed: 09/09/2011]

Laevastu T, Hayes ML (1981) Fisheries oceanography and ecology. Fishing News Books, Farnham, 239 pp.

Lara RJ (2003) Amazonian mangroves – a multidisciplinary case study in Pará State, North Brazil: Introduction. Wetlands Ecology and Management 11: 217–221. https://doi.org/10.1023/A:1025012914237

Leão ZMAN, Kikuchi RKP, Oliveira MDM, Vasconcelos V (2010) Status of Eastern Brazilian coral reefs in time of climate changes. Pan-American Journal of Aquatic Sciences 5: 224–235.

Legendre P, Legendre LFJ (1998) Numerical ecology. Elsevier, Oxford, 853 pp.

Lekve K, Boulinier T, Stenseth NC, Gjosaeter J, Fromentin J-M, Hines JE, Nichols JD (2002) Spatio-temporal dynamics of species richness in coastal fish communities. Proceedings of the Royal Society B: Biological Sciences 269: 1781–1789. https://doi.org/10.1098/rspb.2002.2058

Lessa R, Araújo B, Bezerra J Jr, Nóbrega M (2009) Dinâmica das Frotas Pesqueiras da Região Nordeste do Brasil. Programa REVIZEE – Score Nordeste, Fortaleza, 164 pp.

Lessa R, Nóbrega M (2000) Programa REVIZEE/Score-NE: Guia de Identificação de peixes marinhos da região nordeste. Recife, Universidade Federal Rural de Pernambuco, Departamento de Pesca, 128 pp. Available on line at: http://www.mma.gov.br/estruturas/revizee/_arquivos/guiaiden.pdf [Accessed: 19/03/2011]

Lessa RP, Monteiro A, Duarte-Neto PJ, Vieira AC (2009) Multidimensional analysis of fishery production systems in the state of Pernambuco, Brazil. Journal of Applied Ichthyology 25: 256–268. https://doi.org/10.1111/j.1439-0426.2009.01264.x

Lira AKF, Teixeira SF (2008) Ictiofauna da praia de Jaguaribe, Itamaracá, Pernambuco. Iheringia, Série Zoologia, 98: 475–480. https://doi.org/10.1590/S0073-47212008000400010

Longhurst AR, Pauly D (1987a) Ecology of Tropical Oceans. Academic Press, San Diego, 407 pp.

Longhurst AR, Pauly D (1987b) Geography of the Tropical Oceans. In: Longhurst AR, Pauly D (Eds) Ecology of Tropical Oceans. Academic Press, San Diego, 5–26. https://doi.org/10.1016/B978-0-12-455562-4.50005-7

Lowe-McConnell RH (1987) Ecological studies in tropical fish communities. Cambridge University Press, Cambridge, 390 pp. https://doi.org/10.1017/CBO9780511721892

Macêdo SJ, Muniz K, Montes MJF (2004) Hidrologia da região costeira e plataforma continental do estado de Pernambuco. In: Eskinazi-Leça E, Neumann-Leitão S, Costa MF (Eds) Oceanografia: um cenário tropical. Recife, Bagaço, 255–286.

Mafalda PO Jr, Sinque C, Muelbert H (2006) Associações de larvas de peixes na costa norte da Bahia Atlântica 28: 5–11.

Magurran AE (2004) Measuring biological diversity. John Wiley & Sons, Cornwall, 256 pp.

Melo AS (2008) O que ganhamos "confundindo" riqueza de espécies e equabilidade em um índice de diversidade. Biota Neotropica 8: 21–27. https://doi.org/10.1590/S1676-06032008000300001

Menezes N, Figueiredo JL (1980) Manual de Peixes Marinhos do Sudeste do Brasil. IV. Teleostei (3) São Paulo, Museu de Zoologia, Universidade de São Paulo, 96 pp.

Mitchell SB, West JR, Arundale AMW, Guymer I, Couperthwaite JS (1999) Dynamics of the turbidity maxima in the upper Humber estuary system, UK. Marine Pollution Bulletin 37: 190–205. https://doi.org/10.1016/S0025-326X(98)00178-7

Moyle PB, Cech JJ (1988) Fishes: An Introduction to Ichthyology. Prentice Hall, Englewood Cliffs, 559 pp.

Muehe D, Garcez DS (2005) A plataforma continental brasileira e sua relação com a zona costeira e a pesca. Mercator 4: 69–88.

Nelson JS (2006) Fishes of the World. John Wiley & Sons, New York, 601 pp.

Nielsen LA, Johnson DL (1983) Fisheries techniques. American Fisheries Society, Bethesda, 468 pp.

Passos CVB, Fabré NN, Malhado ACM, Batista VS, Ladle RJ (2016) Estuarization increases functional diversity of demersal fish assemblages in tropical coastal ecosystems. Journal of Fish Biology 89: 847–862. https://doi.org/10.1111/jfb.13029

Pet-Soede C, Van Densen WLT, Hiddink JG, Kuyl S, Machiels MAM (2001) Can fishermen allocate their fishing effort in space and time on the basis of their catch rates? An example from Spermonde Archipelago, SW Sulawesi, Indonesia. Fisheries Management and Ecology 8: 15–36. https://doi.org/10.1046/j.1365-2400.2001.00215.x

Rangely J, Fabré NN, Tiburtino C, Batista VS (2010) Estratégias de pesca artesanal no litoral marinho alagoano (Brasil). Boletim do Instituto de Pesca 36: 263–275.

Schroeder FA, Castello JP (2010) An essay on the potential effects of climate change on fisheries in Patos Lagoon, Brazil. Pan-American Journal of Aquatic Sciences 5: 148–158.

Servain J, Legler DM (1986) Empirical orthogonal function analyses of tropical Atlantic sea surface temperature and wind stress: 1964–1979. Journal of Geophysical Research: Oceans 91: 14181–14191. https://doi.org/10.1029/JC091iC12p14181

Souza UP, Costa RC da, Martins IA, Fransozo A (2008) Relationships among Sciaenidae fish (Teleostei: Perciformes) and Penaeoidea shrimp (Decapoda: Dendrobranchiata) biomass from the north coast of São Paulo State, Brazil. Biota Neotropica 8: 83–92. https://doi.org/10.1590/S1676-06032008000100011

Tubino RA, Monteiro-Neto C, Moraes LES, Paes ET (2007) Artisanal fisheries production in the coastal zone of Itaipu, Niterói, RJ, Brazil. Brazilian Journal of Oceanography 55: 187–197. https://doi.org/10.1590/S1679-87592007000300003

Vendel AL, Lopes SG, Santos C, Spach HL (2003) Fish assemblages in a tidal flat Brazilian. Archives of Biology and Technology 46: 233–242. https://doi.org/10.1590/S1516-89132003000200015

Walther G-R, Post E, Convey P, Menzel A, Parmesan C, Beebee TJC, Bairlein F (2002) Ecological responses to recent climate change. Nature 416: 389–395. https://doi.org/10.1038/416389a

Whitfield AK (1999) Ichthyofaunal assemblages in estuaries: a South African case study. Reviews in Fish Biology and Fisheries 9: 151–186. https://doi.org/10.1023/A:1008994405375

Whittaker RH (1960) Vegetation of the Siskiyou mountains, Oregon and California. Ecological Monographs 30: 279–338. https://doi.org/10.2307/1943563

Whittaker RJ, Willis KJ, Field R (2001) Scale and species richness: towards a general, hierarchical theory of species diversity. Journal of Biogeography 28: 453–470. https://doi.org/10.1046/j.1365-2699.2001.00563.x

Wootton RJ (1990) Ecology of teleost fishes. Springer Netherlands, Dordrecht, 404 pp. https://doi.org/10.1007/978-94-009-0829-1

Zavala-Camin LA (1983) Caracterização das espécies Brasileiras da familia Scombridae (Perciformes) Boletim Instituto de Pesca 10: 73–94.

Author Contributions: CDS, VSB and NNF contributed equally to this article.

Competing Interests: The authors have declared that no competing interests exist.

15

Biology, distribution and geographic variation of loliginid squids (Mollusca: Cephalopoda) off southwestern Atlantic

Rodrigo Silvestre Martins[1], **Marcelo Juanicó**[2]

[1] *Laboratório de Ciências da Pesca, Instituto do Mar, Universidade Federal de São Paulo. Rua Dr Carvalho de Mendonça 144, Encruzilhada, 11070-100 Santos, SP, Brazil.*
[2] *Juanicó – Environmental Consultants Ltd. Building 130, Ram On, 19205, Israel.*
Corresponding author: Rodrigo Silvestre Martins (rsmartins@unifesp.br)

ABSTRACT. The biology and ecology of southwestern Atlantic loliginid squids have been intensively researched in the last few decades, mostly off the Brazilian southern coast. However, information gathered by scientific research cruisers, either past or recent, is limited. Three species of loliginid squids – the warm-tempered *Doryteuthis sanpaulensis* (Brakoniecki, 1984), plus the tropical *D. pleii* (Blainville, 1823) and *Lolliguncula brevis* (Blainville, 1823) – were sampled along 16 degrees of latitude on the southwestern Atlantic (22–38°S). The samples were obtained mostly from oceanographic surveys, but also included squids caught by commercial fisheries, and a few specimens from museum collections. Squid response to abiotic variables, morphological variation and circadian behaviour were surveyed in detail. *Doryteuthis sanpaulensis* can be divided into at least six distinguishable geographical groups, which do not form a cline. In contrast, *D. pleii* can be divided into only two morphological groups that are very similar. *Doryteuthis* spp. were heterogeneously distributed on the shelf, whereas *L. brevis* was confined nearshore. Our data extended the southernmost distribution range of *D. pleii* by at least nine degrees of latitude, owing to specimens obtained at ~38°S (Mar del Plata, Argentina). Small, immature *D. sanpaulensis* were sampled inside the Patos Lagoon estuary (~32°S). The morphologically similar *Doryteuthis* spp. apparently avoid direct competition by concentrating at different depths, displaying different thermal preferences, and inverse circadian levels of activity. The information reported herein may be regarded as a "snapshot" of the ecology of sympatric squids in a marine environment that has not been deeply affected by climate change.

KEY WORDS. Competition, morphology, Myopsida, population biology, reproduction.

INTRODUCTION

Loliginid squids are typical inhabitants of coastal and shelf waters (Jereb et al. 2010). Many species are widely distributed geographically, and for this reason they experience significant shifts in environmental conditions such as water temperature and food availability (Pelc and Jackson 2008). In addition to the geographic location, depth and oceanographic regimes also affect the environmental conditions of these mollusks. Loliginid squids display noticeable variations in body proportions, some of which result in cryptic speciation, or distinct population groups that are clearly discernible based on their morphology, (Jackson and Moltschaniwskyj 2001, Herke and Foltz 2002, Sales et al. 2013, 2014, 2017, van der Vyver et al. 2016). Those groups may differ not only in body proportions, but also in other biological aspects, such as body condition, adult size, fecundity, reproductive cycle and lifespan (Pelc and Jackson 2008).

The veined squid, *Loligo forbesii* Steenstrup, 1856 from the Azores Islands form a distinct population, with individuals that are much larger (and robust) than their con-specifics from the European and north-western African Atlantic continental shelves (Pierce et al. 1994). Despite these differences, which include DNA microsatellite divergences, veined squid from the Azores are conspecific with *L. forbesii* distributed elsewhere (Shaw et al. 1999). Similar (and sometimes almost identical) phenotypes of the Atlantic brief squid *Lolliguncula brevis* (Blainville, 1823) are found scattered along the Brazilian coast (Perez and Zaleski 2013). Interestingly, distinct groups inhabiting nearby areas are often separate by indistinguishable groups of this squid located far apart (Zaleski et al. 2012). In addition, a recent genetic analysis of *L. brevis* has confirmed cryptic speciation within the species (Sales et al. 2014), as previously suggested from the morphometrics that distinguish populations from the North and South Atlantic (Zaleski et al. 2012).

Three loliginid squid species are sympatric and are commonly found in shelf waters off southern Brazil: the "tropical arrow squid" *Doryteuthis pleii* (Blainville, 1823), the "São Paulo squid" *Doryteuthis sanpaulensis* (Brakoniecki, 1984) and the "Atlantic brief squid" *Lolliguncula brevis* (Blainville, 1823) (Haimovici and Perez 1991a). *Doryteuthis pleii* is currently the best-studied species, mostly due to its importance for industrial and artisanal fisheries (e.g. Martins and Perez 2007, Arkhipkin et al. 2015). *Doryteuthis sanpaulensis* was intensively studied during the 1980s owing to its high abundance in the upwelling zone of Cabo Frio (22°S, Rio de Janeiro State) and along the coast of the state of Rio Grande do Sul (29–32.5°S) (Vidal et al. 2013). *Lolliguncula brevis* is the least-studied species thus far, perhaps because the species has little or no commercial value (Coelho et al. 2010, Perez and Zaleski 2013, Vaske et al. 2014).

Of interest, these three western Atlantic loliginids inhabit the so-called Paulista Province (22–33°S), a marine transitional biogeographical zone that shares biota with the Caribbean and Northern Patagonian Provinces, and has a relatively high level of endemism (Palacio 1977). The regional physiography and environmental conditions of the Paulista Province vary widely due to heterogeneous coastal features (including sedimentation patterns, oceanic sandy beaches, rocky shores, sheltered bays and bights, estuaries, mangroves, islands and reefs, and shelf width), oceanographic variability (circulation and water masses), climatic seasonality, and levels of biological productivity. This province is limited to the north and to the south by the 23.0 °C isotherm, and the area between 33–38°S acts as a barrier for the tropical fauna, mostly due temperature fluctuations (Palacio 1982).

The Paulista Province is regarded as the southernmost distributional limit for tropical species such as *D. pleii* and *L. brevis*, whereas the warm-tempered *D. sanpaulensis* extends its southern distributional range towards the Northern Patagonian Province and beyond (Haimovici and Perez 1991a, Vidal et al. 2013). Therefore, these squids are exposed to a suite of environmental conditions (and their fluctuations) in this transitional zone that certainly may affect their biology and ecology (Pierce et al. 1994, Pelc and Jackson 2008).

In Brazil, before the1970s, most studies on cephalopods involved their taxonomy and biogeography. From the 1950s on, the growing importance of squids as fishing targets in the coastal waters of southern Brazil created a demand for studies on their biology and ecology (Yesaki et al. 1976, Juanicó 1980, Gasalla et al. 2005). Even though our knowledge on the various important aspects of the biology of these cephalopods has increased ever since, some basic information on the population structure, reproduction, distribution, and trophic relationships of squids have remained unclear. In addition, there are taxonomic issues pertaining common species that are still unresolved (e.g. Brackoniecki 1984, Vecchione et al. 2005, Sales et al. 2013, 2014, 2017). Despite those gaps mentioned above, Brazilian research on the biology, ecology and fisheries of loliginid squids has developed considerably in the last four decades, mostly on populations off the southeast and south portions of the Brazilian coast (see Perez and Zaleski 2013, Vidal et al. 2013 and Arkhipkin et al. 2015 for comprehensive reviews).

Basic morphometric and sexual maturity studies on the tropical arrow squid, the São Paulo squid and the Atlantic brief squid were published in the late 1970s and early 1980s (Juanicó 1978, 1983). This study comes to complement these efforts by giving the first appraisal on loliginid squid biology and ecology and a first description of loliginid squid population structure, distribution in relation to environmental variables and geographic morphological variation off the southwestern Atlantic, between Rio de Janeiro and Mar del Plata (22–38°S).

MATERIAL AND METHODS

The study area ranges from 22°S (Cabo Frio, Brazil) to 38°S (Mar del Plata, Argentina), covering the entire Paulista Province plus part of the Northern Patagonic Province, encompassing the continental shelf from coastline to shelf break (200 m isobath) (Fig. 1). The seabed within this range is comprised mostly of soft sediments (Mahiques et al. 2004, Acha et al. 2008). Two distinct oceanographic systems are included: the South Brazil Bight (SBB, 22–29°S), and the shelf between Rio Grande do Sul and the northernmost part of the Argentinean shelf (29–38°S).

The SBB is characterized by seasonal interplay between onshore bottom intrusions of the South Atlantic Central Water (SACW) in spring-summer and surface equatorward penetration of the La Plata River Plume in autumn-winter (Pimenta et al. 2005, Castro et al. 2006). South to SBB, the oceanographic regime is influenced by the seasonal oscillation of the Subtropical Convergence (a mesoscale oceanographic feature originated from interactions between Brazil and Falkland currents), and the steady injections of fresh water due to continental drainage of the La Plata River and Patos Lagoon (Garcia 7997). Thus, biotic (biological productivity) and abiotic (temperature, salinity, pH, dissolved oxygen, etc.) variables vary widely throughout the study area.

The bulk of the loliginid samples used in this study were obtained during two oceanographic surveys conducted with the R/V Prof. W. Besnard between 23 and 30°S in September and November of 1975 (FAUNEC Cruisers III and IV) (Figs 1, 2). Squids were caught using a bottom otter trawl (16.7 m headrope fitted with floaters and stretched meshes of 60 mm in the net wings/body and 30 mm in the cod end). Hauls lasted 1.5 h at a velocity of 3.5 knots and were carried out between 10 and 132 m. Thirty-seven hauls were accomplished in the first leg (19–27 September 1975) and 41 in the second leg (07–27 November 1975), totalling 78 oceanographic stations (Fig. 2). Surface and bottom temperatures were obtained *in situ* with inversed thermometers fitted to Nansen bottles. Salinities of surface and bottom water samples were measured onboard using a Kahlsico

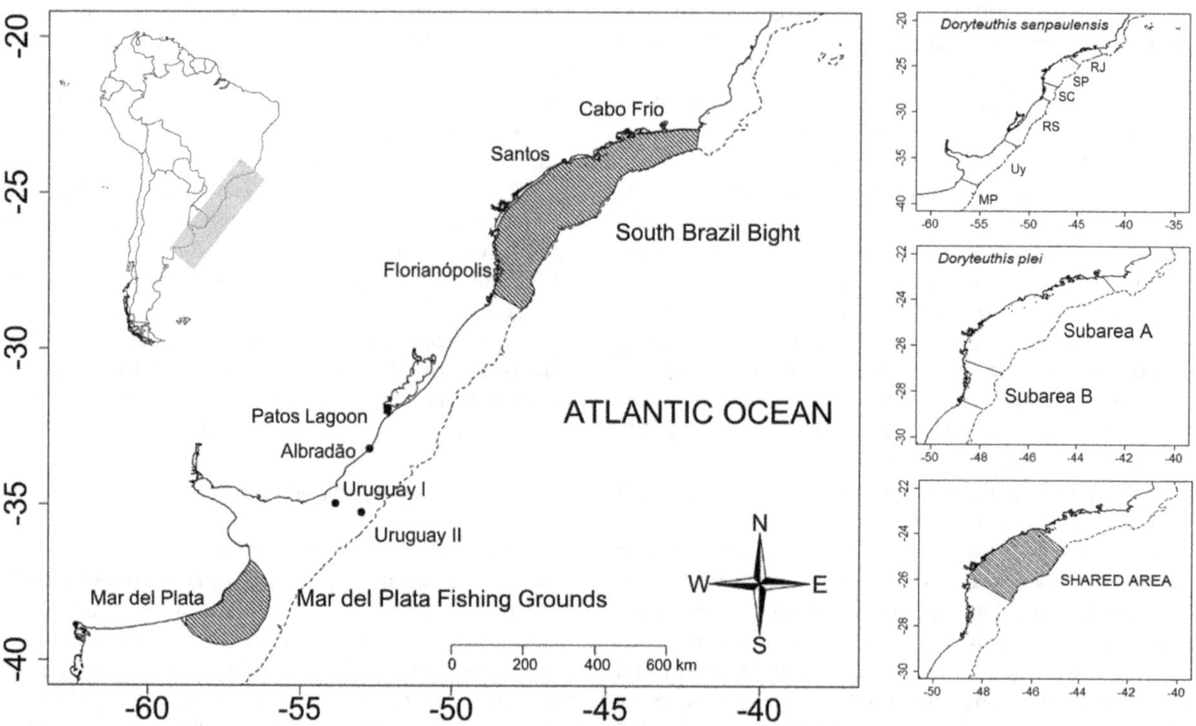

Figure 1. Study area, showing the areas and sites from where squid samples were obtained, also including the subareas used for the geographic differentiation in *Doryteuthis sanpaulensis* and *D. pleii* plus the shared area of co-occurrence of both species. The relative position of the study area on the South American Atlantic coast is boxed. RJ: Rio de Janeiro, SP: São Paulo, SC: Santa Catarina, RS: Rio Grande do Sul, Uy: Uruguay, MP: Mar del Plata. Subarea A: Cabo Frio (23°S) – Paranaguá (26°S), Subarea B: Paranaguá (26°S) – Santa Catarina Island (28°S). Latitudes and longitudes are decimal transformed. The 200 m isobath is showed (dashed line).

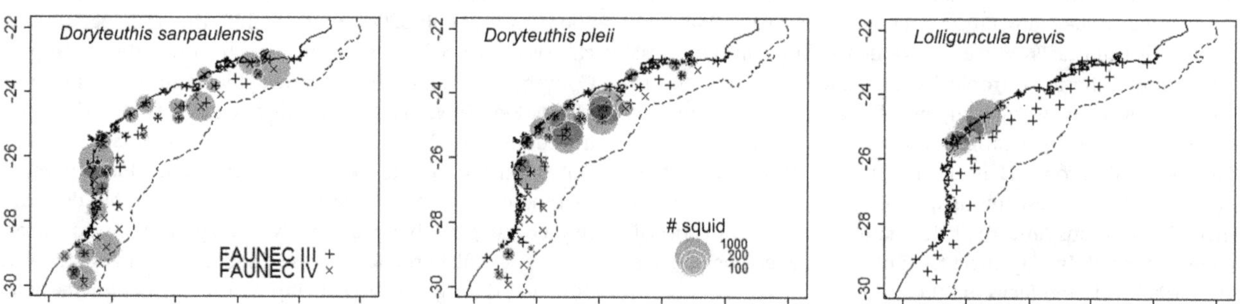

Figure 2. Maps of FAUNEC surveys from where squid where obtained. Abundance data were log-transformed for better visualization. Latitudes and longitudes are decimal transformed. The 200 m isobath is showed (dashed line).

Electrodeless Induction Salinometer (model S/N 21705). Squids were sorted from the remaining catch and were frozen following each haul. After landing, the samples were fixed and preserved in 10% buffered formalin solution in the laboratory.

The sampling effort was further supplemented with squids from Rio Grande do Sul, Uruguay and Mar del Plata (Argentina) coasts (29–38°S) (obtained from scientific surveys, commercial fisheries, and fish markets) to expand the study area domain (Fig. 1). Some few specimens were obtained from the malacological collection of the National Museum located in Rio de Janeiro. Details of the analysed material are given in the electronic Suppl. material 1.

In the laboratory, eight morphometric characters were selected following Cohen (1976) and Clarke (1962): mantle length (ML), fin length (FL), fin width (FW), gill length (GL, right only), raquis width (RW), gladius width (GW) and lower beak rostral length (LBRL). Measurements were taken from formalin-preserved specimens to the nearest millimetre (FL, FW, GL, RW, GW) and to the nearest 0.05 mm (LBRL), with the latter done under a stereoscopic microscope. All measurements (except ML) were taken with a Vernier calliper. The RW and GW were obtained by gently pressing the structure against a flat surface to standardize the lateral angle.

After dissection of the mantle, the sex and maturity stage of specimens were determined according to the five-step maturity scale proposed by Juanicó (1983): juvenile, (A) immature, (B) in maturation, (C) mature and (D) spent (Table 1). The presence of implanted spermatophores in the buccal receptacle (in *Doryteuthis* spp.) or in the inner wall of the mantle cavity, near the tip of gills (in *L. brevis*) was recorded as evidence of mating (Juanicó 1983).

We believe that tissue shrinkage in formalin-preserved squids that had been previously frozen is negligible and did not affect the measurements (Andriguetto and Haimovici 1988). To achieve standardized results , all morphometric measurements, sex identification and maturity scoring were carried out by one person, the second author (Pierce et al. 1994).

Morphometric analyses were conducted for samples collected during FAUNEC surveys. Samples from the Patos Lagoon obtained from the southern coast of Rio Grande do Sul were not used because those samples were collected with different gear and levels of the effort. However, squid samples from Uruguay and Mar del Plata were retained in the analysis to include the southernmost part of the study area (Fig. 1). Small samples ($n \leq 10$) were rejected regardless of their origin.

To address geographic diversification, Juanicó (1978) modified the discriminant method proposed by Royce (1957), originally developed to discriminate fish stocks. The Royce method compares the superposition of normal curves obtained from linear regressions between morphometric characters to discriminate distinct phenotypes, and thus different populations (Royce 1957). However, this method has three inherent problems: (1) it is valid only when the regression slopes are equal, (2) it cannot deal with different variances and (3) it cannot handle large sample sizes. Juanicó (1978) adapted the Royce method (hereafter referred to as the Modified Royce Method – MRM) to resolve these issues. Details on the MRM method are given in Suppl. material 2.

Since the MRM is a statistically based decision method that neither relies on significance levels nor on confidence intervals, it can be applied for paired comparisons among several samples. For the sake of simplicity, a grand mean of superposition ($\bar{\Omega}$) was calculated by averaging all average morphological characters ($\bar{S}$). This value, when multiplied by 100, yields the mean superposition in terms of percentage. Squid were considered morphologically distinct when $\bar{\Omega} < 70$.

For comparative purposes, the study area was divided into subareas according to the species investigated and the squid groups bounded within each subarea were analysed accordingly. Those divisions were based on latitudinal discontinuities in the catches of both species, and were named after geographic references on the coast (six subareas for *D. sanpaulensis*) and latitudinal intervals (two subareas for *D. pleii*) (Fig. 2).

The size structure of males, females and juveniles of the three loligind species was ascertained by visual analysis of ML histograms depicting the percentage of squids of each sex and juveniles in pooled 10 mm intervals. Histograms were made for each of the geographic divisions used in the morphometric analysis of *Doryteuthis* spp. (see Fig. 2). All *L. brevis* individuals obtained were analysed regardless their source due the small number of individuals sampled (see Suppl. material 1).

A hypothesis regarding the parity of sexes (i.e., 1 female: 1 male) was tested by means of a two-tailed chi-square test with Yates correction for continuity (Zar 2009). The sex ratio was further examined by subarea (*D. sanpaulensis* only) and depth. The latter analyses were carried out at specific depth strata

Table 1. Juanicó's (1983) loliginid squid maturity scale.

Stages	Females	Males
Juvenile	Reproductive system unrecognizable	Reproductive system unrecognizable
A – Immature	Nidamental glands distinguishable under dissection microscope. The ovary is small and filiform, and lacks a granulose structure	Spermatoforic organ and sac very small and distinguishable only under dissection microscope. The spermatoforic sac lacks spermatophores and testis resemble a thin membrane or a small lobule
B – In maturation	Nidamental glands small to large in size. Early stages differs from stage **A** by a bulged ovary with a clearly granulose structure containing small oocytes or ovules. Latter stages differ from stage **C** by an opaque whitish ovary with a reticular system on the walls of oocytes and ovules (particularly evident in *Doryteuthis sanpaulensis* and *Lolliguncula brevis*)	There are few and small spermatophores in the spermatoforic sac, the testis is clearly distinguishable to the naked eye
C – Mature	Ovary and oviduct full of oocytes, occupying half of the posterior mantle cavity. Nidamental glands swollen and firm. All or most ovules amber in color, semi-transparent, and lacking a reticular system. Mature ovules filling the oviduct, but in some females they may also fill the ovary	Spermatoforic sac full of completely developed spermatophores. Testis at its largest size. When males mature, they produce spermatophores continuously, making it difficult to tell them apart from spent individuals
D – Spent	Large females with flaccid or reduced nidamental glands. Ovary and oviduct flabby with few or no mature ovules, but with some immature ovules and tissue remain	–

for both *Doryteuthis* species, namely <31 m and >55 m for *D. sanpaulensis* and <30 m and >45 m for *D. pleii*. This approach was taken because there were few or no squid of each species caught at the intermediate depths between these bathymetric intervals. All statistical tests were considered significant at a significance level of 0.05.

Size-at-maturity (ML_{50}) was estimated for each species by fitting logistic regression models to the proportion of mature squid (Martins and Perez 2007). Regressions were fitted with generalized linear models (GLMs) using binomial family and logit link functions (Crawley 2007). The ML_{50} was estimated for five out the six subareas established for *D. sanpaulensis* (squid from Uruguay were not included due to decay problems in females), and for the whole SBB for the remaining two species.

The distribution of *D. sanpaulensis* and *D. pleii* along the SBB (FAUNEC surveys) was exanimated according to environmental variables recorded at the oceanographic stations, namely surface and bottom temperature, surface and bottom salinity, time of day and depth strata. Time of day was divided into four intervals of 6 h each (3:00 am – 9:00 am, 9:00 am – 3:00 pm, 3:00 pm – 9:00 pm and 9:00 pm – 3:00 am). Depth was also stratified into four bathymetric intervals: 10–40, 40–70, 70–100 and 100–130 m (10–30, 30–50, 50–70 and 70–90 m in the "shared area", see Fig. 2).

Spatial overlap among the three loliginid species over all samples were using the Fager affinity index (*I*) (Fager and McGowan 1963):

$$I = \left(\frac{J}{\sqrt{N_A \cdot N_B}} \right) - \left(\frac{1}{2\sqrt{N_B}} \right)$$

where *J* is the number of joint occurrences of both species, and N_A and N_B are the numbers of occurrences of species *A* and *B*, where N_B is larger or equal to N_A. Since this index quantifies the overlap of only two species at a time, calculations were done separately for each pair of each species. A minimum value of 0.5 indicates a positive relationship between the pair of species.

Food and feeding were examined only for *D. pleii* from FAUNEC samples. Stomachs were dissected, and their degrees of fullness were scored according the following criteria: (1) empty, (2) 50% full and (3) 100% full or distended. Since stomach contents were considerably digested and mostly unrecognizable (unpublished results), a refined description of the diet was not attempted. However, stomach fullness data were pooled by sex and ontogeny and analyzed in relation to the time of the day to describe the feeding schedule of the species.

RESULTS

Characterization of geographic groups in *Doryteuthis sanpaulensis*

Figures 3 and 4 show the standardized dimensions of each morphological measurement as a result of linear regressions against the ML. Differences can clearly distinguish female and male squid according to the geographic subarea. Despite those differences, however, these groups do not form a cline along the latitudinal gradient.

Among the female squid, individuals from the coast of Santa Catarina were the most distinct, with the largest FL, FW, RW, GW, LBRL and the smallest GL amongst all areas. Squid from Mar del Plata were also very distinct, since all measurements (except FL and LBRL) were the smallest among subareas. Squid from Rio de Janeiro, São Paulo and Rio Grande do Sul were similar, but individuals from São Paulo had the largest GL and those from Rio de Janeiro had the smallest LBRL (Fig. 3).

Morphological differences were also evident in males. Except for the FW, differences were noticeable for the remaining measurements. Squid from Rio de Janeiro had the smallest FL and LBRL, whereas those from São Paulo had the largest GL and a relatively large RW. Individuals from Santa Catarina had the smallest GL and RW, whereas Rio Grande do Sul squid had the smallest FL and largest RW, GW and LBRL. Uruguayan squid were fairly alike to Rio de Janeiro individuals, but differed in RW, GW and LBRL. Squid form Mar del Plata had the largest FL and FW (Fig. 4).

Geographic variation in *Doryteuthis* spp.

The two species of *Doryteuthis* presented distinguishable morphological differences along the latitudinal gradient. In *Doryteuthis pleii*, males and females from the two subareas overlapped >70%, implying that the latitudinal differences were subtle. In contrast, *D. sanpaulensis* can be separated into at least six distinguishable geographical groups (five in the case of female squid) (Fig. 5).

Doryteuthis sanpaulensis females inhabiting Santa Catarina and Mar del Plata diverged morphologically from their immediate counterparts to the north and south (São Paulo and Rio Grande do Sul, $\bar{\Omega}$ = 60.5–67) and to the north (Rio Grande do Sul coast, $\bar{\Omega}$ = 55.8) respectively. Squid from Mar del Plata also overlapped very little in their morphologies within all remaining subareas. Squid from São Paulo and Rio Grande do Sul were very similar ($\bar{\Omega}$ = 75.2). The smallest overlap between groups were between squids from Santa Catarina and Rio de Janeiro ($\bar{\Omega}$ = 49.3) (Fig. 5).

Doryteuthis sanpaulensis males from Santa Catarina also overlapped very little with their closest neighbours to the north and to the south ($\bar{\Omega}$ = 64.1–68.6) and, again, squid from the São Paulo and Rio Grande do Sul coasts were very similar ($\bar{\Omega}$ = 75.9). The Uruguayan group diverged strongly from their immediate northern neighbour (Rio Grande do Sul, $\bar{\Omega}$ = 53) and was very similar to squid from the northernmost geographical area (Rio de Janeiro, $\bar{\Omega}$ = 72.2) (Fig. 5).

Population size structure, maturity, sex ratio and size at maturity (ML_{50})

The size structure histograms constructed for *D. sanpaulensis* were plotted in three figures for better visualization (Figs 6–8). The size of individuals ranged from 14 to 114 mm ML. The

3

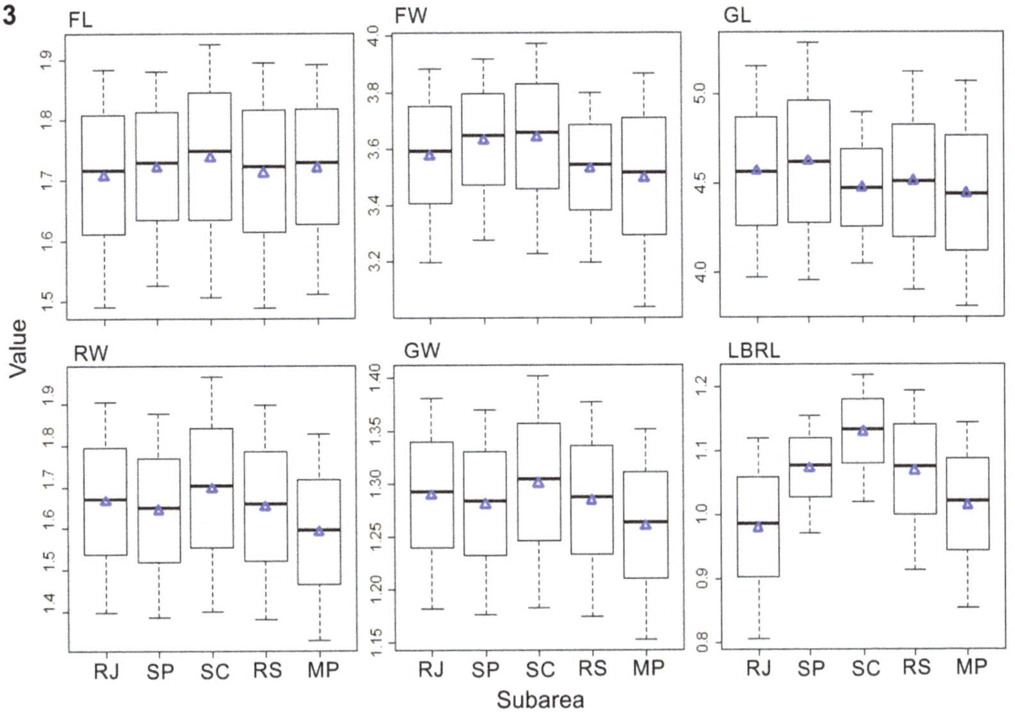

4

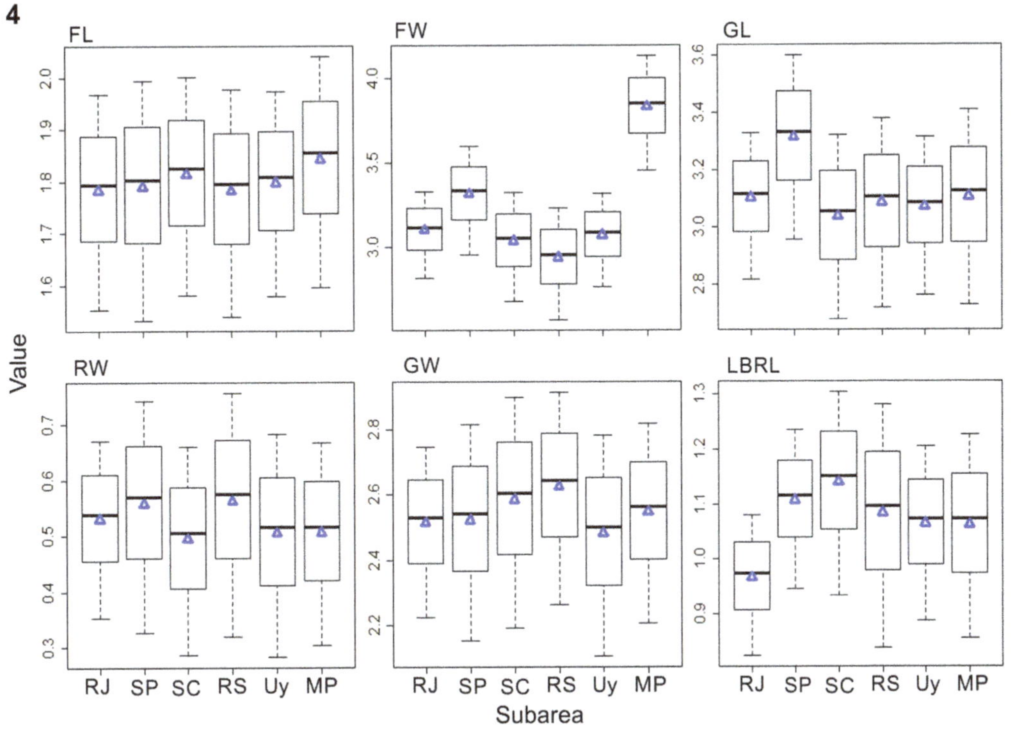

Figures 3–4. Box-Whisker plots of standardized values of attributes for the soft body parts of *Doryteuthis sanpaulensis* females (3) and males (4) from each region surveyed. Blue triangles represent mean values. FL: fin length, FW: fin width GL: gill length, RW: raquis width, GW: gladius width, LBRL: lower beak rostral length. RJ: Rio de Janeiro, SP: São Paulo, SC: Santa Catarina, RS: Rio Grande do Sul, Uy: Uruguay, MP: Mar del Plata.

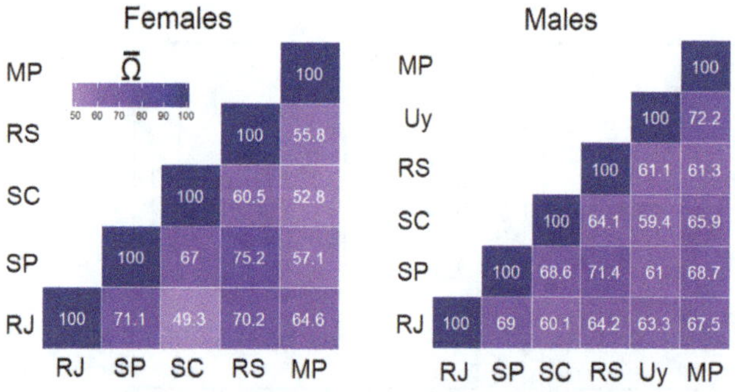

Figure 5. Heatmaps depicting the morphological overlap ($\bar{\Omega}$) among geographic areas in *Doryteuthis sanpaulensis*. RJ: Rio de Janeiro, SP: São Paulo, SC: Santa Catarina, RS: Rio Grande do Sul, Uy: Uruguay, MP: Mar del Plata.

size distribution of juvenile squid was unimodal regardless of the geographic area, but squid from the Patos Lagoon estuary were smaller than squid found north of it (which encompassed a similar size range) (Fig. 6).

The largest and smallest female squid were sampled south of 29°S (at Patos Lagoon estuary and off Uruguay, respectively), and in all cases, their size distribution was unimodal (Figs 7, 8). Females from Santa Catarina were larger than their male counterparts (Fig. 6).

In males, the average ML size and modes tended to decrease from Rio de Janeiro to Santa Catarina, increasing in squid sampled south of 29°S (Rio Grande do Sul) except at the Patos Lagoon estuary, where squid were <40 mm ML (Fig. 7). The largest average ML and modes were found off the Uruguayan coast, decreasing again in squid from Mar del Plata (Fig. 8). The size distribution of male squid from all remaining areas was unimodal, except for exemplars caught off Uruguay (Figs 6–8). Male-biased sex ratios were found in Rio de Janeiro (1 female: 2.2 males), São Paulo (1 female: 1.3 males), Santa Catarina (1 female: 1.9 males) and Mar del Plata (1 female: 1.4 males), and females outnumbered males in Uruguay (1 female: 0.5 males) ($\chi^2 = 2.7–19.6$, $p < 0.05$). Regarding bathymetry, the sex ratio was biased toward males in shallow (<31 m) bottoms (1 female: 1.4 males, $\chi^2 = 21.67$, $p = 3.24 \times 10^{-6}$) and toward females in deeper (>55 m) bottoms (1.3 females: 1 male, $\chi^2 = 4.06$, $p = 0.04$).

The proportion of *D. sanpaulensis* in each maturity stage in relation to the geographic subareas is given in Table 2. Immature females were the most frequent stage in all areas (≥74%), except off Uruguay (Table 2). Mature and spent females were frequent off Santa Catarina (42%) and off Uruguay (89.7%) (Table 2). Spent individuals were mostly recorded off the coast of Santa Catarina (Table 2). Maturing and mature males predominated in all subareas (>67.7%), except at the Patos Lagoon estuary (31.9%) (Table 2). Considering all subareas pooled, implanted

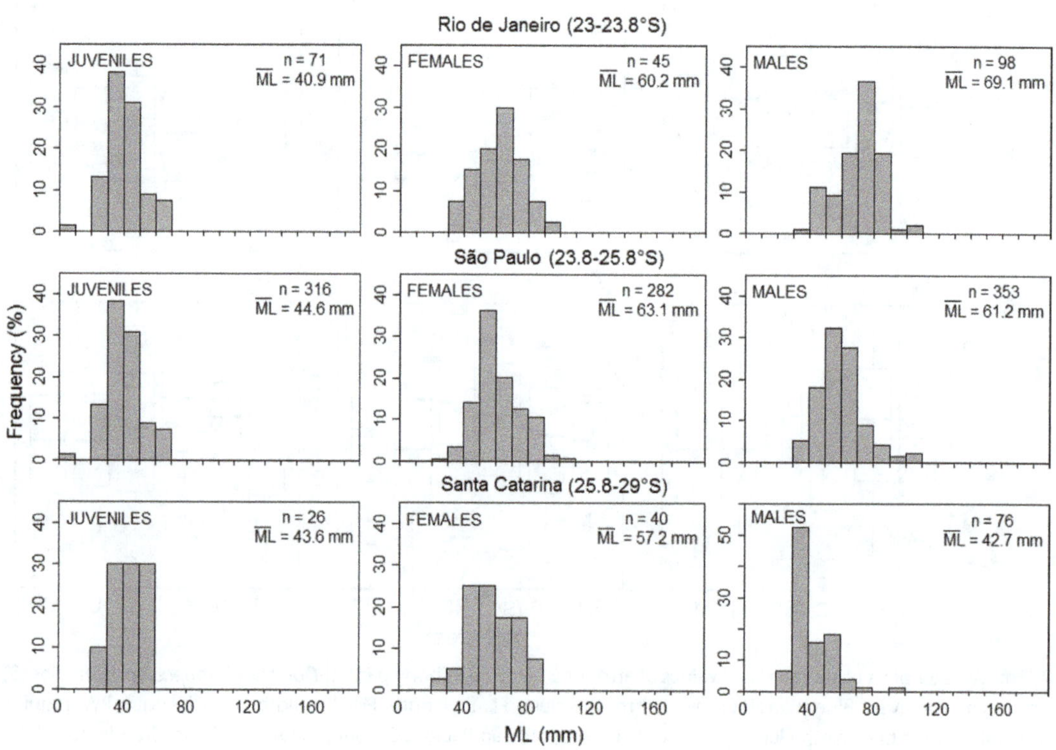

Figure 6. Size structure of *Doryteuthis sanpaulensis* according to the subareas nested within the SBB (22–29°S). Note de different y-axis.

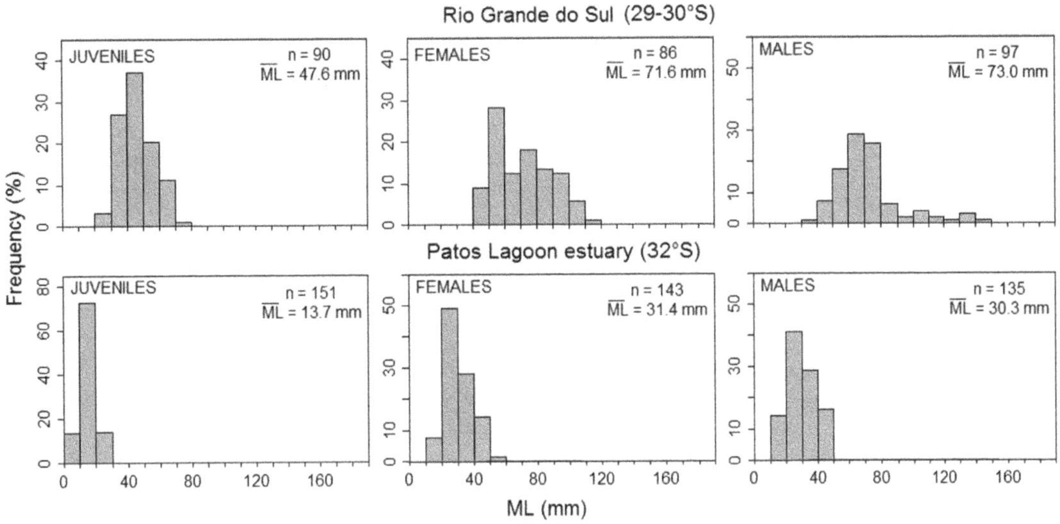

Figure 7. Size structure of *Doryteuthis sanpaulensis* according to the subareas nested within the Rio Grande do Sul coast (29–33.7°S). Note de different y-axis.

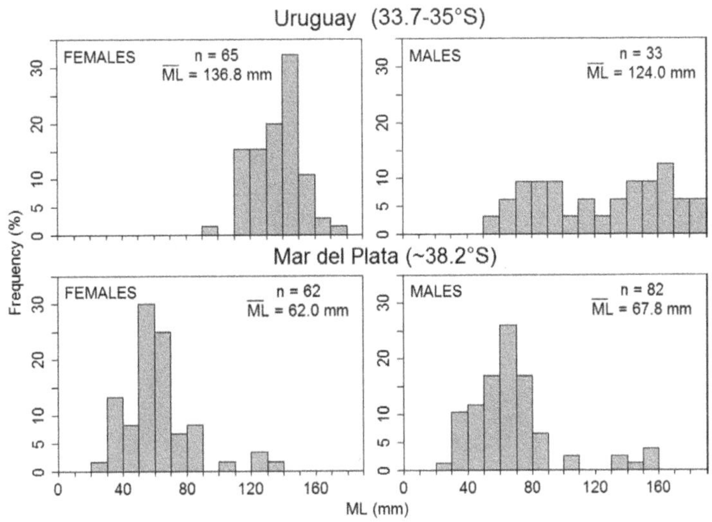

Figure 8. Size structure of *Doryteuthis sanpaulensis* according to the subareas nested within the Uruguayan-Argentinean coast (33.7–38.2°S).

spermatophores in the buccal receptacle occurred more frequently in mature females (n = 19), followed by spent (n = 6) and maturing squids (n = 1).

The size structure of *Doryteuthis pleii* is depicted in Fig. 9. The length of the mantle ranged from 30 to 247 mm. Since the morphological analysis had detected distinct groups (see above), two histograms were plotted separately (Fig. 9). Female squid between Cabo Frio and Paranaguá (23–26°S) had a normal size distribution with a distinct mode around 90–100 mm ML, whereas those off Santa Catarina (26–28°S) had a mode at smaller sizes (80–90 mm ML) (Fig. 9). The size distribution of male squid from both areas was skewed to the left, with a clear mode at 90–100 mm ML, although individuals between Cabo Frio and Paranaguá encompassed a wider size range (Fig. 9). The size range of juvenile squid was similar in both areas, but the mode obtained for squids sampled from Cabo Frio to Paranaguá was greater (Fig. 9).

In the two areas, the sex ratio of squid was significantly biased toward males (0.78 female: 1 male; $\chi^2 = 4.09$, p = 0.026). There were, however, no differences regarding bathymetric sex ratio (<30 m: 0.9 female: 1 male, $\chi^2 = 1.57$, p = 0.209 and >49 m: 0.88 female: 1 male, $\chi^2 = 1.10$, p = 0.292).

Of the 326 squid sampled for maturity assessment (143 females and 183 males), most (58% of females and 65.6% of

Table 2. Number of females and males of *Doryteuthis sanpaulensis* of each maturity stage per geographical area.

Geographic area	Maturity stages	Females		Males	
		n	%	n	%
Rio de Janeiro (23–23.8°S)	A	36	90.0	25	25.8
	B	1	2.5	21	21.6
	C	1	2.5	51	52.6
	D	2	5.0	–	–
	Total	40	100.0	97	100.0
São Paulo (23.8–25.8°S)	A	78	78.0	62	32.3
	B	5	5.0	58	30.2
	C	14	14.0	72	37.5
	D	3	3.0	–	–
	Total	100	100.0	192	100.0
Santa Catarina (25.8–29°S)	A	18	47.4	17	22.7
	B	4	10.5	23	30.7
	C	11	28.9	35	46.7
	D	5	13.2	–	–
	Total	38	100.0	75	100.0
Rio Grande do Sul (29–30°S)	A	57	74.0	39	43.8
	B	4	5.2	17	19.1
	C	13	16.9	33	37.1
	D	3	3.9	–	–
	Total	77	100.0	89	100.0
Patos Lagoon estuary (32°S)	A	118	82.5	92	68.1
	B	23	16.1	15	11.1
	C	2	1.4	28	20.7
	D	0	0.0	–	–
	Total	143	100.0	135	100.0
Uruguay (33.7–35°S)	A	2	2.9	2	6.5
	B	5	7.4	4	12.9
	C	60	88.2	25	80.6
	D	1	1.5	–	–
	Total	68	100.0	31	100.0
Mar del Plata (~38.2°S)	A	45	72.6	42.0	51.2
	B	12	19.4	29.0	35.4
	C	5	8.1	11.0	13.4
	D	0	0.0	–	–
	Total	62	100.0	82.0	100.0

males) were in an advanced stage of maturity (i.e. stages B, C and D). Nearly 44.1% of the females were mature, whereas 14% were maturing. Only 13.3% of the females were found spent. Immature squid comprised 34.4% and 42% of all males and females, respectively. No spent males were recorded. Mature females had the highest incidence of implanted spermatophores (n = 31), followed by spent individuals (n = 7) and maturing squid (n = 3).

The size of *Lolliguncula brevis* varied between 16 and 56 mm ML, females being larger than males, with a wider size interval. Two distinct modes were observed in female squid, 15–20 and 25–30 mm ML, whereas a well-defined mode at 25–30 mm ML (Fig. 10) was obtained for males. Only four females were found with implanted spermatophores, all of which were inside the mantle cavity.

All mated females were mature (stage C). Maturing and mature females (stages B and C) comprised little more than half of all individuals sampled (53.5%), followed by immature individuals (stage A) (43.6%). Only three spent females were found. Most mature females were caught in September 1979 at depths shallower than 18 m (FAUNEC III survey). In contrast, most male squid were mature (83.1%), and the bulk of them (n = 64) were caught in deeper waters (20–30 m). No spent males were found. The sex ratio was biased towards females (2.24 female: 1 male), although no significant difference was detected ($\chi^2 = 3.71$, p = 0.053).

The sizes of the three loliginids at maturity (ML_{50}) are given in Table 3. Note that calculations were made according to the availability of maturing/mature squid and, in some cases, not all stages were found in reasonable numbers. Size at maturity in female and male *D. sanpaulensis* varied according to the latitudinal gradient, with the largest and smallest ML_{50} found in Rio Grande do Sul and Santa Catarina, respectively, for both sexes

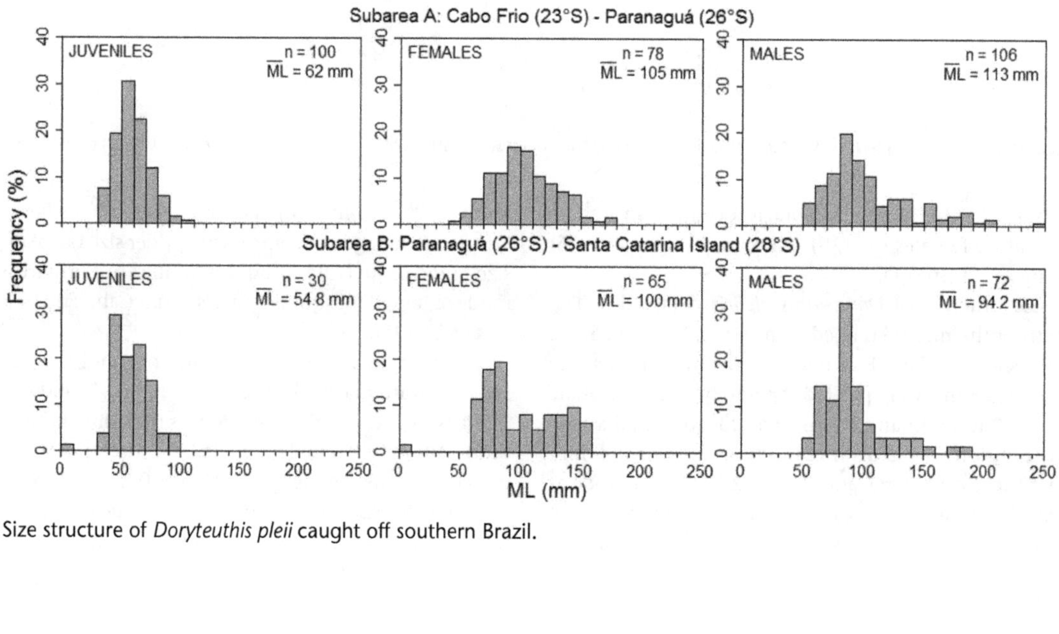

Figure 9. Size structure of *Doryteuthis pleii* caught off southern Brazil.

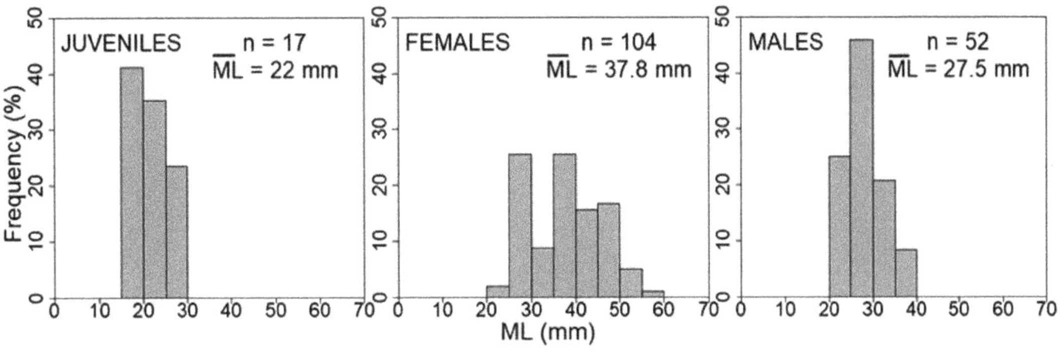

Figure 10. Size structure of *Lolliguncula brevis* caught off southern Brazil (24–26°S).

Table 3. Size-at-maturity (ML_{50}) of the three loliginid squids. Capital letters within brackets indicates the maturity stages used in the calculations (see Table 1). n = sample size.

Species	ML_{50} (mm)				Area
	n	females	n	males	
Doryteuthis pleii	22	110.16 (C + D)	30	86.85 (B + C)	22–29°S
Lolliguncula brevis	5	36.58 (B + C + D)	–	–	24–26°S
	–	–	10	66.66 (C)	Rio de Janeiro (23–23.8°S)
	8	81.56 (C + D)	16	56.55 (B + C)	São Paulo (23.8–25.8°S)
Doryteuthis sanpaulensis	9	61.11 (C + D)	10	56.55 (C)	Santa Catarina (25.8–29°S)
	9	82.52 (C + D)	14	78.17 (C)	Rio Grande do Sul (29–30°S)
	8	71.71(B +C + D)	8	66.66 (B + C)	Mar del Plata (~38.2°S)

(Table 3). Sizes at maturity of 104.66 mm ML and 86.85 mm ML were estimated for females and males of *D. pleii*, respectively, pooled for the whole study area (Table 3). A ML_{50} of 36.58 mm ML was estimated for female *L. brevis*, but in this case, sample size was very small (n = 5), which means that our results should be interpreted with caution. The number of mature males of *L. brevis* was not sufficient for a similar calculation.

Spatial and temporal patterns and response to environmental variables

The three loliginids were present on the SBB (22–29°S). *Doryteuthis sanpaulensis* and *D. pleii* were also recorded up to the southernmost portion of the study area (~38°S, Mar del Plata). The southernmost occurrence of *D. pleii* was based on four large male squid (>200 mm ML) sampled off Mar del Plata in 1972 and deposited at the malacological collection of the National Museum. *Lolliguncula brevis* occurred exclusively nearshore (<30 m) between 24 and 26°S (Fig. 2).

On the wider continental shelf (up to 130 m deep), *Doryteuthis* spp. had patchy distributions (Fig. 2), with mean catch rates of 21 ± 95.3 (SD) squid haul^{-1} for *D. sanpaulensis* (range 0–812 squid haul^{-1}) and 6.23 ± 23.2 (SD) squid haul^{-1} for *D. pleii* (range 0–167 squid haul^{-1}).

An analysis of the catches by depth showed that *D. sanpaulensis* was disproportionally more abundant in shallower depths (<40 m, 1210 squid caught, catch rate: 39 squid haul^{-1}) and moderately abundant at the deepest depth strata (100–130 m), although the highest catch rate was found between 70 and 100 m deep (993 squid haul^{-1}) (Fig. 11). *Doryteuthis pleii* was abundant on midshelf (395 squid caught, catch rate: 15.8 squid haul^{-1}), and the highest catch rate was recorded in the shallowest depth strata (<40 m, 369 squid haul^{-1}) (Fig. 11).

Regarding temporal occurrence, *D. sanpaulensis* was more abundant during mid-afternoon and in the first third of the night (between 3:00 pm and 9:00 pm). In contrast, the highest abundance of *D. pleii* was recorded between late night and the first third of the morning (3:00–9:00 am) (Fig. 11).

Doryteuthis sanpaulensis occurred more often in colder waters (T <19.5 °C), a pattern particularly evident for both bottom and surface temperatures. Likewise, the species was frequent in fresher waters (S <34.5), in both surface and the bottom, although it was found in high bottom salinities as well (Fig. 12). Interestingly, a reasonable number of *D. sanpaulensis* (n = 311) was also caught inside the Patos Lagoon estuary.

The patterns were less clear for *Doryteuthis pleii*, but apparently the species commonly occurs in warmer surface temperatures (T >21.5 °C), sharing with *D. sanpaulensis* the same bottom temperature (i.e. T ~19.0 °C). The response of *D. pleii* to water salinity signatures was also poorly discernible, but the species was apparently more frequent in saltier water (Fig. 12).

Doryteuthis sanpaulensis, *D. pleii* and *L. brevis* occurred together in the same haul only in four oceanographic stations

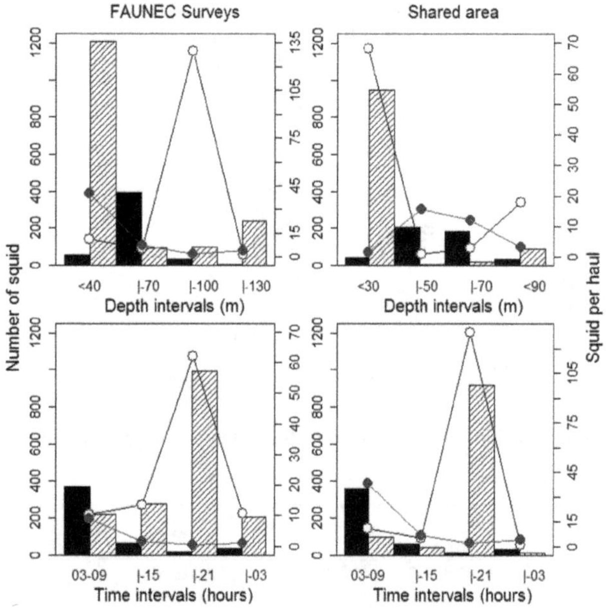

Figure 11. Total number and catch rates of *Doryteuthis sanpaulensis* and *D. pleii* collected in different depths and time of day during FAUNEC surveys and within the shared area off southern Brazil. Both metrics were calculated for stations where at least one of the two species was present (FAUNEC surveys) or where both species were present (shared area). Bars: number of squid. Circles: squid per haul. Crosshatched bars and empty circles: *D. sanpaulensis*. Black bars and dark grey circles: *D. pleii*.

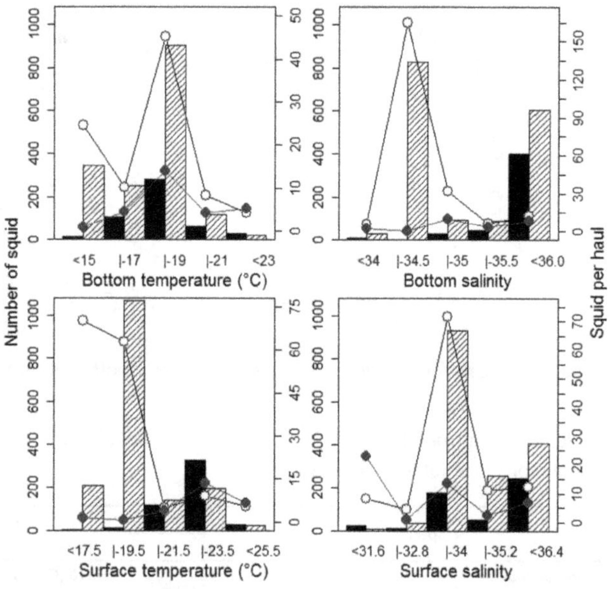

Figure 12. Total number and catch rates of *Doryteuthis sanpaulensis* and *D. pleii* collected in different temperatures and salinities during FAUNEC surveys off southern Brazil. Both metrics were calculated for stations where at least one of the two species was present. Bars: number of squid. Circles: squid per haul. Crosshatched bars and empty circles: *D. sanpaulensis*. Black bars and dark grey circles: *D. pleii*.

conducted at a depth of <15 m. The Fager's affinity indexes calculated for *D. sanpaulensis–L. brevis* and *D. pleii–L. brevis* pairs were 0.28 and 0.31, respectively, indicating a poor association between *L. brevis* and the two *Doryteuthis* species.

Spatial and temporal patterns and response to environmental variables within the shared area

The two species of *Doryteuthis* co-occurred in only 12 out of the 78 oceanographic stations during the FAUNEC III and IV surveys. Those 12 stations were between 23.5 and 27°S ("shared area", comprising 32 oceanographic stations, see Fig. 2). A Fager affinity index of 0.7 was calculated for the two co-ocurrying species, suggesting a strong positive relationship between them. However, we believe that it is possible that this index does not reflect a strong affinity, since it was common to collect many exemplars of one species and only a few of the other in the same batch (Fig. 13).

Catches by depth strata within the shared area mirrored the general pattern for both species (Fig. 11). Catches by time of day differed between the two species: *Doryteuthis pleii* peaked between 3:00 am and 9:00 am (361 squid caught, 40 squid haul^{-1}), whereas *D. sanpaulensis* took peaked between 3:00 pm and 9:00 pm (915 squid caught, 130 squid haul^{-1}) (Fig. 11).

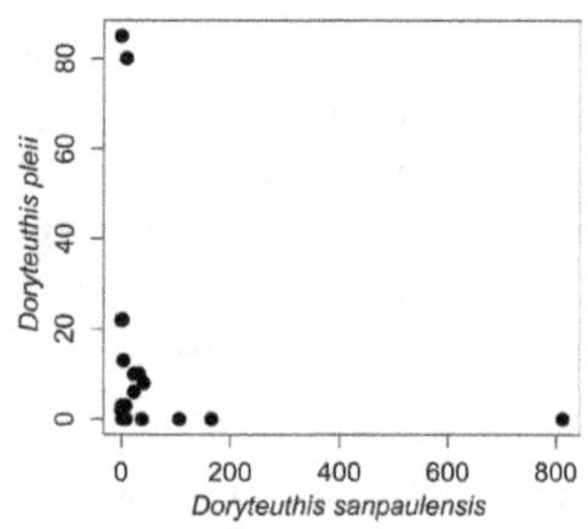

Figure 13. Catches of *Doryteuthis sanpaulensis* and *D. pleii* within the shared area.

Both species were caught in very similar environmental conditions within the shared area, except for the fact that *D. sanpaulensis* was caught in a slightly lower average salinity (surface: 34.90, bottom: 33.98) than *D. pleii* (surface: 35.15, bottom: 33.99). The temperature at the time of capture was the same for the two species (surface: 21.6 °C, bottom: 18.5 °C).

Feeding schedule in *Doryteuthis pleii*

The feeding activity of *D. pleii* was clearly concentrated between late evening and mid-morning (3:00–9:00 am) regardless of the stage of maturity stage and sex of the specimen. There was also an elevated number of empty stomachs found for the same period (Fig. 14).

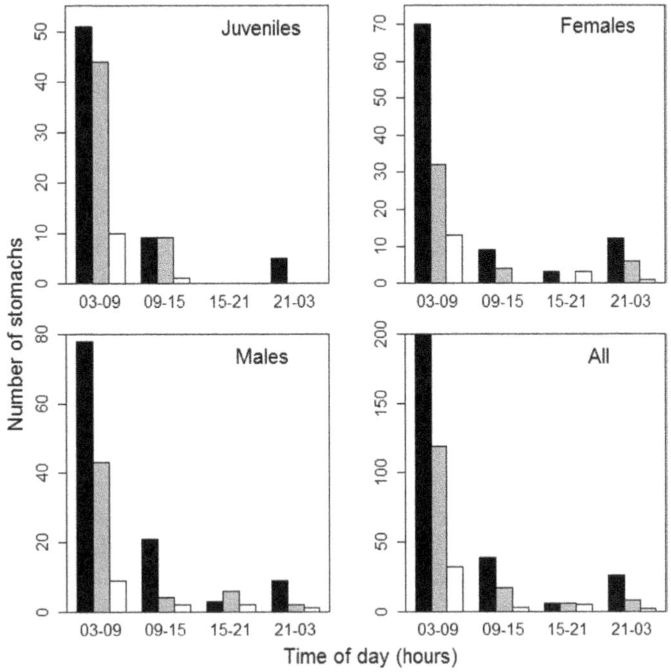

Figure 14. Stomach fullness frequency of *Doryteuthis pleii* collected during FAUNEC surveys per period of the day. Black bars: empty stomachs, grey bars: 50% full stomachs, white bars: 100% or distended stomachs.

DISCUSSION

This study has pioneered squid biology and ecology research in Brazil. For the first time ever, the population biology and ecology of the three most common shelf loliginids (*D. sanpaulensis*, *D. pleii* and *L. brevis*) were surveyed, along with early data on their spatial and bathymetric distributions and responses to environmental variables. The circadian levels of activity were described for the two species of *Doryteuthis*, along with the feeding schedule for *D. pleii*. In addition, a discriminant morphological analysis identified a latitudinal phenotypic variability for *D. sanpaulensis* and *D. pleii*, more remarkable in the first species. Hence, this survey may be regarded as a framework for all subsequent studies on loliginid squid biology and ecology off the southern Brazilian coast (e.g., Perez and Zaleski 2013, Vidal et al. 2013, Arkhipkin et al. 2015).

Geographic variation in *Doryteuthis* spp.

Ideally, studies on morphologic variations in squid populations must be conducted with samples collected simultaneously, to avoid mixed stock effects (Pierce et al. 1994). Even though this recommendation was followed for FAUNEC samples, we were not able to take simultaneous samples from Uruguay and Argentina. Despite the fact that Uruguayan and Argentinean squid were sampled in the same season (austral summer), their samples were obtained on different years, which may have introduced temporal biases to our analysis. Notwithstanding the latter, it was most instructive to keep these samples for comparison purposes, as they expanded our study area southward in nearly six degrees of latitude.

In addition to the limitations evoked above, it is also important to recognize that there were some other issues with our data that deserve mention. Most of our samples were obtained during a very narrow window of time, and since both *Doryteuthis* species produce at least two cohorts per year (Vidal et al. 2013, Arkhipkin et al. 2015), it is unlikely that we ended up with a representative set of environmentally driven morphological variability.

Cryptic squid species are difficult to separate on the basis of morphological analyses only (Herke and Foltz 2002) because squid display great phenotypic plasticity in response to environmental conditions (Pelc and Jackson 2008). Furthermore, even though it is unlikely that the morphological differences found in the squid from FAUNEC result from an artefact in the selectivity of the fishing gear (e.g. Martins and Perez 2007), since the same net was employed in all surveys, the same may not be the case of the samples collected from Uruguay and Argentina. This certainly must be considered when interpreting the results.

In our data, latitudinal morphological variability was very evident in *D. sanpaulensis*, as the dimensions of both soft and hard parts diverged significantly among geographically separated groups. This suggest that *D. sanpaulensis* populations experience different environmental conditions throughout the sampled area. This is in line with the restricted loliginid larval dispersal potential along the SBB (see Martins et al. 2014), which would expose paralarvae and early juveniles to distinct biotic and abiotic conditions. Environmental variability results in different growth patterns, and thus different final dimensions of body parts as squids mature (Pelc and Jackson 2008). In addition, the morphological variability among geographical groups suggests a strong homing behavior in adult squid.

The morphological variability of body parts along with the morphological overlap analysis and size at maturity data found for *D. sanpaulensis* in the study area suggests that there are at least six distinct groups scattered between Cabo Frio (23°S) and Mar del Plata (37.8°S). On the SBB (23–29°S), squid inhabiting ~26–28°S (Santa Catarina coast) had shorter mantle lengths and matured at smaller sizes than squid to the north and to the south of this region. In addition, males were significantly smaller than females, in contrast with the typical sexual dimorphism described for this species (i.e., males larger than females; Vidal et al. 2013). Actually, the squid from the Santa Catarina coast differed from others in almost all morphological characters investigated.

Doryteuthis sanpaulensis dwelling to the north and to the south of the coast of Santa Catarina were very similar in morphology. This suggests that these geographically divided groups experienced similar environmental conditions, at least during some part of their lifetime. Latitudinal gradients in water temperature may be a major driver of phenotypic divergence in *D. sanpaulensis*, since temperature has a direct effect in growth-related parameters (Forsythe 2004). However, variability in other environmental conditions, such as dissolved oxygen and food availability, may also play a role (Jackson and Moltschaniwskyj 2001, van der Vyver 2016).

The most distinct geographic group of *D. pleii* was also found off the coast of Santa Catarina. Nevertheless, the species had just two distinguishable groups, and these were much less distinct than those of the co-generic species. Thus, it is suggested that the phenotype of *D. pleii* in Santa Catarina also results from local environmental conditions. However, due to the subtle morphological differences between those two groups, we hypothesise that *D. pleii* individuals move more between latitudes than *D. sanpaulensis*, perhaps "homogenizing" the environmental conditions experienced during the squid's lifetime, since they actively track suitable temperature fields (see Haimovici and Perez 1991b). In addition, these latitudinal movements may promote strong gene flow among different stocks at the SBB. This hypothetical scenario is consistent with a recent genetic analysis that failed to find any relevant genotypic differences between squid from this area and their counterparts on the rest of the South American Atlantic coast (Sales et al. 2017).

Spatial patterns resulting from environmental variability

In our data, *Doryteuthis* spp. have specific thermal preferences. *Doryteuthis sanpaulensis* was found more often in colder waters, whereas *D. pleii* was associated with tropical (warmer) waters. The two species occurred together only when the water was thermally stratified (when both cold and warm water were present in the same location but in different strata) in the shared area. This ability to actively select water strata that are suitable for them has allowed these two species of squid to exploit locations where they would not normally occur.

We found small and mostly immature representatives of *D. sanpaulensis* at the Patos Lagoon estuary in January, February and March 1979. The presence of these squid inside the estuary during three consecutive summer months does not seem to be by chance. It is worthy of note that *D. sanpaulensis* had not been recorded in estuaries (at least in Brazilian waters), and to the best of our knowledge, this is the first record of a loliginid squid other than *L. brevis* (which has well-described physiologic mechanisms to tolerate low salinities, see Hendrix et al. 1981) inside an estuary.

These small *D. sanpaulensis* may enter the estuary during events of physiologically tolerable environmental conditions to exploit the food-rich, productive estuarine areas. This could be related to landward intrusion of the salt edge into the upper estuary, a near-bottom, high-salinity tongue of seawater, which is particularly strong during austral summer months, when the freshwater discharge is less (Möller et al. 2001). Off the coast Rio de Janeiro, juvenile *D. sanpaulensis* were found entering the warm Guanabara Bay (22°43'S), during flood tide, tracking bottom landward intrusions of cold SACW. In this case, squid seems to use the physiologically suitable near-bottom temperature field to exploit the abundant food resources inside this eutrophized tropical coastal bay (Moraes and Lavrado 2017).

When this study was conducted, our data extended the southernmost distribution range of *D. pleii* (29°S, Palacio 1977) by at least nine degrees of latitude, owing to some specimens obtained at ~38°S (Mar del Plata, Argentina). In fact, latter studies found that *D. pleii* occurs during spring and summer to the south of its normal range over the shelf break between 29 and 34°S, tracking the warm, fast flowing superficial Brazil Current (BC) (Haimovici and Perez 1991b). Recently, *D. pleii* was recorded as far as Península Valdez, Argentina (42°S) (Augusto Crespi-Abril 2017, personal communication). Thus, it is clear that stronger poleward BC penetration in some years may carry vagrant individuals toward higher latitudes during austral warmer seasons.

Competition avoidance between *Doryteuthis* spp.

According to our results, the two species of *Doryteuthis*, which are morphologically similar, employ several mechanisms to avoid direct competition in the wider continental shelf. For instance, the diel activity pattern was inverse, with *D. pleii* being more active during daylight, whereas *D. sanpaulensis* was more active at night. Furthermore, both species had clearly different thermal preferences: *D. sanpaulensis* was found more often in colder temperatures than *D. pleii*. In addition, the most common depths where each species was found were different. The feeding schedule in *D. pleii* peaks between late night and early morning, whereas *D. sanpaulensis* feeds more intensively between mid-morning and late afternoon (Andriguetto and Haimovici 1997). This temporal mismatch between feeding activity schedules may be regarded as an additional mechanism that prevents direct competition.

Concluding remarks

This study revisits data from a time when climate change had not yet been perceived as an agent of significant environmental variability in the ocean (Abran et al. 2016). Thus, the information reported here may be regarded as a "snapshot" on the ecology of sympatric squids in a colder and less acidic marine environment (see Nagelkerke and Connell 2015). It seems that the abundance of the commonest loliginid squids off the southern Brazilian coast has shifted in the last few decades. *Doryteuthis pleii* apparently replaced *D. sanpaulensis* as the most abundant squid in some areas where the latter species used to be dominant, such as off the coast of northern Rio de Janeiro (~23°S) (e.g., Costa et al. 2017). Whether this apparent shift was caused by climate change or not is currently unknown. Accordingly, our results may be particularly important for a better understanding of more recent studies on the biology and ecology of loliginid squids in the southwestern Atlantic Ocean and elsewhere.

ACKNOWLEDGEMENTS

The FAUNEC Project (Brazilian Continental Shelf Nectonic Fauna) was funded by the Conselho Nacional de Desenvolvimento Científico e Tecnológico (CNPq). MJ acknowledges a PhD scholarship granted by the Brazilian Ministry of Foreign Affairs as part of the cultural exchange program celebrated between Brazil and Uruguay in the late 1970s. We extend our gratitude to José Angel Alvarez Perez (UNIVALI) and an anonymous referee for the careful observations and thoughtful comments and suggestions that have improved greatly the final version of the manuscript.

LITERATURE CITED

Abran NJ, McGregor HV, Tierney JE, Evans MN, McKay NP, Kaufman DS (2016) Early onset of industrial-era warming across the oceans and continents. Nature 536: 411–418. https://doi.org/10.1038/nature19082

Acha EM, Mianzan H, Guerrero R, Carreto J, Giberto D, Montoya N, Carignan M (2008) An overview of physical and ecological processes in the Rio de la Plata Estuary. Continental Shelf Research 28: 1579–1588. https://doi.org/10.1016/j.csr.2007.01.031

Andriguetto JM, Haimovici M (1988) Effects of fixation and preservation methods on the morphology of a loliginid squid (Cephalopoda, Myopsida). American Malacological Bulletin 6: 213–217.

Andriguetto JM, Haimovici M (1997) Feeding habits of *Loligo sanpaulensis* Brakoniecki, 1984 (Cephalopoda, Loliginidae) in southern Brazil. Nerítica 11: 63–76.

Arkhipkin AI, Rodhouse PGK, Pierce GJ, Sauer WHH, Sakai M, Allcock L, Arguelles J, Bower JR, Castillo G, Ceriola L, Chen C-S, Chen X, Diaz-Santana M, Downey N, Gonzalez AF, Amores JG, Green CP, Guerra A, Hendrickson LC, Ibañez C, Ito K, Jereb P, Kato Y, Katugin ON, Kawano M, Kidokoro H, Kulik VV, Laptikhovsky VV, Lipiński MR, Liu B, Mariategui L, Marin W, Medina A, Miki K, Miyahara K, Moltschaniwskyj N, Moustahfid H, Nabhitabhata J, Nanjo N, Nigmatullin ChM, Ohtani T, Pecl G, Perez JAA, Piatkowski U, Saikliang P, Salinas-Zavala CA, Steer M, Tian Y, Ueta Y, Vijai D, Wakabayashi T, Yamaguchi T, Yamashiro C, Yamashita N, Zeidberg LD (2015) World squid fisheries. Reviews in Fisheries Science and Aquaculture 23: 29–252. https://doi.org/10.1080/23308249.2015.1026226

Brackoniecki TF (1984) A full description of *Loligo sanpaulensis* (Brackoniecki 1984) and a redescription of *Loligo gahi* (d'Orbigny 1835), two species from the southwest Atlantic. Bulletin of Marine Science 34: 435–448.

Castro BM, Lorenzzetti JA, Silveira ICA, Miranda LB (2006) Estrutura termohalina e circulação na região entre o Cabo de São Tomé (RJ) e o Chuí (RS) In: Rossi-Wongtschowski CLDB, Madureira LAS-P (Coords) O ambiente oceanográfico da plataforma continental e do talude na região sudeste-sul do Brasil. EDUSP, São Paulo, 11–20.

Clarke M (1962) The identification of cephalopod "beaks" and the relationships between beak size and total body weight. Bulletin of the British Museum (Natural History) Zoology 8: 421–497.

Coelho LI, Muto EY, Marian JEAR, Soares LSH (2010) Contribuição ao conhecimento da dieta, atividade alimentar e reprodução de *Lolliguncula brevis* (Blainville, 1823) na região costeira de Santos (Estado de São Paulo). Boletim do Instituto de Pesca 36: 225–236.

Cohen A (1976) The systematics and distribution of *Loligo* (Cephalopoda, Myopsida) in the western tropical Atlantic, with descriptions of two new species. Malacologia 15: 299–367.

Costa PAS, Braga AC, Martins AS, Haimovici M, Mincarone MM (2017) Padrões de distribuição das assembleias de peixes e cefalópodes demersais na plataforma e talude continental da Bacia de Campos. In: Curbelo-Fernandez MP, Braga AC (Eds) Comunidades Demersais e Bioconstrutores: caracterização ambiental regional da Bacia de Campos, Atlântico Sudoeste. Elsevier, Rio de Janeiro, 87–110. ttps://doi.org/10.1016/B978-85-352-7295-6.50004-X

Crawley MJ (2007) The R Book. John Wiley and Sons Ltd, West Sussex, 1076 pp.

Fager EW, McGowan JA (1963) Zooplankton species groups in the North Pacific. Science 140: 453–461. https://doi.org/10.1126/science.140.3566.453

Forsythe JW (2004) Accounting for the effect of temperature on squid growth in nature: from hypothesis to practice. Marine and Freshwater Research 55: 331–339. https://doi.org/10.1071/MF03146

Garcia CAE (1997) Hydrographic Characteristics In: Seeliger U, Odebrecht C, Castello JP (Eds) Subtropical Convergence Environments: The Coast and Sea in the Southwestern Atlantic. Springer-Verlag, Berlin, 18–19.

Gasalla MA, Postuma FA, Tomás ARG (2005) Captura de lulas (Mollusca: Cephalopoda) pela pesca industrial desembarcada em Santos: comparação após 4 décadas. Brazilian Journal of Aquatic Science and Technology 9: 5–8. https://doi.org/10.14210/bjast.v9n2.p5-8

Haimovici M, Perez JAA (1991a) Coastal cephalopod fauna of southern Brazil. Bulletin of Marine Science 49: 221–230.

Haimovici M, Perez JAA (1991b) Abundância e distribuição de cefalópodes em cruzeiros de prospecção pesqueira demersal na plataforma externa e talude continental do sul do Brasil. Atlântica 13: 189–200.

Hendrix Jr JP, Hulet WH, Greenberg MJ (1981) Salinity tolerance and the responses to hypoosmotic stress of the bay squid *Lolliguncula brevis*, an euryhaline cephalopod mollusc. Comparative Biochemistry and Physiology – Part A: Molecular and Integrative Physiology 69: 641–648. https://doi.org/10.1016/0300-9629(81)90150-X

Herke SW, Foltz DW (2002) Phylogeography of two squid (*Loligo pealei* and *L. plei*) in the Gulf of Mexico and northwestern Atlantic Ocean. Marine Biology 140: 103–115. https://doi.org/10.1007/s002270100680

Jackson GD, Moltschaniwskyj NA (2001) Temporal variation in growth rates and reproductive parameters in the small near-shore tropical squid *Loliolus noctiluca*: is cooler better? Marine Ecology Progress Series 218: 167–177. https://doi.org/10.3354/meps218167

Jereb P, Vecchione M, Roper CFE (2010) Family Loliginidae In: Jereb P, Roper CFE (Eds) Cephalopods of the world: an annotated and illustrated catalogue of cephalopod species kwon to date. FAO, Rome, FAO Species Catalogue for Fishery Purposes 4, vol. 2, 38–117.

Johntson J (1977) Métodos econométricos. Atlas Editora, São Paulo, 531 pp.

Juanicó MR (1978) Analisis estadístico de caracteres morphometricos: comentários sobre el Método de Sobreposicion de Royce. In: V Simpósio Latinoamericano de Oceanografia Biológica, São Paulo, Instituto Oceanográfico, Universidade de São Paulo, 33–34.

Juanicó MR (1980) Developments in South American squid fisheries. Marine Fisheries Review 12: 10–14.

Juanicó MR (1983) Squid maturity scales for population analysis In: Caddy JF (Ed.) Advances in Assessment of World Cephalopod Resources. FAO, Rome, FAO Fishery Technical Paper 231, 341–378.

Mahiques MM, Tessler MG, Ciotti AM, Silveira ICA, Sousa SHM, Figueira CL, Tassinari CG (2004) Hydrodynamically-driven patterns of recent sedimentation in the shelf and upper slope off southeast Brazil. Continental Shelf Research 24: 1685–1697. https://doi.org/10.1016/j.csr.2004.05.013

Martins RS, Perez JAA (2007) The ecology of loliginid squid in shallow waters around Santa Catarina Island, southern Brazil. Bulletin of Marine Science 80: 125–145.

Martins RS, Camargo R, Gasalla MA (2014) The São Paulo shelf (SE Brazil) as a nursery ground for *Doryteuthis plei* (Blainville, 1823) (Cephalopoda, Loliginidae) paralarvae: a Lagrangian particle-tracking Individual-Based Model approach. Hydrobiologia 725: 57–68. https://doi.org/10.1007/s10750-013-1519-4

Möller Jr OO, Castaing O, Salomon J-P, Lazure P (2001) The influence of local and non-local forcing effects on the subtidal circulation of Patos Lagoon. Estuaries 24: 297–311. https://doi.org/10.2307/1352953

Moraes MCM, Lavrado HP (2017) Distribution of loliginid squids in a eutrophicated tropical coastal bay. Marine Biology Research 13: 330–341. https://doi.org/10.1080/17451000.2016.1262041

Nagelkerken I, Connell, SD (2015) Global alteration of ocean ecosystem functioning due to increasing human CO_2 emissions. Proceedings of the National Academy of Sciences of the United States of America 112: 13272–13277. https://doi.org/10.1073/pnas.1510856112

Palacio FJ (1977) A study of cephalopods from Brazil with a review of Brazilian zoogeography. PhD Thesis, University of Miami, Miami. https://goo.gl/VWd3mC

Palacio FJ (1982) Revisón zoogeográfica marina del sur del Brasil. Boletim do Instituto Oceanográfico 31: 69–92. https://doi.org/10.1590/S0373-55241982000100006

Pelc GT, Jackson GD (2008) The potential impacts of climate change on inshore squid: biology, ecology and fisheries. Reviews in Fish Biology and Fisheries 18: 373–385. https://doi.org/10.1007/s11160-007-9077-3

Perez JAA, Zaleski T (2013) *Lolliguncula brevis*, Western Atlantic Brief Squid. In: Rosa R, O'Dor RK, Pierce G (Eds) Advances in Squid Biology, Ecology and Fisheries. Part I: Myopsid Squids. Nova Science Publishers, New York, 301–322.

Pierce GJ, Thorpe RS, Hastie LC, Brierley AS, Guerra A, Boyle PR, Jamieson R, Ávila P (1994) Geographic variation in *Loligo forbesi* in the Northeast Atlantic Ocean: analysis of morphometric data and tests of causal hypotheses. Marine Biology 119: 541–547. https://doi.org/10.1007/BF00354316

Pimenta FM, Campos EJD, Miller J, Piola ARJ (2005) A numerical study of the Plata river plume along the Southeastern South American Continental Shelf. Brazilian Journal of Oceanography 53: 129–146. https://doi.org/10.1590/S1679-87592005000200004

Royce W (1957) Statistical comparison of morphological data. US Fish and Wildlife Service, Special Scientific Report – Fisheries, 208: 7–28. https://spo.nmfs.noaa.gov/SSRF/SSRF208.pdf

Sales JBL, Shaw PW, Haimovici M, Markaida U, Cunha DB, Ready J, Figueiredo-Ready WM, Schneider H, Sampaio I (2013) New molecular phylogeny of the squids of the family Loliginidae with emphasis on the genus *Doryteuthis* Naef, 1912: Mitochondrial and nuclear sequences indicate the presence of cryptic species in the southern Atlantic Ocean. Molecular Phylogenetics and Evolution 68: 293–299. doi:10.1016/j.ympev.2013.03.027

Sales JBL, Markaida U, Shaw PW, Haimovici M, Ready JS, Figueredo-Ready WM, Angioletti F, Carneiro MA, Schneider H, Sampaio I (2014) Molecular phylogeny of the genus *Lolliguncula* Steenstrup, 1881 based on nuclear and mitochondrial DNA sequences indicates genetic isolation of populations from North and South Atlantic, and the possible presence of further cryptic species. PLoS ONE 9(2): e88693. https://doi.org/10.1371/journal.pone.0088693

Sales JBL, Rodrigues-Filho LFS, Ferreira YS, Carneiro J, Asp NE, Shaw PW, Haimovici M, Markaida U, Ready J, Schneider H, Sampaio I (2017) Divergence of cryptic species of *Doryteuthis plei* Blainville, 1823 (Loliginidae, Cephalopoda) in the Western Atlantic Ocean is associated with the formation of the Caribbean Sea. Molecular Phylogenetics and Evolution 106: 44–54. https://doi.org/10.1016/j.ympev.2016.09.014

Shaw PW, Pierce GJ, Boyle PR (1999) Subtle population structuring within a highly vagile marine invertebrate, the veined squid *Loligo forbesi*, demonstrated with microsatellite DNA markers. Molecular Ecology 8: 407–417. https://doi.org/10.1046/j.1365-294X.1999.00588.x

van der Vyver JSF, Sauer WHH, McKeown NJ, Yemane D, Shaw PW, Lipiński MR (2016) Phenotypic divergence despite high gene flow in chokka squid *Loligo reynaudii* (Cephalopoda: Loliginidae): implications for fishery management. Journal of the Marine Biological Association of the United Kingdom 96: 1507–1525. https://doi.org/10.1017/S0025315415001794

Vaske Jr T, Knoeller JMS, Viliod MCL (2014) Occurrence and biological aspects of the Atlantic brief squid *Lolliguncula brevis* in the estuary of São Vicente-SP, southeastern Brazil. Unisanta Bioscience 3: 66–77. Available online at: http://periodicos.unisanta.br/index.php/bio/article/view/172/208

Vecchione M, Shea E, Bussarawit S, Anderson FE, Alexeyev D, Lu C-C, Okutani T, Roeleveld M, Chotiyaputta C, Roper CFE, Jorgensen E (2005) Systematics of Indo-West Pacific loliginids. Phuket Marine Biological Center Research Bulletin 66: 23–26. http://tolweb.org/tree/ToLdocuments/loliginidsystematics.pdf

Vidal ÉAG, Marian JEAR, Martins RS (2013) *Doryteuthis sanpaulensis*, São Paulo Squid. In: Rosa R, O'Dor RK, Pierce G (Orgs) Advances in Squid Biology, Ecology and Fisheries. Part I: Myopsid Squids. Nova Science Publishers, New York, 257–300.

Yesaki M, Rahn E, Silva G (1976) Sumário das explorações de peixes de arrasto de fundo ao largo da costa sul do Brasil. PDP (FAO/PNUMA, SUDEPE), Rio de Janeiro, Documentos Técnicos 19, 44 pp.

Zaleski T, Perez JAA, Gandara-Martins AL (2012) Morphological and morphometric variability of the squid *Lolliguncula brevis* (Mollusca: Cephalopoda) in Brazilian waters: evidence for two species in the western Atlantic? Anais da Academia Brasileira de Ciências 84: 1015–1028. https://doi.org/10.1590/S0001-37652012000400016

Zar JA (2009) Biostatistical Analysis (5th edn). Pearson, Essex, 960 pp.

Author Contributions: RSM conducted statistical analysis and wrote the paper; MJ conducted fieldwork, biological sampling, manuscript review and statistical analysis.

Competing Interests: The authors have declared that no competing interests exist.

16

Morphological evidence for the taxonomic status of the Bridge's Guan, *Penelope bridgesi*, with comments on the validity of *P. obscura bronzina* (Aves: Cracidae)

Oscar Diego Evangelista-Vargas[1], **Luís Fábio Silveira**[1]

[1] *Seção de Aves, Museu de Zoologia da Universidade de São Paulo, São Paulo, Brasil. Avenida Nazaré 481, Ipiranga, 04263-000 São Paulo, SP, Brazil.*
Corresponding author: Diego Evangelista-Vargas (diegoevangelistav@gmail.com)

ABSTRACT. *Penelope obscura* Temminck, 1815 is a forest guan found in Brazil, Paraguay, Uruguay, Argentina and Bolivia. Three subspecies are currently recognized: *Penelope o. obscura* Temminck, 1815, *P. o. bridgesi* Gray, 1860, and *P. o. bronzina* Hellmayr, 1914. The limits between *Penelope* taxa are poorly understood since few studies have evaluated their differences in plumage, distribution and taxonomy. Based on 104 specimens deposited in ornithological collections we studied the variations in the plumage of *P. obscura*, including all characters that have been used to describe the included subspecies. Our results show that the plumage of these birds is extremely variable in southern and southeastern Brazil. Without any morphological and morphometric characters to support *P. o. bronzina* as a valid taxon, we synonymized it with *P. obscura*. Conversely, *P. o. bridgesi*, which occurs in the Yungas and the Chaco, is a distinct taxon and should be treated as a separate species from *P. obscura*.

KEY WORDS. Distribution, morphology, *Penelope*, taxonomy.

INTRODUCTION

Penelope obscura Temminck, 1815 is distributed from eastern Minas Gerais and northern Espírito Santo in Brazil, to southeastern Paraguay, northeastern Argentina (extending to central Bolivia) and Uruguay (Del Hoyo and Kirwan 2016). As it is often the case among members of the genus, *P. obscura* is syntopic with other species throughout its distribution: on the eastern slopes of the Andes of Bolivia and northwestern Argentina, this species shares its habitat with *P. montagnii* and *P. dabbenei*, and in southeastern Brazil, eastern Paraguay and northeastern Argentina, it is usually found with *P. superciliaris* (Vaurie 1968).

The body plumage of *P. obscura* is usually dark brown; neck, mantle, upper wing coverts and breast are edged with dull white, as the forehead and eyebrow. The tarsus is blackish brown or black in life (Vaurie 1968). It presents sexual dimorphism in the color of the iris, red to orange in males and brown or dark brown in females, which is already well differentiated in the first year of life (Sick 1997, Chalukian 1997, Nacinovic 2012).

Based on the "Yacuhú" of Azara (1805), Temminck (1815) described *P. obscura*. Paraguay is considered as its type locality. Gray (1860) described *Penelope bridgesi*, type locality in Bolivia (restricted to Villa Montes, Tarija by Vaurie 1966). Ogilvie-Grant (1893) suggested that *P. bridgesi* should be considered a synonym of *P. obscura*; however, Hartert and Venturi (1909) considered *P. bridgesi* as a subspecies of *P. obscura*, a placement that is currently accepted. Finally, Hellmayr (1914) described *P. obscura bronzina* based on a specimen collected in Santa Catarina (Colonia Hansa, currently Corupá). The author also delimited the geographic distribution of the three subspecies. Hellmayr's hypotheses (1914) have been adopted up to now, and no revisions of *Penelope* have been published since the taxonomic study of Vaurie (1968).

Penelope obscura bronzina is an endemic subspecies of the Brazilian Atlantic Forest and has a wide distribution, occurring from east of Minas Gerais to Santa Catarina through Espírito Santo, Rio de Janeiro, São Paulo and Paraná (Sick and Teixeira 1979). The nominate form is distributed in southern Brazil, in the states of Santa Catarina and Rio Grande do Sul, also including the forested regions in Paraguay, Uruguay and northeastern

Argentina (in the provinces of Misiones, Chaco, Corrientes, Entre Ríos and Santa Fe); the exact range of this subspecies in Brazil is unclear, but it is most likely restricted to the forests of the Paraná and Uruguay Rivers. Lastly, *P. o. bridgesi* occurs in western South America, being found in the Bolivian Yungas in the provinces of Cochabamba, Santa Cruz, Chuquisaca and Tarija, and towards the south in Argentina in the forests of eastern slopes of the Andes in the provinces of Jujuy, Salta, Tucumán and Catamarca. The gap between the ranges of this and the two other subspecies may have been caused by an increase of the aridity in the Chaco, which has isolated this taxon in the Andean foothills (Vaurie 1968).

Vaurie (1966) recognized that the nominate subspecies differs from *P. o. bronzina* by having the feathers of the head entirely blackish with only a suggestion of light edgings on the forehead and brows. *Penelope o. bronzina*, according to this author, has the head with long and bordered feathers, especially on the anterior half of the crown, with silvery-white. Also, the race *obscura* would be smaller in size and lighter, with plumage browner and throughout duller. Other authors such as Delacour and Amadon (2004) documented that *P. o. bridgesi* is similar to *P. o. bronzina*, being larger and having the crest feathers less edged with white and little or no trace of eyebrow stripe, with the plumage duller and browner.

Del Hoyo and Kirwan (2016) mentioned that plumage differences among races suggest that more than one species may be involved in the *P. obscura* complex. They also pointed out that the possibility of contact and intergradation of the two eastern races, *P. o. bronzina* and *P. o. obscura*, required detailed study. Vaurie (1968) suggested that these races do not intergrade, and Del Hoyo and Motis (2004) noted a gap between the two forms in the state of Santa Catarina, Brazil, and also recognized the need of studies to evaluate the taxonomy of this complex.

MATERIAL AND METHODS

We analyzed and measured 80 museum specimens of all taxa currently considered as belonging to the *P. obscura* complex (36 males, 29 females and 15 of unknown sex), all of which are distributed in Brazil, Argentina, Paraguay and Bolivia. These specimens are housed in the following institutions: Museu de Zoologia da Universidade de São Paulo (MZUSP, including the topotype of *P. o. bronzina*), São Paulo, Brazil; Museu Nacional da Universidade Federal do Rio de Janeiro (MNRJ), Rio de Janeiro, Brazil; Museu de História Natural do Capão da Imbuia (MHNCI), Curitiba, Brazil; Museu de Ciências Naturais da Fundação Zoobotânica do Rio Grande do Sul (MCNRS), Porto Alegre, Brazil; and Museo Argentino de Ciencias Naturales "Bernardino Rivadavia" (MACN), Buenos Aires, Argentina. Photographs of the specimens were taken against a white background using 10 watt LED spotlights, always in the same place to standardize the lightening. We also examined photographs of 24 specimens, including the holotypes of *P. o. bronzina* and *P. o. bridgesi*, deposited in the following institutions: Museum für Naturkunde (MFN), Berlin, Germany; Field Museum of Natural History (FMNH), Chicago, USA; Rijksmuseum van Natuurlijke Historie (RNHL), Leiden, Netherlands; Zoologische Staatssammlung München (ZSM), Munich, Germany; and the Natural History Museum (BMNH), Tring, UK (Fig. 1).

Localities and geographic coordinates were obtained from the labels attached to the specimens. When the coordinates were not available they were obtained by consulting ornithological gazetteers (Paynter and Traylor 1991, Paynter 1992, 1995, Vanzolini 1992). The maps were designed using the software QGis 2.6.0, showing the divisions of countries, Brazilian states and the biogeographic provinces proposed by Morrone (2009) to differentiate the areas where the taxa occur. In order to perform detailed analysis of geographic variations in body size and plumage coloration we defined nine geographic groups considering morphological homogeneity, geographic proximity and ecological similarity. Specimens identified as juveniles by distinct plumage and smaller sizes were excluded from the morphometric and plumage analyses, and were considered only to compile the distribution of the taxa.

We measured the length of the exposed culmen, bill from nostril, tarsus, middle toe, wing and tail using calipers with precision of 0.01 mm following Baldwin et al. (1931). Morphometric data were analyzed to test the sexual dimorphism and intra- and inter-geographic group variations. Shapiro-Wilk and Lilliefors (Kolmogorov-Smirnov) tests were performed to test the normality of data. Sexual dimorphism in each group was evaluated through the Student's t test, testing the differences between sexes for each selected character. For multivariate analyses, the data requires log-transformed variables as input so that all of them have the same relevance for the analysis. Multivariate analyses are based on multiple variables that are measured in the same individual and are not independent from each other, and we tested the general morphology of *P. obscura* with all the morphometric variables taken together. To corroborate the sexual dimorphism we performed the Hotelling's T square test, the multivariate counterpart of Student's t, testing the differences between the multivariate means. A Lineal Discriminant Analysis (LDA) was also used; this test identifies all the elements that belong to each geographic group (including each measure for each variable) and generates a linear combination of variables that can find the pattern within the data and effectively classify it. It represents in percentages the possibility of individuals belonging to the pre-determined group or meeting the pattern of another group. The Canonical Variates Analysis (CVA) was performed to check whether the groups form significantly distinct clusters morphometrically. The univariate part of the tests was conducted in R software statistical package v. 3.2.1 and the multivariate analysis was carried out in Past v. 2.15.

The plumage was evaluated according to the Munsell Color Guide (1994), based on traits usually considered as diagnostic for the three subspecies: supercilium, crown, throat, breast, abdomen, back, tail and wing coverts. We also analyzed

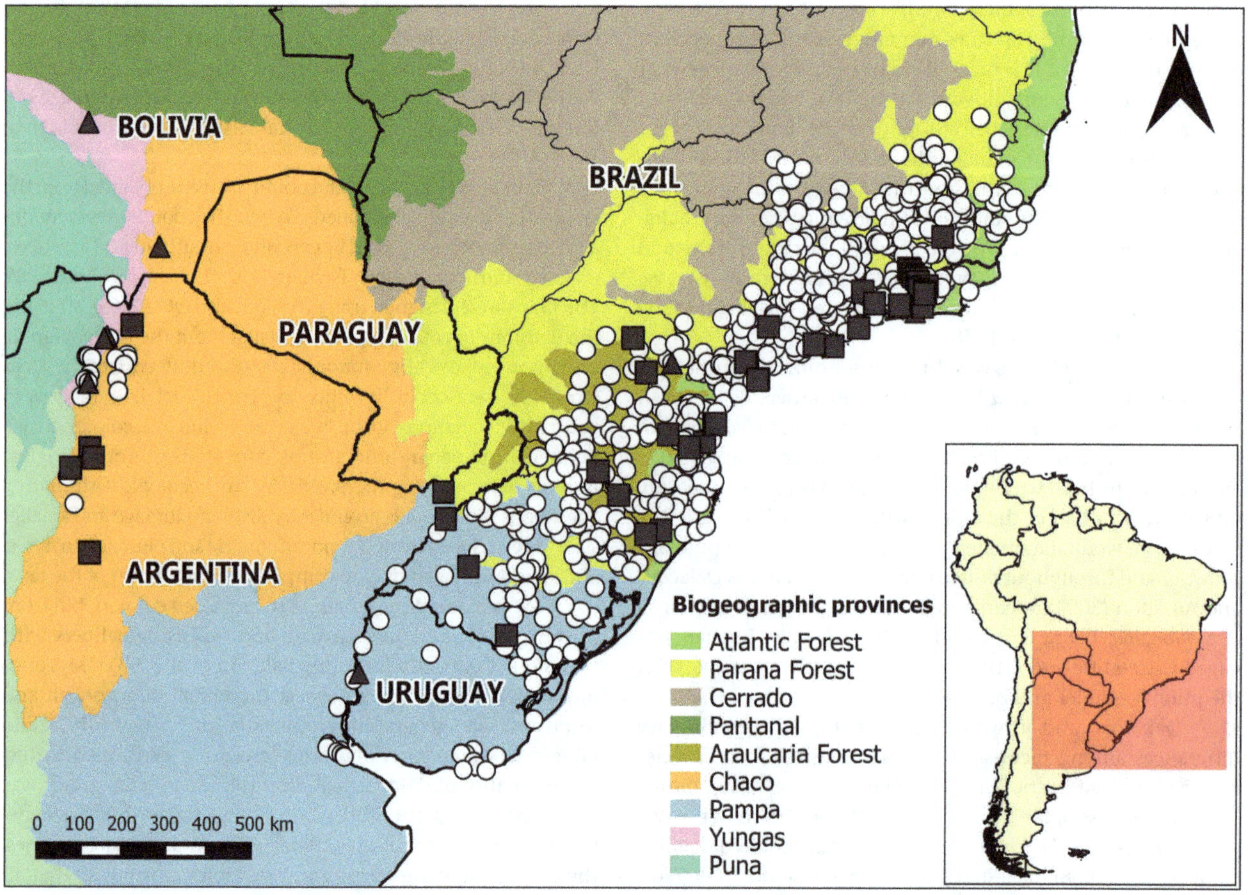

Figure 1. Distribution of *Penelope obscura*. Black squares: specimens analyzed personally; Black triangles: specimens analyzed by photographs; White circles: living individuals (photos).

photographs of living individuals available in the online databases Wiki Aves (http://www.wikiaves.com.br) and Ecoregistros (http://www.ecoregistros.org). From WikiAves were reviewed 4.342 images from the states of Espírito Santo (122), Minas Gerais (976), Rio de Janeiro (711), São Paulo (1.582), Paraná (333), Santa Catarina (274) and Rio Grande do Sul (344) in Brazil; and 100 images from Ecoregistros, for the records in Uruguay (18) and northeast (31) and northwest (51) of Argentina (Fig. 1). Some of the examined photographs were excluded from our analysis because they had been taken from a long distance, were very dark, or portrayed partially or wrongly identified taxa.

RESULTS

Statistical analyzes: nine geographic groups were defined: three from the Atlantic forest, two from the Araucaria forest, and one from the Parana forest, Pampa, Chaco and Yungas (Table 1). For statistical analyses, we examined 77 adult specimens distributed in the aforementioned biogeographic provinces (Fig. 2). Table 2 presents the descriptive statistical analysis including mean, standard deviation and minimum and maximum values.

Univariate analysis of normality suggests that data of the six variables meet the criteria of the parametric tests. Student's t-test was not performed for the groups with small sample number; it was used to test the null hypothesis that the mean value of each variable does not differ between the two sexes for each group. Based on the test, the vast majority of the means of the two sexes do not differ significantly; only in five specific cases was there a significant difference between the sexes, with males being larger than females (Table 3). The T^2 Hotelling test has as null hypothesis that the vectors of the means of the sexes are equal, and shows that there is no sexual dimorphism in the same geographic groups tested for Student's t-test ($p > 0.05$).

Table 4 presents the results of Lineal Discriminant Analysis and shows that everything in the off-diagonal are the frequencies of individuals that also meet the pattern of other groups. Only 33 (44 %) of the individuals were correctly classified within their pre-determined group. The MANOVA/CVA explained

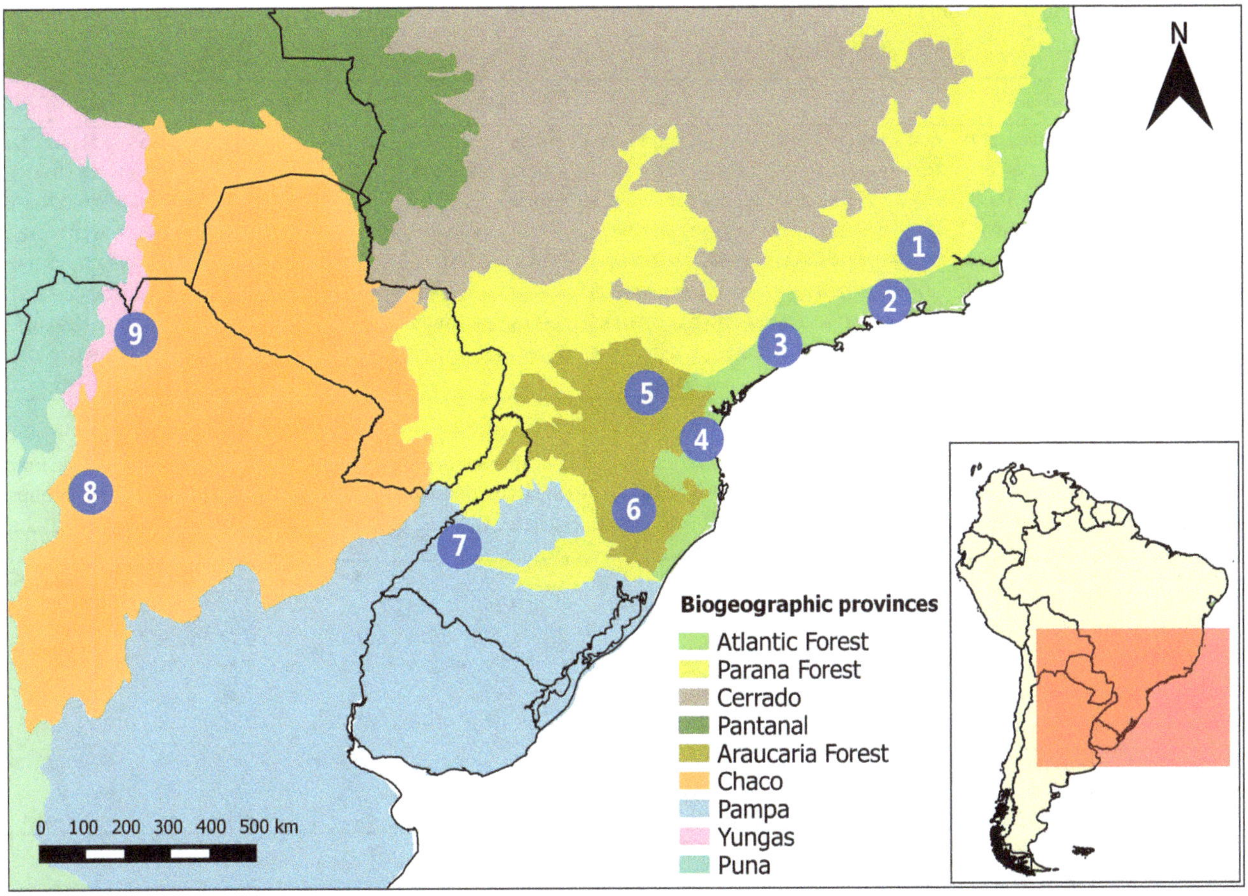

Figure 2. Geographic groups of *Penelope obscura* separated by biogeographic provinces: Parana forest (1), Atlantic Forest (2,3,4), Araucaria Forest (5,6), Pampa (7), Chaco (8) and Yungas (9).

Table 1. Geographic groups and respective number of skins used for statistical analysis.

Geographic groups	Number and gender	Localities
1. Parana Forest	8 (4 males, 4 females)	Southeast of Minas Gerais (Brazil)
2. Atlantic Forest 1	24 (14 males, 8 females, 2 unknown)	Northeast of São Paulo (Brazil)
3. Atlantic Forest 2	11 (5 males, 6 females)	East of São Paulo (Brazil)
4. Atlantic Forest 3	5 (1 male, 4 unknown)	East of Paraná and Northeast of Santa Catarina (Brazil)
5. Araucaria Forest 1	5 (1 male, 2 females, 2 unknown)	Central Paraná and North of Santa Catarina (Brazil)
6. Araucaria Forest 2	4 (1 male, 3 females)	Center Santa Catarina and Northeast of Rio Grande do Sul (Brazil)
7. Pampa	7 (4 males, 2 females, 1 unknown)	Southwest of Rio Grande do Sul (Brazil) and Misiones (Argentina)
8. Chaco	11 (6 males, 4 females, 1 unknown)	Tucumán and Catamarca (Argentina)
9. Yungas	1 (unknown)	Salta (Argentina)

the variations between the geographic groups in the first two axes, and most of them could not be differentiated by any of the morphometric variables. Groups 6 and 7 from southern Brazil and northeastern Argentina (Araucaria forest and Pampa, respectively) differed slightly from the others in the length of the tail (Fig. 3).

Specimens collected in eastern South America showed continuous distribution with no gap in the state of Santa Catarina in Brazil. Their plumage showed no sexual dimorphism in any geographic group. The characters used to diagnose the subspecies *P. o. obscura* were found in specimens from various parts of its wide range. For instance, the whitish edges of head feathers are wide (MZUSP 31577, from Varjão de Guaratuba, São Paulo, BR), thin (MZUSP 34130, Maromba, Rio de Janeiro, BR) or almost imperceptible (MACN 44036, Chantecler, Misiones, ARG), without any apparent geographic trend. The overall color of the crown feathers affects the appearance of the head, which can be black, gray, or brown. In addition, the prominence of the white

Table 2. Mean, standard deviation, minimum and maximum values of body measurements (mm) from the geographic groups of *Penelope obscura*.

Geographic groups	Exposed culmen	Nostril	Tarsus	Middle Toe	Wing	Tail
1 (n = 8)	29.04 ± 2.62 (26.2–33.1)	25.41 ± 1.83 (23.3–27.6)	75.41 ± 5.83 (70.2–88.6)	62.38 ± 2.39 (59–67)	314.75 ± 14.51 (292–338)	357.38 ± 10.25 (342–372)
2 (n = 24)	30.24 ± 2.70 (24.9–35.1)	24.52 ± 1.66 (21.6–27.3)	76.35 ± 4.10 (68.4–83.1)	66.75 ± 3.35 (62–73)	324.3 ± 15.89 (295–360)	356.4 ± 19.23 (321–410)
3 (n = 11)	32.16 ± 1.49 (30–34.7)	25.21 ± 1.22 (23.7–27.8)	80.73 ± 2.07 (76.2–82.8)	69.54 ± 1.86 (67–72)	324.7 ± 25.68 (293–362)	346.2 ± 23.62 (322–385)
4 (n = 5)	32.1 ± 2.41 (29–35)	25.68 ± 1.70 (23.1–27.2)	84.32 ± 3.54 (78.7–87.2)	71 ± 4.74 (63–75)	326.8 ± 14.20 (306–343)	346 ± 16.98 (321–368)
5 (n = 5)	27.42 ± 2.74 (23.2–30.4)	22.66 ± 1.98 (20.5–24.8)	79.96 ± 3.90 (77–86.7)	66.8 ± 4.15 (61–72)	320.2 ± 20.90 (299–347)	356.6 ± 19.77 (330–384)
6 (n = 4)	25.95 ± 0.87 (25.2–27.2)	22.3 ± 0.22 (22–22.5)	77.65 ± 3.13 (74–81.2)	64.75 ± 2.5 (62–68)	311.25 ± 16.68 (287–324)	311.5 ± 16.98 (288–327)
7 (n = 7)	25.81 ± 1.23 (23.9–27.7)	22.7 ± 0.81 (21.4–23.7)	79.01 ± 2.33 (77.1–83.3)	65.86 ± 2.91 (61–70)	322.4 ± 17.35 (297–345)	323 ± 15.06 (299–339)
8 (n = 11)	27.16 ± 1.03 (25.9–28.5)	22.66 ± 1.18 (21.1–24.5)	76.86 ± 3.32 (71.2–81.4)	63.4 ± 3.96 (58.2–70)	329.5 ± 18.13 (295–346)	355.3 ± 15.03 (328–376)
9 (n = 1)	26.1	23.3	81.8	68	346	368

Table 3. Student t-test comparing males and females of each geographic group with more than four individuals for each sex. Asterisks indicates significant differences between both sexes at $p < 0.05$.

Geographic group	Exposed culmen	Nostril	Tarsus	Middle Toe	Wing	Tail
1 (♂: 4, ♀: 4)	t = -1.730 p = 0.134	t = -2.006 p = 0.091	t = -0.118 p = 0.91	t = 0.137 p = 0.895	t = -0.971 p = 0.369	t = 0.921 p = 0.392
2 (♂: 14, ♀: 8)	t = -1.448 p = 0.163	t = -0.354 p = 0.727	t = -1.132 p = 0.27	t = -3.248 p = 0.004*	t = -2.388 p = 0.027*	t = -1.047 p = 0.307
3 (♂: 5, ♀: 6)	t = -2.342 p = 0.044*	t = 0.116 p = 0.910	t = -0.409 p = 0.692	t = -0.084 p = 0.935	t = -1.160 p = 0.276	t = -1.430 p = 0.186
8 (♂: 6, ♀: 4)	t = -0.468 p = 0.652	t = -1.756 p = 0.117	t = -0.533 p = 0.608	t = -0.620 p = 0.552	t = -3.367 p = 0.010*	t = -2.968 p = 0.018*

Table 4. Frequency of percentages of predictive performance from the Lineal Discriminant Analysis (LDA) for all the geographic groups.

Predicted	Original							
	1	2	3	4	5	6	7	8
1	50.0	12.5	–	–	–	–	–	9.1
2	37.5	62.5	27.3	–	20	–	28.6	36.4
3	–	8.3	54.5	80	40	–	–	–
4	12.5	–	9.1	–	–	–	–	–
5	–	4.2	–	–	–	–	–	9.1
6	–	–	–	–	–	25	28.6	–
7	–	8.3	9.1	20	–	75	28.6	–
8	–	4.2	–	–	40	–	14.2	45.4

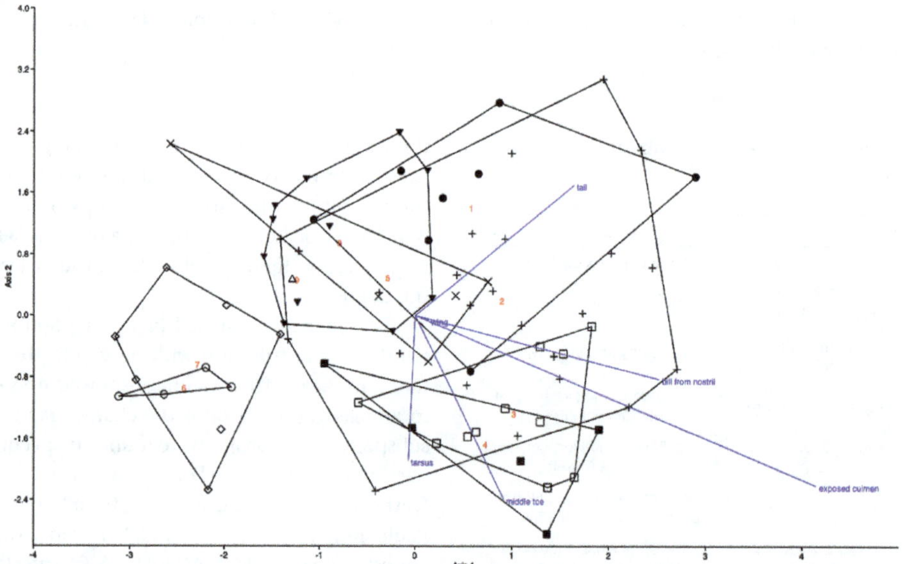

Figure 3. CVA Plot of geographic groups according to morphometric measurements. Black circles: Parana Forest (group 1); Plus: Atlantic Forest 1 (group 2); White squares: Atlantic Forest 2 (group 3); Black squares: Atlantic Forest 3 (group 4); Exes: Araucaria Forest 1 (group 5); White circles: Araucaria Forest 2 (group 6); White diamonds: Pampa (group 7); Inverted black triangles: Chaco (group 8); White triangle: Yungas (group 9).

Figure 4. From top to bottom, dorsal view of *P. o. bronzina* (MZUSP 49384, São Paulo, southeastern BR), *P. o. obscura* (MACN 48384, Corrientes, northeastern ARG), and *P. o. bridgesi* (MACN 8148a, Tucumán, northern ARG). Scale bars: 10.0 cm.

hue in the edge of the feathers appears to vary considerably. Specimens from Potreirinhos in Rio Grande do Sul (MCNRS 682, BR) to the south often have less noticeable whitish edges and tend to have a blackish general coloration, but this pattern is also recorded in birds from the northern portion of the range, which present general blackish coloration and feathers with white edges reduced or absent (WA700594, WA823706). On the other hand, specimens in the south of the range may present feathers with noticeable white edges and general grayish coloration (WA914588, WA1763524, WA1782826; ER10072, ER10141).

The plumage of the back also showed considerable variation, without any apparent geographical pattern. Across the entire range of the species we found individuals with the back greenish-brown (MZUSP 32906, Teresópolis, Rio de Janeiro, BR), brownish (MZUSP 7021, Fazenda Monte Alegre, Paraná, BR) or blackish (MZUSP 49384, Rocha, São Paulo, BR). Apparently, specimens from the northern part of the range have the white streaks on the back more noticeable when compared to the southern specimens. However, we also found specimens in the south with wide white streaks (MCNRS 1765, Bagé, Rio Grande do Sul), and in the north the white streaks in some specimens are almost absent (MN 45542, Simão Pereira, Minas Gerais, BR) (Fig. 4).

Some specimens showed a contrasting dark chest with a paler belly (MN 47672, Matias Barbosa, Minas Gerais; MHNCI 204, Tibagi, Paraná, BR), whereas other specimens had a uniformly colored chest with either a dark belly (MN 49619, Levi Gasparian, Rio de Janeiro; MZUSP 49382, Morretinho, São Paulo; MCNRS 1765, Bagé, Rio Grande do Sul, BR) or a paler, uniform belly (MHNCI 399, Vale do Ivaí, Paraná, BR). Both phenotypes

Figure 5. From top to bottom: lateral view of *P. o. bronzina* (MZUSP 49384, São Paulo, southeastern BR), *P. o. obscura* (MACN 48384, Corrientes, northeastern ARG), and *P. o. bridgesi* (MACN 8148a, Tucumán, northern ARG). Scale bars: 10.0 cm.

were found in photographs taken at the same locality (e.g. WA1380991 and WA1423978, from Afonso Cláudio, Espírito Santo, BR). The color of the belly was also variable, ranging from brown (MZUSP 1911, Colonia Hansa, Santa Catarina, BR, topotypical) to orangish (MZUSP 78284, Bananal, São Paulo, BR).

It is common to find individuals with different overall coloration in the same place, demonstrating the existence of a large individual variation (e. g. WA1681568, Pouso Alegre, Minas Gerais; WA450514, Nova Friburgo, Rio de Janeiro; WA1518511, Itatiba, São Paulo; WA1084879 and WA1090972, Tibagi, Paraná, BR). As mentioned before we located specimens with plumage characters supposedly diagnostic for the nominate subspecies in specimens found in the range of *P. o. bronzina* (WA1765142, Bocaina de Minas, Minas Gerais, BR) and specimens with characters considered diagnostic for *P. o. bronzina* in the geographic range of nominate subspecies (ER10141; Ribera del Norte, Buenos Aires, Argentina).

On the other hand, all specimens attributed to *P. o. bridgesi* lacked olive brown coloration with a green tinge to the upperparts (unlike *P. o. obscura* and *P. o. bronzina*) (Fig. 4). Instead, specimens of *P. o. bridgesi* showed chestnut upperparts contrasting with the black head. The underparts were consistently uniform and slightly paler than the back, whereas in *P. o. obscura* and *P. o. bronzina* the coloration of the upperparts was highly variable. The wing coverts of *P. o. bridgesi* show broad white margins, a character state not found in *P. o. obscura* and *P. o. bronzina*, which have indistinct white margins to the wing coverts (Fig. 5). Overall, *P. o. bridgesi* shows less pronounced individual variation in plumage coloration than *P. o. obscura* and *P. o. bronzina*.

DISCUSSION

Misidentifications of *Penelope obscura* are still common, especially where it is sympatric with *P. superciliaris*. The range of *P. obscura* in eastern South America is continuous, without any gap between the distribution of the nominate race and *P. o. bronzina* in southern Brazil (Fig. 1). We suspect that the absence of specimens collected between the states of Santa Catarina (from north, called *P. o. bronzina*) and Rio Grande do Sul (to south, *P. o. obscura*), deposited in North American and European museums may have led to the assumption that there was a large geographic gap between the ranges of *P. o. obscura* and *P. o. bronzina*. As no detailed specimen-based study has been conducted to date, including analyses of individual and geographic variation, the validity of *P. obscura bronzina* remained uncontested for many decades in the ornithological literature. The same applies to the taxonomic rank of *P. o. bridgesi*, an isolated taxon occurring in the biogeographic provinces of Yungas and Chaco on the eastern slopes of the Andes in Bolivia and Argentina, which also was never studied in detail.

The eastern population (*P. o. obscura* and *P. o. bronzina*) showed considerable variation in plumage coloration, with many individuals presenting characters previously considered as diagnostic for one of the subspecies in the range of the other. The plumage in *P. o. bridgesi* was much less variable throughout its range than that of *P. o. obscura/bronzina*. Statistical analyses showed no morphological differences between males and females within the groups, and the few differences among some geographic groups do not represent a pattern of biogeographic variation in the eastern or western populations. We cannot be sure that there is a marked difference in size between the aforementioned subspecies of *P. obscura*, as was previously mentioned by Vaurie (1968). Total size measurements attached to some specimens' labels suggest that males are larger than females, but the color of the iris is the only character that effectively differentiates the two sexes.

The occurrence of *P. obscura* is still questionable in some regions. Sousa (1999) mentioned the presence of this species in Elísio Medrado (13°S, 39°45'W) in the state of Bahia (Brazil), being the north limit of its distribution. However, as this record was based on a personal communication from a third part, and there are no records from naturalists in the 19th and 20th centuries, when the habitat was undisturbed, we consider unlikely that *P. obscura* occurs in the state of Bahia since there is no reliable evidence for it. In Paraguay, Hayes (1995) mentioned *Penelope obscura* as a hypothetical species. It was visually reported in the Chaco (President Hayes, Boquerón) and Orient (Central, Alto Paraná and Ñeembucú), but unfortunately those reports had no other details. The only specimens mentioned by the author are deposited at the National Museum of Natural History, Smithsonian Institution (USNM 256875, 58994), and they lack precise information on collecting dates and localities, and were obtained at a time when Paraguay included parts of what is now Brazil and Argentina. Smith and Derna (2015) recently reviewed the records in the country and showed that museum skins attributed to *P. obscura* actually represent members of *P. superciliaris*. Recent records from Paraguay are photographs taken in the Refugio de Vida Silvestre Atinguy, in Misiones department (27°20'S, 56°41'W), claimed by the authors to be wild-origin birds.

After the review of a large series of specimens of *P. o. bridgesi*, and by comparing it with *P. o. obscura* and *P. o. bronzina*, we find that what is currently defined in the literature as *Penelope obscura* does not include the characters observed in the individuals of *P. o. bridgesi*. When Ogilvie-Grant (1893) proposed *bridgesi* as a synonym of *Penelope obscura*, he only reviewed seven specimens (four from Brazil, one from Uruguay, one from Paraguay, and the holotype from Bolivia), excluding the characters of *bridgesi* cited by Gray (1860) that are diagnosable to considered it as a species; the same situation was also observed in Hartert and Venturi (1909), which considered it as a subspecies of *P. obscura*. Our results clearly show that *Penelope bridgesi* is a taxon that can be differentiated from all others without overlapping with any other *Penelope*. It is geographically isolated and should be treated as a valid species (see below). The populations in eastern and southern Brazil are continuously distributed and show considerable individual variation in plumage, without any clear geographic trend. There are any morphological or morphometric characters to support the continued existence of more than one taxon in the region. As *P. o. bronzina* cannot be distinguished from the nominate subspecies, we here treat *Penelope obscura bronzina* Hellmayr, 1914 as a synonym of *Penelope obscura* Temminck, 1815. The treatment proposed here is consistent with the current proposals found for the other species of *Penelope*.

TAXONOMY

Penelope obscura Temminck, 1815

Holotype. The description of the species is based on the 'Yacuhú', made by Azara (1805). Type locality: Paraguay.

Common names. Portuguese: Jacuaçu. Spanish: Pava oscura. English: Dusky-legged Guan.

Diagnosis. Distinguished from *P. bridgesi* by olive brown coloration with green tinge in upperparts, and generally darker overall coloration. Underparts very variable, unlike *P. bridgesi*. White margins of wing coverts markedly reduced. Head, hindneck, and back coloration more uniform than in *P. bridgesi*.

Re-description. No sexual dimorphism found in plumage coloration. Whitish edges extension of head feathers varies from wide to almost imperceptible. Some individuals show a moderately distinct superciliary stripe formed by short feathers with white edges. Overall color of crown feathers variable, ranging from black, gray to brown. White hue of the edge of the feathers varies individually. Brown back may show greenish, brownish or blackish tinge. Neck, breast, upper back, and shoulders conspicuously edged with white. Birds in the northern parts of the species' range tend to have more noticeable white streaks on the back, but streaks also may be lacking. The coloration of the underparts is variable and may show a contrastingly darker chest than belly, but underparts may also be uniform dark or lighter than the breast. The belly feathers have contrasting horizontal scaling, which may range from brown to orangish, and may vary in intensity. Tail feathers are dark brown with bronze or purplish sheen. Bare skin of face slaty, legs dusky gray to blackish, and wattle red.

Distribution. Southeastern and southern Brazil (from the east of Minas Gerais and north of Espírito Santo to Rio Grande do Sul), northeastern Argentina (Misiones, Chaco, Corrientes, Entre Ríos, Santa Fe and Buenos Aires), and the forested regions of Uruguay, with recent photographic records for Paraguay. Inhabits the biogeographic provinces of Atlantic Forest, Cerrado, Paraná Forest, Araucaria Forest and Pampa.

Examined material (n = 84). BRAZIL. *Minas Gerais*: Muriaé (1 ♀- MNRJ 36791); Juiz de Fora Km.118 (1 ♂- MNRJ 45036); Juiz de Fora Km.788 (1 ♂- MNRJ 48346); Matias Barbosa Km.795 (1 ♂- MNRJ 47672); Matias Barbosa Km.801 (1 ♀- MNRJ 48384); Simão Pereira Km.807 (1 ♀- MNRJ 49558); Simão Pereira Km.817 (1 ♀- MNRJ 45542, 1 ♂- MNRJ 48997); Simão Pereira Km.825 (1 ♀- MNRJ 48383); Estação do Túnel (2 ♀- BMNH 1901.3.28.65, BMNH 1901.3.28.64); *Rio de Janeiro*: 2-BMNH 1889.6.1.215, RNHL no number, 1 ♂- 1895.4.1.3, 1 ♀- 1895.4.1.4; Rio de Janeiro (1 – ZSM no number); Levi Gasparian Km.792 (1 ♂- MNRJ 49619); Levi Gasparian (1 ♀- MNRJ 48338); Tres Rios Km.27 (1 ♀- MNRJ 49080); Areal Km.44 (1 ♂- MNRJ 49081); Petropolis Km.52 (1 ♂- MNRJ 49574); Petropolis Km.64 (1 ♂- MNRJ 44389); Teresopolis Km.52 (1 ♀- MNRJ 47743); Teresopolis Km.82 (1 ♀- MNRJ 48354); Teresopolis (2 – MNRJ 9321, MNRJ 9322, 3 ♂- MZUSP 32905, MZUSP 32906, MZUSP 32907, 2 ♀- MZUSP 32904, MZUSP 32908; Guapimirim Km.96 (1 ♂- MNRJ 48337); Parnaso (1 ♂- MNRJ 48336); Inga (1 ♂- MNRJ 47552); Morro Redondo (1 ♀- MNRJ 9405); Maromba (1 ♀- MZUSP 34129, 1 ♂- MZUSP 34130); Paraty (1 ♂- MNRJ 36927); *São Paulo*: 1 – BMNH 1889.6.1.203; Bananal (1 ♂- MZUSP 78284, 1 ♀- MZUSP 78285); Varjão de Guaratuba (3 ♀- MNRJ 31062, MZUSP 31576, MZUSP 43763, 1 ♂- MZUSP 31577); Cabreúva (1 ♂- MZUSP 93098); Morretinho e Rocha (2 ♂- MZUSP 49382, MZUSP 49383, 1 ♀- MZUSP 49384); Rio Ribeira (1 ♀- MZUSP 66890, 1 ♂- MZUSP 66891); Icapara (1 ♀- MZUSP 73215); *Paraná*: Vale do Ivai (1 – MHNCI 399); Santa Barbara (1 ♀- MHNCI 251); Fazenda Monte Alegre (1 ♂- MZUSP 7021); Tibagi (1 ♀- MHNCI 2041); Guaratuba-Caiobá (1 ♂- MHNCI 1189, 2 – MHNCI 256, MHNCI 2519); Santa Catarina: Rio Negro (1 – MHNCI 5776); Joinville (1 ♂- RNHL 1292-3, 1 – MZUSP 4863); Colonia Hansa (1 ♂- ZSM 12415 *Penelope obscura bronzina* HOLOTYPE, 2 – MZUSP 1911, MFN 2759); Catanduvas (1 ♂- MNRJ 36928); Anita Garibaldi (1 ♀- MCNRS 2768); Jaguariaiva, Fazenda Morungava (1 – FMNH no number, 1 ♂- 110740); *Rio Grande do Sul*: Potreirinhos (1 ♀- MCNRS 682); Bom Jesus (1 ♀- MCNRS 2544); São Francisco de Assis (1 – MCNRS 1975); Bagé (1 ♂- MCNRS 1765). URUGUAY. 1 ♂- BMNH 1878.1.25.1; Payssandu: 1 – BMNH 1901.2.10.24. PARAGUAY. 1 – BMNH no number. ARGENTINA. *Misiones*: Chantecler (2 ♂- MACN 44034, MACN 44036, 1 ♀- MACN 44035); Corrientes: Rincón Mercedes (1 ♂- MACN 48384, 1 ♀- MACN 48385).

Mentioned photographs. Wikiaves.com.br (n = 13): BRAZIL. 700594 (Guarapuava, Paraná); 823706 (Caputira, Minas Gerais); 914588 (Taquara, Rio Grande do Sul); 1763524 (Aceguá, Rio Grande do Sul); 1782826 (São Martinho da Serra, Rio Grande do Sul); 1380991 (Afonso Cláudio, Espirito Santo); 1423978 (Afonso Cláudio, Espirito Santo); 1681568 (Pouso Alegre, Minas Gerais); 450514 (Nova Friburgo, Rio de Janeiro); 1518511 (Itatiba, São Paulo); 1084879 (Tibagi, Paraná); 1090972 (Tibagi, Paraná); 1765142 (Bocaina de Minas, Minas Gerais).

Ecoregistros.org (n = 2). ARGENTINA: 10072 (Reserva Ecológica de Ribera del Norte – San Isidro, Buenos Aires); 10141 (Reserva Ecológica de Ribera del Norte – San Isidro, Buenos Aires.

Penelope bridgesi Gray, 1860

Holotype. BMNH 1846.9.9.42 (sex unknown, Bolivia; examined).

Common name. Spanish: Pava de Bridge. English: Bridge's Guan.

Diagnosis. Distinguished from all members of *Penelope* by having chestnut-olive brown coloration without green tinge, and overall coloration browner than *P. obscura*. Underparts are uniform and slightly paler than the back, unlike *P. obscura*. Wing coverts with contrasting large white margins. Black head and hindneck contrast with brownish back, being distinct of *P. obscura*.

Re-description. No sexual dimorphism found in plumage coloration. General coloration chestnut, with limited individual variation. Crest feathers dark with whitish margins, most strongly marked on the forehead; little or no trace of an superciliary stripe. Upperparts chestnut, contrasting with the black head and hindneck. Lower parts uniform dull reddish brown and slightly paler than the back. The belly feathers have contrasting horizontal scaling, which may range from brown to orangish and may vary in intensity. Feathers of the neck, mantle and chest are margined with white, wider on the mantle and wing coverts. Tail feathers dark brown with brownish sheen. Bare skin of face slaty, legs dusky gray to blackish, and wattle red.

Distribution. Forests of the eastern slopes of the Andes, from central Bolivia (Cochabamba, Chuquisaca and Tarija) to northwest Argentina (Jujuy, Salta, Tucumán and Catamarca). Found in the biogeographic provinces of the Yungas and the Chaco.

Examined material (n = 20). ARGENTINA, *Jujuy*: Cerro Calilegua (1 ♂- ZSM 21118); Valle Grande, Ledesma (1 ♀- ZSM 59536); Salta: Río Santa María (1 – MACN 30575); Salta (1 ♀- BMNH 1899.1.27.214); *Tucumán*: Vipos (1 ♀- MACN 8148a); Tafí Viejo (1 ♂- MACN 8884); Sierra del Aconquija (1 ♀- MACN 42153, 4 ♂- MACN 42156, MACN 42157, MACN 42158, MACN 42159, 1 – MACN 42160); Aconquija (1 ♀- MACN 32840); *Catamarca*: El Suncho (1 ♂- MACN 42154, 1 ♀- MACN 42155). BOLIVIA: *Penelope bridgesi* HOLOTYPE (BMNH 1846.9.9.42); Cochabamba: Tin-Tin (1 ♂- FMNH 413837); Tarija: Villa Montes (2 – ZSM 27977, ZSM 27978, 1 ♀- MFN 313633).

ACKNOWLEDGMENTS

We thank Fundação de Amparo à Pesquisa do Estado de São Paulo (FAPESP), Conselho Nacional de Desenvolvimento Científico e Tecnológico (CNPq) and Coordenação de Aperfeiçoamento do Pessoal de Nível Superior (CAPES) for the grants received and for the financial support; Idea Wild for the equipment donated to DEV; the photographers who posted the photos in EcoRegistros and WikiAves websites. The curators and staff of the following institutions kindly welcomed us and allowed us to study the specimens under their care: MFN and ZSM (Germany); FMNH (USA); RNHL (Netherlands); BMNH (UK); MACN (Argentina); MNRJ, MHNCI and MCNRS (Brazil). We also thank Instituto Chico Mendes de Conservação da Biodiversidade (ICMBio/SISBIO) for issuing collecting permits. LFS is supported by a CNPq research productivity fellowship. George Sangster kindly reviewed the manuscript and offered important contributions to this text.

LITERATURE CITED

Azara F de (1805) Caracteres generales de los Yacús. Apuntamientos para la historia natural de los páxaros del Paraguay y Rio de La Plata 3. Doña Manuela Ibarra, Madrid, 69–77.

Baldwin SP, Oberholser HC, Worley LG (1931) Measurements of Birds. Scientific Publications of the Cleveland Museum of Natural History, Cleveland, 165 pp. https://doi.org/10.5962/bhl.title.60247

Chalukian S (1997) Estudio preliminar de La Pava de Monte (*Penelope obscura*) en el Parque Nacional El Rey, Argentina. In: Strahl SD, Beaujon S, Brooks DM, Begazo AJ, Sedaghatkish G, Olmos F (Eds) The Cracidae, Their Biology and Conservation. Hancock House, USA, 64–70.

Delacour J, Amadon D (2004) Curassows and Related Birds. Second edition. Lynx Edicions, The National Museum of Natural History, New York, 476 pp.

Del Hoyo J, Kirwan GM (2016) Dusky-legged Guan (*Penelope obscura*). In: Del Hoyo J, Elliot A, Sargatal J, Christie DA, De Juana E (Eds) Handbook of the Birds of the World Alive. http://www.hbw.com/node/53291 [Accessed: 15/11/2016]

Del Hoyo J, Motis A (2004) Update chapter. In: Delacour J, Amadon D (Eds) Curassows and related birds. Lynx Edicions, Barcelona, 322–476.

Gray GR (1860) *Penelope bridgesi*. Proceedings of the Zoological Society of London 28. Academic Press, London, 270.

Hartert E, Venturi S (1909) Notes sur les oiseaux de la république Argentine. In: Rothschild W, Hartert E, Jordan K (Eds) Novitates Zoologicae XVI. Zoological Museum of Tring, London and Aylesbury, 159–267.

Hayes FE (1995) Status, Distribution and Biogeography of the Birds of Paraguay. American Birding Association, Colorado, 230 pp.

Hellmayr CE (1914) *Penelope obscura bronzina* new subspecies. In: Rothschild W, Hartert E, Jordan K (Eds) Novitates Zoologicae XXI. Zoological Museum of Tring, London and Aylesbury, 178.

Morrone JJ (2009) Evolutionary Biogeography: An Integrative Approach with Case Studies. Columbia University Press, New York, 304 pp.

Munsell AH (1994) Soil Color Charts. Macbeth Division of Kollmorgan Instruments Corporation, New York, 10 pp.

Nacinovic JB (2012) Uma contribuição iconográfica para a identificação dos Jacus (Aves, Cracidae, *Penelope*) que ocorrem no Brasil. Agência 2A Comunicação, Rio de Janeiro, 80 pp.

Ogilvie-Grant WR (1893) Catalogue of the Birds in the British Museum 22. Order of the Trustees, London, 585 pp.

Paynter RA (1992) Ornithological Gazetteer of Bolivia. Museum of Comparative Zoology of Harvard University, Cambridge, 185 pp. https://doi.org/10.5962/bhl.title.14591

Paynter RA (1995) Ornithological Gazetteer of Argentina. Museum of Comparative Zoology of Harvard University, Cambridge, 1043 pp. https://doi.org/10.5962/bhl.title.14627

Paynter RA, Traylor MA (1991) Ornithological Gazetteer of Brazil. Museum of Comparative Zoology of Harvard University, Cambridge, 788 pp.

Sick H (1997) Ornitologia Brasileira. Nova Fronteira, Rio de Janeiro, 912 pp.

Sick H, Teixeira DM (1979) Notas sobre aves brasileiras raras ou ameaçadas de extinção. Publicações Avulsas do Museu Nacional 62: 1–39.

Smith P, Derna R (2015) On the status of Dusky-legged guan *Penelope obscura* Temminck, 1815 (Aves: Cracidae) in Paraguay. Papéis Avulsos de Zoologia do Museu de Zoologia da Universidade de São Paulo 55(14): 211–216. https://doi.org/10.1590/0031-1049.2015.55.14

Sousa D (1999) Novos registros de espécies de aves no estado de Bahia. Atualidades Ornitológicas 88: 6.

Temminck CJ (1815) Histoire naturelle générale des pigeons et des gallinacés 3. Sepp & Fils, Amsterdam, 757 pp.

Vanzolini PE (1992) A supplement to the ornithological gazetteer of Brazil. Museu de Zoologia da Universidade de São Paulo, São Paulo, 252 pp.

Vaurie C (1966) Systematic notes on the bird family Cracidae. 5, *Penelope purpurascens*, *Penelope jacquaçu*, and *Penelope obscura*. American Museum Novitates 2250: 1–23.

Vaurie C (1968) Taxonomy of the Cracidae (Aves). Bulletin of the American Museum of Natural History 138(3): 131–260.

Author Contributions: LFS conceived the study. LFS and DE-V collected the data; DE-V analyzed the data. LFS and DE-V wrote the ms and gave final approval for publication.

Competing Interests: The authors have declared that no competing interests exist.

17

A new species of *Cellaria* (Bryozoa: Cheilostomata) from northeastern Brazil, with a tabular identification key to the Atlantic species

Ana C.S. Almeida[1,2], Facelúcia B.C. Souza[2], Leandro M. Vieira[1]

[1]*Laboratório de Estudos de Bryozoa, Departamento de Zoologia, Centro de Biociências, Universidade Federal de Pernambuco. 50670-810 Recife, PE, Brazil.*
[2]*Museu de Zoologia, Universidade Federal da Bahia. 40170-115 Salvador, BA, Brazil.*
Corresponding author: Ana Carolina S. Almeida (carol.salmeida@gmail.com)

ABSTRACT. A new species of the erect cheilostome bryozoan *Cellaria* Ellis & Solander, 1786 is described from Bahia, NE Brazil. *Cellaria oraneae* **sp. nov.** is the first formally characterized species of the genus reported from Northeastern coast of Brazil, distinguished from all congeners by the combination of hexagonal autozooids and rhomboid fertile zooids, hexagonal interzooidal avicularium with sagittate foramen, completely immersed ovicell with oval aperture and proximal rectangular lip. A brief discussion of the diversity of *Cellaria* from the Atlantic Ocean and a tabular identification key to these species are also provided.

KEY WORDS. Bryozoan, Cellariidae, taxonomy, Western Atlantic.

INTRODUCTION

Marine bryozoans are found in all oceans, from the intertidal to abyssal depths. The vast majority of some 5,689 living species (Bock and Gordon 2013) belongs to Order Cheilostomata Busk, 1852, which shows a wide variety of colonial growth forms, from encrusting laminar to erect colonies (Taylor and James 2013). Encrusting colonies seem to be more abundant in tropical regions, whereas erect bryozoans are particularly diverse in temperate and polar waters. (Barnes and Whittington 1999, Taylor and James 2013).

Among the erect bryozoans, *Cellaria* Ellis & Solander, 1786 is a speciose genus, comprising more than 50 living species (Hayward and Thorpe 1989, Bock and Hayward 2014). Colonies have articulated skeletal elements (internodes) joined by chitinous joints, with each internode containing multiple zooids (Taylor and James 2013). *Cellaria* species may occur from less than 20 to over 1000 m deep in all oceans. They are common in shallow, coastal waters, attached to hard substrata (McKinney and Jaklin 2000, 2001, Hayward and McKinney 2002). Most *Cellaria* are known from the Indo-Pacific regions, Antarctica and Northeast Atlantic, being apparently limited to their expected native distribution (Hayward and Thorpe 1989). Only a few species have been documented from the Western Atlantic (Winston and Woollacott 2009, Bock and Hayward 2014). However, in the past few years, five new species were described from this region – *Cellaria brasiliensis* Winston, Vieira & Woollacott, 2014, *Cellaria riograndensis* Ramalho & Calliari, 2015 and *Cellaria subtropicalis* Vieira, Gordon, Souza & Haddad, 2010 from Brazil; *Cellaria novanglia* Winston & Hayward, 2012 and *Cellaria louisorum* Winston & Woollacott, 2009 from United States and Barbados, respectively. This indicates that the diversity of *Cellaria* from the Atlantic is yet to be described.

Over the past decade, knowledge of the diversity of marine bryozoans in Brazil has grown significantly and more than 100 new species were described (e.g., Vieira et al. 2010, 2012, 2014, Winston and Vieira 2013, Winston et al. 2014, Almeida et al. 2015a, Ramalho and Calliari 2015). Although *Cellaria* have been documented in the Brazilian waters since the 1960s (Braga 1967, Vieira et al. 2008), the first species from Brazil, *C. subtropicalis*, was only recently formally characterized and described (Vieira et al. 2010). Since then, two other new species were described, *C. brasiliensis* from Rio de Janeiro (Winston et al. 2014), and *C. riograndensis*, from Rio Grande do Sul (Ramalho and Calliari 2015). These three valid species from the Brazilian coast are from South and Southeastern Brazil, and there is a record of a new species from Bahia (Vieira et al. 2008, Almeida et al. 2015b). This species was named *Cellaria bahiensis* in a doctoral dissertation but never formally described (Vieira et al. 2008). Therefore, according to the International Code of Zoological Nomenclature Article 9.7 and Article 11.1 (ICZN 1999), the name *Cellaria bahiensis* Souza is a *nomen nudum* and cannot be considered valid, as pointed by Vieira et al. (2008).

Here we formally describe this new taxon from state of Bahia, Northeast Brazil, and provide a diagnosis for it based on specimens from the same locality (i.e. Todos os Santos Bay) as those studied by Souza (Vieira et al. 2008).

MATERIAL AND METHODS

All type and non-type specimens analyzed in this study are deposited in the Museu de Zoologia of Universidade Federal da Bahia (UFBA), Salvador, Brazil. The specimens were collected by dredge in different localities between 1997 and 2006 along the coast of Bahia, northeast Brazil, from 15 to 50 m (Fig. 1). Some specimens were collected by Laboratório de Malacologia e Ecologia de Bentos (LAMEB) from Universidade Federal da Bahia (UFBA) on different years but at the same locality, and were deposited in a single lot for each locality. Selected specimens were coated with gold-palladium alloy and examined under a scanning electron microscopy (JEOL JSM-6390LV) at Centro de Pesquisa Gonçalo Moniz (FIOCRUZ/BA), Bahia, Brazil. Measurements were taken from digital SEM images of the holotype using ImageJ® software, including range (minimum and maximum), mean, and standard deviation.

TAXONOMY

Order Cheilostomata Busk, 1852
Suborder Neocheilostomina d'Hondt, 1985
Cellariidae Fleming, 1828
Cellaria Ellis & Solander, 1786

Cellaria oraneae sp. nov.

Figs 2–7

Cellaria sp. 2: Vieira et al. 2008: 21. [Brazil: Bahia]
Cellaria sp.: Almeida et al. 2015a: 4. [Brazil: Bahia]

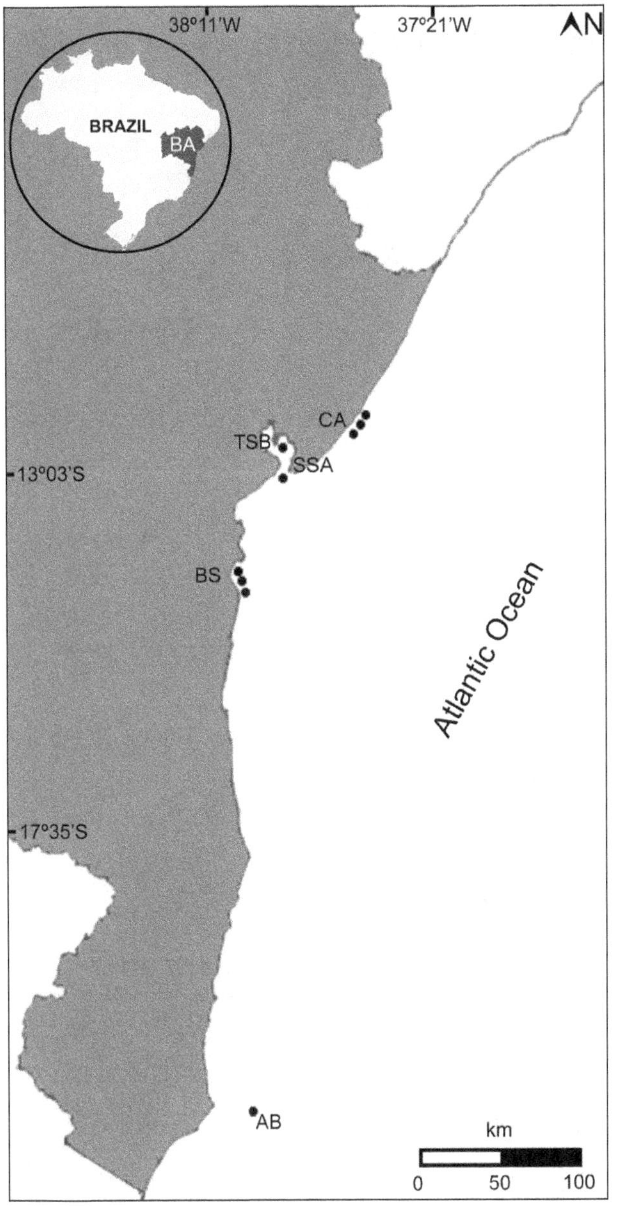

Figure 1. Distribution map of *Cellaria oraneae* sp. nov. in state of Bahia, northeast Brazil. (BA) Bahia, (CA) Camaçari, (SSA) Salvador, (TSB) Todos os Santos Bay, (BS) Baixo Sul, (AB) Abrolhos Bank.

Diagnosis. *Cellaria* with hexagonal autozooids, opesia semi-elliptical with distal rim tuberculate and slightly convex proximal rim with two rounded condyles; interzooidal avicularia hexagonal, with longitudinal ridges forming a spear-shaped area, with sagittate foramen and placed only between non-fertile zooids; fertile zooids rhomboid and ovicell completely immersed with oval aperture and a proximal rectangular lip.

Description. Colony erect, cylindrical, jointed, symmetrically bifurcating and attached at the base by chitinous rhizoids. Branches formed by 6–20 zooidal series, 0.353–0.372 mm in diameter when formed only by infertile autozooids, but larger in regions of fertile zooids, about 0.544–0.574 mm in diameter (Fig. 2). Autozooids hexagonal in outline, about 0.358–0.409 mm (0.395 ± 0.016) long and 0.164–0.231 mm (0.211 ± 0.019) wide, limited by raised lateral walls, comprising a pair of longitudinal ridges extending from lateral walls forming a somewhat lanceolate area (Fig. 3). Frontal wall with a sunken granular cryptocyst. Opesia semi-elliptical, about 0.056–0.068 mm (0.62 ± 0.003) long and 0.072–0.095 mm (0.085 ± 0.006) wide, distal rim tuberculate and arched, proximal rim smooth

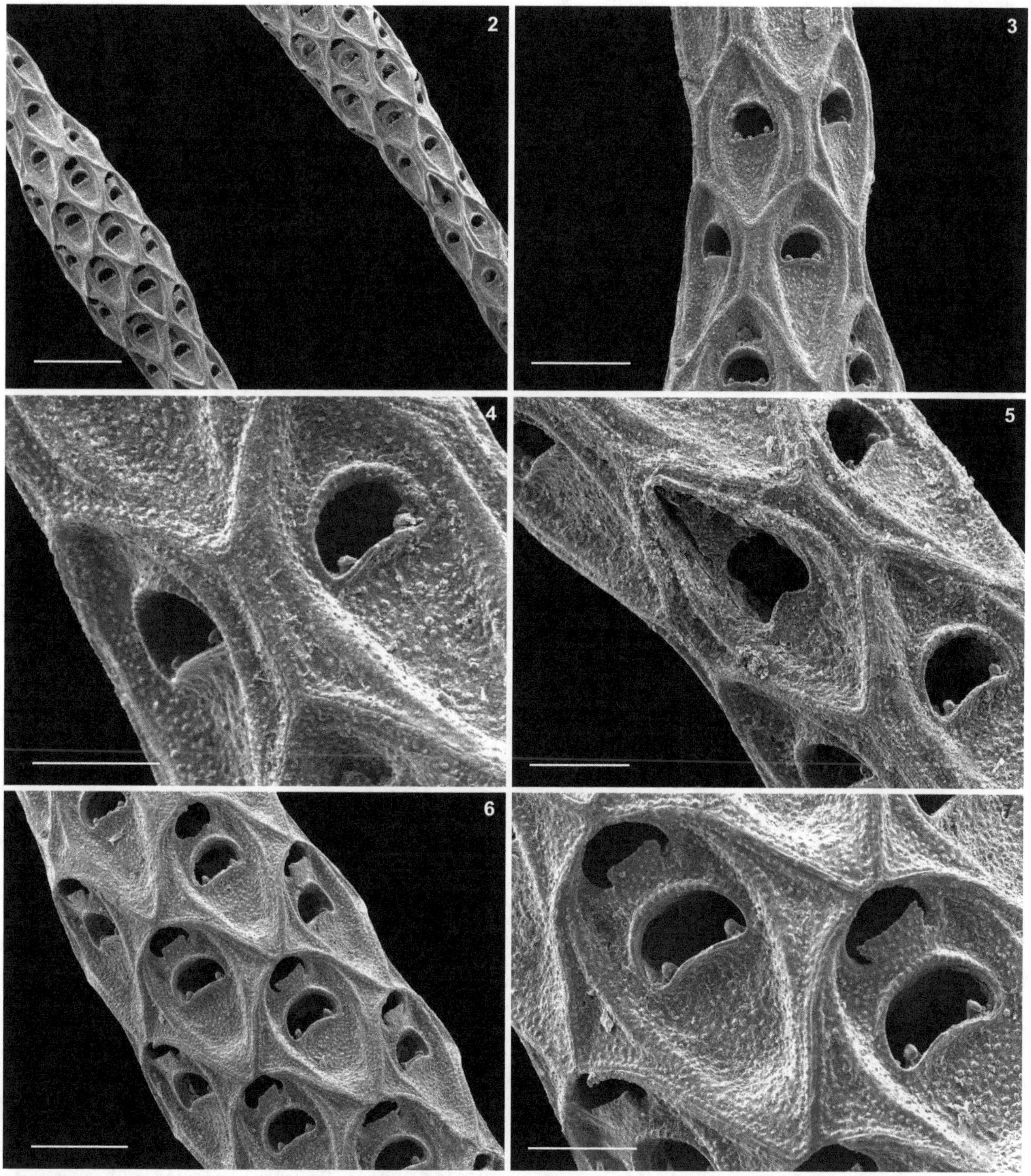

Figures 2–7. *Cellaria oraneae* sp. nov., UFBA 280, holotype, Bahia, Brazil: (2) general aspect of the branches; (3) close-up of subhexagonal autozooids; (4) close-up of opesia showing tuberculate distal rim and proximal condyles; (5) close-up of hexagonal interzooidal avicularia with longitudinal ridges forming a spear-shaped area and sagittate foramen; (6) close-up of fertile rhombic zooids; (7) close-up of immersed ovicells showing oval aperture with proximal rectangular lip. Scale bars: 2 = 500 μm; 3, 6 = 200 μm; 4, 5, 7 = 100 μm.

and slightly convex, two rounded denticles placed near proximal corners (Fig. 4). Interzooidal avicularium sometimes present, about 0.340–0.367 mm (0.354 ± 0.010) long and 0.186–0.210 mm (0.198 ± 0.009) wide, with same cryptocystal calcification and longitudinal ridges as autozooids, forming a spear-shaped area; foramen sagittate, tapered distal edge, convex proximally, forming a broad lip, placed only between infertile zooids (Fig. 5). Fertile zooids wider than autozooids, rhomboid in outline, about 0.316–0.374 mm (0.355 ± 0.016) long and 0.200–0.320 mm (0.285 ± 0.038) wide; opesia semi-elliptical, about 0.050–0.069 mm (0.60 ± 0.004) long and 0.078–0.115 mm (0.103 ± 0.009) wide, same cryptocystal calcification and longitudinal ridges as autozooids (Fig. 6). Ovicells completely immersed, aperture oval with a proximal rectangular lip (Fig. 7).

Material examined. Holotype: Brazil, *Bahia*: Todos os Santos Bay (12°51'62"S, 38°39'78"W, 43 m), 1 specimen, April 1997, Orane Alves leg., UFBA 280. Paratypes: Brazil, *Bahia*: Todos os Santos Bay (12°49'62"S, 38°37'38"W), 1 specimen, May 1997, Orane Alves leg., UFBA 2263; Salvador (12°58'79"S, 38°33'41"W, 50 m), 1 specimen, UFBA 2262. *Additional specimens*: Brazil, *Bahia*: Abrolhos Bank (18°20'80"S, 38°55'97"W, 41 m), 10 specimens, April 1997, Zelinda Leão leg., UFBA 1446; Maraú (Baixo Sul, 14°06'49"S, 38°57'31"W), 5 specimens, August 2004, LAMEB-UFBA leg., UFBA 2261; Cairú (Baixo Sul, 13°27'01"S, 38°46'34"W, 48 m), 2 specimens, 2002, José Dominguez leg., UFBA 2260; Ituberá (Baixo Sul, 13°44'02"S, 38°48'58"W, 47 m), 6 specimens, 2002, José Dominguez leg., UFBA 2259; Maraú (Baixo Sul, 14°05'00"S, 38°53'55"W, 30 m), 6 specimens, 2002, José Dominguez leg., UFBA 2258; Itacaré (Baixo Sul, 14°11'01"S, 38°58'59"W, 15 m), 8 specimens, 2002, José Dominguez leg., UFBA 2241; Camaçari (12°44'12"S, 38°05'12"W, 23 m), 4 specimens, 2002-2004, LAMEB-UFBA leg., UFBA 2257; Camaçari (12°50'00"S, 38°10'06"W, 37 m), 6 specimens, 2002-2006, LAMEB-UFBA leg., UFBA 2256.

Distribution. Atlantic: Brazil (Bahia); infralittoral, from 15 to 50 m.

Etymology. The species epithet is in honor of Orane Alves (Universidade Federal da Bahia), in recognition of her contribution to the knowledge of Brazil's marine biodiversity.

Remarks. *Cellaria oraneae* sp. nov. can be distinguished from all congeners by the combination of hexagonal autozooids and rhomboid fertile zooids, hexagonal interzooidal avicularia with longitudinal ridges forming a spear-shaped area and with sagittate foramen, ovicell completely immersed with aperture oval and a proximal rectangular lip. We analyzed 50 internodes of the new species, among which 30 were formed only by autozooids (infertile internodes) and 20 were fertile (with autozooids and ovicelled zooids). Avicularia were found in both fertile and infertile internodes, but no avicularia were found between fertile zooids. Therefore, we believe that interzooidal avicularia of *C. oraneae* sp. nov. are placed only between autozooids.

Among *Cellaria* species already recorded from the Atlantic Ocean, *C. oraneae* sp. nov. resembles *C. riograndensis* in having cryptocyst mostly granular, hexagonal interzooidal avicularia with longitudinal ridges forming a spear-shaped area, with no condyles and completely immersed ovicell. *Cellaria oraneae* sp. nov. can be distinguished from *C. riograndensis* by having autozooids hexagonal and fertile rhomboid zooids (in *C. riograndensis*, both autozooids and fertile zooids have the same shape), avicularia placed only between autozooids and with sagittate foramen (avicularia is placed between either autozooids and fertile zooids and with triangular foramen in *C. riograndensis*), and ovicell aperture with a proximal rectangular lip (without lip in *C. riograndensis*).

Among other *Cellaria* distributed worldwide, *C. oranae* sp. nov. resembles *C. tenuirostris* (Busk, 1852) in the hexagonal autozooids and rhomboid fertile zooids, granular cryptocyst, hexagonal interzooidal avicularia with longitudinal ridges forming a spear-shaped area and completely immersed ovicell. *Cellaria oranae* sp. nov. is distinct from *C. tenuirostris*, however, by having avicularium with sagittate foramen and no condyles (*C. tenuirostris* have an avicularium with triangular foramen and well-developed condyles), avicularium placed only between autozooids (in *C. tenuirostris* avicularia are placed between either autozooids and fertile zooids), and ovicell aperture with a proximal rectangular lip (without lip in *C. tenuirostris*).

DISCUSSION

The morphological characters commonly used to discriminate among *Cellaria* species include the shape of autozooids and fertile zooids (assigned as hexagonal when formed by six sides or rhomboid when with four sides; Hastings 1947), opesial condyles, type of avicularia and shape of its foramen, ovicell and its aperture. Species of *Cellaria* show a wide range of morphologies, which include species with condyles at the distal opesial rim or at both distal and proximal opesial rim; avicularium large, interzooidal (i.e. replacing one autozooid; Hayward and Ryland 1998) or small, fistulose-like (i.e. interposed in the longitudinal series of autozooids; Hayward and Ryland 1998 and Berning 2013); and ovicells slightly prominent or completely immersed, with aperture varying from a single entire fenestra or ornamented with a proximal lip. The foramen of the avicularia, either interzooidal or fistulose, shows several morphologies, including bell-shaped (i.e. forming an inverted U-shaped curve; López de La Cuadra and García-Gómez 1996); triangular (i.e. three sides with pointed tips); subtriangular (i.e. three sides with rounded edges); semicircular (i.e. as long as wide); semielliptical (i.e. longer than wide); subcordate (i.e. somewhat heart-shaped; López de La Cuadra and García-Gómez 1996); mushroom-shaped (i.e. resembling the profile of a mushroom; Winston and Hayward 2012) and sagittate (i.e. shaped like an arrowhead, Table 1). As known in other bryozoan genera (e.g., Vieira et al. 2013, Almeida et al. 2014), the analysis and characterization the morphology of the ovicell/ooecia are needed to make a taxonomic assignment. In the absence of fertile colonies, it is difficult to assign specimens to any known species using only autozooidal characteristics.

Table 1. Tabular identification key to Atlantic species of *Cellaria*: present (+) and absent (–), avicularia interzooidal (I), avicularia fistulose (F), unknown states (?).

Species	Autozooid		Opesia			Avicularia				Dimorphism in fertile zooid	Ovicell/Aperture
	Shape	Cryptocyst	Distal rim	Proximal rim	Distal condyles	Type	Shape	Foramen	Condyles		
C. atlantida Cook, 1967	rhombic	granular	beaded	convex	–	I	rhombic	bell-shaped	–	–	immersed/with rectangular lip
C. bassleri Hastings, 1947	rhombic	smooth	smooth	convex	–	I	hexagonal	subtriangular	–	–	immersed/ semicircular
C. bafouri Matsuyama, Titschack, Baum & Freiwald, 2015	hexagonal	smooth	beaded	convex	–	–	–	–	–	–	immersed/with rectangular lip
C. brasiliensis Winston, Vieira & Woollacott, 2014	hexagonal	granular	beaded	convex	–	I	hexagonal	subtriangular	+	–	immersed/oval to circular
C. clavata (Busk, 1884)	rhombic	granular	beaded	convex	+	F	semicircular	semicircular	–	+ (hexagonal)	immersed/ circular, with rectangular lip in later astogeny
C. cookae Lopez de la Cuadra & García-Gómez, 1996	hexagonal	granular	beaded	convex	–	I	hexagonal	subcordate	–	–	immersed/with rectangular lip
C. crassa Wood, 1844	hexagonal	granular	?	convex	+	I	rhombic	subtriangular	–	+ (rhombic)	immersed/ semicircular
C. diffusa Robertson, 1905	hexagonal	granular	beaded	convex	+	F	quadrangular	semicircular	–	–	immersed/oval
C. elongatoides Bassler, 1936	hexagonal	granular	?	convex	–	F	circular	semicircular	–	+ (with larger opesial condyles)	immersed/with semicircular lip
C. fistulosa (Linnaeus, 1758)	rhombic/ hexagonal	granular	beaded	convex	–	F	quadrangular to rounded	semielliptical	+	–	immersed/ circular, with rectangular lip in later astogeny
C. harmelini d'Hondt, 1973	hexagonal	smooth	beaded	convex	–	F	semicircular	semicircular	-	–	immersed/with rectangular lip
C. louisorum Winston & Woollacott, 2009	hexagonal	granular	beaded	convex	–	I	hexagonal	triangular	+	+ (with concave proximal rim)	immersed/circular
C. normani (Hastings, 1947)	hexagonal	granular	beaded	convex	–	I	hexagonal	semicircular	+	–	immersed/oval to circular
C. novanglia Winston & Hayward, 2012	hexagonal	granular	beaded	convex	–	F	quadrangular	mushroom-shaped	+	–	immersed/with rectangular lip
C. oraneae sp.	hexagonal	granular	beaded	convex	–	I	hexagonal	sagittate	–	+ (rhombic)	immersed/with rectangular lip
C. ornata d'Orbigny, 1842	rhombic	granular	beaded	convex	?	I	?	triangular	?	?	?
C. paradoxa Hayward & Cook, 1979	hexagonal	granular	smooth	convex	+	–	–	–	–	+ (rhombic, with larger opesia)	immersed/ semicircular
C. riograndensis Ramalho & Calliari, 2015	rhombic/ hexagonal	granular	beaded	convex	–	I	hexagonal	triangular	–	–	immersed/circular
C. salicornioides Lamouroux, 1816	hexagonal	granular	beaded	convex	+	I	hexagonal	semicircular	+	–	immersed/circular
C. scoresbyi Hastings, 1947	hexagonal	granular	beaded	convex	-	I	hexagonal	triangular	+	–	immersed/with rectangular lip
C. sinuosa (Hassall, 1840)	hexagonal	granular	smooth	straight	+	F	circular	semicircular	+	–	immersed/with rectangular lip
C. sobrionoi Lopez de la Cuadra & García-Gómez, 2000	hexagonal	granular	beaded	convex	–	I	rhombic	subtriangular	+	–	prominent/ triangular
C. subtropicalis Vieira, Gordon, Souza & Haddad, 2010	rhombic/ hexagonal	granular	beaded	convex	–	I	hexagonal	subtriangular	+	–	Immersed/ semielliptical to subtriangular
C. triangulata Canu & Bassler, 1925	rhombic	smooth	?	straight	–	?	?	?	?	+ (with larger opesia)	immersed/ triangular

At least 58 species of *Cellaria* are recognized worldwide (Bock and Hayward 2014). So far, 23 species have been recorded from the Atlantic Ocean and, among these, 12 from the Western Atlantic. *Cellaria oraneae* n. sp. is the fourth species of the genus described from Brazil, representing the only species from the Northeast. Since there are few studies on the continental shelf and slope of the Brazilian Northeast, where the genus seems to be common, it is likely that the diversity of *Cellaria* from Brazil has been underestimated. This seems to be the case with other Brazilian bryozoan taxa (e.g., Vieira et al. 2010, Almeida and Souza 2014, Winston et al. 2014, Almeida et al. 2014, 2015a). The real diversity of the genus around the world is also difficult to estimate, since many records have been attributed to the widely-known *C. tenuirostris*, restricted to the Indo-Pacific Oceans (Hastings 1947, Winston 2005). Additionally, some species assigned to *Cellaria* are only known from their original descriptions (e.g., *Cellaria ornata* d'Orbigny, 1842, *Cellaria triangulata* Canu & Bassler, 1925). Without more recent morphological characterization or comparison with related taxa, the validity of some of these may be questionable. Comprehensive studies using SEM are still lacking, preventing the elucidation of the identity of some species as *Cellaria clavata* (Busk, 1884), *Cellaria salicornia* (Pallas, 1766) and *Cellaria normani* (Hastings, 1947) (Hastings 1947, Hayward and Thorpe 1989, López de La Cuadra and García-Gómez 1996). As pointed out by Cook (1967), a review of these taxa is recommended.

ACKNOWLEDGMENTS

This study is part of A.C.S. Almeida's PhD thesis supported by PROTAX-CNPq (440620/2015-5) through the Graduate Program in Animal Biology (Programa de Pós-Graduação em Biologia Animal) of the Departamento de Zoologia, UFPE. We are grateful to Centro de Pesquisa Gonçalo Moniz (FIOCRUZ/BA) for SEM images, Orane Alves (Laboratório de Geoecologia de Sedimentos Marinhos, UFBA), Marlene Peso-Aguiar (Laboratório de Malacologia e Ecologia de Bentos, UFBA) and Ulisses Pinheiro (Laboratório de Porifera, UFPE) for logistical support.

LITERATURE CITED

Almeida ACS, Souza FBC (2014) Two new species of cheilostome bryozoans from the South Atlantic Ocean. Zootaxa 3753: 283–290. https://doi.org/10.11646/zootaxa.3753.3.7

Almeida ACS, Souza FBC, Menegola CMS, Sanner J, Vieira LM (2014) Taxonomic review of the family Colatooeciidae Winston, 2005 (Bryozoa, Cheilostomata), with description of seven new species. Zootaxa 3868: 001–061. https://doi.org/10.11646/zootaxa.3868.1.1

Almeida ACS, Souza FBC, Sanner J, Vieira LM (2015a) Taxonomy of recent Adeonidae (Bryozoa, Cheilostomata) from Brazil, with the description of four new species. Zootaxa 4013: 348–368. https://doi.org/10.11646/zootaxa.4013.3.2

Almeida ACS, Alves O, Peso-Aguiar M, Dominguez J, Souza FBC (2015b) Gymnolaemata bryozoans of Bahia State, Brazil. Marine Biodiversity Records 8: e120. https://doi.org/10.1017/S1755267215000743

Barnes DKA, Whittington M (1999) Biomechanics and mass mortality of erect bryozoans on a coral reef. Journal of the Marine Biological Association of the United Kingdom 79: 745–747. https://doi.org/10.1017/S0025315498000903

Berning B (2013) New and little-known Cheilostomata (Bryozoa, Gymnolaemata) from the NE Atlantic. European Journal of Taxonomy 44: 1–25. https://doi.org/10.5852/ejt.2013.44

Bock PE, Gordon DP (2013) Phylum Bryozoa Ehrenberg, 1831. Zootaxa 3703: 67–74. https://doi.org/10.11646/zootaxa.3703.1.14

Bock PE, Hayward PJ (2014) World Register of Marine Species, World List of Bryozoa, Genus *Cellaria* Ellis & Solander, 1786. In: Bock PE, Gordon D (Eds) The Bryozoa Home Page. http://www.marinespecies.org/aphia.php?p = taxdetails&id = 110869

Braga LM (1967) Notas sobre alguns briozoários marinhos brasileiros coletados pelo Navio Oceanográfico Almirante Saldanha. Instituto de Pesquisa da Marinha, Notas Técnicas, 2: 1–12.

Cook PL (1967) Polyzoa (Bryozoa) from West Africa. The Pseudostega, the Cribrimorpha and some Ascophora Imperfecta. Bulletin of the British Museum of Natural History 15: 321–351.

Hastings AB (1947) Notes on Polyzoa (Bryozoa). III. On some species of *Cellaria*, with remarks on G. Busk's grouping of the species of this genus. Annals and Magazine of Natural History 11: 217–241. https://doi.org/10.1080/00222934608654548

Hayward PJ, Thorpe JP (1989) Membraniporoidea, Microporoidea and Cellarioidea (Bryozoa, Cheilostomata) collected by Discovery investigations. Journal of Natural History 23: 913–959. https://doi.org/10.1080/00222938900771071

Hayward PJ, Ryland JS (1998) Cheilostomatous Bryozoa. Part 1. Aeteoidea - Cribrilinoidea. In: Barnes RSK, Crothers JH (Eds) Synopses of the British Fauna. Field Studies Council, Shrewsbury, 1–366.

Hayward PJ, McKinney FK (2002) Northern Adriatic Bryozoa from the vicinity of Rovinj, Croatia. Bulletin of the American Museum of Natural History 270: 1–139. https://doi.org/10.1206/0003-0090(2002)270<0001:NABFTV>2.0.CO;2

International Commission on Zoological Nomenclature (1999) International code of zoological nomenclature. London, The International Trust for Zoological Nomenclature.

López de la Cuadra C, García-Gómez JC (1996) The species of *Cellaria* (Bryozoa, Cheilostomatida) with large avicularia from West Africa. Journal of Natural History 30: 153–161. https://doi.org/10.1080/00222939600770101

McKinney FK, Jaklin A (2000) Spatial niche partitioning in the *Cellaria* meadow epibiont association, northern Adriatic Sea. Cahiers de Biologie Marine 41: 1–17.

McKinney FK, Jaklin A (2001) Sediment accumulation in a shallow-water meadow carpeted by a small erect bryozoan. Sedimentary Geology 145: 397–410. https://doi.org/10.1016/S0037-0738(01)00156-7

Ramalho LV, Calliari L (2015) Bryozoans from Rio Grande do Sul Continental Shelf, Southern Brazil. Zootaxa 3955: 569–587. https://doi.org/10.11646/zootaxa.3955.4.8

Taylor PD, James NP (2013) Secular changes in colony-forms and bryozoan carbonate sediments through geological history. Sedimentology 60: 1184–1212. https://doi.org/10.1111/sed.12032

Vieira LM, Migotto AE, Winston JE (2008) Synopsis and annotated checklist of recent marine Bryozoa from Brazil. Zootaxa 1810: 1–39.

Vieira LM, Gordon DP, Souza FBC, Haddad MA (2010) New and little-known cheilostomatous Bryozoa from the south and southeastern Brazilian continental shelf and slope. Zootaxa 2722: 1–30.

Vieira LM, Winston JE, Fehlauer-Ale KH (2012) Nine new species of *Bugula* Oken (Bryozoa: Cheilostomata) in Brazilian shallow waters. PLoS One 7: e40492. https://doi.org/10.1371/journal.pone.0040492

Vieira LM, Spencer jones ME, Winston JE (2013) *Cradoscrupocellaria*, a new bryozoan genus for *Scrupocellaria bertholletii* (Audouin) and related species (Cheilostomata, Candidae): taxonomy, biodiversity and distribution. Zootaxa 3707: 1–63. https://doi.org/10.11646/zootaxa.3707.1.1

Vieira LM, Migotto AE, Winston JE (2014) Ctenostomatous Bryozoa from São Paulo, Brazil, with descriptions of twelve new species. Zootaxa 3889: 485–524. https://doi.org/10.11646/zootaxa.3889.4.2

Winston JE (2005) Redescription and revision of Smitt's "Floridan Bryozoa" in the Collection of the Museum of Comparative Zoology, Harvard University. Virginia Museum of Natural History Memoir 7: 1–150.

Winston JE, Woollacott RM (2009) Scientific results of the *Hassler* expedition. Bryozoa. No. 1. Barbados. Bulletin of the Museum of Comparative Zoology 159: 239–300. https://doi.org/10.3099/0027-4100-159.5.239

Winston JE, Hayward PJ (2012) The Marine Bryozoans of the Northeast Coast of the United States: Maine to Virginia. Virginia Museum of Natural History Memoir 11: 1–180.

Winston JE, Vieira LM (2013) Systematics of interstitial encrusting bryozoans from southeastern Brazil. Zootaxa 3710: 101–101. https://doi.org/10.11646/zootaxa.3710.2.1

Winston JE, Vieira LM, Woollacott RM (2014) Scientific results of the Hassler Expedition. Bryozoa. 2. Brazil. Bulletin of the Museum of Comparative Zoology at Harvard College 161: 139–239. https://doi.org/10.3099/MCZ14.1

Author Contributions: ACSA FBCS and LMV analyzed, measured and described the specimens and wrote the paper.

Competing Interests: The authors have declared that no competing interests exist.

18

A new species of burrowing snake (Serpentes: Dipsadidae: *Apostolepis*) from the state of Mato Grosso, Central-West region of Brazil

Fernanda Martins dos Santos[1], Omar Machado Entiauspe-Neto[2], Jailini da Silva Araújo[3], Moisés Barbosa de Souza[3], Thales de Lema[4], Christine Strüssmann[5], Nelson Rufino de Albuquerque[1,6]

[1]*Laboratório de Zoologia, Campus do Pantanal, Universidade Federal de Mato Grosso do Sul. Avenida Rio Branco 1270, 79304-902 Corumbá, MS, Brazil.*
[2]*Laboratório de Vertebrados, Instituto de Ciências Biológicas, Universidade Federal do Rio Grande. Avenida Itália Km 8, 96203-900 Rio Grande, RS, Brazil.*
[3]*Laboratório de Herpetologia, Centro de Ciências Biológicas e da Natureza, Universidade Federal do Acre. Rodovia BR 364, Km 4, 69915-900 Rio Branco, AC, Brazil.*
[4]*Laboratório de Sistemática de Vertebrados, Museu de Ciências e Tecnologia, Pontifícia Universidade Católica do Rio Grande do Sul. Avenida Ipiranga 6681, 90619-900 Porto Alegre, RS, Brazil.*
[5]*Faculdade de Medicina Veterinária, Universidade Federal de Mato Grosso. Avenida Fernando Correia da Costa 2367, 78060-900 Cuiabá, MT, Brazil.*
[6]*Programa de Pós-Graduação em Biologia Animal, Instituto de Biociências, Universidade Federal de Mato Grosso do Sul. Avenida Costa e Silva, Cidade Universitária, 79070-900 Campo Grande, MS, Brazil.*
Corresponding author: Nelson Rufino de Albuquerque (nelson_rufino@hotmail.com)

ABSTRACT. During a faunal rescue conducted at a hydroelectric power station constructed in a Cerrado savanna area in the state of Mato Grosso, a sample of five small stripe-patterned individuals of snakes of the genus *Apostolepis* Cope, 1862 document the existence of an undescribed species, which is named herein. The new species can be distinguished from its congeners by a combination of scale counts, number of maxillary teeth and color pattern. The new species is most similar to *Apostolepis borellii* Peracca, 1904, *A. lineata* Cope, 1887, *A. nelsonjorgei* Lema & Renner, 2004, *A. nigroterminata* Boulenger, 1896, *A. serrana* Lema & Renner, 2006 and *A. underwoodi* Lema & Campbell, 2017 in its coloration pattern. However, it is distinguished from these species by having a pair of triangular blotches covering portions of the third to sixth supralabials, a white nuchal collar, the shape of the fourth supralabial and the shape of the tip of tail, the number of supralabials in contact with parietals, the size of the anterior chinshields, the color pattern of the paraventral side, parietal and terminal scales, the width of dorsal stripes, and a distinct number of subcaudals. The new species occurs in areas within the Cerrado biome.

KEY WORDS. Biodiversity, Cerrado, Chapada dos Guimarães, Elapomorphini, taxonomy.

INTRODUCTION

The taxonomic history of the South American Dipsadinae genus *Apostolepis* Cope, 1862 has been rather confusing, and researchers have long recognized the lack of a comprehensive review of taxonomy and phylogenetic relationships in this genus (Ferrarezzi et al. 2005, Nogueira et al. 2012). The genus presently comprises 38 species (Curcio et al. 2011, Lema 2016, Cabral et al. 2017, Lema et al. 2017 and Uetz et al. 2018) distributed in five groups – *dimidiata, dorbignyi, flavotorquata, longicaudata* and *nigroterminata* (Nogueira et al. 2012). Additional groups were referred to by Lema (2003a), Martins and Lema (2015) and Lema and Renner (2016) – including *ambinigra, assimilis, lineata, nigrolineata, phillipsi, polylepis, quinquelineata* and a *borellii* group, which presently includes only *A. borellii* Peracca, 1904 and *A. underwoodi* Lema & Campbell, 2017. However, with the exception of the *ambinigra* (Lema and Martins 2016), *assimilis* (Ferrarezzi et al. 2005) and *dimidiata* groups (Nogueira et al. 2012), the remaining groups have not yet been formally described and we refrain from considering them.

Apostolepis is particularly speciose in Brazil, where 32 species have already been recorded (Costa and Bérnils 2018). Contributions to the taxonomy of these burrowing snakes in Central-West region of Brazil were initially made by Cope (1887), who described two species – *Apostolepis lineata* and *A. vittata* – from the municipality of Chapada dos Guimarães, state of Mato Grosso. Subsequently, Koslowsky (1898) described *A. intermedia* from Miranda, state of Mato Grosso do Sul. A few years later, Peracca (1904) described *A. borellii* based on a specimen from Urucum massif, near Corumbá, Mato Grosso do Sul; Amaral (1925) described *A. rondoni* based on a specimen from "Matto Grosso" (=the former name of a huge region which presently comprises three Brazilian states: Mato Grosso do Sul, Mato Grosso, and Rondônia), and Prado (1942) described *A. goiasensis* based on a specimen from Rio Verde, Goiás. More recently, Lema (2002a,b, 2003a), and Ferrarezzi et al. (2005), respectively, described *A. albicollaris* from Brasília, Distrito Federal, *A. christineae* from Barra do Bugres, presently Porto Estrela, Mato Grosso, *A. cerradoensis* from Minaçú, Goiás, and *A. ammodites* from Palmas, Tocantins, whereas Lema and Renner (2006) and Lema (2016) described two new species from Serra do Roncador, Mato Grosso, *A. serrana* and *A. roncadori*, respectively. Additionally, Martins and Lema (2015) and Lema and Renner (2016) removed *A. borellii* from the synonymy of *A. nigroterminata*.

During a faunal rescue conducted at a hydroelectric power station constructed in a Cerrado savanna area in the state of Mato Grosso, a sample of five small stripe-patterned individuals of snakes of the genus *Apostolepis* document the existence of an undescribed species, which is named herein.

MATERIAL AND METHODS

Institutional acronyms are as listed in Sabaj (2016), except for UFAC (Coleção Herpetológica da Universidade Federal do Acre, Rio Branco, Acre, Brazil) and UFMT (Coleção Zoológica de Vertebrados da Universidade Federal de Mato Grosso, Cuiabá, Mato Grosso, Brazil). Specimens of the new species were compared with 85 museum specimens of 15 species of *Apostolepis* (see Appendix) and data from the literature (e.g., Amaral 1925, Harvey et al. 2001, Lema 2003b, 2004, 2016, Lema and Renner 2004, 2016, Curcio et al. 2011, Albuquerque and Lema 2012, Nogueira et al. 2012, Borges-Nojosa et al. 2016, Cabral et al. 2017, Lema and Campbell 2017).

All measurements were made to the nearest 0.1 mm using digital calipers, except for snout-vent (SVL) and tail (TL) lengths, which were taken with a flexible ruler to the nearest 1.0 mm. Ventral scales were counted according to Dowling (1951). Bilateral variation is reported as right/left. When no everted hemipenis was available, the sex of each specimen was determined by making a post-cloacal incision between the eighth and tenth subcaudals to verify the presence of the hemipenes. The hemipenial description is based on right organs from two preserved specimens (UFMT-R 1933 and MCP 14524), which were prepared according to the method described in Pesantes (1994) and Zaher and Prudente (2003). UFMT-R 1933 was not expanded or inflated since it was irreversibly damaged during filling of the hemipenial body. On the other hand, MCP 14524 was fully inflated with red petroleum jelly. Terminology for hemipenial morphology followed Dowling and Savage (1960) and Zaher (1999). The distribution map was made using the free Quantum GIS software. We use the names *A. quinquelineata* and *A. nigrolineata* in the sense of Lema (1997) and Lema and Renner (1998), respectively (but see Curcio et al. 2011).

TAXONOMY

Apostolepis kikoi sp. nov.

Figs 1–3, Table 1

Apostolepis sp. – Strüssmann 2000: 163. [Cresonymy]
Apostolepis sp. 1 – Martins and Lema 2015: 102; Lema and Renner 2016: 71. [Cresonymy]
Apostolepis sp. 3 – Martins and Lema 2015: 102 (*partim*). [Cresonymy]
Apostolepis aff. *borellii* – Lema and Campbell 2017: 28 (*partim*). [Cresonymy]

Holotype. A female (MCP 12096) collected in 2000 at the Manso multi-use reservoir and hydroelectrical power plant – locally known as APM Manso – constructed at the confluence of the rivers Manso and Casca, Chapada dos Guimarães (15°27'39"S, 55°45'00"W; 811 m.a.s.l.), Mato Grosso, Brazil, by the faunal rescue team. Paratypes. three males and one female, same locality as the holotype: MCP 14524 (male), MCP 14525 (male) and MCP 11372 (female), date of collection unknown, collected by the faunal rescue team, and UFMT-R 1933 (male) collected on 1 December 1999 by the faunal rescue team.

Diagnosis and comparison with other species. *Apostolepis kikoi* sp. nov. can be distinguished from all other *Apostolepis* by the combination of having five dorsal stripes (vs. dorsal stripes absent in *A. ambiniger*, *A. ammodites*, *A. assimilis*, *A. breviceps*, *A. cearensis*, *A. dorbignyi*, *A. flavotorquata*, *A. multicincta*, *A. roncadori* and *A. tertulianobeui*; the presence of seven stripes on the dorsum in *A. gaboi* and *A. niceforoi*; three stripes on the dorsum in *A. cerradoensis*, *A. goiasensis*, *A. nigrolineata*, *A. quirogai* and *A. tenuis*; a pair of narrow lateral stripes in *A. barrioi*; a pair of wide lateral stripes in *A. albicollaris*, *A. dimidiata* and *A. polylepis*); the presence of a white nuchal collar (vs. white nuchal collar absent in *A. ambiniger*, *A. barrioi*, *A. breviceps*, *A. christineae*, *A. goiasensis*, *A. intermedia*, *A. lineata*, *A. longicaudata*, *A. niceforoi*, *A. nigrolineata*, *A. polylepis*, *A. quinquelineata*, *A. rondoni*, *A. serrana*, *A. striata*, *A. thalesdelemai* and *A. vittata*); the presence of a triangular blotch covering portions of the third, fourth, fifth and sixth supralabials (vs. a light lateral spot below the eye, usually occupying the third and fourth supralabials in *A. mariae*); 15 scale rows at midbody (vs. 17 in *A. polylepis*); six supralabials (vs. five in *A. breviceps*, *A. christineae*, and *A. vittata*); second and third supralabials entering orbit (vs. only the third supralabial entering orbit in *A. breviceps*); preocular contacting nasal (vs. nasal and preocular separated by prefrontal in *A. ammodites*, *A. arenaria*, *A. assimilis*, *A. breviceps*, *A. cearensis*, *A. dorbignyi*, *A. gaboi*, *A. goiasensis*, *A. intermedia*,

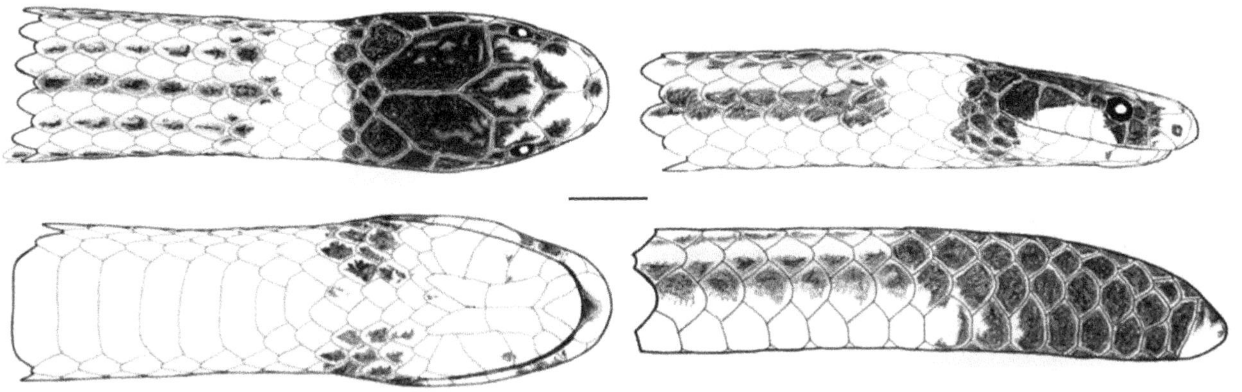

Figure 1. Line drawings from holotype of *Apostolepis kikoi* sp. nov. (MCP 12096) from the reservoir of Manso hydroelectrical power plant, Chapada dos Guimarães, Mato Grosso, Brazil. Scale bar: 2 mm (Eduardo Melloni Lucchesi del.).

A. multicincta, A. phillipsi, A. polylepis, A. quirogai, A. tenuis and *A. tertulianobeui*); temporals absent (usually 0 + 1 in *A. ammodites, A. assimilis, A. cearensis, A. mariae, A. niceforoi, A. nigrolineata, A. quirogai, A. tertulianobeui* and *A. thalesdelemai*, and 1 + 1 in *A. flavotorquata* and *A. quinquelineata*); seven infralabials (vs. five in *A. breviceps*; five to six in *A. nelsonjorgei* and *A. vittata*; six in *A. christineae, A. intermedia, A. multicincta*; eight in *A. gaboi* and *A. quirogai*); four infralabials contacting the first pair of chinshields (vs. three infralabials contacting the first pair of chinshields in *A. dorbignyi, A. intermedia, A. multicincta, A. tenuis* and *A. vittata*); a higher number of ventrals than *A. arenaria* and *A. striata* (203–209 vs. 168–181 and 202, respectively); fewer ventrals than *A. christineae, A. intermedia, A. longicaudata, A. nelsonjorgei, A. niceforoi, A. phillipsi, A. polylepis, A. serrana, A. tertulianobeui, A. thalesdelemai* and *A. vittata* (203–209 vs. 211–248 in the latter eleven species); fewer subcaudals than *A. borellii, A. intermedia, A. longicaudata, A. nelsonjorgei, A. serrana* and *A. tertulianobeui* (26–30 vs. 32–55 in the latter six species); a higher number of subcaudals than *A. lineata, A. niceforoi* and *A. polylepis* (26–30 vs. 24, 23 and 20–25, respectively) and fewer maxillary teeth than *A. longicaudata* and *A. phillipsi* (4 + 2 vs. 5 + 2 in the latter two species).

Preserved specimens of *Apostolepis kikoi* sp. nov. are most similar to *A. borellii, A. lineata, A. nelsonjorgei, A. nigroterminata, A. serrana* and *A. underwoodi* in its general pholidosis and coloration pattern. However, the new species is distinguished from *A. borellii* by having its paraventral sides cream and unblemished (vs. blackish), a pair of triangular blotches covering portions of the third, fourth, fifth and sixth supralabials (vs. a small, trapezoidal blotch covering only the posterior half of the third and the entire fourth supralabial); tip of the tail conical (vs. rounded) and fewer subcaudals (vs. 26–30 vs. 32). *Apostolepis kikoi* sp. nov. can be distinguished from *A. nigroterminata* by having a triangular blotch covering portions of the third, fourth, fifth and sixth supralabials (vs. an irregular blotch on the posterior margin of the third to the anterior margin of the fourth supralabial – see also Harvey 1999: 401, fig. 7 in Lema and Renner 2016, and Figs 4–5 below), a blackish blotch on rostral scale adjacent to anterior border of prefrontals (vs. blotch absent), darker parietals (vs. light blotches irregularly distributed on parietal scales) and a distinct number of maxillary teeth (4 + 2 vs. 3 + 2). Further, it differs from *A. nigroterminata* in having its background color beige (in living specimens) (vs. background color red-orange); paravertebrals distinct (vs. paravertebrals indistinct); first and fifth stripes wider, covering upper half of third and lower half of fourth row on each side (vs. first and fifth stripes also wider, but covering first and about 50% of fourth row of scales on each side). *Apostolepis kikoi* sp. nov. can be distinguished from *A. underwoodi* by having a vertebral stripe one scale wide (vs. vertebral stripe narrow, running on the medial line of each vertebral scale); paravertebral stripes covering sixth row on each side (vs. paravertebral stripes covering half of fifth and half of sixth rows). Finally, *Apostolepis kikoi* sp. nov. is distinguished from *A. lineata, A. nelsonjorgei* and *A. serrana* by having its fourth supralabial scale rectangular (vs. triangular in *A. nelsonjorgei*), 4–6 supralabials contacting parietals (vs. 5–6 contacting parietals in *A. nelsonjorgei*), terminal scale black dorsolaterally (vs. terminal scale entirely white in *A. lineata* and *A. nelsonjorgei*), anterior chinshields longer than posterior (vs. anterior and posterior chinshields of about the same size in *A. nelsonjorgei*), the presence of a white nuchal collar (vs. white nuchal collar absent in *A. lineata* and *A. serrana*) and a distinct number of subcaudals (27–30 vs. 40–46 in *A. nelsonjorgei* and 33 in *A. serrana*).

Description of holotype (Figs 1–3). A small female, possibly juvenile, SVL 262 mm, TL 26 mm (9.92% of SVL). Body subcylindrical. Tail very short with tip conical and laterally compressed. Terminal scale pointed. Head slightly distinct from neck, narrower than diameter of midbody. Head length from quadrate-articular jaw joint to tip of snout (in lateral view) 6.56 mm (2.5% of SVL), 4 mm at widest point (60.9% of head length). Snout rounded in dorsal and lateral views, slender and slightly projected beyond jaws; snout length from tip of snout to anterior margin of right orbit 2.37 mm (36.12% of head

Figure 2–5. (2) Holotype of *Apostolepis kikoi* sp. nov. (MCP 12096) from the reservoir of Manso hydroelectrical power plant, Chapada dos Guimarães, Mato Grosso, Brazil (photo by Marcos Di Bernardo). (3) Lateral view of holotype (preserved) of *Apostolepis kikoi* sp. nov. (MCP 12096) from the reservoir of Manso hydroelectrical power plant, Chapada dos Guimarães, Mato Grosso, Brazil (photo by Marcos Di Bernardo). Dorsal (4) and lateral (5) views of the specimen of *Apostolepis nigroterminata* (UFAC 504) from Campus Universitário da Universidade Federal do Acre, Rio Branco, Acre, Brazil (photo by Danyella Paiva da Silva).

length). Rostral wider than high, visible from above; portion of rostral trapezoidal from above, 1.52 mm wide; length of rostral visible from above 0.63 mm. Prefrontal-rostral contact broadly separating nasals. Suture between prefrontals 58.4% of length of frontal. Prefrontals almost as wide as long. Frontal hexagonal, longer (2.09 mm) than wide (1.42 mm at widest point). Nostril-orbit distance in right lateral view 1.23 mm (16.49% of head length). Interorbital width (shortest distance between dorsomedial margins of orbits) 2.51 mm (66.57% of head width). Nasal entire, contacting preocular. Preocular rectangular, about 34% of length of nasal. Loreal absent. Parietal almost twice as long (3.08 mm) as wide (1.44 mm at widest point); suture between parietals 1.97 mm; length of suture between parietals 94.2% length of frontal. Supraoculars subretangular, almost twice as long as wide. Maximum diameter of eye 0.70 mm, pupils round. Postoculars pentagonal, higher than wide. Temporals absent. Five occipitals wider than long; median occipital positioned between posterior tips of parietals, smaller than adjacent vertebral; two pairs of lateral occipitals twice as large as dorsals, contacting posterior edges of sixth supralabials. Supralabials 6/6, 1 contacting rostral, 1–2 contacting nasal, 2 contacting nasal and preocular, 2–3 entering orbit, 3–4 contacting single postocular, 4–6 contacting parietal. Supralabials in ascending order of size, with sixth supralabial higher and longer (1.46 mm long, 1.29 mm high) than remaining supralabials. Infralabials 7/7, 1–4 contacting anterior chinshields, 4–7 contacting posterior chinshields. Mental subtriangular, wider (1.14 mm) than long (0.82 mm), separated from anterior chinshields by contact between first infralabials. Anterior chinshields elongated, longer than posterior chinshields. Suture between chinshields 1.82 mm. Chinshields separated from ventrals by four gulars and two preventrals. Gulars in four rows between last supralabial and

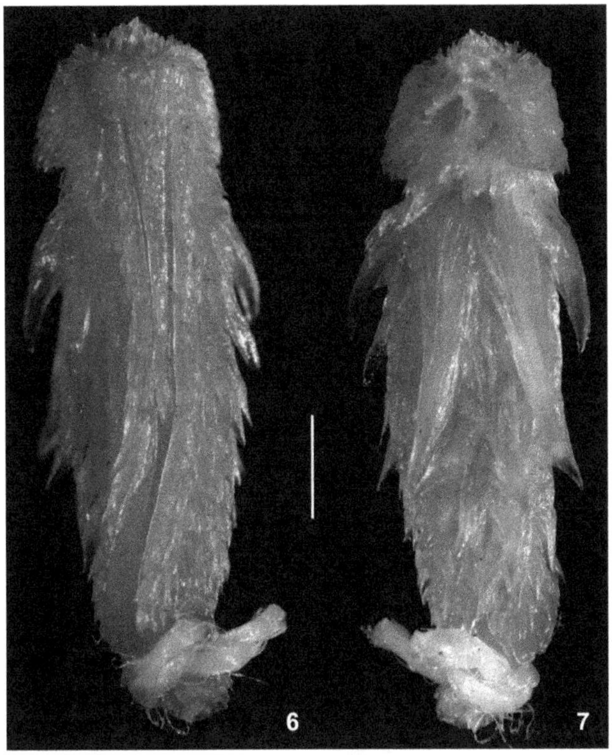

Figures 6–7. Hemipenis of *Apostolepis kikoi* sp. nov. (MCP 14524) from the reservoir of Manso hydroelectrical power plant, Chapada dos Guimarães, Mato Grosso, Brazil, showing sulcate (6) and absulcate (7) sides (photo by Fernanda Martins dos Santos). Scale bar: 1 mm.

first preventral. Dorsal scales smooth in 15 rows at midbody. No apical pits. 205 ventrals. 26 paired subcaudals (tail complete). Anal scale divided. 4 + 2 maxillary teeth.

Coloration of holotype in life (Fig. 2). Head dorsally blackish with light blotches irregularly distributed on rostral, nasal, prefrontal, frontal, supraocular and parietal scales. Suture between prefrontals brownish. Dorsally, rostral scale has a blackish blotch adjacent to anterior border of prefrontals; ventrally, rostral scale has a blackish spot in its convex portion. Posterior upper margin of first and fourth supralabials blackish. Second supralabial blackish. Anterior portion and upper half of third supralabial blackish. Blackish head cap covers upper half of fifth and sixth (except for its lower anterior portion) supralabials. Posterior portion of nasal, preocular and postocular scales blackish. Infralabials and chinshields cream, except for small blackish spots on third and fourth scales. Brown pigment of throat region restricted to more laterally positioned gular scales (i.e., evidencing an incomplete gular band). White nuchal collar two and half to three dorsals long. Black nuchal collar absent. Background color beige, with five black stripes; vertebral stripe one scale wide; paravertebral stripes covering sixth and tenth rows, less distinct than remaining dorsal stripes; first and fifth stripes wider, covering upper half of third and lower half of fourth row on each side. Paraventral sides and venter unblemished. Black band on tail extends for nine scales dorsally; seven subcaudals are black. Terminal scale with band of melanophores dorsolaterally and entirely clear ventrally.

Color of holotype in preservative (Fig. 3). Head dorsally brownish with light blotches irregularly distributed on rostral, nasal, prefrontal, frontal, supraocular and parietal scales. Dorsally, rostral scale brown pigmented; this blotch covers anterior border of prefrontals; ventrally, rostral scale with brownish spot on its convex portion. Anterior and posterior parts of prefrontals brown. Anterior and posterior parts of suture between prefrontals brown. Posterior upper margin of first and fourth supralabials brown. Second supralabial brown. Anterior portion and upper half of third supralabial brown. Brown head cap covers upper half of fifth and sixth (except for its lower anterior portion) supralabials. Posterior portion of nasal, preocular and postocular scales brown. Infralabials and chinshields cream, except for small brown spots on third and fourth scales. Brown pigment of throat region restricted to more laterally positioned gular scales (i.e., evidencing an incomplete gular band). White nuchal collar two and half to three dorsals long. Black nuchal collar absent. Background color light brown, with five brown stripes; paravertebral stripes light brown and less distinct than remaining dorsal stripes. Venter immaculate. Dark brown band on tail extends for nine scales dorsally; seven subcaudals are dark brown. Terminal scale with band of melanophores dorsolaterally and entirely clear ventrally.

Variation. Measurements and morphological variation are summarized in Table 1. Largest male (MCP 14524) SVL 310 mm, TL 38 mm. Largest female (MCP 12096, holotype) SVL 262 mm, TL 26 mm. Smallest specimen measured is female (MCP 11372) with SVL 152 mm and TL 16 mm. Fifth and sixth supralabials in

Table 1. Variation in characters of the type series of *Apostolepis kikoi* sp. nov. Supralabials (SL), infralabials (IL), snout-vent length (SVL), tail length (TL), snout length (SL), head length (HL), preventrals (PV), ventrals (VE), subcaudals (SC), number of dorsal scales of black caudal band (DBB) and number of subcaudal scales of black caudal band (SBB). Measurements in mm.

	MCP 11372	MCP 12096 (holotype)	MCP 14524	MCP 14525	UFMT-R 1933
Sex	F	F	M	M	M
SL	6/6	6/6	6/6	6/6	6/6
IL	7/7	7/7	7/7	7/7	7/7
SVL	152	262	310	304	263
TL	16	26	38	37	35
SL	1.74	2.37	2.70	2.61	2.46
HL	5.83	6.56	8.01	7.56	7.44
PV + VE	2 + 207	1 + 205	1 + 209	2 + 208	2 + 203
SC	28	26	29	29	30
DBB	10	9	7	8	9
SBB	6	8	5	3	5

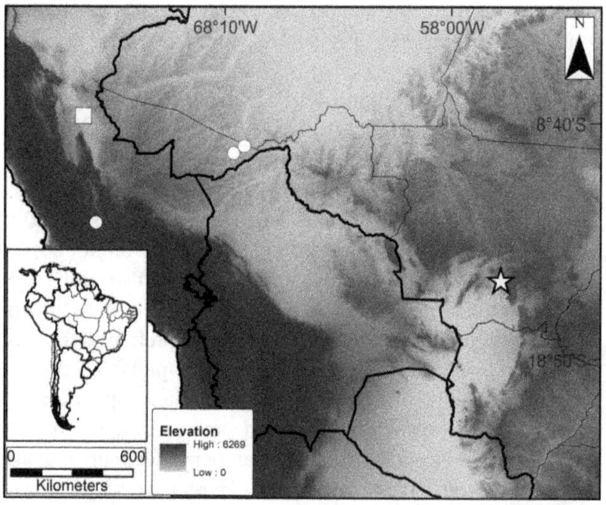

Figure 8. Map of South America, illustrating the type locality (star) of *Apostolepis kikoi* sp. nov. – reservoir of Manso hydroelectrical power plant – Chapada dos Guimarães (15°27′39″S, 55°45′00″W, 811 m a.s.l.), Mato Grosso, Brazil, and locality records for *Apostolepis nigroterminata* (circles) in the Brazilian state of Acre and Peru based upon the material examined. Square indicates the type locality of *Apostolepis nigroterminata*.

contact with parietal in MCP 14524, MCP 14525 and UFMT-R 1933; fourth, fifth and sixth supralabials in contact with parietal in MCP 11372. Head length ranges from 7.44–8.01 mm ($\bar{x}$ = 7.7 ± 0.3) in males, 5.83–6.56 mm ($\bar{x}$ = 6.2 ± 0.5) in females. Number of ventral scales ranges from 203–209 ($\bar{x}$ = 206.7 ± 3.2, n = 3) in males and 205–207 ($\bar{x}$ = 206 ± 1.4, n = 2) in females. Subcaudals range from 29–30 ($\bar{x}$ = 29.3 ± 0.6, n = 3) in males and 26–28 in females ($\bar{x}$ = 27 ± 1.4, n = 2). Preventrals 1 (n = 3) or 2 (n = 2). First infralabial fused with first chinshield in MCP 14524 on right side. Suture between prefrontals entirely pigmented by brown blotch in UFMT-R 1933. White nuchal collar two and a half to three dorsals long (MCP 11372, MCP 14524 and UFMT-R 1933) or two to three dorsals long (MCP 14525). Black nuchal collar vestigial in MCP 11372, MCP 14524 and MCP 14525 and apparent in UFMT 1933. First dorsal stripe, vertebral stripe and fifth dorsal stripe brown; paravertebral stripes light brown and less distinct than remaining dorsal stripes in all specimens. Caudal band extends for length of 7–10 scales dorsally and 3–8 subcaudals. In MCP 11372 fourth supralabial on each side entirely clear; it has a narrow, vestigial band of melanophores on the anterior portion of the terminal scale (dorsally) and small black blotches can be observed on its 22nd, 24th, 25th, 26th and 27th pair of subcaudal scales. Terminal scale with band of melanophores dorsolaterally and entirely clear ventrally in MCP 11372, MCP 14524 and UFMT-R 1933.

Hemipenial morphology (Figs 5–7). Retracted organs extend for length of nine subcaudals. Everted hemipenes subcylindrical, unilobed, unicapitate and noncalyculate. Basal region on sulcate side bears numerous spines of similar size. Several moderate-sized spines present on lateral region of hemipenial body in its absulcate side. Sulcus spermaticus bifurcates about two-thirds before end of organ; branches – which extend centrolineally – reach distal tip of lobe. Basal region on absulcate side also bears numerous spines of similar size, but these are abruptly replaced by two larger spines in middle region of hemipenial body, larger than those disposed on lateral region of hemipenial body. Capitulum confined to sulcate side. Distal region of absulcate side bears transverse papillate flounces, without calyces, whereas a small number of papillae are concentrated above flounces.

Distribution (Fig. 8). All individuals of the type series of *Apostolepis kikoi* sp. nov. were obtained in the area presently occupied by the Manso multi-use reservoir and hydroelectrical power plant (APM Manso), in most part situated in the municipality of Chapada dos Guimarães, Mato Grosso, Brazil. With nearly 428 km^2, the reservoir of APM Manso inundated – from December 1999 – many different Cerrado physiognomies (see Conceição 2000), established over terrains up to 287 m.a.s.l. from the confluence of the Casca and Manso rivers (approximately at 14°52′ S, 55°48′ W) upwards. Manso is the main tributary to the Cuiabá River, a major tributary of the left bank of the Paraguay River. The local climate is generally hot and semi-humid (classified as "Aw climate" in the Köppen's climate classification map for Brazil, see Alvares et al. 2014), with a well-marked seasonality and rains concentrated in the summer, from October/November to April/May. Mean annual precipitation is 1350 mm; mean annual temperature is around 26 °C. There are four to five months of drought (May to September), and relative air humidity may drop to less than 30% from July to September (Strüssmann and Mott 2009). The Manso River basin is included in the morphostructural domain of the Paraná River sedimentary basin. It is cut across two distinct lithostratigraphic units in nonconformity: the Permeable Mesozoic sandstones belonging to the São Bento (or Botucatu) Group and the inclined Precambrian phyllites and gneisses of the Cuiabá Group, frequently with a lateritic horizon at or near the surface (Barros et al. 1982; Vieira Jr. et al. 2012).

Etymology. The specific epithet honors Francisco Luís Franco ("Kiko"), a specialist in Brazilian snakes, as a tribute to his relentless friendship, dedication and enthusiasm as curator of Herpetological Collection Alphonse Richard Hoge of Instituto Butantan, São Paulo, Brazil (partially and tragically destroyed by fire on 15 May 2010).

Remarks. MCP 12096 was selected as the holotype because the general color in life of *A. kikoi* was described from the live holotype before it was euthanized.

Apostolepis nigroterminata was described by Boulenger (1896) after a single specimen from "Cayaria" (= Callaria, Departamento de Ucayali), eastern Peru. Harvey's (1999) placement of *A. borellii* in the synonymy of *A. nigroterminata* resulted in the inclusion of the latter species in a former Brazilian list of reptiles (see Costa and Bérnils 2015). However, it should be noted that Harvey's (1999) nomenclatural act was based on the analysis of the holotype of

Table 2. Variation in characters of five specimens of *Apostolepis nigroterminata*. Snout-vent length (SVL), tail length (TL), ventrals (VE) and subcaudals (SC). The UFAC specimens presumably are males; the FMNH 39646 specimen is a male. *Holotype, data from Lema and Renner (2016: 67). Measurements in mm.

	BMNH 1946.1.9.77*	FMNH 39646	UFAC 383	UFAC 397	UFAC 504
Sex	M	M	M	M	M
SVL	199	301	210	261	238
TL	18	26	17	19	18
VE	213	234	209	216	219
SC	26	28	24	27	25

A. borellii (a specimen from Urucum massif, Mato Grosso do Sul, Brazil), one specimen collected at the confluence of Rio Araguaia and Tapirapé, Tapirapé Village, Mato Grosso do Sul (AMNH 87942) – that was subsequently re-identified as *A. phillipsi* by Martins and Lema (2015) – and several other specimens from Bolivia and Peru. Recently, Lema and Renner (2016) removed *A. borellii* from the synonymy of *A. nigroterminata* and restricted the distribution of the latter to some localities in Peru. However, these authors also listed a specimen of *A. nigroterminata* from "Brazil: Acre: Rio Branco" in Appendix as "UFAC w/n". The species was also included in an updated list of Brazilian reptiles and referred to occur in the states of Acre, Mato Grosso, and Pará (Costa and Bérnils 2018). Besides the unvouchered mention to Acre in Lema and Renner (2016), Lema et al. (2017) presented a picture of a specimen from Acre without a clear locality description or voucher number, which also provides little evidence for the occurrence of this species in Brazil. The record for Pará is also unvouchered (Maschio et al. 2012), while the specimen (UFMT 10672) from Nobres, Mato Grosso, referred to the species by Santos et al. (2011) was examined and is reidentified here as *Apostolepis* sp. Therefore, we argue that the specimens examined herein (UFAC 383, UFAC 397, UFAC 504) represent not only three locality records of *A. nigroterminata* for the state of Acre, Brazil, but also the first documented record of the species for the country. The specimens collected in Rio Branco (UFAC 397, UFAC 504) extend the geographic distribution of *A. nigroterminata* about 760 km northeastward from Callaria. In particular, the specimen depicted in Figs 4–5 matches the original description of Boulenger (1896) and that given by Lema and Renner (2016) in most details of scalation and color pattern of this species (see Table 2).

Two other species of *Apostolepis* – *A. lineata* and *A. vittata* – were also described from Chapada dos Guimarães (Cope 1887). Although the only existing syntype of *A. lineata* is in very bad condition, Harvey's (1999) redescription is sufficiently complete to allow it to be unambiguously distinguished from *Apostolepis kikoi* sp. nov. Together with *A. assimilis* (Fig. 9) and *A. vittata* (Fig. 10), the description of *Apostolepis kikoi* sp. nov. increases the number of species of *Apostolepis* reported to occur sympatrically in the Manso reservoir area to three.

Figure 9. *Apostolepis assimilis*. A specimen deposited at the Universidade Federal de Mato Grosso (UFMT-R 11088) from district of São Vicente, Santo Antonio do Leverger, Mato Grosso, Brazil (photo by Christine Strüssmann).

Figure 10. *Apostolepis vittata*. A specimen deposited at the Universidade Federal de Mato Grosso (UFMT-R 12259) from the National Park of Chapada dos Guimarães, Cuiabá, Mato Grosso, Brazil (photo by Christine Strüssmann).

An attempt was made to identify all the specimens of *Apostolepis kikoi* sp. nov. using Nogueira et al.'s (2012) key. However, the specimens could not be characterized beyond couplet 1, because of the many overlapping characters presented in the couplets. The assignment of *Apostolepis kikoi* sp. nov. into a formal group should await a more comprehensive phylogenetic arrangement than is available for the genus.

ACKNOWLEDGEMENTS

We thank each of the curators and collection managers for sending us specimens on loan or giving permission to examine specimens in their care, and Eduardo Melloni Lucchesi for the drawings. Tainá Figueras Dorado Rodrigues helped with the construction of the map. We thank also Felipe Grazziotin and three anonymous reviewers for their constructive comments, which helped us to improve the manuscript, and Renato Bérnils providing us detailed feedeback and references that report the occurrence of *Apostolepis nigroterminata* in Brazil. Conselho Nacional de Desenvolvimento Científico e Tecnológico (CNPq) provided a fellowship to FMS during his graduate studies at the Universidade Federal de Mato Grosso do Sul and to OME-N during his undergraduate studies at the Universidade Federal do Rio Grande, and a productivity research grant to CS (CNPq process 309541/2012-3).

LITERATURE CITED

Albuquerque NR, Lema T (2012) Description of the second known specimen of *Apostolepis intermedia* (Serpentes, Colubridae, Xenodontinae). Zootaxa 3325: 53–58.

Alvares CA, Stape JL, Sentelhas PC, Gonçalves JLM, Sparovek G (2014) Köppen's climate classification map for Brazil. Meteorologische Zeitschrift 22(6): 711–728. https://doi.org/10.1127/0941-2948/2013/0507

Amaral A (1925) Ophidios de Matto Grosso (Contribuição II para o conhecimento dos ophidios do Brasil). Commissão de Linhas Telegraphicas Estrategicas de Matto Grosso ao Amazonas. História Natural, Zoologia 84(5):1–29.

Barros AM, Silva RH, Cardoso ORFA, Freire FA, Souza JJ Jr, Rivetti M, Luz DS, Palmeira RCB, Tassinari CCG (1982) Geologia. In: Brasil. Projeto RADAMBRASIL. Rio de Janeiro, Ministério das Minas e Energia, 25–192.

Borges-Nojosa DM, Lima DC, Bezerra CH, Harris DJ (2016) Two new species of *Apostolepis* Cope, 1862 (Serpentes: Elapomorphini) from brejos de altitude in northeastern Brazil. Revista Nordestina de Zoologia 10(2): 74–94.

Boulenger GA (1896) Catalogue of the Snakes in the British Museum (Natural History). London, British Museum (N.H.), vol. 3, xiv+727 pp.

Cabral H, Lema T, Renner MF (2017) Revalidation of *Apostolepis barrioi* (Serpentes: Dipsadidae). Phyllomedusa 16(2): 243–254. https://doi.org/10.11606/issn.2316-9079.v16i2p243-254

Conceição PN (2000) Os hábitats da área de influência do APM Manso. In: Alho CJR (Ed.) Fauna silvestre da região do rio Manso – MT. Brasília, IBAMA, 31–127.

Cope ED (1887) Synopsis of the Batrachia and Reptilia obtained by H. H. Smith in the Province of Mato Grosso, Brazil. Proceedings of the American Philosophical Society 24: 24–60.

Costa HC, Bérnils RS (2015) Répteis brasileiros: lista de espécies 2015. Herpetologia Brasileira 4(3): 75–93.

Costa HC, Bérnils RS (2018) Répteis do Brasil e suas Unidades Federativas: Lista de espécies. Herpetologia Brasileira 7(1): 11–57.

Curcio FF, Nunes PMS, Harvey MB, Rodrigues MT (2011) Redescription of *Apostolepis longicaudata* (Serpentes: Xenodontinae) with comments on its hemipenial morphology and natural history. Herpetologica 67(3): 318–331. https://doi.org/10.1655/HERPETOLOGICA-D-10-00043.1

Dowling HG (1951) A proposed standard system of counting ventral scales in snakes. British Journal of Herpetology 1: 97–99.

Dowling HG, Savage JM (1960) A guide to the snake hemipenis: a survey of basic structure and systematic characteristics. Zoologica 45(2): 17–28.

Ferrarezzi H, Barbo FE, Albuquerque CE (2005). Phylogenetic relationships of a new species of *Apostolepis* from Brazilian Cerrado with notes on the *assimilis* group (Serpentes: Colubridae: Xenodontinae: Elapomorphini). Papéis Avulsos de Zoologia 45(16): 215–229. https://doi.org/10.1590/S0031-10492005001600001

Harvey M (1999) Revision of Bolivian *Apostolepis* (Squamata: Colubridae). Copeia 1999(2): 388–409. https://doi.org/10.2307/1447485

Harvey M, Gonzales AL, Scrocchi G J (2001) New species of *Apostolepis* (Squamata: Colubridae) from the Gran Chaco in Southern Bolivia. Copeia 2001(2): 501–507. https://doi.org/10.1643/0045-8511(2001)001[0501:NSOASC]2.0.CO;2

Koslowsky J (1898) Ofidios de Matto-Grosso (Brasil). Revista del Museo de La Plata 8: 25–32.

Lema T (1997) A redescription of the tropical Brazilian serpent *Apostolepis nigrolineata* (Peters, 1869), (Colubridae: Elapomorphinae), synonymous with *A. pymi* Boulenger 1903. Studies on the Neotropical Fauna and Environment 32(4): 193–199. https://doi.org/10.1080/01650521.1997.11432422

Lema T (2002a) Nova espécie de *Apostolepis* do grupo *lineata* do sudoeste do Brasil (Serpentes, Elapomorphinae). Facena 18: 41–52.

Lema T (2002b) Nova espécie de *Apostolepis* Cope do grupo *dimidiata* do Cerrado do Brasil (Serpentes: Elapomorphinae). Comunicações do Museu de Ciências e Tecnologia da PUCRS, Série Zoologia 15(2): 227–238.

Lema T (2003a) Descrição de nova espécie de *Apostolepis* Cope do cerrado do Brasil, pertencente ao grupo *dimidiata* (Serpentes, Elapomorphinae). Acta Biologica Leopoldensia 25(1): 123–131.

Lema T (2003b) Reencontro de *A. goiasensis* Prado, 1942, com a redescrição da espécie (Serpentes: Elapomorphinae). Comunicações do Museu de Ciências e Tecnologia da PUCRS, Série Zoologia 16(2): 199–209.

Lema T (2004) Description of a new species of *Apostolepis* Cope, 1861 (Serpentes, Elapomorphinae) from Brazilian Cerrado. Acta Biologica Leopoldensia 26(1): 155–160.

Lema T (2016) Description of new species of *Apostolepis* (Serpentes: Dipsadidae: Xenodontinae: Elapomorphini) from Serra do Roncador, Central Brazil. Caderno de Pesquisa, Série Biologia 28(1): 1–11. Available online at: https://online.unisc.br/seer/index.php/cadpesquisa/article/download/7854/4993

Lema T, Martins LA (2016) The complexity of the *Apostolepis dimidiata* group, with partition in two groups (Serpentes: Dipsadidae: Xenodontinae: Elapomorphini). Caderno de Pesquisa, Série Biologia 28(2): 20–29.

Lema T, Renner MF (1998) O status de *Apostolepis quinquelineata* Boulenger, 1896, *A. pymi* Boulenger, 1903 e *A. rondoni* Amaral, 1925 (Serpentes, Colubridae, Elapomorphini). Biociências 6(1): 99–121.

Lema T, Renner MF (2004) New specimens of *Apostolepis vittata* (Cope 1887) (Serpentes, Elapomorphinae). Caderno de Pesquisa, Série Biologia 16(1): 51–56.

Lema T, Renner MF (2006) A new species of *Apostolepis* with striped pattern from Mato Grosso, Brazil (Serpentes, Elapomorphinae). Ciência em Movimento 8: 13–18.

Lema T, Renner MF (2016) Status of *Apostolepis borellii* Peracca 1904 (Serpentes, Xenodontinae), with restriction of the *A. nigroterminata* concept. Neotropical Biology and Conservation 11(2): 62–71. https://doi.org/10.4013/nbc.2016.112.02

Lema T, Campbell P (2017) New species of *Apostolepis* (Serpentes, Dipsadinae, Elapomorphini) from Bolivia, from the *Apostolepis borellii* group. Journal of Zoological Sciences 5(1): 19–28.

Lema T, Queiroz AN, Martins L (2017) Color variation in *Apostolepis nigrolineata* (Serpentes, Colubridae: Dipsadinae: Elapomorphini), and contribution to the knowledge of the *nigrolineata* group. Cuadernos de Herpetología 31(2): 93–101.

Martins LA, Lema T (2015) Elapomorphini (Serpentes, Xenodontinae) do Brasil sudoeste. Neotropical Biology and Conservation 10(2): 93–102. https://doi.org/10.4013/nbc.2015.102.05

Maschio G, Galatti U, Neckel-Oliveira S, Gordo M, Bitar YOC (2012) Répteis. In: Martins FD, Castilho AF, Campos J, Hatano FM, Rolim SG (Eds) Fauna da Floresta Nacional de Carajás: estudos sobre vertebrados terrestres. Nitro Imagens, São Paulo, 84–101.

Nogueira C, Barbo FE, Ferrarezzi H (2012) Redescription of *Apostolepis albicollaris* Lema, 2002, with a key for the species groups of the genus *Apostolepis* (Serpentes: Dipsadidae: Elapomorphini). South American Journal of Herpetology 7(3): 213–225. https://doi.org/10.2994/057.007.0303

Peracca MG (1904) Viaggio del Dr. A. Borelli nel Matto Grosso Brasiliano e nel Paraguay, 1899. IX. Rettili ed amfibii. Bollettino del Museo Regionale di Scienze Naturali di Torino 19(460): 1–15.

Pesantes OS (1994) A method for preparing the hemipenis of preserved snakes. Journal of Herpetology 28(1): 93–95. https://doi.org/10.2307/1564686

Prado A (1942) Notas ofiológicas. 14. Comentários acerca de algumas serpentes opistóglifas do gênero *Apostolepis*, com a descrição de uma nova espécie. Memórias do Instituto Butantan 16: 7–12.

Sabaj MH (2016) Standard symbolic codes for institutional resource collections in herpetology and ichthyology: an online reference. Washington, DC, American Society of Ichthyologists and Herpetologists, version 6.5. http://www.asih.org [Accessed: 30/5/2017]

Santos MM, Ávila RW, Kawashita-Ribeiro RA (2011) Checklist of the amphibians and reptiles in Nobres municipality, Mato Grosso state, central Brazil. Herpetology Notes 4: 455–461.

Strüssmann C (2000) Herpetofauna. In: Alho CJR (Ed.) Fauna Silvestre da Região do Rio Manso – MT. Brasília, IBAMA, 153–189.

Strüssmann C, Mott T (2009) Sympatric amphisbaenids from Manso Dam region, Mato Grosso State, western Brazil, with the description of a new two-pored species of *Amphisbaena* (Squamata, Amphisbaenidae). Studies on Neotropical Fauna and Environment 44(1): 37–46. https://doi.org/10.1080/01650520802628295

Uetz P, Freed P, Hošek J (2018) The Reptile Database. http://www.reptile-database.org [Accessed: 26/04/2018]

Vieira Jr HT, Moraes JM, Paula TLF (2012) Geoparque Chapada dos Guimarães (MT): proposta. In: Schobbenhaus C, Silva CR (Eds) Geoparques do Brasil: propostas. Rio de Janeiro, CPRM, vol. 1, 283–316. http://www.cprm.gov.br/publique/media/guimaraes.pdf [Accessed: 13/05/2016]

Zaher H (1999) Hemipenial morphology of the South American xenodontine snakes, with a proposal for a monophyletic Xenodontinae and a reappraisal of colubroid hemipenes. Bulletin of the American Museum of Natural History (240): 1–168.

Zaher H, Prudente ALC (2003) Hemipenis of *Siphlopis* (Serpentes, Xenodontinae) and techniques of hemipenial preparation in snakes: a response to Dowling. Herpetological Review 34(4): 302–307.

APPENDIX 1

Additional specimens examined

Apostolepis sp. Brazil, *Mato Grosso*: Nobres, Gruta da Cantina (UFMT-R 10672).

Apostolepis assimilis. Brazil, *São Paulo*: Barueri (MPEG 19172, MPEG 19173). *Mato Grosso*: Chapada dos Guimarães (MCP 13282). *Mato Grosso do Sul*, Campo Grande, Jardim Autonomista (ZUFMS-REP 256); Campo Grande, Vila Carvalho (ZUFMS-REP 94, ZUFMS-REP 95, ZUFMS-REP 233, ZUFMS-REP 234).

Apostolepis ammodites. Brazil, *Maranhão*: Carolina, U.H.E de Estreito (MPEG 24395). *Minas Gerais*: Pirapora, margem direita do Rio São Francisco (MPEG 18347). *Tocantins*: Barra do Ouro (MPEG 23633), Filadélfia (MPEG 24697).

Apostolepis borellii. Brazil, *Mato Grosso do Sul*: Urucum Massif (MZUT 962, holotype).

Apostolepis aff. *borellii*. Brazil, *Mato Grosso*: Serra do Amolar (UFMT-R 1171, UFMT-R 1173, UFMT-R 1174, UFMT-R 1179, UFMT-R 1182, UFMT-R 1184, UFMT-R 1186, UFMT-R 1214, UFMT-R 1216).

Apostolepis cearensis. Brazil, *Ceará*: Fortaleza (MPEG 18219, MPEG 18220). *Maranhão*: Urbano Santos (MPEG 20550, MPEG 20551, MPEG 20552). *Piauí*: Castelo do Piauí (MPEG 22760, MPEG 22762, MPEG 22810, MPEG 22814).

Apostolepis dimidiata. Brazil, *Mato Grosso do Sul*: Aquidauana (ZUFMS-REP 96, ZUFMS-REP 97, ZUFMS-REP 235); Aquidauana, Camisão (ZUFMS-REP 100); Aquidauana, Vila Cidade Nova (ZUFMS-REP 98); Campo Grande (ZUFMS-REP 1732); Campo Grande, Jardim Autonomista (ZUFMS-REP 256); Chapadão do Sul, Campus da UFMS (ZUFMS-REP 2170); Sidrolândia, Estância Belém (ZUFMS-REP 101). *Mato Grosso*: Chapada dos Guimarães (MCP 13282).

Apostolepis lineata. Brazil, *Mato Grosso*: Chapada dos Guimarães (ANSP 11212, syntype).

Apostolepis longicaudata. Brazil, *Pará*: Portel, Fazenda Riacho Monte Verde (MPEG 22729).

Apostolepis nigrolineata. Brazil, *Pará*: Barcarena, arredores de Barcarena (MPEG 17292); Vila São Francisco, antiga sede de Barcarena (MPEG 16339); Benevides, Estrada da Belágua, Maguari (MPEG 21079); Canaã dos Carajás (MPEG 26453); Itaituba, A.P.A. do Tapajós, Mina do Tocantinzinho (MPEG 24569); Óbidos, ESEC Grão-Pará (MPEG 23737); Oriximiná, Serra do Acarai (MPEG 23681); Santa Bárbara do Pará, Parque Ecológico do Gunma (MPEG 21330, MPEG 21333); Santarém, comunidade Tapari (MPEG 26512, MPEG 26553); Santo Antônio do Tauá (MPEG 3940); Viseu, Curupati (MPEG 10010, MPEG 10884, MPEG 10886, MPEG 10887, MPEG 13260); Viseu, km 220 da BR-316, antigo km 74 de Capanema (MPEG 3581, MPEG 8192, MPEG 10841, MPEG 10851); Viseu, Rio Gurupi, Colônia Nova, próximo do rio, BR-316 (MPEG 11487); Vitória do Xingu, U.H.E. de Belo Monte (MPEG 26189, MPEG 26190, MPEG 26503).

Apostolepis nigroterminata. Brazil, *Acre*: Km-80 da BR-317, Fazenda da Patroa, near Boca do Acre (UFAC 383), Campus Universitário da UFAC, Rio Branco (UFAC 504), Parque Zoobotânico da UFAC, Rio Branco (UFAC 397). Peru: Ayacucho, La Mar, in Sivia, Apurimac River, 760 m.a.s.l. (FMNH 39646); Loreto (MUSM 005); "Cayaria" (= Callaria, Departamento de Ucayali, Peru) (BMNH 1946.1.9.77, holotype).

Apostolepis polylepis. Brazil, *Maranhão*: Estreito, U.H.E. de Estreito (MPEG 63444, MPEG 23645, MPEG 24713). *Tocantins*: Darcinópolis (MPEG 24173), Palmeiras do Tocantins, U.H.E. de Estreito (MPEG 24712, MPEG 24714, MPEG 24715).

Apostolepis serrana. Brazil, *Mato Grosso*: Serra do Roncador, Rio das Mortes (BMNH 1972.430, holotype).

Apostolepis striata. Brazil, *Rondônia*: Vilhena (CHUNB 12794, holotype).

Apostolepis tenuis. Bolivia, *Santa Cruz*: Buenavista (UMMZ 64436, holotype).

Author Contributions: TL, CS and NRA conceived the project; FMS, TL, CS and NRA contributed to the acquisition of data of *Apostolepis* specimens, including those of the type series of *A. kikoi*; JSA and MBS analyzed the specimens of *A. nigroterminata*; FMS, CS and NRA wrote the paper, with contributions from OME-N and TL.
Competing Interests: The authors have declared that no competing interests exist.

19

Temperature and circadian effects on metabolic rate of South American echimyid rodents, *Trinomys setosus* and *Clyomys bishopi* (Rodentia: Echimyidae)

Ana Paula Fabio-Braga[1], Wilfried Klein[1]

[1] *Departamento de Biologia, Faculdade de Filosofia, Ciências e Letras de Ribeirão Preto, Universidade de São Paulo. Ribeirão Preto, SP, Brazil.*
Corresponding author: Wilfried Klein (wklein@usp.br)

ABSTRACT. Basal metabolic rate (BMR) represents the lowest level of metabolic activity capable to sustain homeostasis in an endotherm and is an important tool to compare metabolic rates of different species. Echimyidae is the most specious family within caviomorph rodents, however, little is known about the biology of its species, such as *Trinomys setosus* (Desmarest, 1817) and *Clyomys bishopi* (Ávila-Pires & Wutke, 1981), a ground and an underground dwelling echimyid, respectively. The ambient temperature and circadian effects on metabolic rate were evaluated through closed-system respirometry for these two species, as well as the circadian effects on CO_2 production and respiratory exchange ratio (RER). *Trinomys setosus* and *C. bishopi* showed the lowest metabolic rates (0.56 ± 0.02 $mLO_2.h^{-1}.g^{-1}$ and 0.53 ± 0.03 $mLO_2.h^{-1}.g^{-1}$, respectively) at 32 °C and during the light phase. Under laboratory conditions, *T. setosus* showed metabolic rate variation compatible with nocturnal activity, whereas *C. bishopi* activity cycle remains unclear. Both species showed BMR lower than expected by allometric regressions for rodents.

KEY WORDS. Basal metabolic rate, fossoriality, neotropical, oxygen consumption, thermoneutral zone.

INTRODUCTION

Animals expend energy for numerous purposes such as maintenance of homeostasis, foraging, food digestion, growth, and reproduction. The timing of those activities is fundamental to increase survival. Deviations from the standard biological rhythm can be strongly selected against in nature (DeCoursey et al. 2000, Spoelstra et al. 2016) and most species will display activity cycles entrained to light and dark phases (Koukkari and Sothern 2006).

The basal metabolic rate (BMR) represents the lowest level of metabolic activity necessary to sustain homeostasis in endotherms (Hulbert and Else 2004, Careau et al. 2008), and has been an important tool to compare metabolic rates among different species (Hulbert and Else 2004). Although the BMR is widely used to answer physiological, ecological and behavioral questions, a reliable BMR measurement depends on ensuring that experimental trials were performed using an adult animal at rest in a postabsorptive (non-digestive) and non-reproductive state and in a thermoneutral environment (Hulbert and Else 2004).

Variations in BMR are mainly explained by body mass (Kleiber 1932), but many other factors may relate to BMR variation such as climate (Lovegrove 2003) and ecological habits (McNab 2008). Although body mass and BMR are strongly related, a single scaling exponent has not been determined to date (Hulbert and Else 2004, Capellini et al. 2010). Nevertheless, fossorial mammals are believed to present lower than expected BMR for a given body mass (McNab 1966, 1979, Vleck 1979, 1981). Two hypotheses try to explain this reduced BMR, the thermal-stress hypothesis (McNab 1966, 1979) and the cost-of-burrowing hypothesis (Vleck 1979, 1981). The former assumes that BMR is reduced due to heat dissipation limitations underground, while the latter assumes that this feature is related to the high energetic costs of excavating tunnels for foraging. Nevertheless, fossorial rodents will display different strategies to cope with underground constrains and lower than expected BMR could have been selected by different environmental pressures. In *Octodon degus* Molina, 1782, net primary productivity explains the intraspecific BMR variation (Bozinovic et al. 2009). On the other hand, net primary productivity is not a good predictor for BMR in fossorial rodents in general, and ambient temperature was pointed out as the major determinant of residual BMR variation among underground rodents (Luna et al. 2017).

Echimyidae is the most specious family within caviomorph rodents (Upham and Patterson, 2012); however, little is known about the biology of its species. *Trinomys setosus* (Desmarest, 1817) is known to be a ground dwelling echimyid, but the exact distribution within the endangered Brazilian Atlantic rainforest is unknown (MMA 2012). Echimyidae also includes *Clyomys*, a fossorial genus with only scarce physiological information available (Nowak 1999). *Clyomys bishopi* (Ávila-Pires & Wutke, 1981) is restricted to the Brazilian savanna – called Cerrado, and presents morphological features related to a fossorial habit, such as, well-developed claws on the forefeet and enlarged bullae in the cranium (Nowak 1999). The only physiological data available for *C. bishopi* is resting metabolic rate (Barros et al. 2004), however, neither species have data available regarding thermoneutral zone nor biological rhythmicity. BMR data of these two closely related species are therefore not available and could contribute to our knowledge of metabolic physiology of echimyid species. These data can also provide information about the relationship between metabolic rate and fossoriality in this representative rodent taxon (Fabre et al. 2012).

This study aims to report for the first time BMR, CO_2 production and the respiratory exchange ratio (RER) of *T. setosus* and *C. bishopi*. To achieve this, ambient temperature and circadian rhythm effects will be considered. We hypothesize that both species will present the lowest metabolic rate within the light phase, as echimyids are usually nocturnal (Emmons and Feer 1997), and that BMR of *T. setosus* will fit the expected from allometric relationships (see Kleiber 1932, McNab 2008), whereas *C. bishopi* will present a lower than predicted BMR due to its fossorial life style.

MATERIAL AND METHODS

Trinomys setosus (316.6 ± 29.1 g) was caught in Brazilian Atlantic rainforest (13°00'S, 38°01'W) and *C. bishopi* (348.3 ± 23.8 g) was caught in Savanna-like environment (Ecological Station of Itirapina: 22°14'S, 47°52'W). Animals were housed in plastic opaque cages (30 x 40 x 16 cm) and exposed to an inverted 12:12 light:dark cycle. Food (NUVILAB CR1) and water were provided *ad libitum*. Animal manipulation was carried out in accordance to the guidelines of American Society of Mammologists (Sikes and Ganon 2011) and current Brazilian laws for capture (SISBIO: 43334-1), handling and care of mammals in captivity (CEUA:13.1.866.53.3).

Oxygen consumption ($\dot{V}O_2$) and carbon dioxide production ($\dot{V}CO_2$) were obtained by a closed respirometry system (see Barros et al. 1998, Barros et al. 2004). Animals were individually kept inside a chamber (4.55L) within a temperature controlled cabinet (Q315F, QUIMIS). Measurements were conducted with chambers sealed for 10 minutes (600 samples) while circulating air samples through a CO_2 (CA-10A, Sable Systems) and an O_2 analyzer (PA-10, Sable System). CO_2 was measured during all experimental trials to ensure that CO_2 levels inside the chambers were below 1 %, a concentration unlikely to affect VO_2 measurements (Barros et al. 2004). Between measuring intervals, the chambers were flushed with room air using a high flow rate (≈ 1000 $mL.min^{-1}$). Animals were placed in the experimental apparatus three hours before experimental trials and were fasted for eight hours prior to measurements to ensure a post-absorptive state. To ensure acclimation, animals were previously submitted to experimental conditions two times with no measures recorded.

The same individuals were used in two experimental protocols to study the effects of ambient temperature (T_a) and circadian cycle on BMR. First, the effect of T_a on metabolic rate was evaluated through measurements of $\dot{V}O_2$ for each animal (N = 5) at T_a ranging from 20 to 40 °C for *T. setosus* and from 20 to 36 °C for *C. bishopi*, using 4 °C increments. Animals were acclimated at room temperature (25 °C) and experimental trials were conducted during the light phase in an attempt to minimize interference of other activities in the measurements taken. Animals were submitted to each T_a for 30 minutes and $\dot{V}O_2$ was determined on the last 10 minutes of each temperature exposure. Second, to access circadian effect on BMR, animals (N = 8) were measured in trials lasting 6 hours (starting in dark phase and ending in light phase), where each animal was measured every 40 minutes. Acclimation and experimental trials were conducted at the T_a indicated by the first protocol (32 °C, see Results). Circadian effects on $\dot{V}CO_2$ and RER are also reported.

$\dot{V}O_2$ and $\dot{V}CO_2$ were obtained based on the regression of gas variation inside chambers and corrected to STPD. These two variables were used to calculate respiratory exchange ratio (RER). Segmented linear regression was used to evaluate the effects of T_a on metabolic rate. Among the different functions obtained, we selected the one that maximized the statistical coefficient of explanation (see Baldo et al. 2015). Circadian effect on $\dot{V}O_2$ was evaluated using an analysis of variance (repeated measures ANOVA, α = 0.05) between the lowest $\dot{V}O_2$ of dark and light phases. $\dot{V}O_2$, $\dot{V}CO_2$ and RER measures of different daytimes were compared within the inactive phase (repeated measures ANOVA, α = 0.05). All values represent mean and standard error (mean ± SE). Measures within the thermoneutral zone and during the inactive phase were considered to represent basal metabolic rate (BMR). Statistical analyses were conducted using PRISM 6 (GraphPad Software, San Diego, USA).

RESULTS

The lowest metabolic rate of *T. setosus* was 0.60 ± 0.01 $mLO_2.h^{-1}.g^{-1}$ at 32 °C (Fig. 1). The best fitting model gave the lower limit of the thermoneutral zone at T_a = 25.7 °C. Effect of T_a on $\dot{V}O_2$ below the thermoneutral zone was described by segmented regression as $\dot{V}O_2$ ($mLO_2.h^{-1}.g^{-1}$) = -0.018.T_a + 1.06. The

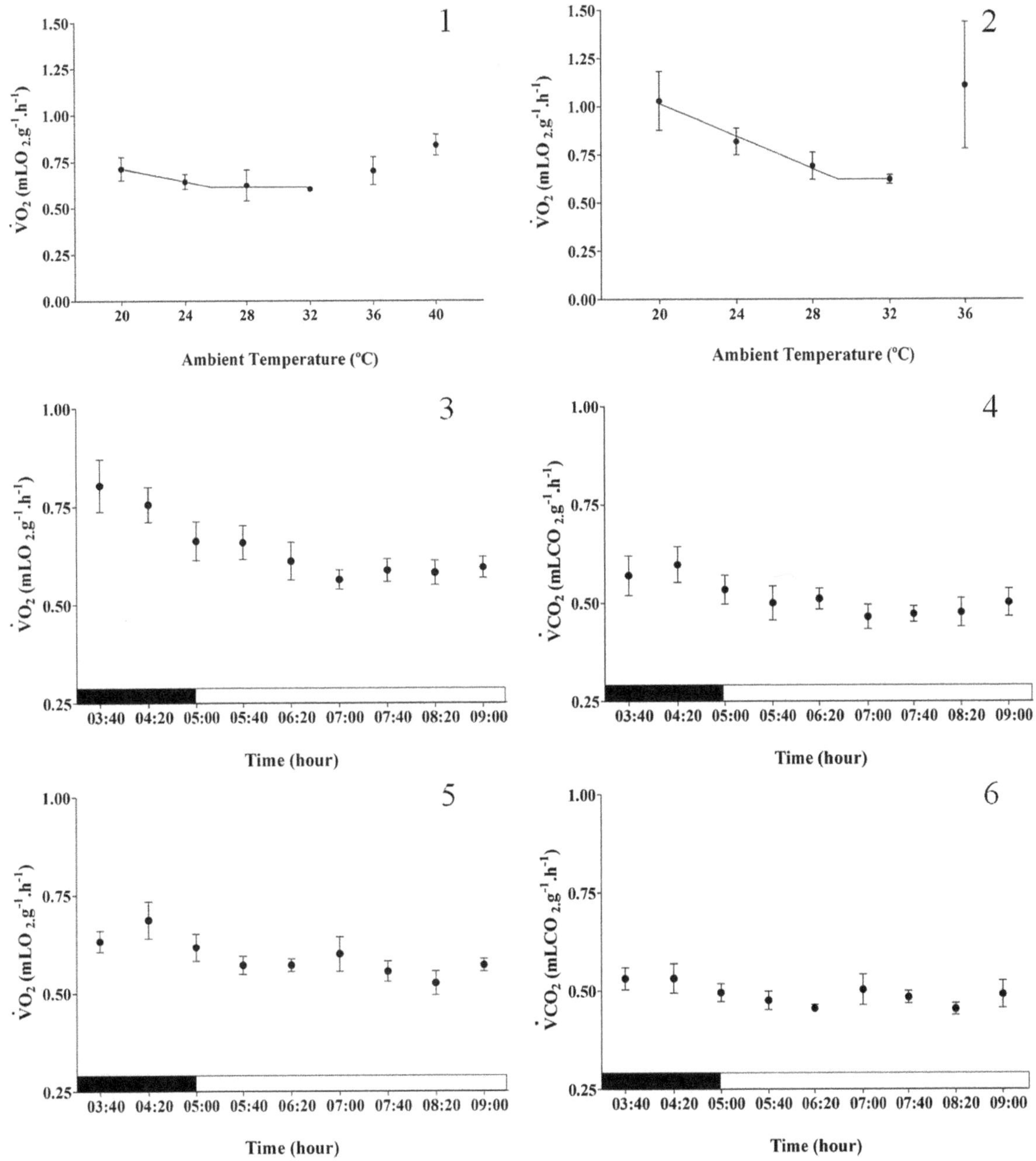

Figures 1–6. Relationship between O_2 consumption ($mLO_2.g^{-1}.h^{-1}$) and environmental temperature (°C) in *T. setosus* (1) and *C. bishopi* (2), as well as relationship between O_2 consumption ($mLO_2.g^{-1}.h^{-1}$) and time of day in *T. setosus* (3) and *C. bishopi* (5) and relationship between CO_2 production ($mLCO_2.g^{-1}.h^{-1}$) and time of day in *T. setosus* (4) and *C. bishopi* (6). Bars indicate the respective phase of the inverted photocycle, dark (black) or light (white). Values represent mean and standard error (mean ± SE)

mean of $\dot{V}O_2$ at minimum T_a of 20 °C was 0.71 ± 0.06 $mLO_2.h^{-1}.g^{-1}$ and at the maximum T_a of 40 °C was 0.84 ± 0.06 $mLO_2.h^{-1}.g^{-1}$, representing 116 % and 138 % of the $\dot{V}O_2$ of the thermoneutral zone, respectively.

Clyomys bishopi's lowest metabolic rate was found at 32 °C with 0.62 ± 0.02 $mLO_2.h^{-1}.g^{-1}$ (Fig. 2). The lower limit of the thermoneutral zone was found at T_a = 29.3 °C. The increase of $\dot{V}O_2$ due to decreasing T_a can be described as $\dot{V}O_2$ ($mLO_2.h^-$

$^{-1}.g^{-1}$) = -0.042.T_a + 1.86. The mean of $\dot{V}O_2$ at minimum T_a of 20 °C was 1.03 ± 0.15 $mLO_2.h^{-1}.g^{-1}$ and at the maximum T_a of 36 °C was 1.11 ± 0.33 $mLO_2.h^{-1}.g^{-1}$, representing 156% and 179% of $\dot{V}O_2$ of the thermoneutral zone, respectively. $\dot{V}O_2$ at T_a of 40 °C was not obtained because the first tested animal was not able to survive at this T_a, and therefore we refrained from exposing the other animals to 40 °C.

Trinomys setosus' lowest metabolic rate measurements during the animal's dark and light phase were found at 04:20 am (0.76 ± 0.04 $mLO_2.h^{-1}.g^{-1}$) and 07:00 am (0.56 ± 0.02 $mLO_2.h^{-1}.g^{-1}$), respectively (Fig. 3). $\dot{V}O_2$ was lower in the light phase than in the dark phase (p = 0.01) and did not vary within the light phase (p = 0.19); therefore the mean $\dot{V}O_2$ of the light phase at 32 °C was $\dot{V}O_2$ = 0.61 ± 0.04 $mLO_2.h^{-1}.g^{-1}$. The same pattern was observed for $\dot{V}CO_2$ (Fig. 4), the lowest $\dot{V}CO_2$ being found during the light phase at 07:00 am (0.46 ± 0.03 $mLCO_2.h^{-1}.g^{-1}$) and not varying within the light phase (p = 0.38), resulting in a mean $\dot{V}CO_2$ during the light phase of 0.49 ± 0.03 $mLCO_2.h^{-1}.g^{-1}$. RER (0.81 ± 0.04) remained constant throughout the light phase (p = 0.48).

Clyomys bishopi's lowest metabolic rate measurements during the animal's dark and light phase were found at 03:40 am (0.63 ± 0.03 $mLO_2.h^{-1}.g^{-1}$) and 08:20 am (0.53 ± 0.03 $mLO_2.h^{-1}.g^{-1}$), respectively (Fig. 5). $\dot{V}O_2$ was lower during the animals light phase than the dark phase (p = 0.02) and did not vary within the light phase (p = 0.39), giving a mean $\dot{V}O_2$ of the light phase at 32 °C as 0.57 ± 0.03 $mLO_2.h^{-1}.g^{-1}$. An identical pattern was observed for $\dot{V}CO_2$ (Fig. 6), the lowest $\dot{V}CO_2$ being found in the animals light phase at 08:20 am (0.45 ± 0.02 $mLCO_2.h^{-1}.g^{-1}$), without variation within the light phase (p = 0.62), giving a mean $\dot{V}CO_2$ for the light phase of 0.48 ± 0.02 $mLCO_2.h^{-1}.g^{-1}$. RER (0.84 ± 0.04) remained constant during the light phase (p = 0.70).

DISCUSSION

The intervals for the thermoneutral zones obtained in the present study are in accordance with data available for other caviomorph terrestrial rodents such as *O. degus* (27–35 °C), *Thrichomys apereoides* Lund, 1839 (25–35 °C), *Chinchilla laniger* Bennett, 1829 (22–35 °C), *Dazyprocta azarae* Lichtenstein, 1823 (18–35 °C) (Arends and Mcnab 2001) and fossorial rodents such as *Ctenomys talarum* Thomas, 1898 (25–35 °C) (Bush 1989). *Trinomys setosus* showed a relatively broad thermoneutral zone with a lower limit at T_a = 25.7 °C extending at least to T_a = 32 °C (Fig. 1), which is compatible with its environment's mean T_a = 25 °C (Hijmans et al. 2005). The increases of metabolic rate below (116%) and above (138%) the limits of the thermoneutral zone were small in comparison to other caviomorph rodents such as *C. laniger* and *O. degus* (Arends and McNab 2001), however measurements in lower temperatures could reveal that these animals may be more affected by a lower T_a.

Clyomys bishopi showed a narrow thermoneutral zone with its lower limit at T_a = 29.3 °C extending to T_a = 32 °C (Fig. 2), which is higher than its mean environmental T_a of 20 °C (Hijmans et al. 2005). However, it is possible that the thermoneutral zone in this species is not related to above ground T_a, but to the amplitude in burrow temperature (Burda et al. 2007). *Clyomys bishopi*'s metabolic rate was more affected by experimental temperatures below (156%) and above (179%) the limits of the thermoneutral zone. Moreover, *C. bishopi* seems not be able to cope well with experimental T_a = 40 °C, since mortality was observed in the only tested animal at T_a = 40 °C. *Ctenomys talarum* shows a broader thermoneutral zone (25–35 °C) than *C. bishopi*, but this species was also not able to survive temperatures above its thermoneutral zone at T_a = 40 °C (Bush 1989). Such a thermoregulatory constrain may be expected in fossorial rodents as long as burrow temperature variation is predictable and animals are able to use behavioral strategies to avoid overheating, i.e. timing their digging activity according to burrow temperatures or moving to cooler parts of the burrow (Burda et al. 2007). Moreover, it is known that other fossorial rodents will be exposed to only small temperature variations inside burrows (McNab 1966, Baldo et al. 2015). Further studies could address daily and annual temperature ranges within burrows and evaluate the whole set of behavioral strategies to cope with temperature variation.

The majority of echimyids forage during the dark phase (Emmons and Feer 1997) and, as expected, *T. setosus* and *C. bishopi* showed the lowest metabolic rates during the light phase. The variation observed in metabolic rate of *T. setosus* is compatible with a nocturnal species with its circadian cycle entrained to day-night phases (Fig. 3) (Koukkari and Sothern 2006). The same pattern has been observed in another species of the same genus, *Trinomys yonenagae* Rocha, 1995 (Marcomini and Spinelli 2003). Concurrently, although *C. bishopi* presented the lowest metabolic rate during the light phase and is usually not seen on the surface during the day, the metabolic rate variation through time does not provide strong evidence that *C. bishopi* is nocturnal (Fig. 5). Still, it is possible that this species presents a circadian cycle entrained to light and dark phases or synchronized with ambient temperature (Šklíba et al. 2014), food availability (Nelson et al. 1975) or social environment (Crowley and Bovet 1980). Field data on activity patterns are needed to clarify which environmental cues could be regulating *C. bishopi* activity under natural conditions. The absence of variation in RER values was expected as no change in metabolic fuel use (i.e., carbohydrates, lipids or proteins) was imposed and RER values were compatible with animals in a post-absorptive state, during which mixed fuels are generally used (Wang et al. 2006).

BMR data of these species are important due to the poor representation of the Echimyidae family in other physiological studies. Echimyidae include 95 species of rodents distributed in 21 genera, however, only few data are available regarding metabolic rate (Table 1). Moreover, some species' metabolic rate was measured with no consideration of ambient temperature and circadian effects, which is not compatible with BMR conditions (Hulbert and Else 2004, Connolly and Cooper 2014). BMR of

Table 1. Available data of metabolic rate for Echimyidae species.

	Body mass (g)	Metabolic rate ($mLO_2.h^{-1}.g^{-1}$)	Reference
Clyomys bishopi[b]	343.2	0.53	This work
Trinomys setosus[b]	316.6	0.56	This work
Trinomys yonenagae[a]	137.6	0.63	Barros et al. (1998)
Trinomys iheringi[a]	223.9	0.83	Barros et al. (1998)
Proechimys semispinosus[b]	498.0	0.63	Arends and McNab (2001)
Thrichomys apereoides[b]	323.0	0.64	Arends and McNab (2001)
Myocastor coypus[b]	4325.0	0.70	Segal (1978)

[a]RMR conditions, [b]BMR conditions.

C. bishopi and *T. setosus* were both lower than predicted by body mass when applying Kleiber's regression of BMR (kcal.day^{-1}) = 73.3 × $M_B^{0.74}$ (Kleiber 1932), representing 68% and 70% of the predicted values, respectively. When applying the allometric regression given by McNab (2008) for rodents, BMR ($mLO_2.h^{-1}$) = 3.87 × $M_B^{0.71}$, *C. bishopi* and *T. setosus* BMRs are equivalent to 76 % and 79 % of predicted values, respectively. Usually such reduction in metabolic rate would be attributed to fossoriality. However, as *T. setosus* is a terrestrial species, this result does not support the hypothesis of reduced BMR being related to fossoriality.

Our results illustrate the apparent complexity of traits influencing BMR in this group. It is possible that the reduced basal metabolic rate for both species may be explained by a phylogenetic effect as suggested for some echimyids (Roberts et al. 1988) or by climatic variables such as high ambient temperature during these species evolution (Lovegrove 2003). In any case, additional data and broader comparisons are needed to understand BMR variation within Echimyidae species.

ACKNOWLEDGEMENTS

We would like to thank Elisabeth S. de Oliveira for maintaining the animals during the study and Michael Minicozzi for reviewing the manuscript. Ana Paula F. Braga was supported by Conselho Nacional de Desenvolvimento Científico e Tecnológico (134608/2013-5) master's fellowship within the Programa de Pós-Graduação em Biologia Comparada (FFCLRP, Universidade de São Paulo) and Wilfried Klein received support through the Instituto Nacional de Ciência e Tecnologia em Fisiologia Comparada (CNPq: 573921/2008-3: FAPESP: 2008/57712-4).

LITERATURE CITED

Arends A, McNab BK (2001) The comparative energetics of 'caviomorph' rodents. Comparative Biochemistry & Physiology A 130: 105–122. https://doi.org/10.1016/S1095-6433(01)00371-3

Baldo MB, Antenucci CD, Luna F (2015) Effect of ambient temperature on evaporative water loss in the subterranean rodent *Ctenomys talarum*. Journal of Thermal Biology 53: 113–118. https://doi.org/10.1016/j.jtherbio.2015.09.002

Barros RCH, Oliveira ES, Rocha LB, Branco LG (1998) Respiratory and metabolic responses of the spiny rats *Proechimys yonenagae* and *P. iheringi* to CO_2. Respiration Physiology 111: 23–231. https://doi.org/10.1016/S0034-5687(97)00118-7

Barros RCH, Abe AS, Cárnio EC, Branco LG (2004) Regulation of breathing and body temperature of a burrowing rodent during hypoxic–hypercapnia. Comparative Biochemistry & Physiology A 138: 97–104. https://doi.org/10.1016/j.cbpb.2004.03.011

Bozinovic F, Rojas JM, Broitman BR, Vásquez RA (2009). Basal metabolism is correlated with habitat productivity among populations of degus (*Octodon degus*). Comparative Biochemistry and Physiology Part A 152: 560–564. https://doi.org/10.1016/j.cbpa.2008.12.015

Burda H, Šumbera R, Begall S (2007) Microclimate in Burrows of Subterranean Rodents. In: Begall S, Burda H, Schleich CE (Eds) Subterranean rodents: news from underground. Springer-Verlag, Berlin–Heidelberg, 21–31. https://doi.org/10.1007/978-3-540-69276-8_3

Bush C (1989) Metabolic rate and thermoregulation in two species of tuco-tuco, Stenomys talarum and Ctenomys australis (Caviomorpha, Octodontidae). Comparative Biochemistry & Physiology A 93: 345–347. https://doi.org/10.1016/0300-9629(89)90048-0

Capellini I, Venditti C, Barton RA (2010) Phylogeny and metabolic scaling in mammals. Ecology 91: 2783–2793. https://doi.org/10.1890/09-0817.1

Careau V, Thomas D, Humphries MM, Réale D (2008) Energy metabolism and animal personality. Oikos 117: 641–653. https://doi.org/10.1111/j.0030-1299.2008.16513.x

Connolly MK, Cooper CE (2014) How do measurement duration and timing interact to influence estimation of basal physiological variables of a nocturnal rodent? Comparative Biochemistry & Physiology A 178: 24–29. https://doi.org/10.1016/j.cbpa.2014.07.026

Crowley M, Bove J (1980) Social synchronization of circadian rhythms in deer mice (*Peromyscus maniculatus*). Behavioral Ecology and Sociobiology 7: 99–155. doi:10.1007/BF00299514

DeCoursey PJ, Walker JK, Smith SA (2000) A circadian pacemaker in free-living chipmunks: essential for survival? Journal of Comparative Physiology. A 186: 169–180. https://doi.org/10.1007/s003590050017

Emmons LH, Feer F (1997) Neotropical rainforest mammals: a field guide. Chicago University Press, Chicago,.

Fabre PH, Haultier L, Dimitrov D, Douzery EJ (2012) A glimpse on the pattern of rodent diversification: a phylogenetic approach. BMC Evolutionary Biology 12: 117–134. https://doi.org/10.1186/1471-2148-12-88

Hijmans RJ, Cameron SE, Parra JL, Jones PG, Jarvis A (2005) Very high resolution interpolated climate surfaces for global land areas. International Journal of Climatology 25: 1965–1978. https://doi.org/10.1002/joc.1276

Hulbert AJ, Else PL (2004) Basal metabolic rate: history, composition, regulation, and usefulness. Physiological & Biochemical Zoology 77: 869–876. https://doi.org/10.1086/422768

Kleiber M (1932) Body Size and Metabolism. Hilgardia: Journal of Agricultural Science 6: 315–356. https://doi.org/10.3733/hilg.v06n11p315

Koukkari WL, Sothern RB (2006) Introducing Biological Rhythms: A Primer on the Temporal Organization of Life, with Implications for Health, Society, Reproduction and the Natural Environment. Springer, New York,.

Lovegrove BG (2003) The influence of climate on the basal metabolic rate of small mammals: a slow-fast metabolic continuum. Journal of Comparative Physiology B 173: 87–112, 2003. https://doi.org/10.1007/s00360-002-0309-5

Luna F, Naya H, Naya DE (2017). Understanding evolutionary variation in basal metabolic rate: An analysis in subterranean rodents. Comparative Biochemistry and Physiology Part A 206: 87–94. https://doi.org/10.1016/j.cbpa.2017.02.002

Marcomini M, Spinelli EO (2003) Activity Pattern of Echimyid Rodent Species from the Brazilian Caatinga in Captivity. Biological Rhythm Research 34: 157–166. https://doi.org/10.1076/brhm.34.2.157.14491

McNab BK (1966) The metabolism of fossorial rodents: a study of convergence. Ecology 47: 712–733. https://doi.org/10.2307/1934259

McNab BK (1979) The influence of body size on the energetics and distribution of fossorial and burrowing mammals. Ecology 60: 1010–1021. https://doi.org/10.2307/1936869

McNab BK (2008) An analysis of the factors that influence the level and scaling of mammalian BMR. Comparative Biochemistry & Physiology A 151: 5–28. https://doi.org/10.1016/j.cbpa.2008.05.008

Ministério do Meio Ambiente (2008–2009) Monitoramento do Desmatamento nos Biomas Brasileiros por Satélite. Available online at: http://www.mma.gov.br/estruturas/sbf_chm_rbbio/_arquivos/relatorio_tcnico_mata_atlantica_2008_2009_72.pdf [Accessed: 13/01/2016]

Nelson W, Scheving L, Halber F (1975) Circadian Rhythms in Mice Fed a Single Daily Meal at Different Stages of Lighting Regimen. Journal of Nutrition 105: 171–184. https://doi.org/10.1093/jn/105.2.171

Nevo N (1995) Mammalian evolution underground. The ecological-genetic-phenetic interfaces. Acta Theriologica 3: 9–31. https://doi.org/10.4098/AT.arch.95-43

Nowak RM (1999) Walker's Mammals of the World. JHU Press, Baltimore,.

Roberts MS, Thompson KV, Canford JA (1988) Reproduction and Growth in Captive Punare (*Thrichomys apereoides* Rodentia: Echimyidae) of the Brazilian Caatinga with Reference to the Reproductive Strategies of the Echimyidae. Journal of Mammalogy 69: 542–551. https://doi.org/10.2307/1381346

Sikes RS, Gannon WL, the Animal Care and Use Committee of the American Society of Mammalogists (2011) Guidelines of the American Society of Mammalogists for the use of wild mammals in research. Journal of Mammalogy 92: 235–253. https://doi.org/10.1644/10-MAMM-F-355.1

Šklíba J, Lövy M, Hrouzková E, Kott O, Okrouhlik J, Šumbera R (2014) Social and Environmental Influences on Daily Activity Pattern in Free-Living Subterranean Rodents: The Case of a Eusocial Bathyergid. Journal of Biological Rhythms 29: 203–214. https://doi.org/10.1177/0748730414526358

Spoelstra K, Wikelski M, Daan S, Loudon AS, Hau M (2016) Natural selection against a circadian clock gene mutation in mice. Proceedings of the National Academy of Sciences 113: 686–691. doi: https://doi.org/10.1073/pnas.1516442113

Upham NS, Patterson, BD (2012) Diversification and biogeography of the Neotropical caviomorph lineage Octodontoidea (Rodentia: Hystricognathi) Molecular Phylogenetics and Evolution 63: 417–29. https://doi.org/10.1016/j.ympev.2012.01.020

Vleck D (1979) The energy cost of burrowing by the pocket gopher Thomomys bottae. Physiological Zoology 52: 122–136. https://doi.org/10.1086/physzool.52.2.30152558

Vleck D (1981) Burrow structure and foraging costs in the fossorial rodent, Thomomys bottae. Oecologia 49: 391–396. https://doi.org/10.1007/BF00347605

Wang T, Hung CC, Randall DJ (2006) The comparative physiology of food deprivation: from feast to famine. Annual Review of Physiology 68: 223–251. https://doi.org/10.1146/annurev.physiol.68.040104.105739

White CR, Kearney MR (2013) Determinants of inter-specific variation in basal metabolic rate. Journal of Comparative Physiology B, 183: 1–26. https://doi.org/10.1007/s00360-012-0676-5

Author Contributions: APFB and WK contributed equally to this work.

Competing Interests: The authors have declared that no competing interests exist.

20

Thermal biology of *Amphisbaena munoai* (Squamata: Amphisbaenidae)

Nathalia Rocha Matias[1], Laura Verrastro[1]

[1]*Programa de Pós-Graduação em Biologia Animal, Laboratório de Herpetologia, Departamento de Zoologia, Universidade Federal do Rio Grande do Sul. Avenida Bento Gonçalves 9500, prédio 43435, sala 105, 91501-970 Porto Alegre, RS, Brazil.*
Corresponding author: Nathalia Rocha Matias (nathimatias@hotmail.com)

ABSTRACT. Studies on the thermal biology of fossorial reptiles that examine the relationship between the body temperature and thermal environment are needed to determine the extent of their thermoregulation abilities. This study assessed the thermal biology of *Amphisbaena munoai* Klappenbach, 1969 in the rocky fields of the Rio Grande do Sul and in the laboratory. The body temperature of most individuals was between 24 and 30 °C, both in the field (n = 81) and laboratory (n = 19). More individuals were caught in winter (n = 55) and spring (n = 60) than in summer (n = 25) and fall (n = 45), and in spring, individuals showed similar nocturnal and diurnal activities. In the laboratory, we found individuals with body temperatures up to 5 °C higher than the ambient temperature (n = 4), suggesting that some physiological mechanisms participate in the thermoregulation of these animals. *Amphisbaena munoai* is a thigmothermic species that is capable of actively regulating its temperature by selecting microhabitats such that its various activities occur within an ideal temperature range. This study is the first to evaluate the effect of seasonality and diurnal and nocturnal variations on the thermoregulation of an amphisbaenid.

KEY WORDS. Amphisbaenians, microhabitat, seasonal, thermoregulation.

INTRODUCTION

All organisms employ strategies that allow them to cope with the heterogeneity in the thermal environment. These strategies include various combinations of thermosensitivity and thermoregulation (Vitt and Caldwell 2014). Thus, organisms may vary between thermal specialists and thermal generalists and between perfect thermoregulators and perfect thermoconformists (Angilletta 2009).

When active, Squamata maintain their body temperatures within a certain range, which varies among species (Avery 1982, Adolph and Porter 1993). The idea that different species of lizards maintain a certain "optimal" temperature was predominant until the mid-20th century; however, it has since then been shown that there is no single optimal temperature but rather several optima depending on the activity (Rocha et al. 2009). The extent of thermoregulation is also known to vary intraspecifically due to various factors, such as altitude (Gvoždík 2002), seasonality (Huey and Pianka 1977), habitat structure (Sears and Angilletta 2015), sex (Ortega et al. 2016), or even competition with other species (Žagar et al. 2015). Moreover, an individual may be a thermal specialist at specific times but a thermal generalist overall throughout the year (Angilletta 2009).

Fossorial reptiles have a lower thermal preference than most basking reptiles (Brattstrom 1965, Templeton 1970, Rocha et al. 2009). Consequently, amphisbaenians exhibit lower body temperatures than heliotherm diurnal lizards during their periods of activity, and their body temperatures are related to variations in the surrounding substrate (López et al. 2002).

To determine whether fossorial reptiles have the ability to thermoregulate or are conformists, it is necessary to study the thermal biology of species with regard to their body temperature and the thermal environment (López et al. 2002). Furthermore, preferred body temperatures should be measured under artificial conditions that allow an individual to select between equally accessible microclimates, whereby the average of the selected temperatures is considered the preferred temperature and the range of selected temperatures is considered the preferred range (Angilletta 2009).

To date, nine studies have been conducted on the thermal ecology of amphisbaenians, encompassing three of the six extant families (Uetz 2012). These have shown that *Amphisbaena mertensii* Strauch, 1881 (Abe, 1984), *Blanus cinereus* (Vandelli, 1797) (Martín et al. 1990, 1991, Gil et al. 1993, López et al. 1998), *Diplometopon zarudny* Nikolsky, 1907 (Al-Johany 1999), and *Trogonophis wiegmanni* Kaup, 1830 (Gatten and McClung 1981, López et al. 2002, Civantos et al. 2003) thermoregulate and select particular microhabitats. However, only one study has investigated the thermal biology of the family Amphisbaenidae Gray, 1825, which was conducted by Abe (1984) and determined the preferred temperature of *A. mertensii* Strauch, 1881 in Brazil under field and laboratory conditions.

Previous studies have shown that lizards modify their body temperatures and the selection of microhabitats according to the season (Díaz et al. 2006, Basson and Clusella-Trullas 2015, Ortega and Pérez-Mellado 2016), highlighting the importance of considering this effect in studies of thermal ecology. However, the importance of seasonality with respect to the thermal biology of amphisbaenids has never been verified.

Balestrin and Cappellari (2011) reported a high abundance of amphisbaenids in the rocky fields of the Rio Grande do Sul, providing an excellent opportunity for studying the ecology of *Amphisbaena munoai* Klappenbach, 1969. This species occurs in the south of Brazil and in Uruguay (Gans 2005). Two studies have investigated the diet and reproduction of this species in the Shield of Rio Grande do Sul (Bernardo-Silva et al. 2006, Balestrin and Cappellari 2011). Both studies showed that the diet is diverse but mainly consists of termites, whereas Balestrin and Cappellari (2011) found that this species has a seasonal reproductive cycle between September and November.

The purpose of the present study was to describe the thermal biology of *A. munoai* in the Shield of Rio Grande do Sul, identifying this species' behavior regarding heterogeneity of the thermal environment. This is the first study of this kind to be conducted in the south of Brazil, a region that is subjected to severe, low-temperature winters.

Building on current knowledge on this subject, this study addressed the following questions: 1) Is *A. munoai* a thermoregulator or thermoconformist species? 2) What is the set point range of temperatures for *A. munoai*? 3) Is this set point range similar to that of other amphisbaenas? 4) Is this species able to select microhabitats for thermoregulation? 5) What source of environmental heat is most related to the body temperature of *A. munoai*? 6) Do the answers to these questions differ between seasons?

MATERIAL AND METHODS

The study area is known as the Shield of Rio Grande do Sul or the southeastern mountain range. This area has granite intrusions that form the Dom Feliciano Belt (Rambo 2005), also known as the rocky fields of the Pampa biome. It is located at the São Jerônimo municipality in Rio Grande do Sul, Brazil (30°21′55″S, 51°54′35″W) at an altitude of approximately 500–600 m. The area is covered with typical steppe vegetation (gallery forest) and semi-deciduous seasonal forest (Semi-deciduous Tropical Forest). The Pampa biome is under the influence of the temperate mesothermal super-humid climate without droughts, with an average temperature of 10–15 °C (IBGE 2010).

We studied the thermal ecology of *A. munoai* daily throughout the annual study period. The field study took place between March 2011 and October 2012, during which time monthly data were collected from 8:00 am to 6:00 pm for two consecutive days. Nocturnal field studies were also conducted between January and October 2012, during which time observations were made from 6:00 pm to 8:00 am for four consecutive days in each season. Thus, the behavior of *A. munoai* in the field was observed over a full 24 hours period in every season of the year.

A method of active searching was used, whereby the same transect was covered each month. This transect was 870 m long and covered an area of 60 ha, calculated on the basis of a polygon. This area encompassed the largest possible number of habitats (rocks, tree trunks, termite mounds, and anthills) for amphisbaenians. The following environmental parameters were measured every hour: air temperature (Ta); substrate temperature (Ts); substrate temperature 1 cm above the ground at the same location as Ts [Ts (1 cm)]; substrate temperature 5 cm below ground (Tsub); temperature under a rock in contact with the substrate (Tur); and relative humidity of the air. Each time an individual was found and captured, its body temperature (Tb) was taken by inserting a thermometer into the cloaca and recorded (up to 15 s after first contact with the individual), its snout-vent length (SVL) was measured in centimeters with a measuring tape, and all of the abovementioned microhabitat temperatures were recorded at the capture site. All temperatures were measured with a K-Type Minipa® MT-450 thermometer (0.1% + 1 °C) over an approximate time of 5 s or until the thermometer stabilized (up to 15 s). Body temperature was not measured in individuals an SVL of < 7 cm to prevent injuries. All measurements were made away from direct sunlight, ensuring that the collector did not interfere with the animal's temperature.

The height and circumference of the rock under which the animal was found were also measured in centimeters with a measuring tape. Relative humidity was measured with a portable weather station (Portable Digital Hygro-Thermo-Anemometer-Light Meter, model THAL-300; Instrutherm).

To determine the preferred temperature and behavior of *A. munoai* over a 24 hours period under laboratory conditions, 14 individuals were captured from the study site in January and March 2013. Individuals with an SVL of < 9.5 cm were deemed immature according to the study by Balestrin and Cappellari (2011), indicating that this sample contained seven juveniles and seven adults.

These individuals were taken to the Herpetology Laboratory of the Federal University of Rio Grande do Sul, Porto Alegre, Brazil, where they were immediately placed in a glass terrarium (89.5 × 50.0 × 45.0 cm high) containing a 5-cm layer of soil from the collection site, a 100-W "Sun Glo" lamp to simulate sunlight, a 15-W "Night Glo" lamp to simulate moonlight, three natural rocks of different sizes collected in the field (length × height: Rock 1, 60.0 × 17.0 cm; Rock 2, 29.0 × 6.0 cm; and Rock 3, 37.5 × 5.5 cm), an artificial thermal rock (length × height: Rock 4, 36.0 × 3.0 cm), and a thermal mat (length × width: 18 × 10 cm). This terrarium simulated the natural environment by containing various equally accessible microclimates and microhabitats.

Two experiments were conducted: one in January 2013 with four amphisbaenas and another in March 2013 with 10 amphisbaenas. In both cases, these individuals were captured two days before the observations. The "Sun Glo" lamp was turned on at 7:00 am and turned off at 5:30 pm, whereas the "Night Glo" lamp was turned on at 7:00 pm and turned off at 6:00 am, maintaining the accessible temperatures to the amphisbaenas between 23 and 41.3 °C.

The behavior of the amphisbaenas was monitored over 12 hours for two consecutive days, giving a total of 24 hours for each observation. The microhabitats were assessed every two hours, at which time the available rocks were lifted and the underground layer was readily seen because the terrarium was made of transparent glass. During each inspection, all temperatures in the artificial environment were measured, including all rocks, the substrate temperature 3 cm below the ground (thermal mat), and the air temperature inside the terrarium (Ta). Whenever an individual was found, it was captured by hand and its cloacal temperature was measured over a duration of 5–15 s, avoiding any interference with the animal's temperature by the collector. All temperatures were measured using the same K-Type Minipa® thermometer used in the field. After the experiments, all animals were returned to the capture site in the field.

To test the variance in body temperatures of the amphisbaenas, we first tested the homoscedasticity of the seasonal and diurnal/nocturnal data with D'Agostino's K-squared test (Ayres et al. 2007). Data from all seasons except summer ($p < 0.01$) were normally distributed, as were the nocturnal data. However, the diurnal and summer data had a non-normal distribution ($p < 0.01$). Consequently, we performed the Kruskal–Wallis test to analyze the variances in body temperatures.

Multiple linear regression analysis was used to define the relationship between the body temperatures of amphisbaenas and ambient temperatures. Forward stepwise regression analysis was used to determine the relative importance of each thermal source because this allows selection of the predictive variables that are most strongly related to the dependent variable (Ayres et al. 2007). Because the independent variables (ambient temperatures) could be highly correlated (multicollinearity; Graham 2003), we tested the pairwise correlations between these variables with Pearson's correlation and analyzed the main components using principal component analysis (PCA; Hair et al. 1984).

To define the thermal heterogeneity of *A. munoai*, we used two equations. Hertz et al. (1993) developed a way to quantify thermoregulation behavior that combines preferred body temperatures in an artificial thermal gradient with operative temperatures in a natural environment: $E = 1 - db/de$.

This index is called "Effectiveness of thermoregulation" and is calculated from two indexes: the first is known as "Accuracy of thermoregulation" (db), which equals the absolute average deviation of the specimen's body temperature in an artificial gradient, and the second is known as "Thermal quality of the environment" (de), which equals the absolute average deviation of operative temperatures. The value of "E" is the relative intensity of thermoregulation: when $E = 0$, random use of thermal microhabitats; when $E = 1$, thermoregulation is perfect; and when $E = 0.5$, it can be concluded that there is thermoregulation with moderate effectiveness (Hertz et al. 1993, Angilletta 2009). This index can only be used in studies with animals weighing up to 1 kg (Hertz et al. 1993, Seebacher and Shine 2004), which applies to *A. munoai* because it only grows up to 172 mm (Bernardo-Silva et al. 2006); individuals in the present study weighed a maximum of 1.3 g.

Mean absolute deviance of the body temperature of 14 specimens kept in laboratory was used to define (db), and mean absolute deviance of the ambient temperature Ts (1 cm) of the whole field study was used to define (de). In a burrow, where an organism receives no solar radiation, the operative temperature equals the ambient temperature (Chappell and Whitman 1990).

To define the degree of thermal heterogeneity in areas where *A. munoai* can be found (ΔT), it was necessary to calculate the differences between the recorded body temperatures and the temperatures of the microhabitats according to the following equation: $\Delta T = \Delta Tc - \Delta Ts$, where ΔTc represents the variation in body temperatures of amphisbaenas and ΔTs represents the variation in environmental temperatures [Tur, Ts, Ts (1 cm), Tsub, and Ta]. The higher the absolute value of that difference, the higher the degree of thermal regulation (active thermoregulation), as stated by Vrcibradic and Rocha (1998).

Canonical correlation analysis (CCA) was used to test microhabitat selection through rock size and to determine whether the humidity affected the rate at which amphisbaenas were found. This multivariate analysis allows the relationship between two groups of variables to be assessed (Ayres et al. 2007).

RESULTS

Field study

The 15-month-long field study included a total of 1104 collector hours of capture effort, during which time 192 *A. munoai* individuals were found. Between March 2011 and March 2012, 877 collector hours of diurnal capture effort were recorded, during which time 168 *A. munoai* individuals were found. During the warmer months (November and December 2011), the mean rate of amphisbaenas findings decreased. To determine whether this decrease was related to the time of year, nocturnal field studies were also conducted. Hence, 227 collector hours of active nocturnal searching were recorded in January (summer), May (fall), July (winter), and October (Spring) 2012, during which time 24 *A. munoai* individuals were found. All individuals were found under rocks.

We found more individuals in winter and spring than in other seasons (capture rate = 0.24 individuals/collector hour vs. 0.15 individuals/collector hour in fall and 0.08 individuals/collector hour in summer). When the diurnal and nocturnal studies were considered separately, we found that the values were higher for both studies in spring, with a rate of 0.3 individuals/collector hour for the diurnal study and 0.21 individuals/collector hour for the nocturnal study. In contrast, only the diurnal capture rate was higher in winter, with 0.28 individuals/collector hour compared with 0.07 individuals/collector hour at night (n = 3 individuals).

Summer and fall had lower capture rates in the diurnal study and across the entire study. In contrast, all seasons except spring showed similar capture rates in the nocturnal study. During the breeding season (September–November), activity was similar between the day (0.3 individuals/collector hour) and night (0.21 individuals/collector hour), whereas it remained below 0.07 individuals/collector hour in all other seasons in the nocturnal study.

No correlation (CCA) was found between humidity and the rate of finding amphisbaenas across the entire study (Rc = 0, $\chi^2 = 3.5$, gl = 1, $p > 0.06$, n = 351), or in spring (Rc = 0.1, $\chi^2 = 1.8$, gl = 1, $p > 0.1$, n = 77), summer (Rc = 0, $\chi^2 = 0$, gl = 1, $p > 0.9$, n = 91), or fall (Rc = 0.1, $\chi^2 = 1.7$, gl = 1, $p > 0.2$, n = 98). However, there was a positive correlation in winter, the wettest season (Rc = 0.2, $\chi^2 = 5.5$, gl = 1, $p < 0.05$, n = 85) (Table 1).

The average body temperature of *A. munoai* in this study was 23.7 °C (SD = 5.2, range = 12.8–33.8 °C, n = 185), with most individuals exhibiting body temperatures between 24 °C and 30 °C (Fig. 1). Furthermore, most individuals that were found between 8:00 am and 6:00 pm had body temperatures between 28 and 30 °C (Fig. 2), whereas those that were found between 6:00 pm and 8:00 am had body temperatures between 16 and 18 °C (Fig. 3). In the diurnal study, the average body temperature of amphisbaenas was 24.1 °C (SD = 5.2, range = 13.2–33.8 °C, n = 163), which was significantly higher than that in the nocturnal study (mean = 20.5 °C, SD = 4.7, range = 12.8–28 °C, n = 22; Kruskall-Wallis: H = 9.4, gl = 1, $p < 0.01$, Fig. 4). The body temperatures of *A. munoai* (Tb) also significantly differed between seasons (Kruskall-Wallis: H = 59.6, gl = 3, $p < 0.0001$), with the exception of the comparison between spring and fall ($p > 0.05$) (Fig. 5 and Table 2).

According to the index "Effectiveness of thermoregulation," the thermoregulation behavior of *A. munoai* was E = 0.6 in spring and fall and E = 0.5 in winter, indicating thermoregulation with moderate effectiveness. The index was E = 0.4 in summer, indicating random use of microhabitats.

The lowest difference between body temperature and microhabitat temperature (ΔT) for all seasons of the study was ΔTur = 2.1. In the nocturnal study, where ambient temperatures were generally lower than those during the day, all ΔT values were higher, with the exception of ΔTsub. Only in spring, nocturnal ΔT values were lower than diurnal ΔT values (Table 3).

Table 1. Seasonal variation in the relative humidity of the air at São Jerônimo between March 2011 and October 2012.

	Average humidity (%)	SD	Minimum	Maximum	n
Winter	75.5	9.7	53.8	94.1	85
Spring	52.6	18.4	24.6	87.1	77
Summer	60.6	16.7	26.7	94.7	91
Fall	65.7	13.7	32.6	89.4	98
Total	63.9	16.8	24.6	94.7	351

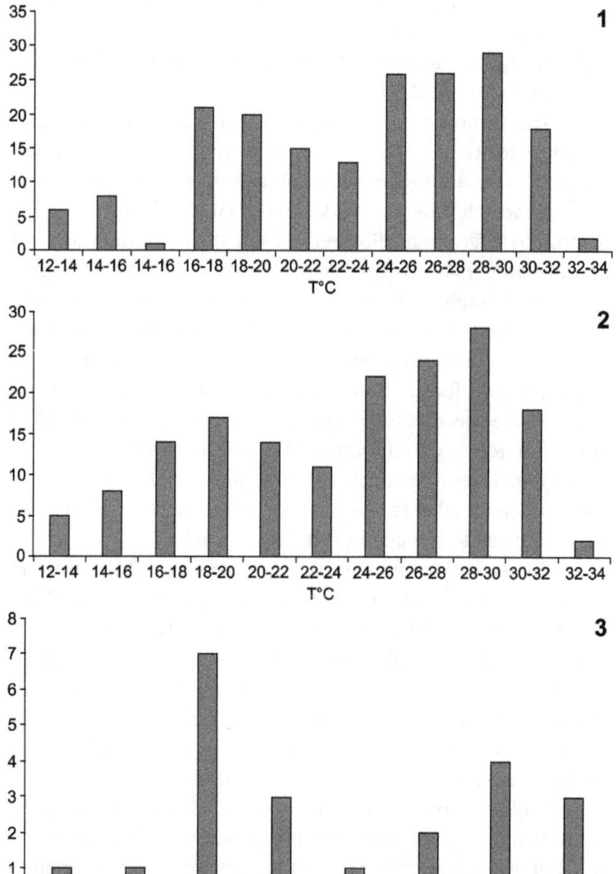

Figures 1–3. Frequency distribution of body temperatures of *Amphisbaena munoai* found at São Jerônimo between March 2011 and October 2012 (n = 185). Body temperatures are shown for: (1) the entire study; (2) the diurnal study; (3) the nocturnal study.

Multiple linear regression showed that there was a significant relationship between body temperature and environmental sources of heat (F = 93.7, $p < 0.0001$, n = 185). However, when we verified the relationship between Tb and each of the environmental variables included in this regression, we only

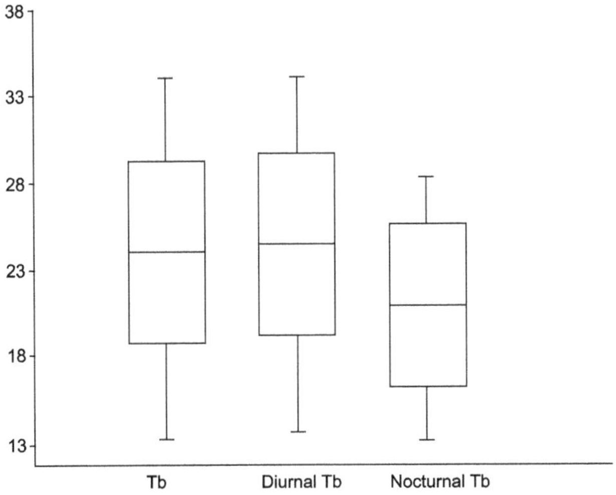

Figure 4. Descriptive statistics (mean and standard deviation) for the body temperatures of *Amphisbaena munoai* found at São Jerônimo between March 2011 and October 2012. Tb: body temperatures for the entire study (n = 183); diurnal Tb: body temperatures for the diurnal study (n = 163); total nocturnal Tb: body temperatures for the nocturnal study (n = 22).

observed a significant relationship with Tur ($p < 0.05$) and Ta ($p = 0.05$), indicating multicollinearity between these variables. We also verified that there was a high correlation between the environmental variables when pairs were analyzed with Pearson's correlation (Table 4). The variable Tur represented 90.3% of the total variance in PCA.

To analyze the relative importance of each thermal source, we conducted forward stepwise regression analyses. Given the high correlation between Ta and Ts (1 cm) ($r = 0.96$, $p < 0.0001$, $n = 185$), we excluded the independent variable Ta from this analysis. We found that the independent variable Ts (1 cm) was most strongly related to the body temperature of amphisbaenas throughout the year (all seasons, $F = 437.8$, $r = 0.84$, $r^2 = 70.41\%$, $gl = 1.2$, $p < 0.01$) and in each individual season (winter, $F = 117.2$, $r = 0.83$, $r^2 = 68.9\%$, $gl = 1.5$, $p < 0.01$; spring, $F = 11.9$, $r = 0.40$, $r^2 = 15.5\%$, $gl = 1.6$, $p < 0.01$; summer, $F = 13.2$, $r = 0.60$, $r^2 = 36.4\%$, $gl = 1.2$, $p < 0.01$; and fall, $F = 34.5$, $r = 0.66$, $r^2 = 44.0\%$, $gl = 1.4$, $p < 0.01$). Tur was then selected as the second variable in all seasons except fall. In all cases, Tsub had the weakest relationship with body temperature.

In the analysis of microhabitat selection through rock size, Tb and SVL formed one group of variables, whereas rock circumference, rock height, and the time at which individuals were

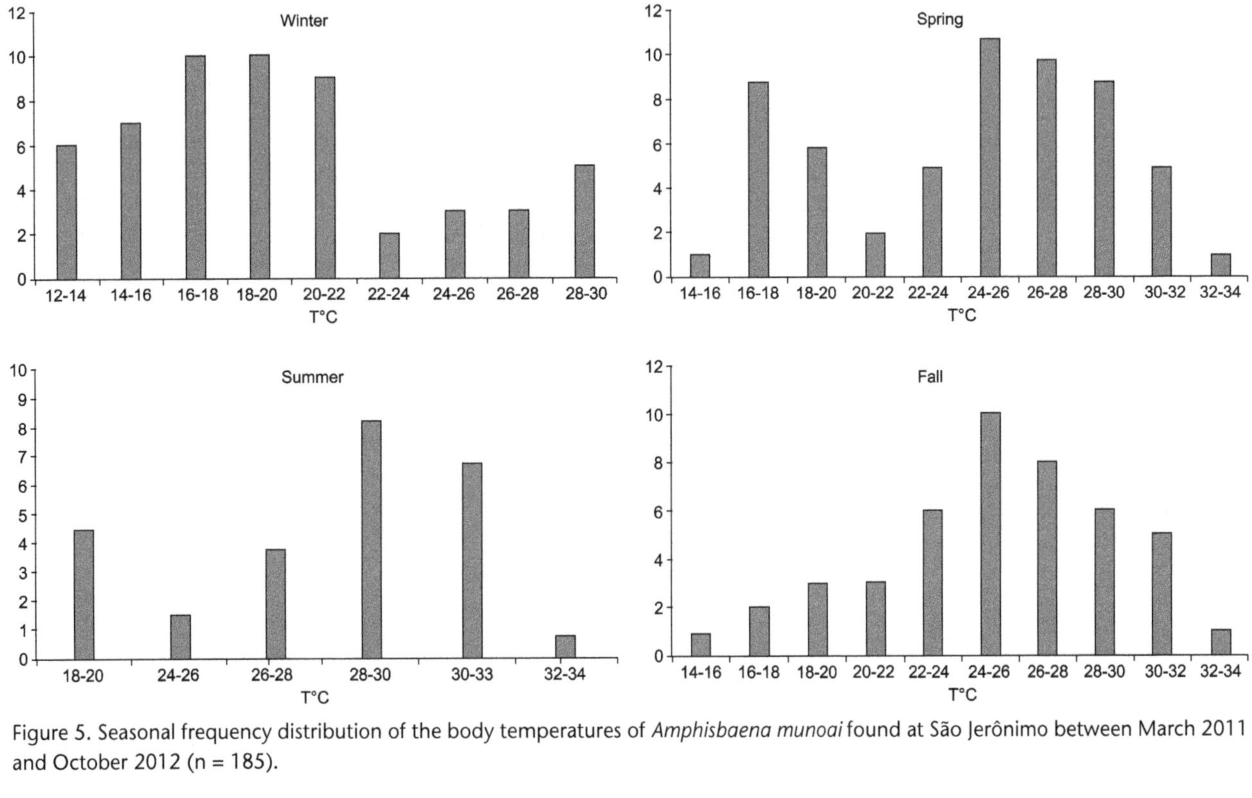

Figure 5. Seasonal frequency distribution of the body temperatures of *Amphisbaena munoai* found at São Jerônimo between March 2011 and October 2012 (n = 185).

Table 2. Seasonal variations in the body temperatures of *Amphisbaena munoai* and the environmental temperatures at São Jerônimo between March 2011 and October 2012.

Winter	Tb	Ta	Ts	Ts (1 cm)	Tsub	Tur
CV (%)	22.9	42.6	29.1	34.3	26.0	29.1
Average (°C)	19.7	13.7	15.1	14.6	15.0	15.3
Minimum (°C)	12.8	2.4	4.5	4.5	6.3	4.7
Maximum (°C)	29.3	25	24.7	26.6	24.0	28.5
SD	4.5	5.8	4.4	5.0	4.0	5.3
n	55	85	85	85	85	85
Spring						
CV (%)	20.1	28.6	31.0	29.8	29.8	34.4
Average (°C)	24.2	19.0	21.7	20.4	21.0	23.5
Minimum (°C)	14.8	9.7	8.8	10.3	9.4	9.4
Maximum (°C)	32.5	31.5	40.3	34.8	38.6	47.3
SD	4.9	5.4	6.7	6.1	6.3	8.1
n	60	77	77	77	77	77
Summer						
CV (%)	9.5	15.2	15.0	15.5	14.7	23.2
Average (°C)	28.6	22.0	25.2	23.7	25.2	26.9
Minimum (°C)	19.0	16.0	19.0	16.5	18.3	16.6
Maximum (°C)	32.0	29.5	34.6	32.9	34.4	44.7
SD	2.7	3.4	3.8	3.7	3.7	6.2
n	25	91	91	91	91	90
Fall						
CV (%)	16.8	34.8	28.3	32.9	24.0	43.9
Average (°C)	25.1	16.8	19.0	17.8	19.0	19.5
Minimum (°C)	14.6	6.9	10.2	7.9	10.2	8.7
Maximum (°C)	33.8	30.0	34.5	33.2	32.2	42.2
SD	4.2	5.8	5.4	5.9	4.6	8.5
n	45	98	98	98	98	98

found formed another group of variables. In the overall analysis, a canonical relationship was found between the two groups (Rc = 0.2, χ^2 = 13.4, gl = 6, p < 0.04); however, when the variables were subjected to individual analysis between groups, there was no significant relationship (Table 5). The seasonal analysis of the canonical relationship showed that there was only a correlation between the two groups in spring (spring, Rc = 0.4, χ^2 = 13.5, gl = 6, p < 0.05, n = 59; summer, Rc = 0.4, χ^2 = 7.2, gl = 6, p > 0.2, n = 25; fall, Rc = 0.4, χ^2 = 11.3, gl = 6, p > 0.07, n = 44; and winter, Rc = 0.3, χ^2 = 6.2, gl = 4, p > 0.1, n = 54). The spring and fall analyses also showed a correlation between rock circumference and both spring Tb (p < 0.05) and fall SVL (p < 0.05).

Laboratory experiment

Microhabitats were surveyed 31 times in the laboratory. In 19 of these surveys, one or more individuals were found in the same or different microhabitats at the same time during the 48 hours observation period. Amphisbaenas activity was observed during both the day (n = 13) and night (n = 19). Amphisbaenas were found at the highest frequency under rocks on the ground (n = 22), followed by in the subterranean environment (n = 9). A displacement was observed above ground near the Petri dish containing water.

Individuals were found under Rock 1, the largest rock in terms of both height and length, at 12 occasions and under artificial Rock 4, which attained the highest temperature, at eight occasions. In addition, individuals were found in the subterranean environment, which was heated by the thermal mat, at nine occasions. In contrast, individuals were only found under Rock 3 at two occasions and were never found under Rock 2, the smallest of the rocks (Table 6).

The average preferred temperature for *A. munoai* in the laboratory was 28 °C (SD = 2.3, range = 24.7 °C–31.8 °C, n = 25), matching the mean activity temperature observed in the natural habitat of this species in summer (28.6 °C). However, unlike the field study, there was no significant difference between day and night in the body temperatures of *A. munoai* in the laboratory (ANOVA: F = 0.1, p > 0.7).

DISCUSSION

Amphisbaenians are a group of reptiles that are specialized for fossorial life, making it difficult to study their biological and ecological characteristics (Martín et al. 2011). In general, the finding of an amphisbaenian under a rock suggests that there is ongoing thermoregulation activity. Although some of the observations made in this study suggest that *A. munoai* are occasionally inactive under rocks, in majority of the cases, activity of this species was confirmed. Therefore, the rate of *A. munoai* findings indicates their diurnal and nocturnal activity during all seasons of the year.

The average body temperature of *A. munoai* found in this study (23.7 °C, SD = 5.2, range = 12.8–33.8 °C, n = 185) is similar to that of other amphisbaenas (15.5–27 °C). The only other species of the Amphisbaenidae that has previously been studied, *A. mertensi*, presented an average body temperature of 21.4 °C (SD = 2.7, range = 15.5–26.0 °C, n = 52) (Abe 1984). Similarly, in studies with *Trogonophis wiegmanni* Kaup, 1830, the range was between 17.6 and 27.6 °C, and thermoregulation and microhabitat selection were confirmed (López et al. 2002, Civantos et al. 2003); in studies with *Blanus cinereus*, the range was 19.6–25.2 °C (Martín et al. 1990), and selection of soil depth and rocks during the day was observed (Martín et al. 1991, López et al. 1998). *Diplometopon zarudny* Nikolsky, 1907, an amphisbaenid with nocturnal habits, has a similar average body temperature (26.4 °C) in Central Saudi Arabia (Al-Johany 1999).

In all cases, higher body temperatures were recorded than the environmental temperatures (Fig. 6), matching the findings of other amphisbaenid thermoregulation studies (Abe 1984, Martín et al. 1990, 1991, Gil et al. 1993, López et al. 1998, 2002, Al-Johany 1999, Civantos et al. 2003). It is not currently known what causes this pattern, but Tossini et al. (1994) suggested that it could be related to foraging, whereby amphisbeanids are se-

Table 3. Difference between the body temperatures of *Amphisbaena munoai* and the thermal environment (ΔT values) for individuals found in São Jerônimo between March 2011 and October 2012 (n = 185). Tur, temperature under a rock; Ts, substrate temperature; Ts (1 cm), substrate temperature 1 cm above the ground; Tsub, substrate temperature 5 cm below the ground; Ta, air temperature.

ΔT/season	Winter				Spring				Summer				Fall			
ΔTur	n	Mean ± SD	Range	ΔTur	n	Mean ± SD	Range	ΔTur	n	Mean ± SD	Range	ΔTur	n	Mean ± SD	Range	ΔTur
Total	55	17.6 ± 4.3	9.6–29.3	2.1	60	21.2 ± 5.0	9.2–30.9	3	25	25.4 ± 3.4	18.6–32.4	3.2	44	20.9 ± 5.0	12.3–36.9	4.2
Diurnal	52	18.0 ± 4.1	12.2–29.3	2	49	22.0 ± 5.0	9.230.9	3.2	20	26.3 ± 2.9	19.7–32.4	2.8	40	21.5 ± 4.8	13.0–36.9	4
Nocturnal	3	10.8 ± 1.2	9.6–12.0	4.8	11	17.4 ± 2.4	14.9–23.0	1.9	5	21.9 ± 3.1	18.6–25.7	5	4	15.0 ± 3.2	12.3–19.3	5.6
ΔTs	n	Mean ± SD	Range	ΔTs	n	Mean ± SD	Range	ΔTs	n	Mean ± SD	Range	ΔTs	n	Mean ± SD	Range	ΔTs
Total	55	17.1 ± 3.6	8.4–24.9	2.6	60	20.2 ± 4.5	10–29.6	4	25	25.0 ± 2.6	20.1–31.3	3.6	44	19.7 ± 3.5	13.0–30.6	5.4
Diurnal	52	17.4 ± 3.4	11.3–24.9	2.6	49	20.6 ± 4.8	10–29.6	4.6	20	25.4 ± 2.6	20.1–31.3	3.7	40	20.1 ± 3.3	14.5–30.6	5.4
Nocturnal	3	11.0 ± 2.3	8.4–13.0	4.6	11	18.3 ± 2.4	16–24.2	1	5	23.3 ± 2.4	20.5–26.2	3.6	4	15.8 ± 2.6	13.0–18.6	4.8
ΔTs (1 cm)	n	Mean ± SD	Range	ΔTs (1 cm)	n	Mean ± SD	Range	ΔTs (1 cm)	n	Mean ± SD	Range	ΔTs (1 cm)	n	Mean ± SD	Range	ΔTs (1 cm)
Total	55	16.7 ± 3.9	6.6–25.1	3	60	19.1 ± 4.9	11.5–30.3	5.1	25	23.6 ± 3.0	18.3–30.1	5	44	19.3 ± 4.0	11.0–28.9	5.8
Diurnal	52	17.1 ± 3.6	11.2–25.1	2.9	49	19.6 ± 5.0	11.5–30.3	5.6	20	24.6 ± 2.4	18.6–30.1	4.5	40	19.9 ± 3.6	13.5–28.9	5.6
Nocturnal	3	9.5 ± 2.5	6.6–11.3	6.1	11	15.7 ± 1.9	16.4–28.1	3.6	5	19.9 ± 1.8	18.3–22.4	7	4	13.2 ± 2.7	11.0–17.2	7.4
ΔTsub	n	Mean ± SD	Range	ΔTsub	n	Mean ± SD	Range	ΔTsub	n	Mean ± SD	Range	ΔTsub	n	Mean ± SD	Range	ΔTsub
Total	55	16.3 ± 3.5	8.7–22.5	3.4	60	19.6 ± 4.3	10.7–28.1	4.6	25	25.3 ± 2.6	21.2–31.0	3.3	44	19.6 ± 3.0	14.5–27.5	5.5
Diurnal	52	16.5 ± 3.3	11.2–22.5	3.5	49	19.6 ± 4.5	10.7–26.8	5.6	20	25.1 ± 2.5	21.2–31.0	4	40	19.8 ± 3.1	14.5–27.5	5.7
Nocturnal	3	11.5 ± 2.7	8.7–14.0	4.1	11	20.0 ± 3.3	16.4–28.1	−0.7	5	25.8 ± 3.3	22.5–30.5	1.1	4	17.4 ± 1.8	15.4–19.0	3.2
ΔTa	n	Mean ± SD	Range	ΔTa	n	Mean ± SD	Range	ΔTa	n	Mean ± SD	Range	ΔTa	n	Mean ± SD	Range	ΔTa
Total	55	15.8 ± 4.8	2.4–24.6	3.9	60	17.7 ± 4.7	9.5–30.9	6.5	25	22.2 ± 3.0	16.3–28.2	6.4	44	18.8 ± 4.0	10–27.6	6.3
Diurnal	52	16.3 ± 4.3	9.6–24.6	3.7	49	18.6 ± 4.8	9.5–30.9	6.6	20	23.1 ± 2.3	18.1–28.2	6	40	19.5 ± 3.5	13.3–27.6	6
Nocturnal	3	6.7 ± 3.7	2.4–9.3	8.9	11	14.0 ± 1.6	12.2–17.8	5.3	5	18.2 ± 2.1	16.3–21.7	8.7	4	12.3 ± 2.4	10.0–15.7	8.3

Table 4. Pairwise Pearson's correlation values (r) for the environmental temperatures at São Jerônimo between March 2011 and October 2012. Tur, temperature under a rock; Ts, substrate temperature; Ts (1 cm), substrate temperature 1 cm above the ground; Tsub, substrate temperature 5 cm below the ground.

	Tur	Ts	Ts (1 cm)	Tsub
Tur	–	–	–	–
Ts	0.94	–	–	–
Ts (1 cm)	0.91	0.89	–	–
Tsub	0.84	0.94	0.79	–
Ta	0.88	0.84	0.96	0.75

lecting higher body temperatures after feeding to aid digestion. In addition, Kamel and Gatten (1983) argued that the selection of higher body temperatures by amphisbaenids was due to the high energetic cost of moving underground. However, even in the controlled environment of the laboratory, we detected individuals with body temperatures up to 5 °C higher than the ambient temperature (n = 4). Thus, it is possible that some physiological mechanism is involved in thermoregulation in these animals, which will be the subject of future studies.

In spring, which is the reproductive season, more *A. munoai* individuals were found in both the diurnal and nocturnal studies, and there was greater conformity with the diurnal environmental temperatures. Furthermore, nocturnal ΔT values were

Table 5. Canonical relationship between two groups of variables: Tb (body temperature of *Amphisbaena munoai*) and SVL (snout-vent length) × time, rock circumference, and rock height for the total study (n = 182).

	SVL	p	Tb	p
Time	−0.044	>0.05	−0.126	>0.05
Circumference	0.142	>0.05	−0.020	>0.05
Height	−0.030	>0.05	0.130	>0.05

Table 6. Variation in the temperatures of microhabitats observed in the laboratory at the Federal University do Rio Grande do Sul (UFRGS) in January and March 2013.

	Average temperature (°C)	SD	Minimum	Maximum	n
Rock 1	26.5	2.4	23.2	32.9	31
Rock 2	26.0	1.8	23.5	30.3	31
Rock 3	26.2	1.8	23.5	29.4	31
Rock 4 (artificial)	32.6	5.4	23.4	41.3	31
Thermal mat	28.3	3.2	23.0	35.2	31
T(ar)	28.1	2.1	23.0	32.0	31

always smaller than the diurnal ΔT values, suggesting either that the preferred temperature for reproductive activity is lower than that for other activities or that the conformity of body tempera-

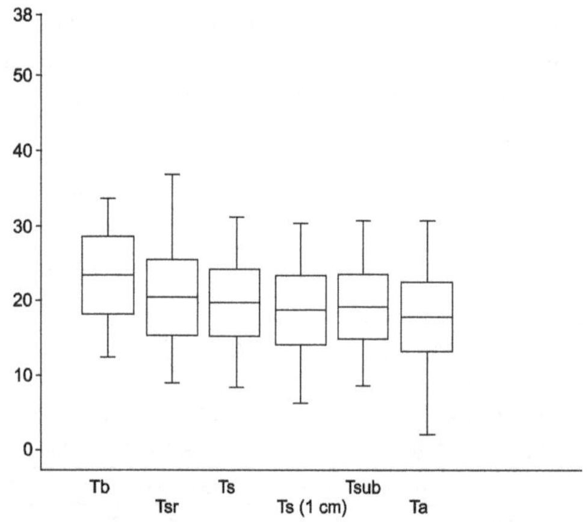

Figure 6. Relationship between ambient temperatures and the body temperatures of *Amphisbaena munoai* found at São Jerônimo between March 2011 and October 2012 (n = 185).

ture to the environment is beneficial for reproductive activity. This was the only season in which low and negative ΔT values occurred, showing conformity with the subterranean environment during the night (Tsub = −0.7). However, CCA showed that in this season, amphisbaenas select their microhabitat based on the size of rocks (Table 5). Therefore, we suggest that *A. munoai* thermoregulate by selecting particular sized rocks to stabilize their body temperatures during the reproductive period.

We found a canonical correlation for total microhabitat selection, but such a correlation could only be found in spring in the seasonal study. However, this result may have been influenced by the lower number of samples when we separated the data by season. In the laboratory, the larger rock (Rock 1) was most used, whereas the smallest rock (Rock 2) was never used, despite there being no significant difference between the temperatures under these rocks (Tukey's test: $p < 0.0001$) (Table 6). This behavior also indicates that amphisbaenas select different rock sizes as well as different thermal microhabitats. A similar situation was found for the amphisbaenas *B. cinereus* (Martín et al. 1991, López et al. 1998) and *Trogonophis wiegmanni* (López et al. 2002, Civantos et al. 2003), both of which thermoregulate and select microhabitats.

In conclusion, *A. munoai* is a thigmothermal species that is able to actively thermoregulate by selecting microhabitats, allowing it to perform its different activities within an ideal temperature range. This study contributes to our knowledge of the thermal ecology of fossorial reptiles because it is only the second study to describe the thermal biology of a species from the largest family of the clade Amphisbaenia. Furthermore, it is the first study to analyze the thermoregulation of an *Amphisbaena* over the entire 24 hours period in the laboratory and the field and to verify seasonal differences in the thermal biology of an amphisbaenid. More studies on the thermal ecology of these fossorial reptiles are required to allow their behavioral patterns to be compared with other similar species.

ACKNOWLEDGMENTS

We would like to thank Gilson Galera for allowing research in the Novos Ares estate and Carlos F. Duarte Rocha for his invaluable suggestions. We also thank the Sistema de Autorização e Informação em Biodiversidade (SISBIO) and the Comitê de Ética na Utilização de Animais (CEUA) by licenses granted SISBIO 12611-1, CEUA 18832. This work was funded by Coordenação de Aperfeiçoamento de Pessoal de Nível Superior (CAPES).

LITERATURE CITED

Abe AS (1984) Experimental and field record of preferred temperature in the neotropical amphisbaenid *Amphisbaena mertensii* Stauch (Reptilia, Amphisbaenidae). Comparative Biochemistry and Physiology A 77: 251–253. https://doi.org/10.1016/0300-9629(84)90056-2.

Adolph SC, Porter WP (1993) Temperature, activity, and lizard life histories. American Naturalist 142: 273–295. https://doi.org/10.1086/285538.

Al-Johany AM (1999) The activity and thermal biology of the fossorial reptile, *Diplometopon zarudnyi* (Amphisbaenia:Trogonophiidae) in Central Saudi Arabia. Asiatic Herpetological Research 8: 1–6.

Angilletta MJ (2009) Thermal Adaptation- A Theoretical and Empirical Synthesis. Oxford University Press, New York,.

Avery RA (1982) Field studies of body temperatures and thermoregulation. Biology of the Reptilia 12: 93–166.

Ayres M, Ayres-Júnior M, Ayres DL, Santos AS dos (2007) BioEstat 5.0: aplicações estatísticas nas áreas das ciências biológicas e médicas. MCT, IDSM, CNPq, Belém.

Balestrin RL, Cappellari LH (2011) Reproduction and feeding ecology of *Amphisbaena munoai* and *Anops kingi* (Amphisbaenia, Amphisbaenidae) in the Escudo Sul-Rio-Grandense, southern Brazil. Iheringia, Série Zoologia 101: 93–102. https://doi.org/10.1590/S0073-47212011000100013.

Basson CH, Clusella-Trullas S (2015) The behavior-physiology nexus: behavioral and physiological compensation are relied on to different extents between seasons. Physiological and Biochemical Zoology 88: 384–394. https://doi.org/10.1086/682010

Bernardo-Silva JS, Von-Muhlen EM, Di-Bernardo MO, Ketterl J (2006) Feeding ecology in the small neotropical amphisbaenid *Amphisbaena munoai* (Amphisbaenidae) in southern Brazil. Iheringia, Série Zoologia, 96: 487–489. https://doi.org/10.1590/S0073-47212006000400014

Brattstrom BH (1965) Body temperature of reptiles. American Midland Naturalist Journal 73: 376–422.

Chappell MA, Whitman DW (1990) Grasshopper thermoregulation. In: Chapman RF, Joern A (Eds) Biology of Grasshoppers. New York, Wiley and Sons, 43–172.

Civantos E, Martín J, López P (2003) Fossorial life constrains microhabitat selection of the amphisbaenian *Trogonophis wiegmanni*. Canadian Journal Zoology 81: 1839–1844. https://doi.org/10.1139/z03-163

Díaz JA, Iraeta P, Monasterio C (2006) Seasonality provokes a shift of thermal preferences in a temperate lizard, but altitude does not. Journal Thermal Biology 31: 237–242. https://doi.org/10.1016/j.jtherbio.2005.10.001

Gatten RE, McClung RM (1981) Thermal selection by an amphisbaenian *Trogonophis wiegmanni*. Journal Thermal Biology 6: 49–51. doi:10.1016/0306-4565(81)90043-7

Gans C (2005) Checklist and bibliography of the amphisbaenia of the World. Bulletin of the American Museum of Natural History 280: 1–130. https://doi.org/10.1206/0003-0090(2005)289<0001

Gil MJ, Guerrero F, Pérez-Mellado V (1993) Observations on morphometrics and ecology in *Blanus cinereus* (Reptilia: Amphisbaenia). Journal of Herpetology 27: 205–209. https://doi.org/10.2307/1564937

Graham MH (2003) Confronting multicollinearity in ecological multiple regression. Ecology 84: 2809–2815. https://doi.org/10.1890/02-3114

Gvoždík L (2002) To heat or to save time? Thermoregulation in the lizard Zootoca vivipara (Squamata: Lacertidae) in different thermal environments along an altitudinal gradient. Canadian Journal of Zoology 80: 479–492. https://doi.org/10.1139/z02-015

Hair JF, Anderson RE, Tatham RL (1984) Multivariate Data Analysis with Readings. MacMillan, New York.

Hertz PE, Huey RB Stevenson RD (1993) Evaluating temperature regulation by fieldactive ectotherms: the fallacy of the inappropriate question. American Naturalist 142: 796–818.

Huey RB, Pianka ER (1977) Seasonal variation in thermoregulatory behavior and body temperature of diurnal Kalahari lizards. Ecology 58: 1066–1075. https://doi.org/10.2307/1936926

IBGE (2010) National atlas of Brazil. In: Climate. 77p. Available on line at: http://www.metadados.geo.ibge.gov.br/geonetwork_ibge/srv/por/metadata.show?id=19597&currTab=simple. Brasília, Instituto Brasileiro de Geografia e Estatística [Accessed: 01/12/2017]

Kamel S, Gatten RE (1983) Aerobic and anaerobic activity metabolism of limbless and fossorial reptiles. Physiological and Biochemical Zoology 56: 419–429. https://doi.org/10.1086/physzool.56.3.30152607

López P, Civantos E, Martín J (2002) Body temperature regulation in the amphisbaenian *Trogonophis wiegmanni*. Canadian Journal of Zoology 80: 42–47. https://doi.org/10.1139/z01-201

López P, Salvador A, Martín J (1998) Soil temperatures, rock selection and the thermal ecology of the amphisbaenian reptile *Blanus cinereus*. Canadian Journal of Zoology 76: 673–679. https://doi.org/10.1139/z97-230

Martín J, López P, Salvador A (1990) Field body temperatures of the amphisbaenid lizard *Blanus cinereus*. Amphibia-Reptilia 11: 87–96. https://doi.org/10.1163/156853890X00492

Martín J, López P, Salvador A (1991) Microhabitat selection of the amphisbaenian *Blanus cinereus*. Copeia 1142–1146. https://doi.org/10.2307/1446118

Martín J, Polo-Cavia N, Gonzalo A, López P, Civantos E (2011) Structure of a population of the amphisbaenian *Trogonophis wiegmanni* in North Africa. Herpetologica 67: 250–257. https://doi.org/10.1655/HERPETOLOGICA-D-11-00007.1

Ortega Z, Pérez-Mellado V (2016) Seasonal patterns of body temperature and microhabitat selection in a lacertid lizard. Acta Oecologica 77: 201–206. https://doi.org/10.1016/j.actao.2016.08.006

Ortega Z, Mencía A, Pérez-Mellado V (2016) Sexual differences in behavioral thermoregulation of the lizard Scelarcis perspicillata. Journal of Thermal Biology 61: 44–49. https://doi.org/10.1016/j.jtherbio.2016.08.006

Rambo B (2005) A fisionomia do Rio Grande do Sul. Editora Unisinos, São Leopoldo.

Rocha CFD, Sluys MV, Vrcibradic D, Kiefer MC, Menezes VA, Siqueira CC (2009) Comportamento de termorregulação em lagartos brasileiros. Oecologia Brasiliensis 13: 115–131.

Sears MW, Angilletta MJ (2015) Costs and benefits of thermoregulation revisited: both the heterogeneity and spatial structure of temperature drive energetic costs. The American Naturalist 185: E94–E102. https://doi.org/10.1086/680008

Seebacher F, Shine R (2004). Evaluating Thermoregulation in Reptiles: The Fallacy of the Inappropriately Applied Method. Physiological and Biochemical Zoology 77: 688–695. 2004. https://doi.org/10.1086/422052

Templeton (1970) Reptiles. In: Whittow GC (Ed.) Comparative Physiology of Thermoregulation. Academic Press, New York, 167–221.

Uetz (2012) Reptile database. Available online at http://www.reptile-database.org [Accessed: 30/06/2016]

Vitt LJ, Caldwell JP (2014) Herpetology: An Introductory Biology of Amphibians and Reptiles. Academic Press, Oxford.

Vrcibradic D, Rocha CFD (1998) The ecology of the skink *Mabuya frenata* in an area of rock outcrops in Southeastern Brazil. Journal of Herpetology 32: 229–237. https://doi.org/10.2307/1565302

Tossini G, Jones S, Avery RA (1994) Effects of feeding on set point temperatures and thermoregulatory behaviour in the lizards *Podarcis muralis* and *Lacerta vivipara*. Amphibia-Reptilia 15: 257–265. https://doi.org/10.1163/156853894X00038

Žagar A, Carretero MA, Osojnik N, Sillero N, Vrezec A (2015) A place in the sun: interspecific interference affects thermoregulation in coexisting lizards. Behavioral Ecology and Sociobiology 69: 1127–1137. https://doi.org/10.1007/s00265-015-1927-8

Author Contributions: NRM and LV contributed equally for the manuscript.

Competing Interests: The authors have declared that no competing interests exist.

Permissions

All chapters in this book were first published in ZOOLOGIA, by Pensoft Publishers; hereby published with permission under the Creative Commons Attribution License or equivalent. Every chapter published in this book has been scrutinized by our experts. Their significance has been extensively debated. The topics covered herein carry significant findings which will fuel the growth of the discipline. They may even be implemented as practical applications or may be referred to as a beginning point for another development.

The contributors of this book come from diverse backgrounds, making this book a truly international effort. This book will bring forth new frontiers with its revolutionizing research information and detailed analysis of the nascent developments around the world.

We would like to thank all the contributing authors for lending their expertise to make the book truly unique. They have played a crucial role in the development of this book. Without their invaluable contributions this book wouldn't have been possible. They have made vital efforts to compile up to date information on the varied aspects of this subject to make this book a valuable addition to the collection of many professionals and students.

This book was conceptualized with the vision of imparting up-to-date information and advanced data in this field. To ensure the same, a matchless editorial board was set up. Every individual on the board went through rigorous rounds of assessment to prove their worth. After which they invested a large part of their time researching and compiling the most relevant data for our readers.

The editorial board has been involved in producing this book since its inception. They have spent rigorous hours researching and exploring the diverse topics which have resulted in the successful publishing of this book. They have passed on their knowledge of decades through this book. To expedite this challenging task, the publisher supported the team at every step. A small team of assistant editors was also appointed to further simplify the editing procedure and attain best results for the readers.

Apart from the editorial board, the designing team has also invested a significant amount of their time in understanding the subject and creating the most relevant covers. They scrutinized every image to scout for the most suitable representation of the subject and create an appropriate cover for the book.

The publishing team has been an ardent support to the editorial, designing and production team. Their endless efforts to recruit the best for this project, has resulted in the accomplishment of this book. They are a veteran in the field of academics and their pool of knowledge is as vast as their experience in printing. Their expertise and guidance has proved useful at every step. Their uncompromising quality standards have made this book an exceptional effort. Their encouragement from time to time has been an inspiration for everyone.

The publisher and the editorial board hope that this book will prove to be a valuable piece of knowledge for researchers, students, practitioners and scholars across the globe.

List of Contributors

Leonardo de Oliveira Cardoso da Silva and Sonia Maria Lopes
Department of Entomology, Museu Nacional, Universidade Federal do Rio de Janeiro, Quinta da Boa Vista, São Cristóvão, 20940-040 Rio de Janeiro, RJ, Brazil

Cristiano Sampaio Costa and Ricardo Pinto-da-Rocha
Departamento de Zoologia, Instituto de Biociências, Universidade de São Paulo. Rua do Matão-Travessa 14, 321, Cidade Universitária, 05508-090 São Paulo, SP, Brazil

Amazonas Chagas-Junior
Departamento de Biologia e Zoologia, Instituto de Biociências, Universidade Federal de Mato Grosso. Avenida Fernando Correa da Costa 2367, Boa Esperança, 78060-900 Cuiabá, MT, Brazil

Daiane Ouvernay and Ildemar Ferreira
Departamento de Biologia Animal, Instituto de Biologia, Universidade Federal Rural do Rio de Janeiro. Rodovia BR-465 km 7, 23897-970 Seropédica, RJ, Brazil

Juan J. Morrone
Museo de Zoología "Alfonso L. Herrera", Departamento de Biología Evolutiva, Facultad de Ciencias, Universidad Nacional Autónoma de México. 04510 Mexico City, Mexico

Thereza de Almeida Garbelotto
Departamento de Ciências Biológicas, Universidade do Sul Catarinense. Avenida José Acácio Moreira 787, 88704-900 Tubarão, SC, Brazil

Ana Paula Leite Kochenborger and Luiz Alexandre Campos
Departamento de Zoologia, Universidade Federal do Rio Grande do Sul. Avenida Bento Gonçalves 9500, 91501-970 Porto Alegre, RS, Brazil

Walter A. Boeger
Laboratório de Ecologia Molecular e Parasitologia Evolutiva, Departamento de Zoologia, Universidade Federal do Paraná. Caixa Postal 19073, 81531-980 Curitiba, PR, Brazil

Taísa Mendes Marques
Laboratório de Ecologia Molecular e Parasitologia Evolutiva, Departamento de Zoologia, Universidade Federal do Paraná. Caixa Postal 19073, 81531-980 Curitiba, PR, Brazil
Departamento de Patologia, Universidade Federal do Paraná, Curitiba, PR, Brazil

Milton P. Ávila, Rafaela Neves Carvalho, Juliana Simião-Ferreira, Lorrane Faria de Morais and Fabrício B. Teresa
Campus de Ciências Exatas e Tecnológicas, Universidade Estadual de Goiás. Rodovia BR 153, 3105, Fazenda Barreiro do Meio, Caixa Postal 459, 75132-903 Anápolis, GO, Brazil

Lilian Casatti
Instituto de Biociências, Letras e Ciências Exatas, Universidade Estadual Paulista. Rua Cristovão Colombo 2265, 15054-000 São José do Rio Preto, SP, Brazil

João Flor dos Santos Neto
Laboratório de Sistemática e Coevolução, Instituto de Estudos Costeiros, Universidade Federal do Pará. Alameda Leandro Ribeiro, 68600-000 Bragança, PA, Brazil
Programa de Pós-Graduação em Biologia Ambiental, Universidade Federal do Pará. Alameda Leandro Ribeiro, 68600-000 Bragança, PA, Brazil

Marcus Vinicius Domingues
Laboratório de Sistemática e Coevolução, Instituto de Estudos Costeiros, Universidade Federal do Pará. Alameda Leandro Ribeiro, 68600-000 Bragança, PA, Brazil
Programa de Pós-Graduação em Biologia Ambiental, Universidade Federal do Pará. Alameda Leandro Ribeiro, 68600-000 Bragança, PA, Brazil
Programa de Pós-Graduação em Biodiversidade e Conservação, Universidade Federal do Pará. Rua Coronel José Porfírio 2515, 68372-040 Altamira, PA, Brazil

Geusivam Barbosa Soares
Laboratório de Sistemática e Coevolução, Instituto de Estudos Costeiros, Universidade Federal do Pará. Alameda Leandro Ribeiro, 68600-000 Bragança, PA, Brazil
Programa de Pós-Graduação em Biodiversidade e Conservação, Universidade Federal do Pará. Rua Coronel José Porfírio 2515, 68372-040 Altamira, PA, Brazil

Ângelo Parise Pinto
Laboratório de Sistemática de Insetos Aquáticos, Departamento de Zoologia, Universidade Federal do Paraná. Caixa Postal 19020, 81531-980 Curitiba, PR, Brazil

Tom Kompier
Schoutenstraat 69, 2596SK, Den Haag, The Netherlands

Cristiano M. Pereira
Programa de Pós-graduação em Biologia Animal, Departamento de Zoologia, Instituto de Biociências, Universidade Federal do Rio Grande do Sul. Avenida Bento Gonçalves 9500, 91501-970 Porto Alegre, RS, Brazil

Denis S. Silva and Gilson R. P. Moreira
Departamento de Zoologia, Instituto de Biociências, Universidade Federal do Rio Grande do Sul. Avenida Bento Gonçalves 9500, 91501-970 Porto Alegre, RS, Brazil

Héctor A. Vargas
Departamento de Recursos Ambientales, Facultad de Ciencias Agronómicas, Universidad de Tarapacá. Casilla 6-D, Arica, Chile

Cristina de Oliveira Dias, Adriana Valente de Araujo and Sérgio Luiz Costa Bonecker
Departamento de Zoologia, Instituto de Biologia, Universidade Federal do Rio de Janeiro. Avenida Prof. Rodolpho Rocco 211, CCS, Bloco A, Laboratório Integrado de Zooplâncton e Ictioplâncton, Sala A0-084, Ilha do Fundão, 21941-590 Rio de Janeiro, RJ, Brazil

Marcela Osorio-Beristain, Ángeles Rodríguez, Cristina Martínez-Garza and Raúl Ernesto Alcalá
Centro de Investigación en Biodiversidad y Conservación, Universidad Autónoma del Estado de Morelos. 62209, Cuernavaca, Morelos, México

Hermes Daros and William Barbosa Dutra
Programa de Pós-graduação em Ecologia de Ecossistemas, Universidade Vila Velha. Campus Boa Vista, 29102-920 Vila Velha, ES, Brazil

Charles Duca
Laboratório de Ecologia de Populações e Conservação, Universidade Vila Velha. Campus Nossa Senhora da Penha, 29102-623 Vila Velha, ES, Brazil

Néstor J. Cazzaniga and Natalia S. Ghezzi
Departamento de Biología, Bioquímica and Farmacia, Universidad Nacional del Sur. San Juan 670, B8000ICN Bahía Blanca, Buenos Aires, Argentina

Julia Pizá
Departamento de Biología, Bioquímica and Farmacia, Universidad Nacional del Sur. San Juan 670, B8000ICN Bahía Blanca, Buenos Aires, Argentina
Instituto de Ciencias Biológicas y Biomédicas del Sur, Universidad Nacional del Sur-CONICET. B8000ICL Bahía Blanca, Buenos Aires, Argentina

Cynthia D. Souza, Vandick S. Batista and Nidia N. Fabré
Instituto de Ciências Biológicas e da Saúde, Universidade Federal de Alagoas. Avenida Lourival Melo Mota, Tabuleiro do Martins, 57072-900 Maceió, AL, Brazil

Rodrigo Silvestre Martins
Laboratório de Ciências da Pesca, Instituto do Mar, Universidade Federal de São Paulo. Rua Dr Carvalho de Mendonça 144, Encruzilhada, 11070-100 Santos, SP, Brazil

Marcelo Juanicó
Juanicó - Environmental Consultants Ltd. Building 130, Ram On, 19205, Israel

Oscar Diego Evangelista-Vargas and Luís Fábio Silveira
Seção de Aves, Museu de Zoologia da Universidade de São Paulo, São Paulo, Brasil. Avenida Nazaré 481, Ipiranga, 04263-000 São Paulo, SP, Brazil

Leandro M. Vieira
Laboratório de Estudos de Bryozoa, Departamento de Zoologia, Centro de Biociências, Universidade Federal de Pernambuco. 50670-810 Recife, PE, Brazil

Ana C. S. Almeida
Laboratório de Estudos de Bryozoa, Departamento de Zoologia, Centro de Biociências, Universidade Federal de Pernambuco. 50670-810 Recife, PE, Brazil
Museu de Zoologia, Universidade Federal da Bahia. 40170-115 Salvador, BA, Brazil

Facelúcia B. C. Souza
Museu de Zoologia, Universidade Federal da Bahia. 40170-115 Salvador, BA, Brazil

Fernanda Martins dos Santos
Laboratório de Zoologia, Campus do Pantanal, Universidade Federal de Mato Grosso do Sul. Avenida Rio Branco 1270, 79304-902 Corumbá, MS, Brazil

Nelson Rufino de Albuquerque
Laboratório de Zoologia, Campus do Pantanal, Universidade Federal de Mato Grosso do Sul. Avenida Rio Branco 1270, 79304-902 Corumbá, MS, Brazil
Programa de Pós-Graduação em Biologia Animal, Instituto de Biociências, Universidade Federal de Mato Grosso do Sul. Avenida Costa e Silva, Cidade Universitária, 79070-900 Campo Grande, MS, Brazil

Omar Machado Entiauspe-Neto
Laboratório de Vertebrados, Instituto de Ciências Biológicas, Universidade Federal do Rio Grande. Avenida Itália Km 8, 96203-900 Rio Grande, RS, Brazil

Jailini da Silva Araújo and Moisés Barbosa de Souza
Laboratório de Herpetologia, Centro de Ciências Biológicas e da Natureza, Universidade Federal do Acre. Rodovia BR 364, Km 4, 69915-900 Rio Branco, AC, Brazil

Thales de Lema
Laboratório de Sistemática de Vertebrados, Museu de Ciências e Tecnologia, Pontifícia Universidade Católica do Rio Grande do Sul. Avenida Ipiranga 6681, 90619-900 Porto Alegre, RS, Brazil

Christine Strüssmann
Faculdade de Medicina Veterinária, Universidade Federal de Mato Grosso. Avenida Fernando Correia da Costa 2367, 78060-900 Cuiabá, MT, Brazil

Ana Paula Fabio-Braga and Wilfried Klein
Departamento de Biologia, Faculdade de Filosofia, Ciências e Letras de Ribeirão Preto, Universidade de São Paulo. Ribeirão Preto, SP, Brazil

Nathalia Rocha Matias and Laura Verrastro
Programa de Pós-Graduação em Biologia Animal, Laboratório de Herpetologia, Departamento de Zoologia, Universidade Federal do Rio Grande do Sul. Avenida Bento Gonçalves 9500, prédio 43435, sala 105, 91501-970 Porto Alegre, RS, Brazil

Index

www.ingramcontent.com/pod-product-compliance
Lightning Source LLC
LaVergne TN
LVHW080906210926
848485LV00004B/603